MODERN PRODUCTION/ OPERATIONS MANAGEMENT

MODERN PRODUCTION/ OPERATIONS

JOHN WILEY & SONS

MANAGEMENT 6/e

ELWOOD S. BUFFA
UNIVERSITY OF CALIFORNIA,
LOS ANGELES

JOHN WILEY & SONS

NEW YORK / CHICHESTER / BRISBANE / TORONTO

PRODUCTION SUPERVISED BY ELLEN P. O'NEILL
DESIGNED BY FERN H. LOGAN
COPYEDITED BY ELAINE MILLER

Library of Congress Cataloging in Publication Data:

Buffa, Elwood Spencer, 1923-
Modern production/operations management.

(The Wiley series in management)
Fifth ed. published in 1977 under title: Modern production management.
Includes index.
1. Production management. I. Title.
TS155.B723 1979 658.5 79-17788
ISBN 0-471-05672-3

Printed in the United States of America

10 9 8 7 6 5 4 3 2 1

To Carl and Jerry

THE WILEY SERIES IN MANAGEMENT

ABOUT THE AUTHOR

Elwood S. Buffa is professor of production and operations management and management science at the Graduate School of Management of the University of California, Los Angeles. He received his B.S. and M.B.A. degrees from the University of Wisconsin, and his Ph.D. from the University of California, Los Angeles. He worked as an operations analyst at the Eastman Kodak Company before entering the teaching profession, and has engaged in consulting activities in a wide variety of settings during the past twenty-five years. He has served as assistant dean and associate dean at the Graduate School of Management, Chairman of UCLA's Academic Senate, and has held visiting appointments at IPSOA in Turin, Italy, and at the Harvard Business School. Professor Buffa has published many research papers in management science and operations management. He is the author of other books published by Wiley, including *OPERATIONS MANAGEMENT: The Management of Productive Systems; OPERATIONS MANAGEMENT: Problems and Models; BASIC PRODUCTION MANAGEMENT;* and, coauthored with James S. Dyer, *MANAGEMENT SCIENCE/ OPERATIONS RESEARCH: Model Formulation and Solution Methods.* In addition, Professor Buffa serves on the board of directors of the University of California Press, and on the board of Planmetrics, Inc.

PREFACE

The sixth edition of *Modern Production/Operations Management* incorporates a title modification to reflect the evolution that has taken place during the last two editions of the book. The addition of "operations" to the title in this edition recognizes what has happened to the book, and to the field as it is taught in colleges and universities.

As the production/operations management field has matured, the technical content has stabilized and the emphasis has returned to P/OM as a functional field of management. As such, there is a reemphasis on "management." Many of the changes in the sixth edition were introduced to strengthen the managerial orientation of P/OM. For example, short cases have been added at the ends of chapters under the heading, "Situations." They are intended to set up situations for analysis and discussion rather than to present problems for solution through analytical models. They culminate in action-oriented questions that require students to think about what managers should do, and how they might trade off one value or consequence against another. These situations emphasize the broader goals and objectives of production and operations systems and their management, and all of them have been classroom tested. Some of the situations are vignettes that come from consulting incidents, and some represent applications reported in the literature. In two instances I have reproduced articles that were reports of important actual applications that may be treated as cases. New sections titled "Implications for the Manager" have been added to the ends of chapters. They summarize the managerial highlights of the chapter and again place more emphasis on the management of operations.

Although there are new and revised materials in all chapters, some are of particular importance. A new section of the book titled "Strategic Planning Decisions" brings together a new chapter on capacity planning with new and revised materials on technology and its impact on productive systems and other long-term decisions, such as location and distribution. I cover these topics early in the sequence to provide a framework within which short-term decisions can be considered. The early emphasis on strategic decisions also helps students to realize how important the operations function is in the success of an enterprise. I have placed the two chapters dealing with the design of productive systems at the end of the book, following the major section on operations planning and control. This arrangement makes it possible to examine the exciting topics of day-to-day operation such as forecasting, aggregate planning, and scheduling of various types of productive systems without delay. Some instructors may prefer to resequence these two chapters and to cover them immediately following the section on strategic planning decisions. The sequence is a matter of personal preference.

A discussion of material requirements planning has been added in a new and separate chapter. The previous edition had substantial materials on MRP, but this important topic now requires a full chapter for appropriate coverage.

There is new and expanded coverage of exponential smoothing with the inclusion of trend and seasonal models. There are new and expanded materials on cyclic scheduling of time-shared equipment, and on personnel scheduling. The former chapter on reliability of productive systems has been condensed and combined with new materials on overhaul and replacement.

The sixth edition continues to take advantage of appendixes to provide coverage of analytical models. Instructors may then assign these materials in any desired sequence in relation to chapters in the main text as review of these materials are needed, or not to assign them if their students' backgrounds are already adequate. The chapter on work measurement that was dropped from the fifth edition has been included in this edition as one of these appendixes.

I have eliminated three previous chapters in this edition. The former chapters on decision making and systems concepts are summarized, and appropriate comments are included in Chapter 1. The former chapter on analytical methods has been dropped, since the appendixes at the end of the book contain all of the material necessary for the study of the basic analytical methods useful in P/OM.

Elwood S. Buffa
Pacific Palisades, California

ACKNOWLEDGMENTS

A book of this length is necessarily based on a wide variety of sources. Although I have made original contributions in some specific areas of analysis and application and in the conceptual framework, the bulk of the material on which *Modern Production/Operations Management* is based comes from original work by scores of colleagues throughout the country. The sources of these materials are cited where the materials are discussed. I hope I have made no omissions.

I have benefited greatly from reviews and comments on previous editions by well-known professors. They include Robert Albanese of Texas A.&M. University; William H. Bolen of Georgia Southern College; Robert W. Boling of the University of Tennessee; John D. Burns of DePaul University; Y. S. Chang of Boston University; Ronald J. Ebert of the University of Missouri-Columbia; Norbert L. Enrick of Kent State University; James A. Fitzsimmons of the University of Texas at Austin; George J. Gore of the University of Cincinnati; Gene K. Groff of Georgia State University; the late S. T. Hardy of Ohio State University; Warren Hausman of the Stanford University; Thomas E. Hendrick of the University of Colorado; Roy Housewright of Western Illinois University; Jarrett Hudnall, Jr., of Louisiana Tech University; Alan Krigline of the University of Akron; Terry Nels Lee of Brigham Young University; James L. McKenney of Harvard University; William T. Newell of the University of Washington; D. Roman of George Washington University; J. A. Sargeant of the University of Toronto; John E. Van Tassel, Jr., of Boston College; Richard J. Tersine of Old Dominion University; and Thomas E. Vollmann of Indiana University.

I thank Professors Louis J. Allain of St. John's University, Joseph D. Blackburn of Boston University, C. W. Dane of the University of Southern California, John P. Dory of New York University, Michael P. Hottenstein of Pennsylvania State University, John P. Matthews of the University of Wisconsin and Michael Summers of Illinois State University for reviewing the proposed revision of the book. They provided both support for the directions of change and many valuable suggestions.

Professors Gene K. Groff of Georgia State University, John P. Matthews of the University of Wisconsin, and Samual Seward of the University of Wyoming reviewed the revised manuscript in detail and helped immeasurably in the development of the final manuscript.

CONTENTS

PART

ONE

INTRODUCTION

CHAPTER 1 The Operations Function

ONE IMPORTANT TRAIL LEFT BY HUMANS FROM THE TIME *HOMO erectus* made the first stone tools is in the systems they created to produce goods and services. Humans set themselves apart from other creatures in developing tools, cooperative systems for hunting and gathering food, cooperative family and larger group systems for sharing tasks, building shelters, and so forth. Through recorded history, humans developed admirable productive systems to build pyramids, roads, aqueducts, and other ancient monuments. They conceived modes of productive systems, beginning with the family handicraft system which evolved into the factory system, culminating in mass production, production lines, and automation concepts.

As the capability to produce goods relieved the threat of starvation and the elements, society became more complex and interdependent. The demand for services in addition to physical products developed. Productive systems were created to deliver education, health care, letters and parcels, fire protection, transportation, financial services, and so on. Thus, productive systems are the unique invention of humankind. They are the means by which we create an endless list of goods and services needed to sustain modern society.

The flow of these goods and services represents the economy as a whole, functioning throughout the world with international flows. There are industries within economies and enterprises within industries, and the *productive system* is the building block on which the entire structure is established. These systems deal with the operations phase of all kinds of enterprises (private and public, manufacturing and service), and their management is called *Production/Operations Management.*

PRODUCTIVE SYSTEMS OF SIGNIFICANCE

It was not long ago that the only productive systems of significance were thought to be manufacturing systems. There seemed to be an insatiable demand for material goods, and it was with the production of goods that we learned the first fundamentals of how to organize resources to produce something effectively. At the dawn of the Industrial Revolution, our attention was focused on the production of goods to satisfy our basic need and desire for material things. Resources were focused to develop manufacturing systems as the productive systems of significance.

In the short span of 200 to 300 years the industrial world developed from handicraft production systems to the highly efficient industrial machine of today. It was in this arena that much of what we know today about the *management* of productive systems was developed.

The fact that manufacturing systems were significant in society focused resources on the solution of problems. Some of the most capable managers put their attention on problems of Production/Operations Management. Production problems attracted the attention of truly outstanding economists, mathematicians, engineers, sociologists, psychologists, and students of the managerial process itself. The result has been a relative abundance of physical goods at low cost, available in a fantastic range of items undreamed of by our ancestors. The result has also been a body of knowledge, experience, and technique dealing with forecasting, design, layout, job design, automation, scheduling models, inventory models, statistical quality control, computers, simulation, waiting line models, mathematical programming, and so on. There is also a sour note—the fantastic production machine also resulted in pollution and many more repetitive, dull jobs.

SERVICE SYSTEMS BECOME SIGNIFICANT

Just as conditions in the past focused resources and attention on manufacturing systems, current conditions have focused attention and resources on new problems. In the past the operations phases of activities, such as health care, education, transportation, and retailing, were carried on at almost the handicraft level. We were suddenly jolted into realizing that while our attention had been fixed on the production of goods, very dramatic changes were developing. Health care and education had grown into huge systems and attracted attention (criticism) when their costs began to increase rapidly.

As the productivity of the economy increased, a reallocation of personal expenditures was taking place, as is so clearly indicated by Figure 1-1. Note that since about 1945 (the end of World War II), services have steadily increased from about 33 percent to nearly 46 percent of personal consumption in 1977, largely at the expense of consumer nondurable goods. Consumer durable goods (dominated by automobiles and household equipment) have remained at an approximately level percentage of personal consumption expenditures since 1950. Of course, the absolute expenditures in all three categories have been increasing, but expenditures for services have been increasing much faster than that for goods.

The same general picture emerges for the United States economy as a whole. Services, as a percentage of Gross National Product (GNP), have increased from about 30 percent in 1948 to about 46 percent currently. The increase in the percentage for services relative to goods in GNP is due both to the reallocation taking place and to inflation in the cost of services. The reasons for the cost increases are undoubtedly complex, but there is general agreement that productivity in the service sector has not increased as it has in manufacturing. During that period, formerly underpaid personnel in professions such as teaching and nursing finally reaped the benefits of increased demand for their services.

The Example of Medical Costs

Medical care is one of the areas that has been the target of critics because of skyrocketing costs. The total U.S. health care bill has more than doubled during the last decade, and in 1971 it stood at $70 billion per year (about $324 per person per year or about 7 percent of the GNP). These expenditures have increased at an average

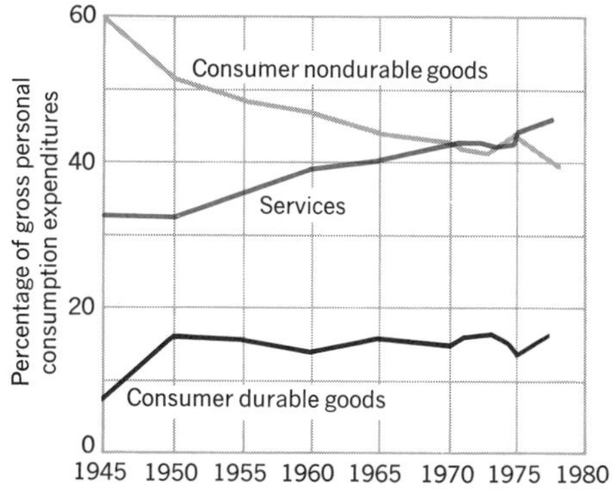

FIGURE 1-1
Relative importance of services and goods in personal consumption expenditures.
SOURCE: *Economic Report of the President, 1978.*

rate of almost 13 percent per year during the last decade (14 percent in the last five years) [Milsum, Turban, and Vertinsky, 1973].*

Figure 1-2 confirms some of the reasons for concern. Since 1950 the price index for medical care has risen from 100 to over 300, while the general index of consumer prices has risen to only 205. While the price index of services has far outstripped the general index, medical care has been the price leader. Of course, during the period from 1950 to the present, there have been substantial increases in productivity to help offset the effect of the general price increases, as is shown in Figure 1-2 (output per worker-hour increased from 100 to 189 between 1950 and 1972). Otherwise, the general price index would undoubtedly have increased faster than it did. In general, services have not benefitted from productivity increases. Medical care cost increases are probably more complex because advancing medical technology has called for huge increases in overhead costs to finance expensive diagnostic and treatment equipment.

The Example of Postal Costs

Nearly everyone has offered advice to the post office, and this steady stream of criticism culminated in a presidential commission. The commisssion reported in 1968 that the problems of the department were due largely to the form of management system used. It recommended that the department be replaced by a largely nonpolitical postal corporation with powers to determine its own policies and procedures. Such a corporation was established, and the new U.S. Postal Service has undertaken a review of all of its activities.

Without attempting to oversimplify post office problems, at least part of the story

* References enclosed in brackets are listed alphabetically at the ends of chapters. When the authors' names are used in a sentence, only the publication date is enclosed in brackets.

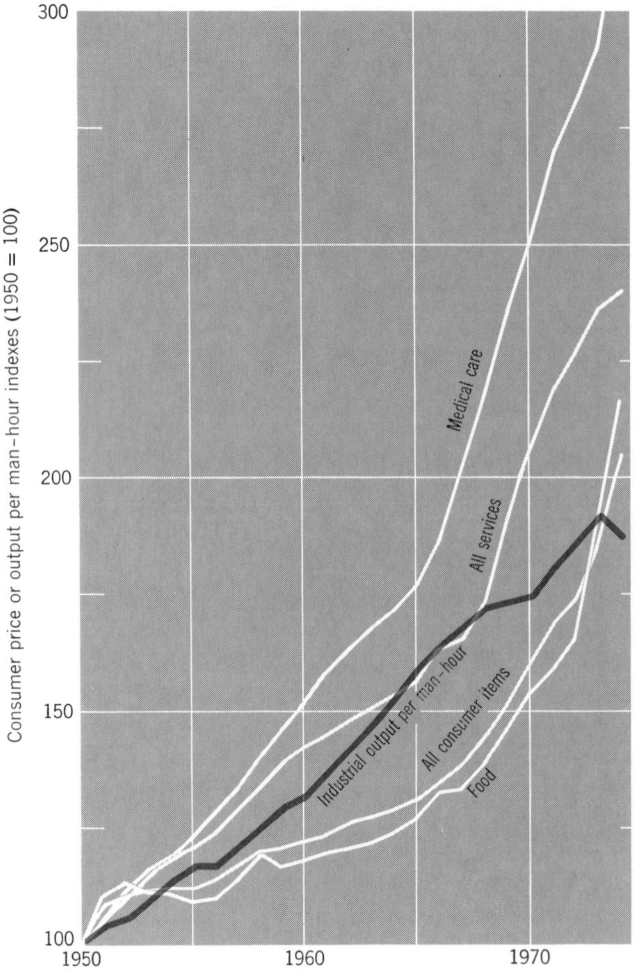

FIGURE 1-2
Comparative price indexes and industrial output per worker-hour index (1950=100).
SOURCE: *Economic Report of the President, 1978.*

is told by Figure 1-3, which gives indexes of salary and productivity trends for the U.S. Postal Service and for United States industry. While average salaries in industry have increased markedly from 1956 to 1967, the effect has been offset substantially by increases in productivity. For the post office, it has been necessary to transmit the salary increases (and other costs) almost directly into price increases to the using public, since productivity has increased only modestly. Meanwhile, the post office must cope with a mail volume growth rate that exceeds the population growth rate. The volume was approximately 80 billion pieces of mail in 1970 and is projected to be 120 billion pieces in 1980.

From the Corner Grocery Store to Significant Productive Systems

Other kinds of nonmanufacturing systems that were once regarded as simple, quaint, and insignificant have undergone great change. The corner grocery store was replaced by the supermarket, which has significant problems in such areas as forecasting, supply, inventory management, facility layout, and material handling. The individual hamburger "joint" is now a franchised mass food preparation service with operations problems that parallel those of the supermarket.

Other franchised operations, such as motels, face significant operations problems

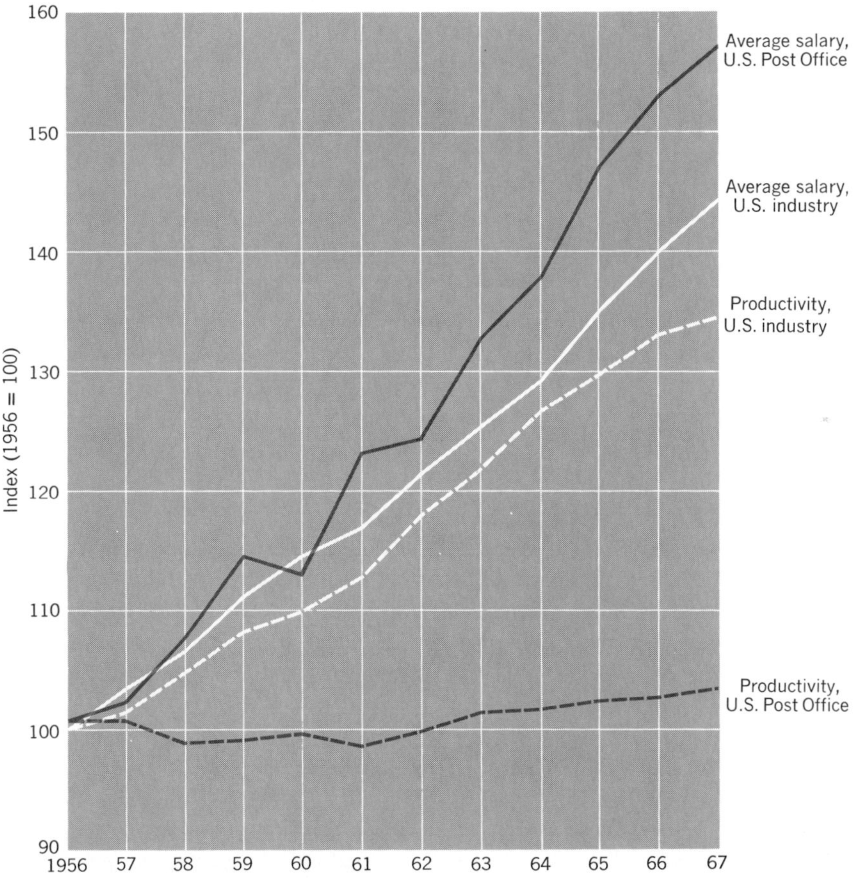

FIGURE 1-3
Productivity and salary trends (1956–1967).
SOURCE: *Post office salary, Post Office Department Annual Report; Post office productivity (weighted), Robert R. Nathan Associates; U.S. industry data, Economic Report of the President, February 1968. Reproduced from C. C. McBride, "Post Office Mail Processing Operations," in* Analysis of Public Systems, *edited by A. W. Drake, R. L. Keeney, and P. M. Morse, MIT Press, Cambridge, Mass., 1972.*

that were not so important when every manager was an individual owner-entrepreneur. Banks have broadened their range of services with branch banking, becoming ardent users of computers and facing large-scale office operations. Other financial institutions, such as insurance companies, face mass information-processing problems of a similar nature.

These kinds of organizations have significant productive systems because they are now significant in our society. Educational and medical systems are currently in high demand, soaking up huge resources. They deserve the attention of the management scientist, the operations researcher, and the industrial engineer. Indeed, society demands that the service sector become as efficient as it can be at converting its input resources to needed services while still maintaining certain quality levels of service. At any rate, these nonmanufacturing systems have taken their place as productive systems of significance.

While we observe that the set of productive systems in the service sector exhibits many very important differences from manufacturing, our thesis is that the two kinds of systems are alike in more ways than they are different, from the point of view of Production/Operations Management. The possible transfer of concept, technique, and experience is certainly not on a one-to-one basis, but significant transfer is possible. We will not attempt to offer expert advice to those who manage these service systems, but we include them for study along with other well-known productive systems. Then we can begin to understand how and why these systems work. Those who have already attempted to transfer concepts, techniques, and experience to service-oriented environments have not found that it was easy to make improvements.

PRODUCTIVE SYSTEMS—A DEFINITION

We define productive systems as *the means by which we transform resource inputs to create useful goods and services.* The productive process is one of transformation or conversion, as shown in Figure 1-4. The resource inputs may take a wide variety of forms. In manufacturing operations, the inputs are various raw materials, energy, labor, machines and facilities, information, and technology. In service-oriented systems, the inputs are likely to be dominated by labor, but depending on the particular system, machines, facilities, and technology may also be important inputs, as in health care, for example. In food service systems, raw materials are an additional important input.

The conversion process itself involves not only the application of the technology but also the adroit management of all the variables that can be controlled. This is where Production/Operations Management is effective in designing and refining or redesigning the system and in planning and controlling operations, as indicated in Figure 1-4.

The essence of effective Production/Operations Management is to see the interrelationships of all the variables and to view the entire process as an integrated system, insofar as possible. When everything works properly, we have outputs of products and services that meet quantity, quality, and cost standards, being available when

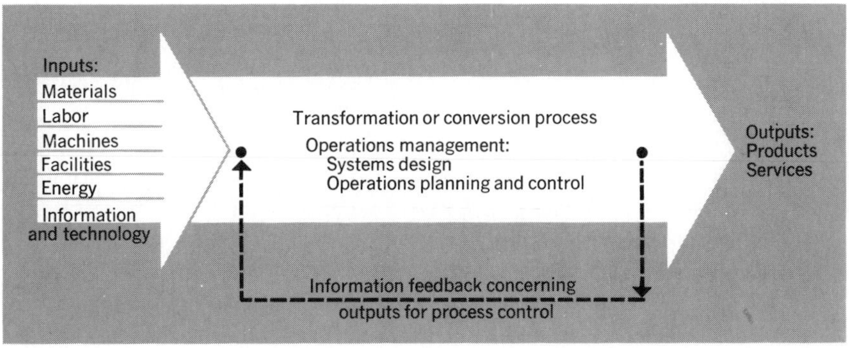

FIGURE 1-4
Productive systems as transformation or conversion processes.

needed. We will see the basic structure shown in Figure 1-4 emerge again and again as we develop the framework for addressing the problems of Production/Operations Management.

HISTORICAL PERSPECTIVE

The Ancients

We do not know who conceived the first productive systems. We do know that the great monuments of the ancient world required both technical know-how and a managerial system that organized resources, made grand plans, and executed those plans with excellent results. Examples are the Egyptian pyramids and sphinxes at Gizeh in about 2500 B.C., the Greek Parthenon in about 440 B.C., the Great Wall of China in about 214 B.C., and the construction marvels of the Roman world—aqueducts, public buildings, roads, and temples—which span a period including at least 400–100 B.C.

During and following the period of the building of the ancient monuments, a wide variety of products was produced through the handicraft system. Production/Operations Management began to develop with the Industrial Revolution, since it was during that period that the factory system evolved out of the handicraft system. A series of changes in industrial technique, and in economic conditions, made possible the development of larger productive units.

The Industrial Revolution

In 1764, James Watt made improvements on the steam engine that made it a practical power source. As a result, external energy sources began to replace muscle power in industry.

During the same period, a rationale for production economics was given its first impetus by the Scottish economist Adam Smith in his book, *The Wealth of Nations* (1776). Under the factory system, division of labor was developing as a commonsense method of production when a relatively large group of workers was brought together to produce in larger quantities. Smith observed that there were three basic economic advantages resulting from division of labor. These were (1) the development of a skill or dexterity when a single task was performed repetitively, (2) a saving in the time normally lost in changing from one activity to the next, and (3) the invention of machines or tools that seemed to follow normally when efforts were specialized on tasks of restricted scope. Smith's book was a milestone in the development of production economics, because a great scholar had recognized that there existed a rationale for production. As we know, the application of division of labor progressed rapidly in this century, to the extent that many people feel that it has gone too far, resulting in boring and dehumanizing work conditions.

An Englishman, Charles Babbage, augmented Smith's observations about division of labor and raised a number of provocative questions about production, organization, and economics. His thoughts were summarized in his book, *On the Economy of Machinery and Manufactures* (1832). Concerning the economic advantages resulting from division of labor, Babbage agreed with Smith, but he observed that Smith had overlooked a most important advantage. Babbage noted the pay scale for different specialties and pointed out that if the shop were reorganized so that each person performed the entire sequence of operations, the wage paid would be dictated by the most difficult or rarest skill required by the entire sequence. Thus the enterprise would pay for the highest or most limiting skill, even when the workers were performing routine tasks. With division of labor, however, just the amount of skill needed could be purchased. Therefore, in addition to the productivity advantages of division of labor cited by Smith, Babbage recognized the principle of limiting skills as a basis for wage payment.

The sources of the productivity increases through 1920, shown in Figure 1-5, were probably dominated by the substitution of machine power and the application of division of labor. But late in this era, the seeds were being sown for dramatic changes.

The Scientific Management Era

Just before the turn of the century, Frederick W. Taylor set in motion a managerial philosophy which he called "scientific management." The practice of the day was to allow workers to decide the means by which production would be achieved. They determined how to produce, according to their skills and past experience. The time required and the cost of production were guided by traditional methods.

"Boondoggling" and spreading of the work were common. Taylor was familiar with these practices, because he entered the industrial system as a worker. He refused to go along with the other workers and instead produced as much as he could. He advanced rapidly and was later in a position to experiment with some of his ideas.

Essentially, Taylor propounded a new philosophy which stated that the scientific

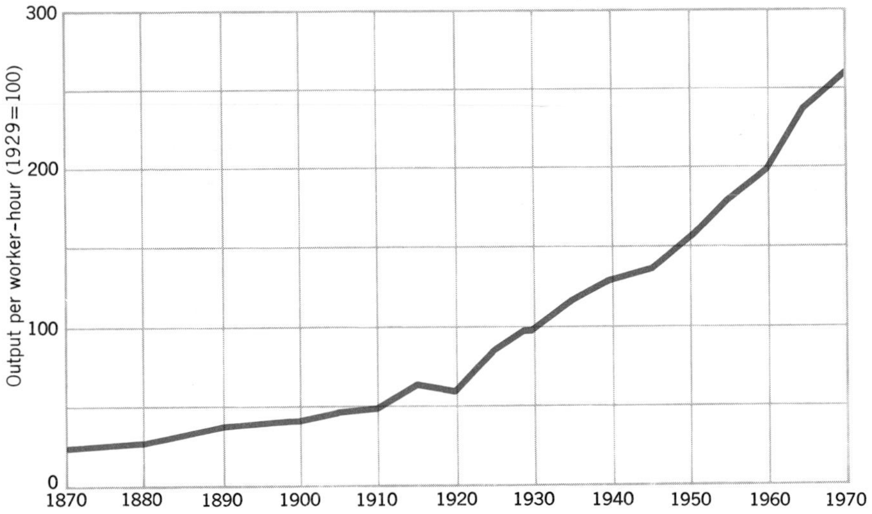

FIGURE 1-5
One hundred years of productivity growth. Output per worker-hour in United States manufacturing, 1870–1972.
SOURCE: *J. W. Kendrick,* Productivity Trends in the United States, *Princeton University Press, Princeton, N.J., 1961, and the Bureau of Labor Statistics.*

method could and should be applied to all managerial problems. He urged that the methods by which work was accomplished should be determined by management through scientific investigation. He listed four new duties of management that may be summarized as follows:

1. The development of a science for each element of work to replace old "rule-of-thumb" methods.

2. The scientific selection, training, and development of workers, instead of the old practice of allowing them to choose their own tasks and to train themselves.

3. The development of a spirit of hearty cooperation between the workers and management to ensure that the work would be carried out in accordance with the scientifically devised procedures.

4. The division of work between workers and management in almost equal shares, each group taking over the work for which it was best fitted.

These four ideas are so much a part of present-day organizational practice that it is difficult to believe that the situation was ever any different. The science that Taylor envisioned was very slow to develop. The techniques, such as time study, production control boards, and wage incentive plans, came to be thought of as scientific management. To Taylor's chagrin, the broad general philosophy was often glossed over.

The effects of the scientific management era can be seen in the dramatic rate of increase in productivity that developed during and after World War I (see Figure 1-5). Although the scientific management era produced great controversy, its results were to revolutionize managerial thought and practice.

Although Taylor was disappointed in the slow evolution of his science, important roots developed in the era that were a forecast of the modern era. These were the development and application of mathematical and probabilistic concepts. In 1915, F. W. Harris developed the first economic lot size model for a simple situation. In 1931, Walter Shewhart developed and introduced to industry statistical quality control concepts. Finally, in 1934, L. H. C. Tippett developed a sampling procedure to determine standards for work delays and work time called "work sampling." These early applications of mathematical and statistical technique foretold future events.

The Modern Era

The current rapid development of concept, theory, and technique began shortly after World War II. Research in war operations by the armed forces produced new math-

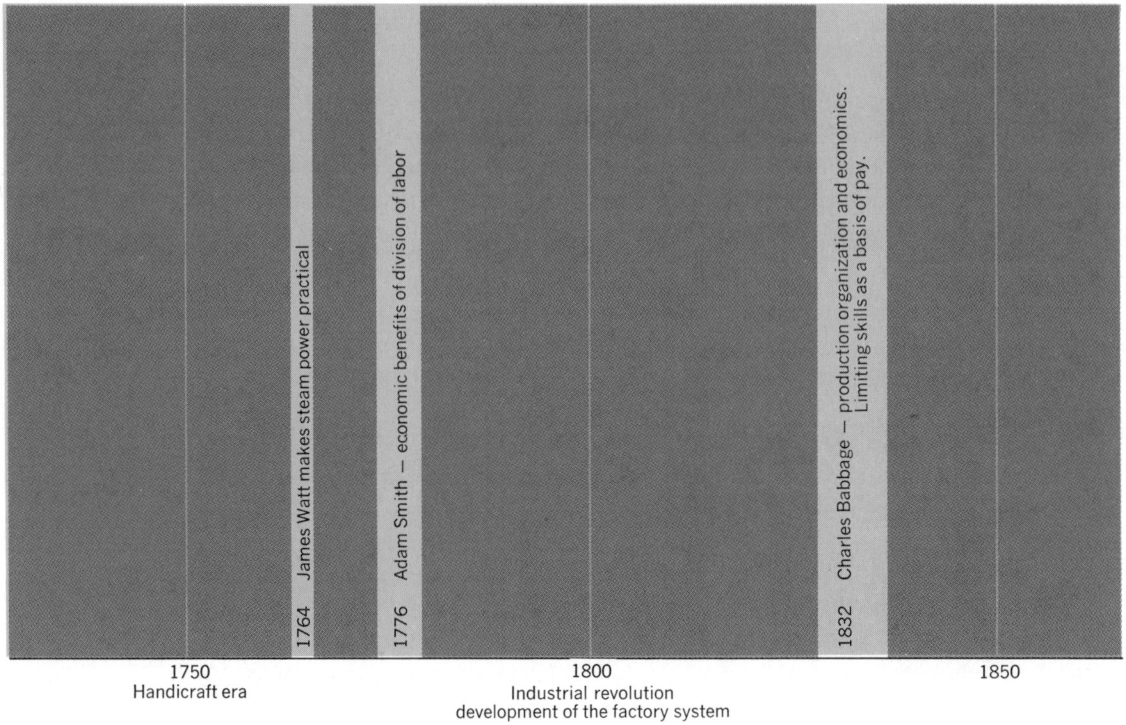

FIGURE 1-6
Milestones in the development of productive systems and operations management.

ematical and computational techniques that were applied to war operations problems. These problems seemed to parallel problems that occurred in productive systems, and the approach to war problems began to trickle into industrial use.

The original proposals for the applications of operations research to industry revolved around a broad systems approach to managerial problems. The techniques that were proposed for use were conceived of as modes of implementation. For a considerable period of time, however, the emphasis was put on the new and powerful analytical and computational techniques, and much of the broad systems view of problems became lost. Research and application became centered on such things as models of inventory control, mathematical programming, PERT/CPM, scheduling techniques, simulation, and waiting line models. The broad philosophical framework of systems analysis seemed to be something to talk about instead of something to do.

As with scientific management, there probably are good reasons why the current era has emphasized technique. It is much easier to work on a smaller problem of restricted scope than to grapple with the large-scale problems. The systems concept requires models that take account of higher-order interactions between organizational

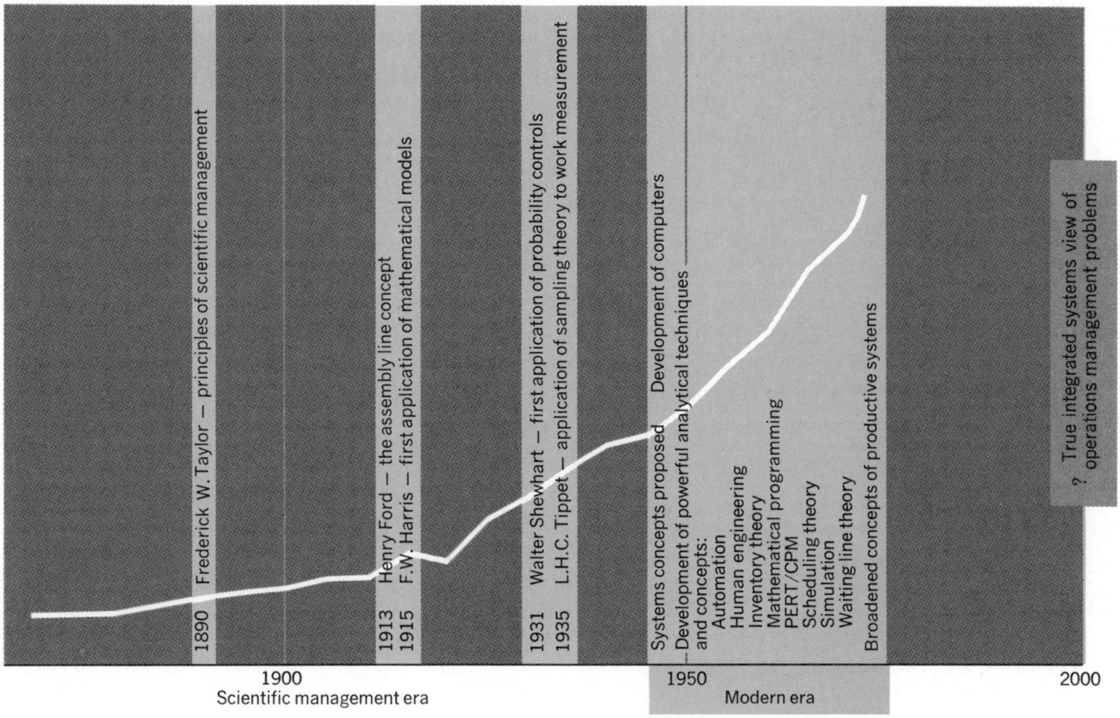

units. This is not only difficult to accomplish in a conceptual sense but may approach the impossible in an organizational sense.

Currently there is a resurgence of interest in the broad conceptual framework of systems, as well as a broadened view of what constitutes a productive system. At some point in the future, we can expect to achieve a truly integrated systems view of operations problems.

Figure 1-6 places the milestone events in perspective with the curve of productivity increase. The average annual rate of increase during the modern era is 5.3 percent, although this average rate was not maintained during the 1973–1975 recession. The modern era is characterized by an acceleration of the substitution of machine power and the use of machines for computation and control. The specialization concept has been broadened and applied at all levels including work, specialization within organizations, and industry. However, the element that is unique about the modern era is science in management within the general framework of systems concepts.

THE FUTURE OF INDUSTRIALIZED SOCIETY

The modern era with its systems approach and powerful new tools of analysis will be tested severely in the future. Present trends of population growth, natural resource drain, and pollution are on a collision course with industrialized society. Some predict disaster for the existence of life on this planet. It is clear that the world population is increasing at rates for which adequate support systems may reach limits and cause a reversal and decay of living standards. At the same time, the effects of finite natural resource limits are being recognized, and indeed the methods of systems analysis have been used to analyze these complex problems and to predict the impact of alternatives.

No one really doubts that the population-resource-pollution complex of problems is of great significance. The delivery systems for all kinds of products and services must be improved and must meet the requirements of a broader set of criteria if we hope to keep pace. This is the role of Production/Operations Management in society.

From a quite different point of view, there will likely be changes in industrialized society. Some say that the postindustrial society has already arrived and that values are changing. The design of productive systems must find an appropriate role for human beings and not treat them simply as links in a giant machine. Thus, even presuming that we survive pollution and gain control of population and limited natural resource problems, industrialized society seems destined for dramatic changes.

PRODUCTION/OPERATIONS MANAGEMENT—THE MANAGEMENT OF PRODUCTIVE SYSTEMS

We pose three questions: What are the long-term strategic decisions in Production/Operations Management that commit major resources and set the course for some time to come? What are the day-to-day, week-to-week, and month-to-month decisions that guide an ongoing operations system? How does one design a productive system, in terms of jobs, the processing required, and the physical flow and arrangement?

Regardless of the nature of the service or product for which the productive system is designed, there is a general rationale. Let us begin by examining a well-known service with these three questions in mind. The following case example covers Production/Operations Management in a nutshell and sets a framework for the entire book.

BURGER—A CASE EXAMPLE

Charles A. Berg has had the feeling that there is a very substantial market in his town for a good-quality hamburger priced under 80 cents. He feels that it can be delivered on demand, together with trimmings and a few limited options such as french fries and soft drinks at similar low prices. He thinks that the phrase "delivered on demand" should be translated to mean that a customer should be able to walk up to the window and place and receive an order within one minute. Even in peak hours, he feels that the customer's total wait in line and time at the window should not exceed four minutes. The quality should be consistently good, but gourmet quality is unnecessary.

Initial Design. After testing the idea informally on friends, Berg decides to set up a pilot operation for an initial experiment. At this stage, he is his own expert on everything and applies good common sense and the cliches of modern business. These include the use of his concept of production line methods and buying raw materials in quantities that seem to strike a balance between inventory costs, ordering costs, and quantity discounts. Berg's whole concept is geared to an attempt to match the output rate to the expected hourly demand rate. The hamburgers, french fries, and other items are produced at these rates and placed under infrared lamps to keep them warm. Any item not sold within five minutes is scrapped.

Everything seems to point to a successful experiment with some rough spots in day-to-day operations. Material costs are higher than anticipated, largely because of scrap. The variability of the demand pattern also affects the amount scrapped. Labor costs are higher than anticipated because the daily demand pattern has large peaks, and proper staffing has resulted in considerable idle time during the off-peak periods. The main thing that he has been able to maintain, however, is his concept of quality and service. Because of this, the volume of business has been excellent.

Berg has come to some conclusions about the workable size for such an operation and the kind of location that is most appropriate. He opens a branch across town in an excellent location and makes some modifications in the way he schedules labor. He also reduces the required labor input slightly by letting customers select their own items and simply pay the cashier. At this point, Berg establishes a new name for the enterprise, simply "BURGER," which he has emblazened high over his two units.

Expansion Plans. The results at the branch are beyond expectations, and Berg is now planning for a large expansion, including possible franchising of the concept. Substantial capital is now available, as is the expertise to proceed. In addition, past experience has provided very important data (e.g., the characteristics of a good location, estimates of the optimal size of a branch, the expected product mix, the expected arrival rate of customers for each hour of the day, and some ideas about how to mechanize somewhat and otherwise improve the processing setup). Berg knows, however, that the current objectives take him well beyond the corner hamburger joint technology. He is now planning a system of BURGER joints, supported by advertising and image building with larger-scale supply, storage, and control problems.

Now let us review the developments at BURGER to extract a rationale for the operations and for the system design. We started conceptually with attempts to make market predictions and long-range plans for the organization. We sharpened our notion of the business we were in, particularly the nature of the food service we provided and the quality level of the food. We made broad studies of the size and location of markets and established a five-year plan for sizes and locations of BURGER branches. These plans became important inputs for specifying the design characteristics of the service, which in turn had an impact on the design of the productive system that we established. Theoretically, the design of the products and services are now fixed, but we shall see. At this point, stop and think about how and why the design of products might change in the future.

We now concentrate our attention on the design of the complex system of jobs and processes where we plan to design a modular branch, and the system of branches. The inputs to the design of the system are from forecasts of demand, analysis of the product or service, technical requirements, and data about human behavior. These forecasts provide data useful in determining a peak and average capacity in the general sense, but the output rates are affected by the processes used, the work flow layout, and the way we decide to put together the work teams.

The first design was based largely on our experience with the most recent branch, but preliminary cost figures suggest that both labor and material costs are too high. We relocate some processes at the central warehouse, because they can be done on a relatively large-scale basis involving some mechanization and automatic material handling (e.g., chopping and extruding hamburger patties and cutting french fried potatoes). This also simplifies operations at each branch and allows the production line concept to be replaced by a rotating team approach to staffing branch systems. The latter concept was recommended after an expert in human behavior analyzed the present setup and found boredom and dissatisfaction among the young people who worked in the branches. The expert also stated that the rotating team concept had other important advantages, since each person on the team learned all aspects of the processes. This broader job structure results in flexibility in case of absence or illness, and the work variety should relieve the boredom problem.

The basic work pattern seemed well developed now, but the work design expert quipped, "The only one idle during this process is the customer for an average of two minutes. What can we have him do that is useful?" An interesting question, but it was passed over for the moment, being kept in reserve. It did point up the fact, however, that we might change the service offered because of impact from the design and the job-process system. Because of the high material cost, another change in the product was considered but rejected—introducing a bit of cereal in the hamburger patty.

Another aspect of the system design centered on the material supply system from the central warehouse. There were two important parts of the supply system. One was concerned with the inventory levels to be maintained at the warehouse and at each branch to ensure continuous service to the customer. Storage adequate to ensure a full week's supply was provided at the warehouse plus two days' supply at each branch. A second problem was to determine an efficient supply routing to all branches each day.

The final overall system design was rendered in terms of a design for the modular branch. The components of the design included a work flow layout and architectural plans, a labor staffing plan that specified just how each operation was to be carried out, including labor time standards, a material flow plan, a customer flow plan and layout, the central warehouse design including storage layout and logistics supply system, the production operations layout, equipment specifications, a schedule for the acquisition of property and the building of branches, and a complete financial analysis.

Operating the Expanded System. In order to manage the ongoing operations of the expanded concept of BURGER, it was recognized that a system of operating plans and controls was needed. These elements were necessary at the broad general level to chart the course for what was becoming a massive operation of a specialized type. Now, however, more detailed plans and controls were also needed to be sure that the details of the operation, which had been so finely tuned, were maintained within current objectives.

It was felt that the heart of the BURGER empire (as it was now being called) was its superb service in a spotless place where one could get consistently good food at a low price. These characteristics of the quality of the food service offered by BURGER had to be maintained. This required a well-trained, highly motivated workforce and reliable equipment. On the other hand, the massive supply problem required good planning, expert buying, close control over perishable products, and a distribution system that ensured that each branch had adequate supplies for sale. In addition, these broader plans had to take account of the rapid expansion in the number of both owned and franchised outlets.

At the outset, a forecasting system geared to the needs of day-to-day, week-to-week, and month-to-month operation was needed. Such a forecasting system had to reflect what was actually happening to the demand for BURGER'S products, including trends and seasonal factors. The key forecasting factor, to which all other sales were tied, was the hamburger, so a statistical forecasting system was developed based on historical data. Each individual monthly forecast took account of current trends and seasonal factors. The forecast was the heart of the entire planning and control system, since the other plans for aggregate levels of operation, materials control, and labor allocation were tied to the forecast of demand.

On the broad aggregate level, a measure of capacity and output was needed in order to make intermediate-term plans and schedules in relation to demand. Such a measure was also needed to form the basis for orderly plans to bring new outlets into the operating system. Since virtually every customer order involved a hamburger, the basic unit of activity chosen was 100 pounds of hamburger sold. An average meal then involved a proportional amount of french fries, beverages, and so forth. Raw material ordering could be geared to this unit, as could other inputs, such as labor. An aggregate planning and scheduling model was then constructed that was driven by forecast of demand, the model yielding the best combination of regular and overtime labor, and provided a basis for advance ordering of raw materials. The aggregate plan and schedule was then allocated to each branch. The aggregate plan was updated each month and presented for management approval and decision.

All raw material ordering was placed on a periodic inventory control system. Each week an order was placed for the amount sold the previous week plus or minus an adjustment for the anticipated sales in the upcoming week, based on data from the aggregate schedule. A buffer inventory of one week's supply was established as a minimum stock to absorb fluctuations in demand and variation in the supply lead time. Branch inventory was brought up to two days' supply each day with daily deliveries, based on current sales rates. Throughout the inventory system, "first-in, first-out" rules were maintained to keep perishable materials as fresh as possible. The entire system was computerized through the management information system (MIS) and updated daily. Each company-owned branch supervisor's performance was partially measured in terms of effective use of the raw materials provided.

While the aggregate plan provided total values of approved regular and overtime work, a separate labor allocation was generated for the warehouse and owned branch outlets, based on branch forecasts and aggregate forecasts. Each supervisor was then given a monthly budget for regular and overtime labor to implement. This budget would allow the supervisor to hire or lay off personnel and use part-time and overtime work to meet the daily, weekly, and monthly fluctuations in labor needs. Each company-owned branch supervisor's performance was partially measured in terms of the effective use of labor.

Quality of product and service was a major concern, as noted previously. The main measures were consistent quality of food (hot items served hot, with meat still moist and bun not soggy, french fries hot and still crisp), customer and kitchen area cleaned continuously, and service times maintained within one minute for service and a maximum customer waiting time of four minutes. With these measures of quality, an inspection brigade regularly made the rounds to all branches on a random-call basis to measure service time and rate the other measures of quality on a rating form. The results were given to branch supervisors as well as to managers of franchised outlets and were summarized in weekly reports generated through the management information system (MIS). Each company-owned branch supervisor's performance was partially measured in terms of the quality record. For franchise outlet managers, the quality measures were related to contract clauses, whereby a franchise could be revoked for consistently poor-quality performance.

Maintenance of plant and equipment was also regarded as an important function.

The cleanliness of the kitchen and customer areas was the responsibility of the individual supervisor. On the other hand, equipment and building maintenance was provided centrally. A regular preventive maintenance program on all equipment was maintained during the night shift, since the equipment was in use continuously every day. A small maintenance crew was used during the day to respond to emergency breakdowns.

Surveillance over the entire operation was conducted through the aggregate plan and a system of budgets and supervisory incentives. Profit performance was the key, with individual components of material utilization, labor cost, and quality performance. Maintaining the delicate balance of operations involved the constant monitoring of the operating plans and control systems. We want to be sure that they represent continuing operations planning models and control reports, particularly since the situation is dynamic.

On a broader scale, there was the attempt to plan and control for the system as a whole through aggregate plans and schedules and the budgets and reports. Actual performance was compared with budget and was tied to a supervisor's and operating manager's incentive bonus plan.

Reassessment of Long-term Goals. As the BURGER system grew, Berg periodically reviewed growth rates in relation to current capacity. His long-term strategy was to maintain a small but discernible gap between perceived future demand in marketing areas and system capacity. Financing new capacity was achieved through a combination of debt and retained earnings.

When new capacity was added, the new units were integrated into the existing systems. Periodically, Berg called for a review of existing systems because he felt that simply adding new units to existing systems could result in suboptimal operation.

PROBLEMS OF PRODUCTION/OPERATIONS MANAGEMENT

Reviewing the BURGER case example, and keeping in mind the three questions posed earlier, we can abstract a fairly typical list of problems.

The key strategic planning decisions were:

1. The selection and design of the service to be offered. It was finalized through assessment of interactions between the original concept, estimated costs of operation, equipment configurations, and alternate job or work crew designs.

2. Capacity planning decisions that also determined the locations for warehouses and branches and a growth plan.

3. A supply, storage, and logistics system.

The key problems and decisions in the day-to-day operations of BURGER were:

1. Forecasting sales as a basis for planning and setting schedules.

2. Aggregate plans and schedules concerning how to allocate productive capacity consistent with demand.

3. Detailed scheduling of personnel and equipment in order to strike some balance between labor costs and the value of good service and scrap. Scheduling of work shifts and the assignment of personnel to the shifts so that the variable load was covered seven days a week.

4. Inventory controls regarding the continuous supply of materials while maintaining reasonable costs of inventory and spoilage.

5. Quality control, setting permissible levels for unacceptable quality, as well as the definition of good quality for products and services. The adroit balance of these factors was regarded as crucial for BURGER.

6. Maintenance of the reliability of the system. With reference to maintenance effort, recognition of the random nature of equipment breakdowns and the fact that equipment downtime may be associated with important costs or loss of sales.

7. Cost control. Although many of the controls were meant to control costs, it is important to note that the control itself was on the activity that may generate cost, such as materials, labor, quality, and scrap. Cost control is then the derivative of activity control.

The mature overall system design was rendered in terms of a design for the modular branch. The components of the design included a work flow layout and architectural plans; a labor staffing plan that specified just how each operation was to be carried out, including labor time standards; a material flow plan; a customer flow plan and layout; the central warehouse design, including storage layout and logistics supply system; the production operations layout; equipment specifications; a schedule for the acquisition of property and the building of branches; and a complete financial analysis.

In reviewing what was done and how it was done, it appears that we employed an iterative design process, involving design and redesign to take account of various interactions. We employed long-range planning concepts, prediction and forecasting techniques, layout planning and work place layout, equipment justification techniques, behavioral work concepts, product and service analyses, and waiting line methods and concepts. The conception of the process and facilities design seemed to represent an integrated view of the conversion process for the system as a whole. In reviewing the longer term strategic decisions that were made at BURGER, we must realize that they were of great significance to the future of the organization. The decisions set the basic approach to supply, distribution, and operations for some time to come and committed the majority of the available capital of the enterprise.

The key decisions that set the design of the productive system were:

1. The design of the service to be offered, finalized through the iterative process.

2. The selection of equipment and processes from among alternate technologies.

3. Job and work crew designs, including a crew schedule plan to take account of daily demand variations.

4. Detailed physical layout to accommodate work flow, equipment, and personnel, as well as the flow of customers.

We have dealt with a food service example in order to discuss the general nature of problems, but we could just as well have used a manufacturing activity or a pure service activity. The relative importance of the problems would have changed considerably with the nature of the system. For example, scheduling of workers and equipment would have been extremely important in a manufacturing system, as would the problem of equipment investment and maintenance. Most service systems are labor intensive, resulting in emphasis on problems of maintaining quality and scheduling labor in work shifts for highly variable demand.

DECISION MAKING

The central role of management is to make decisions that determine the future of the organization. Decision making is complex, because the systems with which we deal are complex and involve multiple criteria. This is why we discuss systems concepts later in this chapter. It is why we will constantly attempt to maintain a systems context, even when we are discussing seemingly separable elements of Production/Operations Management.

Thus, a manager in deciding on the design capacity of a new productive unit cannot arrive at a conclusion based simply on an objective financial analysis. In some situations, the balance of capital and labor costs (low capital, high labor) might suggest that the most economical plan would be to produce according to the dictates of a seasonal demand curve, incurring very little inventory risk. The design capacity would then be set to accommodate the peak demand if the only factors to enter the decision model were the capital and labor costs.

A good manager, however, would recognize that there are a number of possibly interacting criteria. Some criteria of considerable importance might be the state of labor relations with a sharply fluctuating employment level, the impact on the community, and the problems of finding enough qualified people to staff for peak production. Other factors might relate to the maintenance of quality standards, maintaining production controls with widely fluctuating production rates, and widely fluctuating storage needs for raw materials, in process materials, and finished goods.

Complex Decisions Require Multiple Criteria

Why emphasize the points about the complexity of decisions, the importance of both objective and subjective values, and the need for viewing decision problems in a systems context? First, with the growth and development of quantitative methods of analysis, there has been a temptation to make decisions based only on the objective factors. An analytical or simulation model is attractive and persuasive. Although model builders have been very clever in including a wide variety of variables in models, it is not yet possible to include most subjective and behavioral variables, and perhaps it never will be possible.

Decision makers should welcome the objective analysis, but should not assume that all variables have been included. Indeed, perhaps the main reason that most decision makers are people rather than machines is because tradeoffs must be made involving judgment. The second reason for emphasizing complexity of decision making is that the tools to deal with complexity are rapidly being developed. System science is with us, and we must learn to use it. The third reason is related to our framework for decision making. It might be tempting to interpret Figure 1-7 and our comments about decision making only in terms of mathematical models and objective criteria. Therefore, we issue a broad disclaimer. Throughout the book, we assume that the decision maker should consider trade-offs between the objective and subjective criteria.

Later, in chapters where we develop the objective factors of a model, we must recognize that we have for a time closed down our horizon to look at local behavior of certain variables. We may emerge with concepts such as economic order quantities, line balancing, or a hospital admissions rule. These concepts are still out of the context

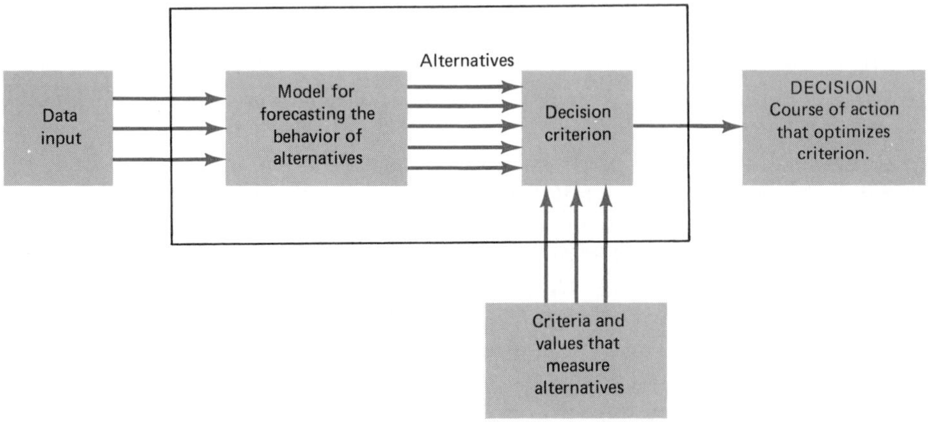

FIGURE 1-7
Decision-making structure.

of a specific organization situation. They are in fact inputs to the broader decision process and subject to trade-off and possibly even dominance by factors not in the model.

Models. Figure 1-7 shows the relationship of major elements entering a decision-making structure. The function of the model is to describe behavior and to forecast behavior of alternatives which might be considered. Using the model, we can generate alternate solutions that are to be appraised by the decision criterion. Here is where we often have trouble, because the model builder often includes a objective function, usually a cost or profit function, in the model. With powerful optimizing methods, such as linear programming, an *optimal* solution may be generated. The difficulty may be with what happens from that point on in the process. A so-called optimal solution to a problem seems to say, "this is the answer," and often has the effect of intimidating the decision maker to accept it as is, or reject it. Too often decision makers reject, because they may not trust the model or legitimately feel that other important factors enter the problem which are not taken into account.

A much more appropriate presentation of results of a model would be to include some range of solutions, including the optimal, and to indicate the costs and other effects associated with each. The decision maker is then in a position to see how trade-offs can be made and see the cost of these trade-offs.

At any rate, Figure 1-7 indicates that the criterion has multiple values and should be thought of as being separate from the model, even when the model contains its own objective function.

PERFORMANCE CRITERIA AND GOALS

The ways that consumers of goods and services evaluate the performance of productive systems is clearly understood. If one were to survey consumers, they would invariably express the degree of their satisfaction in terms of three dimensions: cost or price; availability, that is, delivery time or waiting time; and quality. This is not to say that consumer behavior is simple, for there is ample evidence that it is very complex.

These three dimensions of consumer satisfaction, then become the dominant bases for establishing criteria for the performance of productive systems. While consumers' judgments may be relatively subjective, we must translate them into measures of performance that are as objective as possible. We must design a system to meet certain goals and to make operating decisions that maintain an appropriate balance between competing goals, or improve that balance.

The criteria actually used are surrogates for consumers' judgments. While consumers are thinking of cost to them, we must translate that into labor, materials, and capital costs, and the appropriate balance between these costs. We must think in terms of a larger system of interacting costs and cost trade-offs, as well as cost-price trade-offs. For example, the way that output is programmed for best cost performance may involve accumulating large inventories at considerable cost and risk in order to stabilize and reduce employment costs. Consumers' demand for time performance may be translated into the total time to serve them in a service operation, or delivery time for certain products, or shelf availability for certain other products. Consumers' demand for quality may translate into requirements in the design of products and the quality and consistency of workmanship. These characteristics, however, may finally be measured in such terms as chemical composition of raw materials, or tolerances maintained on critical dimensions of bearings.

To state these measures of performance sounds as though the task is objective and perhaps even simple. It conveys the impression that to improve or even optimize the performance of a productive system, we simply put our efforts into the three areas of cost, on-time delivery or availability, and quality. Some of the reasons why the management of productive systems is neither as objective as many would prefer, or as simple, are that the three dimensions of performance are not independent of each other. The time and quality factors can normally be improved with increased cost. Performance time may itself be a measure of quality as is true with many services. In some instances, extending performance time may improve quality. Therefore, in the decision process, it is the adroit balancing of the criteria that makes for the most effective decisions.

SYSTEMS

Knowledge in all fields has progressed by a "divide and conquer" philosophy that concentrates attention and resources on problems of limited scope. In chemistry and physics, new elements, nuclear fission, and countless other spectacular discoveries have been the results. In the biological sciences, discoveries of RNA and DNA are comparable.

In seeking to understand productive systems, we have also followed the pattern set by successful science research. For at least the past 50 years, efforts have been made to take components of productive systems and to subject them to detailed, even microscopic analysis. Prime examples are in the analysis of inventories, work methods analysis and work place design, maintenance systems, scheduling systems, and quality control systems. For science in general, and certainly also for Production/Operations Management, the past has been characterized by this mechanistic view in which the component was studied and the system as a whole was considered to be the assembly or sum of the parts. There is no question that great progress has been made during this mechanistic era.

More recently, the focus has changed to a study of the system as a whole. In 1947 the idea of a general systems theory was proposed by Bertalanffy. General systems theory rests on the premise that there are properties of systems that do not derive directly from the components, but from the unique combination of components that make up the whole. Furthermore, these properties make the whole (the system) add up to something more than the simple sum of its parts. A man is more than an assembly of cells, tissues, and organs*; an economy is more than a group of industries; a productive system is more than processes, personnel and materials. The behavior and performance of each is an expression of its unity. In each instance, the combination of components with unique relationship among them produces a whole that is unique. To *understand the whole, begin with the whole, not with the components!* This is not to deny the importance of also understanding the components, but it changes the focus on the components. With a systems orientation, we look at a component in terms of its reason for being as a part of the system. Thus systems concepts will take advantage of both the results of looking at the system as a whole and at the results of analyzing the proper role of components within the system.

Because of system goals and objectives, and complex relationships among components, we may need what seems like poor performance from a certain component. For example, in a production-distribution system for a standardized product, we may produce to inventory, thereby incurring a larger inventory cost than would be necessary to meet demand, particularly if demand is seasonal. Looking at the broader system values, the overall result is a more stable workforce size, better employee relations, lower costs of changing production rates, better customer service, but higher inventory costs and possibly higher overtime costs. If one looked at the inventory problem in isolation, there might be a temptation to minimize inventory-connected costs. Note the multiple character of the criteria used to judge performance. Designing the inventory subsystem takes its goals from those of the system as a whole.

PLAN FOR THE BOOK

The main topics of the book divide according to the three basic questions raised in connection with the BURGER example: What are the strategic planning decisions? What are the day-to-day decisions required to operate an ongoing productive system? What decisions set the productive system design? But we are also interested in the interrelationships between topics.

The balance of the book deals with major groups of chapters that center on Strategic Planning Decisions in Part Two, Operations Planning and Control in Part Three, and Productive System Design in Part Four.

There are also seven appendixes that supplement the main thrust of the text. These appendixes support particular topics in the main text and may be used either for review or to add more analytical substance. For example, the materials in Appendix A, Capital Costs and Investment Criteria, are useful at several points in the text where costs are analyzed. Appendixes B and C deal with linear programming and are useful where resource allocation is an issue or where transportation networks are being discussed. Appendix D, Waiting Lines, and Appendix E, Monte Carlo Simulation, are

* Regarding the behavior of whole systems, Buckminster Fuller made the following quip: "It helps explain the fact that an intelligent woman can love $4.98 worth of chemicals arranged into a form called man."

useful in dealing with certain aspects of service systems. Appendix F, Work Measurement, is an important adjunct to job and process design.

REVIEW QUESTIONS

1. Services have become relatively more important in our economy as indicated by Figure 1-1. It is also generally true that the dominant cost in services is labor. What are the future prospects for continued rapid increases in productivity in the United States economy if the conversion to services continues?

2. The rise in medical costs seems to outstrip the price increases in other services, as well as the consumer price index in general. It also continues to rise very much faster than productivity (see Figure 1-2). From what you know of health care systems, what are the most promising areas for cost reduction and productivity increase?

3. If we define productive systems simply as "the means by which we transform resource inputs to useful goods and services," what do you envision as the problems of managing these kinds of processes?

4. Productivity in manufacturing has increased steadily since 1869, as shown by Figure 1-5, and has grown at an increasing rate since about 1920. Rationalize these increases in productivity. What were their sources? What role did operations management play?

5. Postulate an output per worker-hour curve for the service sector of our economy, similar to Figure 1-5. Rationalize the shape of the curve in a way similar to that shown in your answer to question 4.

6. The concept of division of labor has been broadly applied in our society. Give examples at the level of an individual job or task, within organizations, within industries, and within society in general.

7. What did the scientific management era contribute to useful concepts of operations management? How can we measure the value of these contributions?

8. Since both the modern era and the scientific management era profess to apply scientific methods to managerial problems, how can the modern era be distinguished from its predecessor? How can we measure the independent impact of the modern era?

9. Thinking in terms of the BURGER example, outline the problems of operations management. Which of the items in your list of problems do you feel are independent of other problems?

SITUATIONS

10. The vice-president in charge of manufacturing of the Wash-N-Dry Appliance Company is pondering his problems of employment and production scheduling. He has been receiving static from the union, because employment levels have fluctuated considerably. He replied that sales also fluctuate, so why should the union expect employment to be stable? The union president maintained that employment levels actually fluctuate more than sales, and charged that the reason is bad management.

The company board of directors has just held its bimonthly meeting. Among the materials presented to the board were status reports concerning operations, a portion of which are shown in Table 1-1. The board was upset over the rapid growth in finished goods inventories. Finished goods inventories had increased by almost 42 percent in the last eight weeks, and were even 27 percent higher than a year ago. The comparison with the figures for the previous year at this time was particularly unsettling, because current sales levels actually were lower than a year ago. The board chairman instructed the president to take action immediately to reduce finished goods inventories. The president defended himself by pointing to the fact that employees were already being laid off at a rapid rate, and that the employment level was currently almost 26 percent below the level reached only eight weeks ago.

Immediately after the board meeting, the president met with the VP-Manufacturing and "raised hell" (using other colorful language not normally included in textbooks). The president demanded that finished goods inventory be slashed drastically, even if it meant a plant shutdown. "After all, that was how the auto industry reduced its inventories in the 1974-75 depression." The union president walked in at that point, having heard the president's order and justification. The discussions that followed (also censorable) involved strike and boycott threats.

Working late that evening, the VP-Manufacturing resolved to "get a handle on this problem" (or else he might lose his position). He assumed that he could

TABLE 1-1 **Abstract from Operations Report to Board of Directors, Wash-N-Dry Company**

	Current, 80 th week	8 weeks ago	52 weeks ago
Order backlog, aggregate units	5,200	5,300	5,400
Finished goods inventory, units	2,500	1,750	2,250
Finished goods inventory, dollar value	$37,500	$26,250	$33,750
Number of employees	750	1,010	625

muddle through the current crisis, but that he would probably not survive if it happened again. The next morning, he put his staff to work gathering statistics that he hoped might help shed light on the basic problem. He plotted some of these data, as shown in Figure 1-8.

He knew that he had been operating by moving from one crisis to the next, but even so he was startled by Figure 1-8. The union president was correct; employment levels seemed to fluctuate more than sales. But the inventory and backlog curves went off the chart.

a. What is the basic problem?
b. Should the VP-Manufacturing take minimizing of finished goods inventories as his goal? Why or why not?
c. Should the VP-Manufacturing take the stabilization of employment as his goal? Why or why not?

11. The manager of a large branch bank is under heavy pressure from central headquarters to reduce costs of operation. On surveying the items of cost that are under her control, she identifies labor, and materials and supplies as the two main cost components. After examining the materials and supplies costs, however, she finds that virtually all of the forms and other supplies used are specified by headquarters as part of the system. While she could institute a "save paper and pencils" campaign, she is convinced that it would yield very little in terms of tangible cost reduction. She is left with labor cost as the one cost component that may yield cost reductions.

The labor force is made up of tellers and clerical personnel who provide support to the branch officers. The clerical personnel are kept fairly busy, and the manager is certain that it would not be feasible to operate with one less person, even though she observes some idle time.

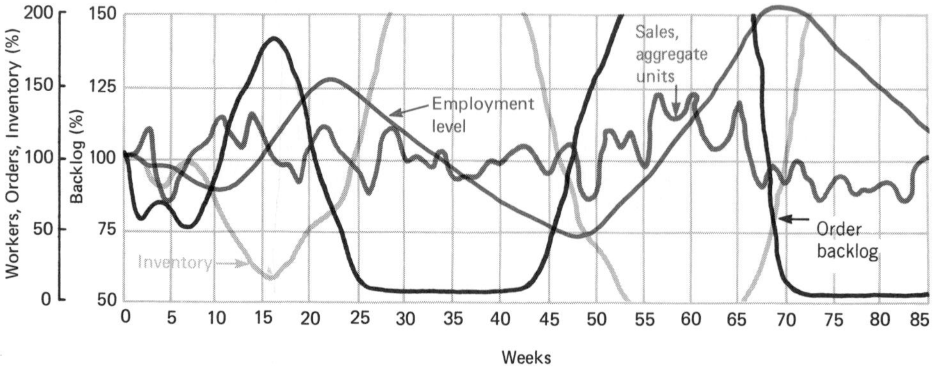

FIGURE 1-8
Indexes of sales, inventory, backlog, and employment for the Wash-N-Dry Company.

The manager has been aware through casual observation that the load at the tellers' windows varies during the day, and from day to day during the week. Her service philosophy has always been to staff the tellers' windows so that bank customers do not have to wait. Perhaps this excellent service policy will have to be fractured a bit during times of peak load. As luck would have it, one teller resigns for personal reasons, and another is granted a request for transfer to another branch in a preferred location. Therefore, the manager has the opportunity to experiment with a lower level of teller staffing without the distasteful aspects of dismissals or other actions to reduce the size of the teller force.

After the first week of operation, the manager has received many complaints from tellers about the workload, and from customers about the poor service. Yet, she has still observed periods where the workload has been light and the tellers relatively idle. She hires one new teller, and ponders her problem of controlling teller labor costs.

How can the branch manager gain control over teller labor costs? What information does she need? What are possible solutions to her problem?

12. An appliance manufacturer is attempting to deal with the range of inventory problems he confronts. He produces a line of small household appliances which are sold in the $20-$75 range, and the demand patterns are seasonal in nature. The raw materials are approximately 10 percent of the manufacturing cost, and the labor is approximately 60 percent.

Labor is semiskilled and can usually be trained to be normally productive within three months. While not paternalistic toward his employees, the owner feels a strong social responsibility toward providing good working conditions and stable employment. Labor turnover, however, has been high in spite of labor being paid higher than average wages for similar work.

The production cycle is relatively short, being approximately two weeks for the average item, and the owner has taken advantage of this fact by keeping finished goods inventories to no more than one month's supply for any time during the seasonal demand pattern. He accomplishes this control by keeping in close touch with market trends and seasonals through a well conceived forecasting system.

The owner has commissioned a study by an outside consultant and has focused attention on minimizing his inventories throughout the process, that is, raw material, in-process, and finished goods inventories. The consultant has manipulated the owner, however, into agreeing that the first phase of the study should be a report which defines the scope and objectives of the study.

Draft a statement of the portion of the consultant's phase I report which defines the statement of the problem and what the scope and objectives of the study should be.

REFERENCES

Bertalanffy, L. V. "General System Theory—A Critical Review," in *Organizations: Systems, Control and Adaptation* (2nd ed.), Vol. II, edited by J. A. Litterer. Wiley, New York, 1969.

Boulding, K. E. "General Systems Theory—The Skeleton of Science," in *Management Systems* (2nd ed.), edited by P. P. Schoderbek. Wiley, New York, 1971.

Buffa, E. S., and J. S. Dyer. *Management Science/Operations Research: Model Formulation and Solution Methods,* Wiley, New York, 1977.

Chase, R. B., and N. J. Aquilano. *Production and Operations Management: A Life Cycle Approach* (rev. ed.). Richard D. Irwin, Homewood, Ill., 1977.

Drake, A. W., R. L. Keeney, and P. M. Morse, editors. *Analysis of Public Systems.* MIT Press, Cambridge, Mass., 1972.

McBride, C. C. "Post Office Mail Processing Operations," in *Analysis of Public Systems,* edited by A. W. Drake, R. L. Keeney, and P. M. Morse. MIT Press, Cambridge, Mass., 1972, pp. 271–286.

Miller, D. W., and M. K. Starr. *Executive Decisions and Operations Research* (2nd ed.). Prentice-Hall, Englewood Cliffs, N.J., 1969.

Milsum, J. H., E. Turban, and I. Vertinsky, "Hospital Admission Systems: Their Evaluation and Management," *Management Science, 19*(6), February 1973, pp. 646–666.

Starr, M. K. *Systems Management of Operations.* Prentice-Hall, Englewood Cliffs, N.J., 1971.

Vollmann, T. E. *Operations Management: A Systems Model-Building Approach.* Addison-Wesley, Reading, Mass., 1973.

PART

TWO

Introduction: STRATEGIC PLANNING DECISIONS

THE CHAPTERS IN PART TWO ARE DIRECTED TOWARD LONG-TERM plans that have significance for the overall strategy. What is the status of technology in both products and processes? Is technology likely to obsolete either products or processes? How large should capacity be five or ten years from now? Where should this capacity be located? Should the capacity be added in steady increments, or periodically in larger units? If the latter, will the initial slack capacity be too large an overhead drain and represent a large risk? The location question is interrelated with the location of distribution points. Where should these distribution points be located? Should the system be tooled up for continuous output, or is a functional arrangement more in keeping with the risks? How will these decisions impact the design of the organization of work, jobs, and detailed physical design?

The preceding questions represent major managerial issues of long-term significance. The decisions commit the organization in ways that are not easily changed without substantial loss. Such decisions commonly require major investments, often the largest of the enterprise assets.

PRODUCT-SERVICE STRATEGIES

At one extreme, we might have products or services that are custom in nature, where the product or service is especially designed to the specifications and needs of the customer or client. Examples are the printing of advertising copy, a prototype spacecraft, many producer goods, or architectural services. The product is not available from inventory because it is one of a kind, and the nature of services are such that they are not inventoriable whether custom or not. The emphasis in the custom product-service strategy is on uniqueness, quality, flexibility, and cost or price is a lesser consideration. If the enterprise is profit making, part of the strategy is to obtain the high profit margins that typically are available for custom designs.

At the other extreme, the product or service is highly standardized. Products of this type are available from inventory—they are "off the shelf" because each unit is identical and the nature of demand is such that availability is an important competitive strategy. There is very little product differentiation, and there is a limited variety in the products. Product examples are automobiles, appliances, standard steel and aluminum shapes, and commodities such as sugar or gasoline. Though not inventoriable, some services are available in highly standardized form such as social security, insurance, motor vehicle registration, and fast food services. The managerial strategies for highly standardized products or services are for dependability of supply and low cost.

Between the extremes of custom design and high standardization of product and services, we have mixed strategies that are sensitive to both variety and some flexibility, and to moderate cost and dependability of supply. In these situations, quality of product or service is an important but not overwhelming criterion. In this middle ground, we have multiple products available, possibly from inventory, or on the basis of order, depending on enterprise strategy and the balance of costs. The great majority of products available today are in the middle category. Most consumer products are available from inventory. Most producer goods are available by order and may be subject to some special design modifications to meet individual needs, though the basic designs are quite standard. Though services are not inventoriable, they may be available in basically standardized modules. For example, treatment of diseases follow standard practices though there are necessarily substantial variations in individual treatments.

PRODUCTIVE SYSTEM TYPES

The basic managerial strategies adopted for the productive system must be related to the product-service strategies. Obviously, it would be inappropriate to use a continuous process capable of producing millions of gallons of product per day to produce an experimental chemical. Again, we think in terms of strategies for the extremes, as well as a middle ground.

A productive system for custom products or services must be flexible. It must have the flexibility to process according to the customer or client needs. For example, an aerospace manufacturer must fabricate special component part designs. The equipment and personnel must be capable of meeting the individual component specifications and of assembling the components in the special configurations of the custom product. The nature of the demand on the productive system results in intermittent use of the facilities, and each component flows from one process to the next intermittently. Physical facilities are organized around the nature of processes, and personnel are specialized by generic process type. For example, in a machine shop we might expect to find milling machine departments, lathe departments, drill departments, and the like. In a general hospital we find X-ray departments, laboratory, surgery, obstetric wards, etc., as generic departments. The flow of the item being processed in intermittent systems is dictated by the individual product or service requirements, so the routes through the productive system are variable. Thus, the intermittent system is flexible as is required by the custom product or service and each generic department and its facilties are used intermittently as needed by the custom orders. Figure II-1 (*a*) and (*b*) show examples of intermittent situations in a machine shop, and in a medical clinic.

By contrast, the nature of the demand on the productive system that produces highly standardized products or services is continuous. Also, material flow may be continuous as in petroleum refining, or approaches continuous flow as with automobile fabrication and assembly. Because of the very high volume requirements of these systems, special processing equipment and entire producing systems can be justified as a productive system strategy. Processing is adapted completely to the product or service. Individual processes are physically arranged in the sequence required, and the entire system is integrated for the single purpose, like a giant machine. The use of inventories may be an important production as well as marketing strategy.

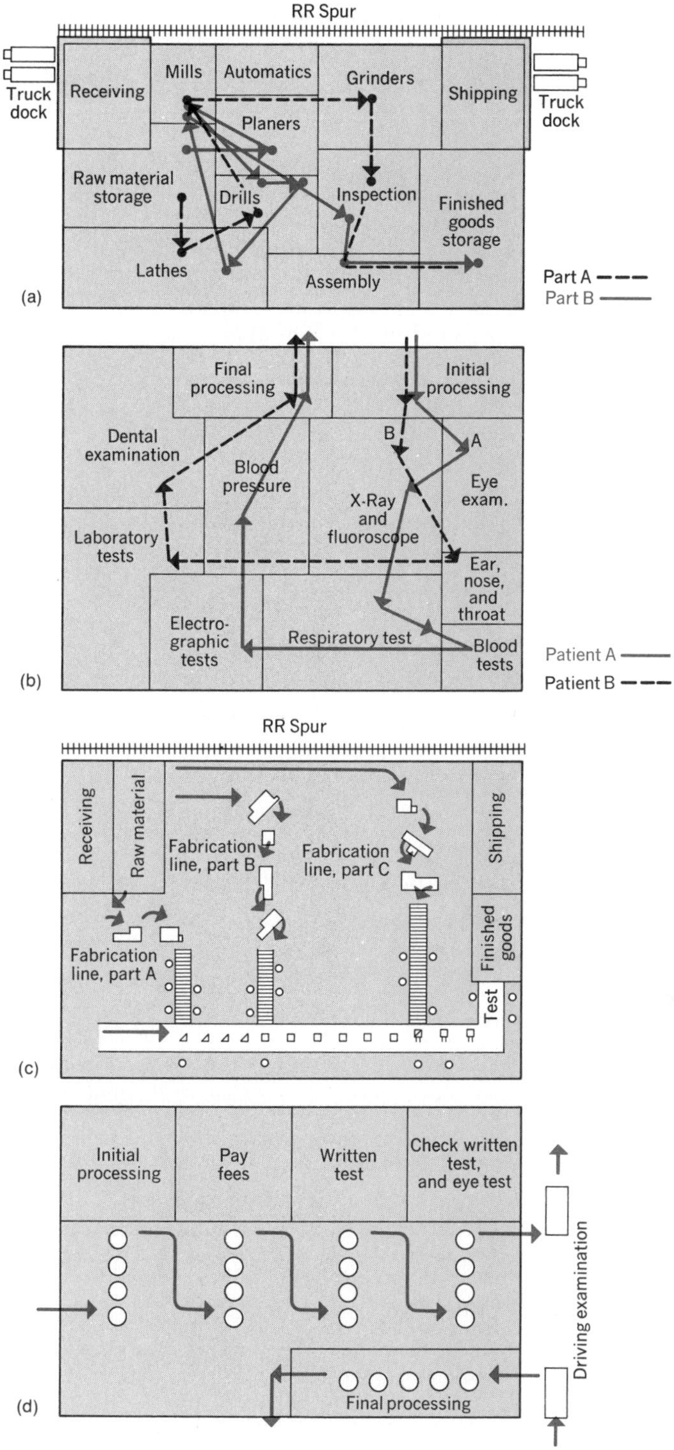

FIGURE II-1
Common examples of intermittent and continuous systems; (a) machine shop, (b) medical clinic, (c) industrial fabrication and assembly lines, and (d) driver's license processing with serial flow.

We also find continuous use of facilities and near continuous productive systems in the service sector. Such mass processing systems are common in some government services such as social security. Also, fast food services approach continuous flow. Figures II-1(*c*) and (*d*) show examples of continuous line situations in manufacturing, and in driver's license processing.

Most productive systems are reflective of the two extremes in that they may be a blend of both. In manufacturing, it is often true that parts fabrication is organized on an intermittent basis with assembly organized on a line or continuous basis. Since output volume may be substantial but not large enough to justify continuous use of facilities, parts and products are produced in economical batches. The inventories resulting from batching again provide an important producing strategy.

To summarize, we have two ways of classifying productive systems, by the intensity of the use of the system, and whether or not inventories are available as a strategy, as shown in Table II-1. First, the nature of the system, being intermittent or continuous, provides the basis for designing operations planning and control systems. If the system is intermittent, planning and control must be based on the individual production order, or its equivalent. We must be concerned with flexibility, variable routes through the system, general purpose equipment, and individual process and quality control. If the system is continuous, planning and control can be somewhat broader or aggregate in nature. The entire system becomes more interdependent and integrated. Failure at any stage within the integrated system may affect the system as a whole. Whether or not the output is inventoriable is the second basis for classifying productive systems. In product systems, inventories can provide flexibility in the alternate producing strategies available. By contrast, non-inventoriable output systems hold their capability in readiness (in inventory), rather than the output itself. This is true of service systems as well as the custom product systems and, in that sense, the custom product system offers a service.

TABLE II-1 **Two-way Classification of Productive Systems**

Type of Productive System (Nature of demand on system)	Output Is Inventoriable	Output Is Noninventoriable (Availability is defined as capability to produce)
Intermittent	Batch processing	Jobbing printer Jobbing machine shop Large-scale projects General hospital Municipal offices
Continuous	Commodity processing Oil refining Sugar refining Auto production Distribution systems	Government services Social Security Mass food services Transportation service

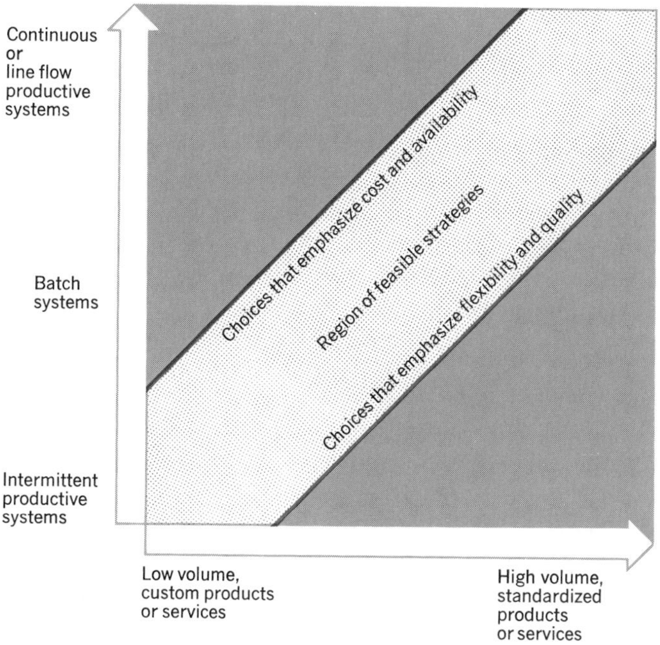

FIGURE II-2
Mapping of productive systems strategies in relation to product-service strategies. Feasible joint strategies occur in region.

PRODUCTIVE SYSTEM, PRODUCT-SERVICE JOINT STRATEGIES

When we examine the product-service and productive system types jointly, it is useful to think of the former as being the independent variable and the latter as the dependent variable, as reprsented by Figure II-2. The feasible joint strategies are shown as a band or range. That is, one cannot say that there is a single correct solution for a given situation; rather, there is a range of choices that may represent alternate joint strategies. Managers might choose a strategy that emphasizes cost and availability by choosing the combinations in the upper part of the band. Similarly, they could choose a strategy that emphasizes flexibility, choice, and quality by choosing combinations in the lower part of the band.

Hayes and Wheelwright* have developed a matrix representation for relating manufacturing process structure as a process life cycle to the product life cycle. They give specific examples of company strategy involving the Lynchburg Foundry, a wholly owned subsidiary of the Mead Corporation.

* Robert H. Hayes and Steven C. Wheelwright, "Link Manufacturing Process and Product Life Cycles," *Harvard Business Review,* January-February 1979, pp. 133–140, and "The Dynamics of Process-Product Life Cycles," *Harvard Business Review,* March-April 1979, pp. 127-136.

Lynchburg has five plants in Virginia and surrounding states. The five plants represent different points in Figure II-2. One plant is a job order shop making one-of-a-kind products, and the joint strategy represents a point in the lower left region of Figure II-2. Two plants are organized to produce a variety of products in batches. These plants involve a strategy in the lower middle range, where emphasis has been placed on the flexibility required by multiple products. A fourth plant is designed as a line-flow setup to produce only a few auto part castings. Thus, the joint strategy is represented by a point in the upper middle range, and in the upper region where cost and availability is emphasized. Finally, the fifth plant is an automated pipe facility producing a highly standardized item in huge quantity on a continuous basis. The joint strategy for the fifth plant is therefore represented by a point in the upper region, to the right.

The Lynchburg examples indicate that an enterprise may need to employ different joint strategies for different product-process situations. The resulting planning and control policies and procedures need to be reflective of these quite different strategies; a uniform set of operations planning and control policies and procedures would be quite inappropriate.

It is unlikely that a strategy can remain static over long periods. As products or services mature in their life cycles, consumer preferences become known, designs become refined, volumes build, and the appropriate joint strategy must reflect these changes. Normally the progression involves a more capital intensive productive process that is more integrated. There is necessarily a loss of flexibility. Abernathy and Wayne* found, for example, in studying the auto industry, that the Ford Motor Company strategy during the Model-T era had become static. The strategy became so completely focused on a standardized product-integrated process strategy that the enterprise was nearly sunk when the competitive arena changed to emphasize product innovation.

PLAN FOR PART TWO

Part Two concentrates on three main topics: technology and its relation to products and process, capacity planning, and location. Chapter 2 presents a background of how technology and innovation percolates through organizations and industries. It raises questions concerning the prediction of technological developments, and how management must somehow relate these predictions to the designs of products and services to remain competitive. The process of product and productive system design are shown to be closely related in an interactive process.

Chapter 3 deals squarely with capacity planning in both stable and risky market situations. These important decisions are analyzed by present value methods because

* W. J. Abernathy and K. Wayne, "Limits to the Learning Curve," *Harvard Business Review*, September-October 1974, pp. 109–119.

the cash flows are likely to occur over a considerable period of time in the future.

Finally, Chapter 4 deals with the location of both producing and distribution units. A rational methodology for plant location is presented that takes account of objective as well as subjective factors, the latter through preference analysis. Distribution is considered within the warehouse location problem for products, and through facility location for service operations.

CHAPTER 2 Technology and the Design of Products and Services

OUR INTEREST IN THE DESIGN OF PRODUCTS AND SERVICES IS primarily in terms of the impact that such designs have on the enterprise strategy for productive system design. We are also interested in the reverse impact that the system may have on the design of the services and products produced. We have little or no interest in the technical or engineering design aspects of products. We care about the interactions that affect the nature of the producing system. Producing capabilities and alternate production costs for alternate designs are information needed to make the best choices among alternatives.

Innovation is the generator of new products and services and new productive processes. Innovation of new products, services, and processes triggers the events that result in the prediction of potential markets, the design of products and services, and the systems to produce them.

TECHNOLOGICAL INNOVATION

In simplest terms we tend to think of the conversion of scientific discovery to application in products, services, and processes as a chain of events: scientific discovery, invention, development, innovation, and application. There are commonly long time lags in this chain, as enabling conditions pace developments. For example, processes for shale oil extraction were known for some time, but economic factors did not justify their costly development. Cheaper sources of crude oil and other energy forms were used first. Given the shortage of conventional energy sources, however, and the resulting price increase, the application of shale oil processes may become a more viable possibility. As another example, all the electronic circuitry required for portable radios was available for many years; however, a true miniature radio was not possible until the development of the transistor. Finally, to take a service item, all the technology necessary for a mass food preparation service has been available for many, many years. Yet enterprises, such as BURGER discussed in Chapter 1, were not possible until our culture had developed a willingness to make trade-offs between time, food quality, and price.

The chain-event model does not help to explain the driving forces that produce the innovations near the end of the chain. Does the process flow from scientific discovery to application rather naturally as a river flows from higher to lower elevation, or is the process somewhat more complex? Do new product innovations stimulate innovations in productive processes, or is that also too simple a concept?

Abernathy and Townsend [1975] suggest that product innovation, process innovation, and changes within a segment of industry appear to feed on each other. "No single external force, such as market factors or technological factors, is dominant in stimulating technological innovation. Sources of stimulation that arise within a productive segment are more frequently the critical factor that sparks technological innovation." They further state, "Historical patterns of development in several productive segments suggest that the efforts of engineers and managers in improving production processes themselves may be a key factor in stimulating technological innovation."

TECHNOLOGICAL INNOVATION IN THE COMPUTER INDUSTRY

It has been commonly held that most innovations are market stimulated and usually applied to new products rather than to production processes. The results of one study in the computer industry are shown in Table 2-1. Note that the largest stimulation source of innovations is in the market and that the greatest application impact is on products.

Abernathy and Townsend, however, analyzed the same data, taking into account the vertical integration structure. Their results are suggestive of a startlingly different conclusion. Figure 2-1 shows the flow of innovations allocated to three levels of industrial integration for the computer industry, based on the data from Table 2-1. The dotted lines with arrows in Figure 2-1 show the source of innovation stimulation, and the solid lines emerging from the "innovation boxes" indicate the frequency and area of application impact. Note that "market factors" represent the primary source of stimulation for 61 percent of the innovations produced by the computer components and supply manufacturers. However, market factors in that industrial segment are the same as the process equipment of the two higher processes in the chain of vertical integration, the computer manufacturers and the computer users. The greatest impact of innovation is upstream.

Abernathy and Townsend note that "*most innovations in the lower levels of the vertical integration chain are product innovations. These are at the same time process innovations for processes at higher levels of vertical integration and as such have*

TABLE 2-1

Source and Impact of Successful Innovations in the Computer Industry

	Components and Supply Manufacturers	Computer Manufacturers
Stimulation source		
Market	47 (61%)	28 (31%)
Production	17 (22%)	32 (36%)
Technical	10 (13%)	20 (22%)
Administrative	3 (4%)	10 (11%)
	77	90
Application impact		
Product	60 (78%)	49 (54%)
Component	10 (13%)	28 (31%)
Process	7 (7%)	13 (14%)
	77	90

SOURCE: S. Meyers and D. Marquis, *Successful Industrial Innovations*, National Science Foundation, Washington, D.C., 1969, pp. 69–70.

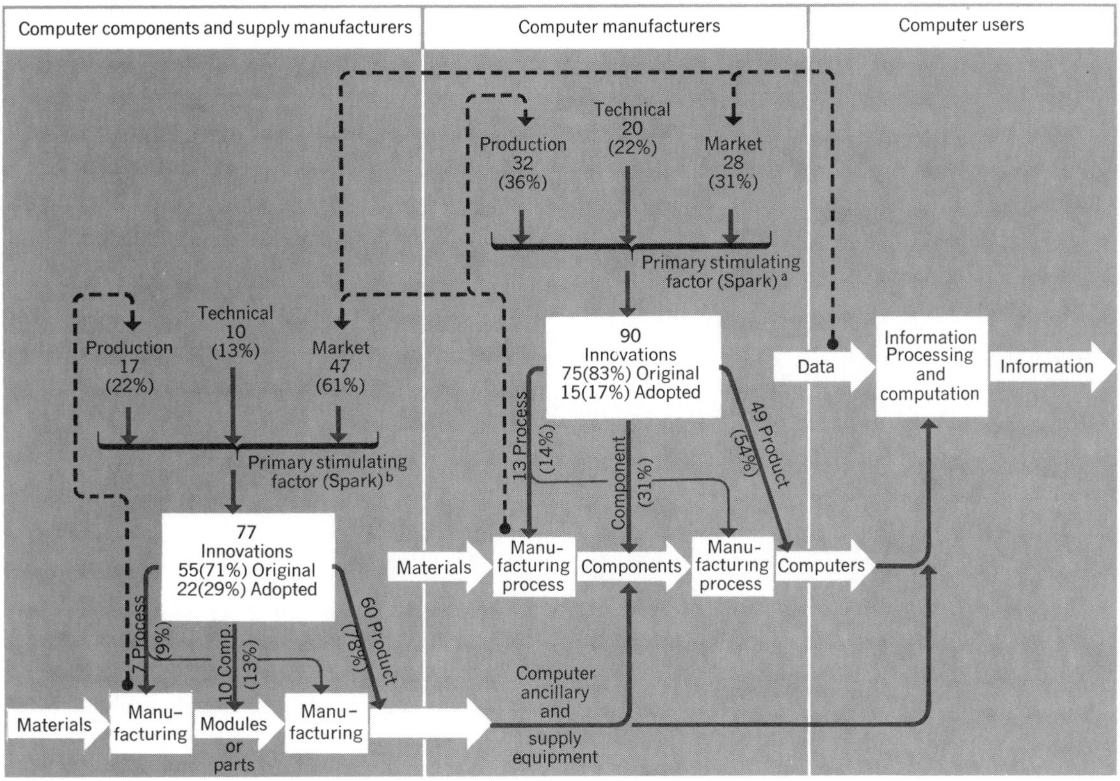

FIGURE 2-1
Technological innovation in the computer industry.
SOURCE: *W. J. Abernathy and P. L. Townsend, "Technology, Productivity and Process Change,"* Technological Forecasts and Social Change 7 *(2), January 1975. Data from Meyers and Marquis, 1969.*

direct productivity implications. In fact, from a strict perspective of the process at the highest level of vertical integration (computer users), all of the innovations considered here are process innovations and all have implications for process productivity." For example, wide-bodied aircraft were product innovations for aircraft manufacturers, but were process innovations to the airlines with very important productivity implications.

Note that the industrial structure culminates in a service industry, that of providing computing service to users. Thus, we see an example of the interlinking between a service industry and manufacturing industries that provide equipment for the processes. Figure 2-1 represents a closed loop system for innovation in the industry, where market stimulation means to be stimulated by an opportunity to serve process needs of an organization higher in the integration structure. A comparable analysis was also performed of the transportation industry, which also culminates in a service, with similar results.

INTERACTION BETWEEN PRODUCT AND PROCESS INNOVATION

We have already alluded to the iterative process between product design and productive system design. Each has an impact on the other in determining the designs of both. This iterative process apparently also takes place on a macro level within industries, as is shown by Figure 2-1, and the concept includes the design of services as well as that of the manufactured product. The nature of services offered is affected by the productive process and vice versa, and so on back through the chain.

A MODEL OF PROCESS AND PRODUCT INNOVATION

Utterback and Abernathy [1975] developed a dynamic model of process and product innovation in firms and tested it on empirical data. The model relates the product and process innovations to three stages of development.

Stage 1

The first stage begins early in the life of products and services and of processes; initially, the innovations are stimulated by needs in the marketplace. Process innovations also are stimulated by the need to increase output rate (see Figure 2-2). In terms of innovation rate, product innovation is high and the initial emphasis is on performance maximization. There may be an anticipation that new capabilities will, in turn, expand requirements in the marketplace.

Although we may think largely in terms of physical products, service innovations are quite comparable: for example, the initial introduction of innovative services such as social security, no-fault auto insurance, comprehensive health services (e.g., Kaiser Permanente), fast-food services, and so on.

Utterback and Abernathy call the first phase "performance maximization" for products and services and "uncoordinated" for processes. High product innovation rates increase the likelihood that product diversity will be extensive. As a result, the productive process is composed largely of unstandardized and manual operations, or operations that rely on general purpose equipment. The productive system is likely to be of the intermittent type, but the characterization—uncoordinated—is probably justified in most instances because the relationships between the required operations are still not clear.

Stage 2

Price competition becomes more intense in the second stage as the industry or product and service group begin to reach maturity. Productive system design emphasizes cost minimization as competition in the marketplace begins to emphasize price. The productive process becomes more capital intensive and more tightly integrated through production planning and control.

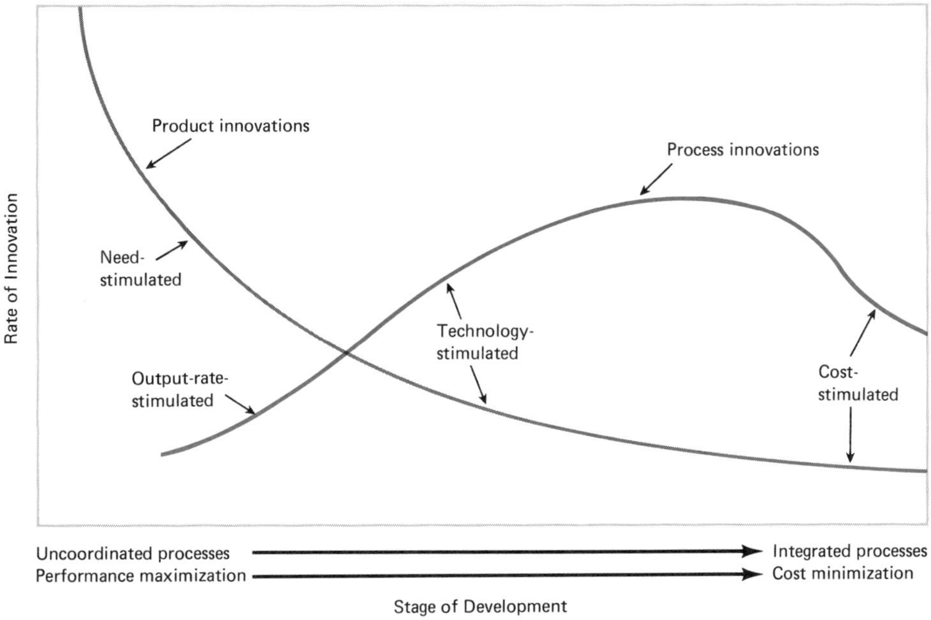

FIGURE 2-2
Relationships of product and process innovations in a dynamic model.
SOURCE: *J. M. Utterback and W. J. Abernathy, "A Dynamic Model of Process and Product Innovation by Firms,"* Omega, *1975.*

At this stage, the production process often is segmented in nature. This is true partly because integration is taking place at a broader level through managerial control systems and partly because the dominant system type is the intermittent system. As shown in Figure 2-2, process innovations dominate; however, both process and product innovations are stimulated by technology.

Stage 3

Finally, as the entire system reaches maturity and saturation, innovations tend to be largely cost stimulated, as indicated in Figure 2-2. Further price competition puts increasing emphasis on cost-minimizing strategies, and the production process becomes even more capital intensive.

The productive process becomes more highly structured and integrated, as illustrated by automotive assembly lines, continuous chemical processes, and such highly automated, large-scale service systems as social security. The productive process becomes so highly integrated that it is difficult to make changes because any change at all creates significant interactions with other operations in the process.

The model of innovation indicates the close relationship between the design and development of products and services and the productive system design. In fact, during the third, or cost-minimizing, stage, the effects of innovation on product cost follow a surprisingly clear pattern. This is indicated in the Ford study which follows.

THE LEARNING CURVE-INNOVATION INTERACTION

One conceptual framework that many manufacturing organizations have used as a basis for a production-marketing strategy is the learning (or experience) curve. The concept is that product costs decline systematically by some fixed percentage with the doubling of volume. For example, a 90 percent learning curve would be one in which the product cost for a doubling of volume would be 90 percent of the former cost. Given such a relationship, managers can increase market share through price competition, depending on costs to decrease according to the learning curve. The learning curve model formalizes the "economics of scale" concept. Abernathy and Wayne [1974] developed a fascinating history of innovation, process change, and organization for the Ford Motor Company, part of which is summarized in Figure 2-3.

Figure 2-3 shows the price decline (in 1958 dollars) of the famous Model T during its long product life cycle. The price decline culminated in a costly conversion to the Model A, and finally, the price increases associated with the annual model changes began in about 1932.

Beginning in 1908, Henry Ford embarked on a conscious policy of price reduction that reduced the price from more than $5000 to nearly $3000. From that point on, the price decline was characterized by an 85 percent learning curve during the Model T era. Market share increased from 10.7 percent in 1910 to a peak of 55.4 percent in 1921. During this spectacular period of stable product design, innovations were largely process or production oriented:

> *The company accomplished savings by building modern plants, extracting higher volume from the existing plant, obtaining economies in purchased parts, and gaining efficiency through greater division of labor. By 1913 these efforts had reduced production throughput times from 21 days to 14. Later, production was speeded further through major process innovations like the moving assembly lines in motors and radiators, and branch assembly lines. At times, however, labor turnover reportedly ran as high as 40 percent per month.*
>
> *Up to this point, Ford had achieved economies without greatly increasing the rate of capital intensity. To sustain the cost cuts, the company embarked on a policy of backward and further forward integration in order to reduce transportation and raw materials costs, improve reliability of supply sources, and control dealer performance. The rate of capital investment showed substantial increases after 1913, rising from 11 cents per sales dollar that year to 22 cents by 1921. The new facilities that were built or acquired included blast furnaces, logging*

operations, sawmills, a railroad, weaving mills, coke ovens, a paper mill, a glass plant, and a cement plant.

Throughput time was slashed to four days and the inventory level cut in half, despite the addition of large raw materials inventories. The labor hours required of unsalaried employees per 1000 pounds of vehicle delivered fell correspondingly some 60 percent during this period, in spite of the additions to the labor force resulting from the backward integration thrust and in spite of substantial use of Ford employees in factory construction.

Constant improvements in the production process made it more integrated, more mechanized, and increasingly paced by conveyors. Consequently, the company felt less need for management in planning and control activities. The percentage of salaried workers was cut from nearly 5 percent of total employment for 1913 to less than 2 percent by 1921; these reductions in Ford personnel enabled the company to hold in line the burgeoning fixed cost and overhead burden. [Abernathy and Wayne, 1974]

Beginning in the middle 1920s, however, General Motors successfully focused the competitive arena in product innovation. The Ford Company was so completely organized to produce a standardized product that the effects of the change in consumer demand nearly sunk the enterprise.

Plotting Innovation Cycles

Even in a period of stable product designs, there appears to be an innovation cycle. Figure 2-4 shows a plot of product and process innovations, and technological transfers over a 38-year period at the Ford Motor Company. Ford-initiated innovations were rated on a scale of 1 to 5 by four independent industry experts. The innovations ranged from the introduction of the plastic steering wheel (average rating 1) in 1921 to the power-driven final assembly line (average rating 5) in 1914. Figure 2-4 shows that new product applications occurred in clusters associated with new models, followed by a decline as the new designs became standardized. Process innovations peaked after the product innovations, presumably to integrate the processes with existing operations and to reduce costs. Technological transfers plotted in Figure 2-4 refer to the transfer of process technology to or from associated industries. These transfers increased as Ford undertook vertical integration.

Managing Technological Change

One of the results of the study was that the need was recognized for balance between a cost-reducing strategy and new product innovations, which were at odds. "The ability to switch to a different strategy seems to depend on the extent to which the organization has become specialized in following one strategy and on the magnitude of change it must face. An extreme in either factor can spell trouble" [Abernathy and Wayne, 1974].

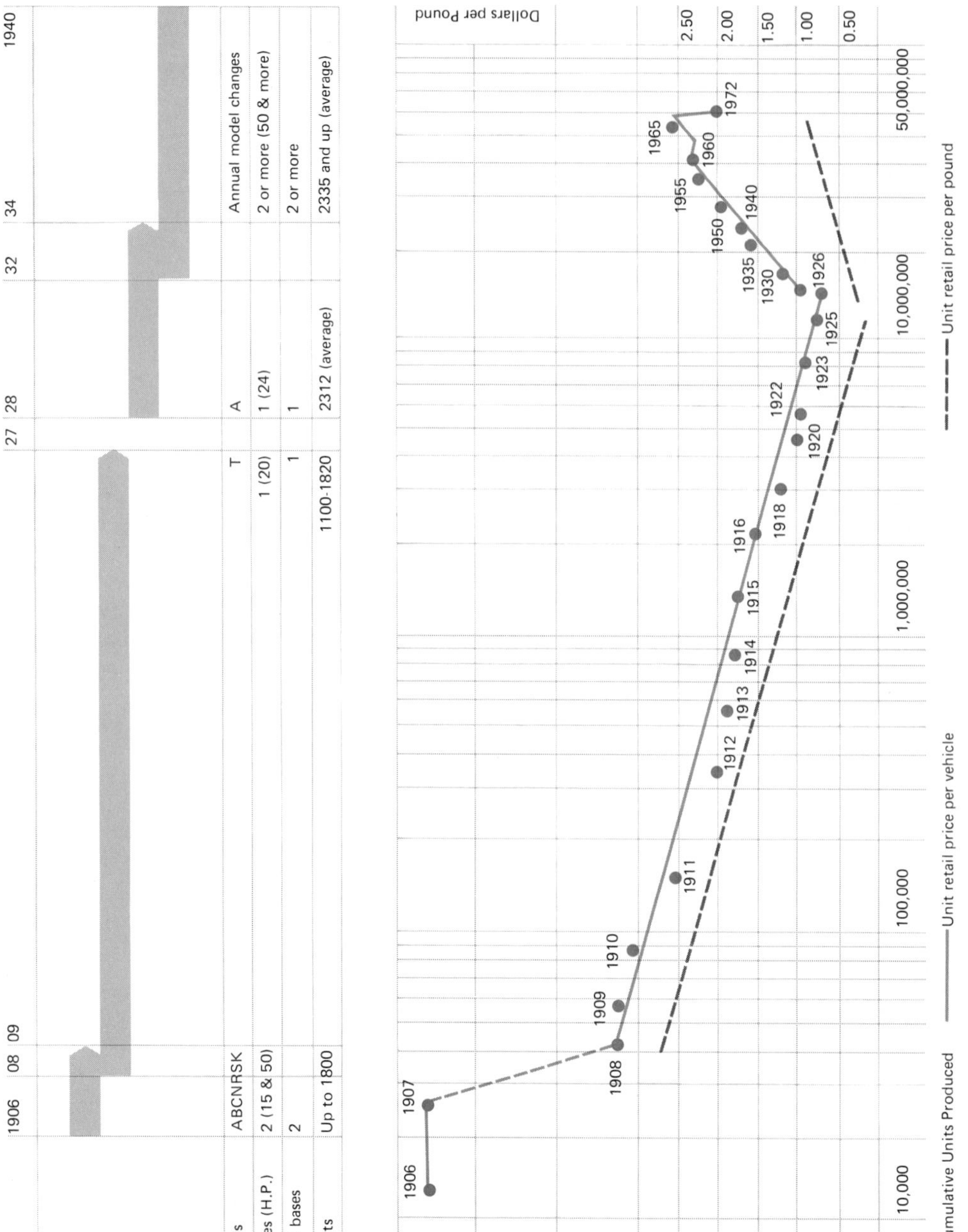

FIGURE 2-3
The Ford experience curve (in 1958 constant dollars).
SOURCE: *W. J. Abernathy and K. Wayne, "Limits of the Learning Curve,"* Harvard Business Review, *September-October 1974, pp. 109–119.*

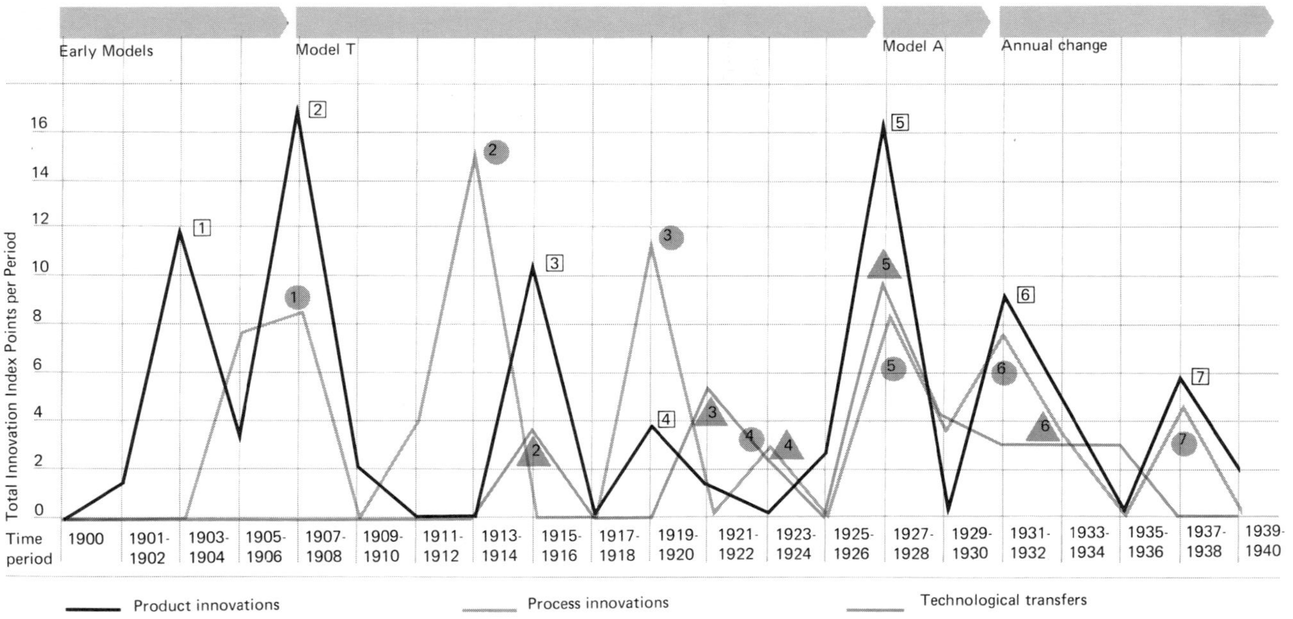

FIGURE 2-4
Innovation and process change at Ford.
SOURCE: *W. J. Abernathy and K. Wayne, "Limits of the Learning Curve,"* Harvard Business Review, *September-October 1974, pp. 109–119.*

This balance can be achieved by periodically inaugurating major product innovations, stressing cost reduction along the learning curve between model changes. An alternate mode of achieving balance is to decentralize within the corporate structure, that is, to have separate organizations follow different strategies with the same general product line. One organization might follow the Ford Model T strategy of cost reduction and volume expansion. Another might develop innovative products or processes which, when developed, might follow the cost reduction strategy, perhaps finally displacing the former products.

PREDICTING MARKETS FOR PRODUCTS AND SERVICES

Rational plans for products and services, including their productive systems, cannot be made without an estimate of the size of the market. This is true whether we are dealing with profit or nonprofit enterprises. If one is determining whether or not a new product or service should be launched, then the data on potential market size is crucial for the decision. If the basic decision has already been made to produce the product or service, then the information is just as crucial to finalizing designs for both the product or service and the productive system.

When we are dealing with an ongoing situation, products and services may be in the growth or even saturation phase of their life cycles. In such cases, we are usually dealing with refinements of the product design and probably additions to physical capacity and/or a relocation or rebalancing of physical capacity. All of these kinds of plans are of a longer-term nature and often require the commitment of large investments. Therefore, insight into the future is required.

PREDICTION AND FORECASTING

We will use two different terms to describe methodologies for estimating future demand, "prediction" and "forecasting." Prediction as a term will have more of the flavor of the "crystal ball." When we predict, we are integrating a great deal of subjective and objective information to form our best estimate of the future. We use prediction methods when we have little experience on which to base estimates of the future. Forecasting, on the other hand, will be used to connote a statistical technique for *casting* the historical record *forward*. Forecasting depends on having enough historical data to be able to describe the record in statistical terms and on reasonably stable market generating factors. There is a place for both prediction and forecasting.

PREDICTION AND FORECASTING METHODS

We will group the methods available into predictive, causal, and time series forecasting models. In general, our use of the term "prediction" applies to the more qualitative methods and "forecasting" to the causal and time series models. Table 2-2 sum-

TABLE 2-2 **Methods of Prediction and Forecasting**

Method	General Description	Applications	Relative Cost	References[a]
Predictive Methods				
Delphi	Expert panel answers a series of questionnaires where the answers of each questionnaire are summarized and made available to the panel to aid in answering the next questionnaire.	Long-range predictions, new products and product development, market strategies, pricing, and facility planning.	Medium-high	Gerstenfeld, 1971; North and Pyke, 1969
Market surveys	Testing markets through questionnaires, panels, surveys, tests of trial products, analysis of time series.	Same as above.	High	Ahl, 1970; Bass, 1969; Bass et al., 1968; Claycamp and Liddy, 1969
Historical analogy and life cycle analysis	Prediction based on analysis of and comparison with growth and development of similar products. Forecasting new product growth based on the S-curve of introduction, growth, and market saturation.	Same as above.	Medium	Bass, 1969; Chambers et al., 1971
Causal Forecasting Methods				
Regression analysis	Forecasts of demand related to economic and competitive factors that control or *cause* demand, through least squares regression equation.	Short- and medium-range forecasting of existing products and services. Marketing strategies, production and facility planning.	Medium	Chambers et al., 1971; Parker and Segura, 1971; Wheelwright and Makridakis, 1977
Econometric models	Based on a system of interdependent regression equations.	Same as above.	High	Wheelwright and Makridakis, 1977

[a] Refer to the reference list at the end of the chapter.

marizes some of the most prominent and useful methods under each of the three main headings. The predictive methods are applicable to new product introductions and longer-range predictions, and the causal methods to short- and medium-range forecasting. Causal and time series models will be covered in Chapter 5.

PREDICTIVE METHODS

In this age of management science and computers, why must we resort to qualitative methods to make some of the most important predictions of future demand for products and services, predictions on which hinge the greatest risks involving large investments in facilities as well as risks in market development? The answer is that where we have no historical record, statistical methods have no validity. What people think, samplings of how they react to market tests, knowledge of consumer behavior, and analogy with similar situations may be the best we can do. Given this situation the most scientific approach is to bring as much order to these kinds of judgments as possible. We cannot create hard demand data that does not exist. The qualitative methods are of considerable significance then, since they provide a basis for some important decisions.

Delphi Methods

"Technological forecasting" is a term used in connection with the longest-term predictions, and the Delphi technique is the methodology often used as a vehicle. The objective of the Delphi technique is to probe into the future in the hope of anticipating new products and processes in the rapidly changing environment of today's culture and economy. In the shortest range of such predictions, it can also be used to estimate market sizes and timing.

The technique draws on a panel of experts in a way that eliminates the possible dominance of the most prestigious, the most verbal, and the best salespeople. The object is to gain the benefit of expert opinion in the form of a consensus instead of a compromise. The result is pooled judgment, with both the range of expert opinion and the reasons for differences of opinion shown.

The Delphi technique was first developed by the RAND Corporation as a means of achieving these kinds of results, in comparison to conferences and panels where the individuals are in direct communication. Thus the undesirable effects of group interaction are eliminated.

The panel of experts can be constructed in various ways and often includes individuals both inside and outside the organization. It may be true that each panel member is an expert on some aspect of the problem, but that no one is an expert on the entire problem. In general, the procedure involves the following:

1. Each expert in the group makes independent predictions in the form of brief statements.

2. The coordinator edits and clarifies these statements.

3. The coordinator provides a series of written questions to the experts that combine the feedback supplied by the other experts.

One of the most extensive probes into the technological future was reported by TRW, Inc. [North and Pyke, 1969]. The project involved the coordination of 15 different panels corresponding to 15 categories of technologies and systems that were felt to have an effect on the comapny's future. Anonymity of panel members was maintained to stimulate unconventional thinking. A Delphi method was then used to question and requestion the experts.

Market Surveys

Market surveys and the analysis of consumer behavior have become quite sophisticated, and the data that result become extremely valuable inputs to predicting market demand. In general, the methods involve the use of questionnaires, consumer panels, and tests of new products and services. The field is a specialty in itself and beyond our scope.

There is considerable literature dealing with the estimation of new product performance based on consumer panels [Ahl, 1970], using analytical approaches [Bass, 1969; Claycamp and Liddy, 1969], as well as simulation and other techniques [Bass, King, and Pessemeier, 1968]. Proposed products and services may be compared with the products and known plans of competitors, and new market segments may be exploited with variations of product designs and quality levels. In such instances, comparisons can be made with data on existing products. These kinds of data are often the best available to refine the designs of products and facilities for new ventures.

Historical Analogy and Life Cycle Analysis

Market research studies can sometimes be supplemented by reference to the performance of an ancestor of the product or service under consideration, applying an analysis of the S-curve. A typical S-curve is shown in Figure 2-5, where demand in the initial phases of market development accelerates to the middle growth period, culminating in market saturation. For example, the assumption was made that color television would follow the general sales pattern experienced with black and white television but that it would take twice as long to reach a steady state [Chambers, Mullick, and Smith, 1971]. Such comparisons provide guidelines during the initial planning phases and may be supplemented by other kinds of analyses and studies as initial actual demand becomes known. Chase and Aquilano [1977] focus their attention on the life cycle of products in studying the problems of production management.

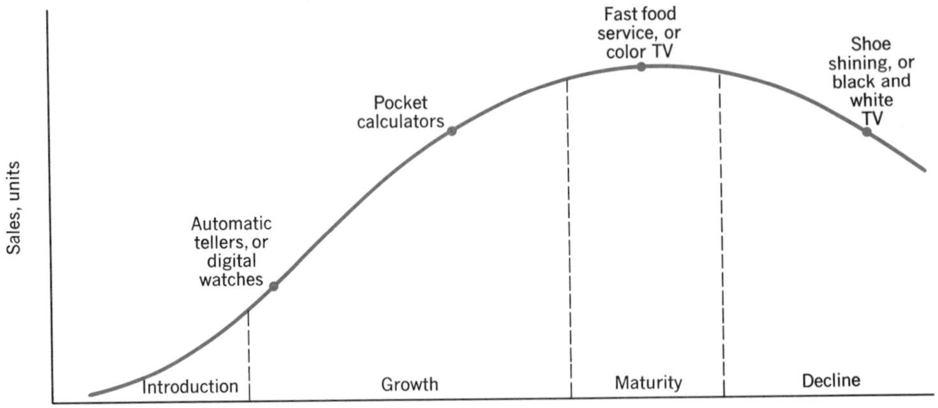

FIGURE 2-5
Typical S-curve of the introduction, growth, market saturation, and decline in the life cycle of products or services.

CAUSAL AND TIME SERIES FORECASTING METHODS

When we have enough historical data and experience, it may be possible to relate forecasts to the factors in the economy that *cause* the trends, seasonal variations, and fluctuations. If we can measure the causal factors and determine their relationships to the product or service of interest, we can then compute forecasts of considerable accuracy.

Time series forecasting models are based on projections or extrapolations from time series data that have been processed by statistical methods. The statistical processing is designed to discount random fluctuations and to take into account trend and seasonal variations.

INTERACTION BETWEEN PRODUCT-SERVICE DESIGN AND PRODUCTIVE SYSTEM DESIGN

On the macro level we saw that innovation in products at lower levels of the vertical integration chain fed into process innovations (productive system design) at higher levels of vertical integration within an industry. At this broad industrywide level, there is an interaction between product-service design and the design of the productive system. Both the designs of services and products may be involved, for example, in both the computer and transportation industries the industry culminates in services rendered instead of a physical product.

A similar process takes place within an enterprise where the design of products and services is partially dependent on the productive system design, and vice versa. The concept is so well recognized in the mechanical industries that a name has been coined for the process of designing products from the point of view of producibility—"production design."

PRODUCTION DESIGN

The producibility and minimum possible production cost of a product are established originally by the product designer. The cleverest production engineer cannot change this situation; he or she can only work within the limitations of the product design. Therefore, the obvious time to start thinking about basic modes of production for products is while they are still in the design stage. This conscious effort to design for producibility and low manufacturing cost is referred to as "production design," as distinct from functional design. To be sure, the product designer's first responsibility is to create something that functionally meets requirements. But once functional requirements are met, there are ordinarily alternate designs, all of which meet functional requirements. Which of these alternatives will minimize production costs? A well-conceived design has already narrowed the available alternatives and specified, for example, a sand casting, if that is appropriate in view of both function and cost considerations.

Given the design, process planning for manufacture must be carried out to specify in careful detail the processes required and their sequence. Production design first sets the minimum possible cost that can be achieved through such factors as the specifications of materials, tolerances, basic configurations, and the methods of joining parts. Final process planning then attempts to achieve that minimum through the specification of processes and their sequence which meet the exacting requirements of the design. Here, process planners may work under the limitations of available equipment. However, if the volume is great or the design stable, or both, process planners may be able to consider special-purpose equipment, including semiautomatic and automatic processes and special purpose layout. In performing their functions, process planners set the basic design of the productive system.

The thesis of a production design philosophy is that alternatives of design that still meet functional requirements nearly always exist. For the projected volume of the product, then, what differences in cost would result? Here we must broaden our thinking, because the possible areas of cost that can be affected by design are likely to be more pervasive than we imagine. There are the obvious cost components of direct labor and materials. But perhaps not so obvious are the effects on equipment costs, tooling costs, indirect labor costs, and the nonmanufacturing costs of engineering.

Indirect costs tend to be hidden, but suppose one design required 30 different parts, whereas another required only 15 (e.g., the reciprocating automobile engine versus the rotary engine). There are differences in indirect costs as a result of greater paper work and the cost of ordering, storing, and controlling 30 parts instead of 15 for each completed item.

DESIGN AND REDESIGN

The design process is an iterative one. In a sense, it is never done. New information feeds in from users, and we find ways to improve designs that reduce production costs, although the quality criterion is often an objective as well.

As an example of production design and redesign, Bright [1958] describes the development of electric light bulb manufacturing during the period from 1908 to 1955. Initially, a batch process was used involving manual operations and simple equipment. The conversion from batch to continuous operation was achieved by adopting systematic layout, standardizing operations, and effecting operation sequence changes. Then, however, the light bulb itself was redesigned a number of times to facilitate process changes and permit mechanical handling. Finally, an evolution took place in which individual standardized manual operations were replaced by mechanical operations, and these operations were in turn integrated to produce a fully automated process.

Also, in manufacturing there are numerous examples that reflect redesign of products from the viewpoints of processes and materials used, methods of joining parts, tolerances, design simplification, and techniques for reducing the amount of processing.

DESIGN OF SERVICES OFFERED

Although no term has been coined to describe it, a process similar to production design goes on and represents the interaction between the design of services to be offered and the productive system design. The motivation for altering the service offered may be accommodations to cost factors and service quality as measured by time performance and other dimensions. Some of the kinds of accommodation in the nature of services offered are as follows:

1. Transfer some of the activity involving the service to the client or customer. This has been one of the most common techniques for lowering costs. In hospitals, supermarkets, and mass-food services, the patient or customer performs some of the activities that were formerly a part of the service. The result is usually a lower labor cost and the elimination of some activities performed by the productive system.

2. Eliminate some aspects of service entirely.

3. Change the mix of services offered.

4. Change the reaction time for service, that is, give poorer service by applying fewer resources to the operation.

5. Change other aspects of service quality, perhaps the range of services offered.

Table 2-3 indicates some of the interactions between the design of services offered and the productive system design for several well-known types of systems. The kinds of accommodations listed for both services and productive systems are not intended to be exhaustive.

Unfortunately, there seems to be a consensus that the quality of services of all kinds has deteriorated. The reasons are probably that most service operations are labor intensive, and labor-saving devices are uncommon in service operations. The result is that rising wage costs are not compensated for by increases in productivity. Recall, for example, the minimal changes in productivity of the post office as compared to United States industry in Figure 1-3.

An Example [Yourdon, 1970]

A simulation study of computing service provides an example. Four different mixes of computing service offered by a service bureau were considered: 100 percent time sharing, 100 percent batch processing, and two alternatives that provided a mix of time sharing and batch processing according to given schedules.

A computer program was written to simulate operations under the four types of service mixes, taking into account assumptions on sales generation, revenue, capacities, and costs of all types. The program calculated profit and loss statements, the time needed to break even, and profitability. Although only one equipment configuration was used, the same program could easily be used to make similar calculations for three or more basic machines, for example, the IBM 360/50, the GE-265, and the XDS-940. The three machines have different capacities and software availabilities and widely varying monthly lease costs. The end result of such a study combines the market estimates and projections, design of the service to be offered, and the interaction with the equipment configuration (productive system design in our terms). The combination yielding maximum profit would presumably be selected.

DECISIONS AND DECISION PROCESSES

We have recognized that the product and productive system design decisions require trade-offs in order to jointly optimize them. Now, what are the prime decision areas, and what decision processes support them? The decision points are focused on the approval of specific designs and product mixes and required alterations to the productive system. There is often the drive to gain advantages through modularity of design and standardization.

Thus, managers are again forced to look at any product-service design decision within a systems context. How will a new or redesigned product or service be received? What impact will it have on the existing product line? What is the best new product mix? What is the impact on the productive system, its design, and its capacities? How will the new design affect operating schedules? And, of course, what are the effects on revenue, costs, and profits? Some of these effects are reflected in forecasts and estimates of incremental revenues and costs, but the manager must make judgmental trade-offs, and the decision framework discussed in Chapter 1 is appropriate.

TABLE 2-3 **Interactions Between the Design of Services Offered and the Productive System Design for Several Types of Services**

Types of Service	Types of Alteration of Service Offered to Accommodate Needs of Productive System	Types of Alteration of Productive System to Accommodate Needs of Service Offered
Nursing care in hospitals	Specialization by levels to reduce costs (e.g., nurses aides).	Changes of layout and activity scheduling.
Food markets	Self-service to reduce cost.	Change in layout and flow. Balance of numbers of check-out stands to maintain waiting time standards.
Postal service	Reduction of services to reduce costs (e.g., number of deliveries per day).	Introduction of some semiautomatic equipment for sorting. System improvement, etc., to improve overall delivery time.
Food service	Elimination of waiters and waitresses to reduce cost, as in cafeterias, and to reduce waiting time and cost in mass food outlets.	Change of layout and flow.
Computing service (e.g., in service bureaus)	Change mix of types of services offered (e.g., time share, batch processing, etc.)	Change mix of equipment (e.g., basic machine and peripheral equipment).

TABLE 2-3 (continued)

Interactions Between the Design of Services Offered and the Productive System Design for Several Types of Services

Types of Service	Types of Alteration of Service Offered to Accommodate Needs of Productive System	Types of Alteration of Productive System to Accommodate Needs of Service Offered
Fire protection	Increase or decrease time to react and provide service.	Relocate stations and/or add or delete stations to maintain a reaction time standard.
Police protection	Increase or decrease time to react and provide service.	Increase or decrease size of police staff. Reallocate staff based on changing crime patterns.
Emergency medical service	Increase or decrease time to react and provide service. Change mix of services available on an emergency basis.	Relocate ambulance stations and/or add or delete stations to maintain a reaction time standard. Change equipment and/or level of training of paraprofessional medical personnel.
Airline cabin service	Change average ratio of passengers to flight attendants. Eliminate services and reduce quality of services offered to reduce cost.	Increase public relations efforts through the media.

The Product Mix Problem

One way of expressing the system of problems surrounding product design is to ask, "What happens to the optimum mix of products when a new product or product line is introduced?"

The product mix problem is a general one throughout industry, and solutions should reflect the most economical allocation of capacity to demand. For example, in oil refining, there are interdependencies in the quantities of different products to be produced. If more of one product is to be produced, such as heating oil, then less of some other products will be produced. The profitability of various products may be different, and there are limits to the markets for each. The result is a complex programming problem to determine the best product mix to produce. The interdependencies always exist in oil refining because the basic raw material, crude oil, can be processed into many different products. An increase or decrease in one always means a change in the quantities of some other products.

In the mechanical industries, the product mix problem is not so obvious. Nevertheless, the mix problem can be acute if we are operating near capacity with time-shared facilities. At this point the interdependencies become critical, and increases or decreases in the amount of one product produced can mean changes in the amounts of other products produced. These problems have been approached through mathematical programming.

The analysis of adding (or deleting) a product can be important in the managerial decision process. In the linear programming format, it focuses managerial attention on the system or products and overall impact on contribution and the impact on the productive system and its capacities. Through sensitivity analysis, the linear programming decision format focuses managerial attention on possible opportunities for selectively enlarging capacity.

IMPLICATIONS FOR THE MANAGER

The design of products and services is of great interest for the manager of operations because the nature of the product or service has great impact on the design of the system to produce it, and vice versa. The manager can often obtain lower costs by exploiting the relationship between the product or service and the productive system.

Innovation is the generator of new products and services. Studies of innovation within industries indicate that a majority of innovations may be classified as product innovations. However, when we look at the vertical integration structure of an industry, we find that product innovations at lower levels in the hierarchy are, in fact, process innovations for the higher levels in the structure. Thus, there is an interaction between product design and productive system design even in these macro terms.

The details of product design, system location, and productive system design all flow from predictions and forecasts of markets. The methods for making market demand estimates are classified as predictive, causal forecasting, and time series forecasting. The manager needs to understand the appropriate use of each of these kinds of methodologies, for the field of application of each is quite different. While predictive methods lack mathematical rigor, they are perhaps the most important to a manager involved in launching a new product, since a historical record does not exist. They are also of value where long-term predictions are needed as a basis for capacity planning and plant locations. Causal methods are quite accurate for short- and medium-range forecasts and are appropriate for operations, as well as for some decisions relating to productive system design. The time series methods are most appropriate for short-range forecasting for inventory and production control. Thus, managers of productive systems are likely to use time series and causal methods appropriately for planning operations in short- to medium-term planning horizons.

The term "production design" is used in industry to connote the important interaction between product and system design. Some of the common interactions involve accommodation in the product design to processes and materials, methods of joining parts, tolerances, design simplifications, and reduced amount of processing.

The accommodations by the design of services offered are most often to cost considerations and may result in the client or customer doing part of the activity. The result is that the client or customer gets less service or a different mix of services or waits longer to obtain service. Managers of service operations need to be sensitive to the nature of their clientele in order to maintain an appropriate definition of the service offered in relation to the cost of providing the service. Unfortunately, improvements in productivity in service operations are less common, and cost reduction objectives are most likely to be achieved through changes in the nature of the service offered.

REVIEW QUESTIONS

1. How does the analysis of the vertical integration structure of an industry alter how one views the nature and source of innovations?

2. Develop an analysis of the flow of innovations through the vertical integration structure, such as Figure 2-1, for another industry. The emphasis here is on the structure, not the innovation data.

3. Contrast the meaning of the terms "prediction" and "forecasting," as they are used in this book.

4. How does the Delphi method gain the benefit of expert opinion in the form of consensus rather than compromise?

5. What kinds of values to the planning base of TRW resulted from the type of Delphi study reported?

6. What kinds of data useful in planning product and productive system designs result from market surveys? From historical analogy and life cycle analysis?

7. What are causal forecasting methods? Time series forecasting methods? What are their fields of application?

8. Discuss the relationship of functional design and production design in determining a product design that meets functional requirements, cost considerations, and the limitations of available processes.

9. Discuss the nature of costs that can be affected by alternate product designs.

10. What feedback loops provide information for the redesign of products and the productive system?

11. What kinds of accommodation in the nature of services offered may result from cost and quality pressures?

12. What kinds of interactions between the design of services offered and the productive system design are likely to occur for fire protection, police protection, emergency medical service, and airline cabin service?

13. How do the chapter materials fit in with the systems approach discussed in Chapter 1?

14. What are the criteria and values that enter the decision process for determining whether or not a new product design should be approved?

15. What kinds of data might result from an analysis of product mix when we are considering the addition or deletion of a product or service? How might these data be of value in making the product decision?

SITUATIONS

16. Nels Jensen started his grocery business 35 years ago in the Lake Tahoe resort community. The combination of good management and a market among well-to-do patrons produced an independent supermarket of unsurpassed quality. One of the hallmarks of the Jensen success model was his emphasis on and definition of the service aspects of his store. Those service aspects were: 10 hours of operation every day and minimum customer hassle and waiting time to obtain the desired purchases and get checked out.

A study of some of the major changes that occurred over the years revealed that both the nature and quality of service and the system design successively affected each other. When the store was small, employees were stationed in areas to help customers select items, check out, bag, and transfer purchases to the parking lot. Later, with a much larger volume of customers, the system transferred virtually all the selection process to the customer, and the system design focused on the check-out stand.

By controlling the number of check-out counters in operation, Jensen set a service standard that he tried to maintain. The standard was set in terms of the time the customer had to wait before being served. Jensen felt that for his clientele, a waiting time of 1.5 minutes was about as long as would be tolerated without complaint. In fact, however, he tried to control waiting time by keeping an eye on the size of the waiting lines. Any time the lines had two or more people waiting, he would open up another check-out counter, even if he had to operate it himself. He then tried to schedule checker shifts to provide capacity for the peak shopping hours. He also trained some of the checkers to do other work, such as pricing and storing stock on display shelves, to provide flexibility in the number of checkers available when peak loads occurred.

Jensen tried several variations with the check-out system. Originally, employees helped unload carts to the checkers, but as labor costs increased, this activity was transferred to the customer. Customer complaints resulted. Then, having changed the design of the check-out stand, Jensen tried a system in which the checker worked directly from the cart. However, the checker complained of backaches from constantly having to lean over to obtain items from the carts.

Jensen then modified the system with a new cart and check-out stand design system. This still enabled the checker to work directly from the cart but did not require leaning over to obtain items. The new cart–check-out stand system raised the working level. The cart had a hinged end, which the checker opened, thus placing the bottom of the cart at the check-out stand working level. In this way, all aspects of service could remain the same in terms of waiting time, what the customer had to do to obtain service, and the check-out time.

The next cycle of system design involved what is called *front-end automation.* Measurements made in a survey by Jensen indicated that the average time for a customer in the check-out process was 5.5 minutes, including 1.5 minutes of waiting time. The 4.0 minutes for actual check-out included 2.5 minutes for ringing up the sale and placing the purchases in bags. The balance was payment, which very often involved check cashing, as well as some chitchat and other miscellaneous activities. With the advent of item scanning systems, Jensen saw the possibility of improving overall service time by reducing the check-out time itself and possibly simultaneously reducing labor costs.

The scanning systems required that a universal code be placed on products (the bar-pattern codes now commonly used on product packages) that would be read by the scanners. A customer's entire order could then be moved over the scanner rapidly. The system required the customer to load purchases on a belt that fed the purchases to the checker, who repositioned them to move past the scanning eye. The scanner read and transmitted the information to a computer, which translated the information into prices and a total bill, including sales tax. The checkers' activities were thus confined to scanning, bagging, collecting, and making change.

While the scanning system had other operating advantages, Jensen was most interested in a possible service improvement, assuming that a productivity increase might justify the scanner's installation on a reasonable basis. The system was installed partially, as a test. Measured results indicated that check-out time, exclusive of waiting time prior to check-out, was reduced to an average of 2.4 minutes. In addition, checker productivity increased from an average of \$252 of sales per hour to \$500, and check-out errors were reduced by 65 percent. Checkers were paid \$4.50 per hour.

Jensen could install the scanning system to cover the present 12 check-out stands for a lease cost of \$7000 per month. Even though only an average of 6 check-out stands were in use (current system), if he were to decide to install the scanners, he would want the entire system to be automated. His concern was that service would be improved, in the sense that the customers' time in a system would be reduced. On the other hand, the scanner system represented a step backward, in that the customers would have to unload their carts and load the conveyor belt. He was not sure how customers of his type of clientele would react.

Should Jensen install the scanner system? Why?

17. History records that the electronic pocket calculator had a product ancestor known as the mechanical desk calculator—first hand powered and later electrically powered. It was a mechanical marvel, prized by those whose jobs required accurate computations and by organizations that needed both accuracy and relatively high productivity in computations not justified for programming on computers.

Calculatron, Inc., was a major manufacturer of desk calculators and had enjoyed long-term profitability. It had a loyal work force of semiskilled and some highly skilled employees. Although product improvements had continued through the years, the basic design of Calculatron's product line was stable during the previous 15 years, and product design changes were carefully implemented to take account of the existing production lines. The market for desk calculators had been an expanding one, and with the advantage of a relatively stable product design, Calculatron had been able to specialize production methods, making continuous improvements in productivity through investments in labor-saving equipment. The productivity increases had helped secure the firm's market position through competitive pricing and produced profitability and security for both the enterprise and its employees. Employees enjoyed high wages and salaries and excellent pension and other benefits. Employees were organized and affiliated with the AFL-CIO and union–management relationships had been generally very good.

Enter electronic minicircuitry, with microcircuitry and the "chip" on the horizon. The first electronic desk calculator had just been announced by a competitor. Calculatron was not far behind. It had employed a staff of electronic engineers two years previously and assigned them the task of producing a revolutionary redesign of the product line. The prototypes had already been tested, and the product and production engineers were at work in the production design phase simultaneously developing preliminary designs of the productive system required to produce the new electronic product line.

Market forecasts indicated a conversion of the former mechanical calculator volume to the electronic, with a "kicker," since a large replacement market was available for the faster, quieter, and more capable electronic machines. The one sour note in the market was the prediction that a pocket-sized calculator would be possible soon if the research and development on microcircuits were to materialize. Reports concerning startling technological innovations in microcircuitry indicated that the probability of a breakthrough was high.

The production engineers are ready to develop final designs of the productive system for the new electronic calculator line. A meeting of the executive committee has been convened to examine preliminary plans for production in relation to short- and longer-term market forecasts and predictions.

a. What kinds of guidelines for the productive system design should the executive committee establish for the production engineers?
b. What plans should Calculatron make for the introduction of the new product line?
c. What kinds of longer-term plans should Calculatron make for future product and process innovation in the calculator field?

18. A company produces refrigerators and is contemplating the addition of air conditioners to the product line. They are currently producing 2000 refrigerators per month. There is idle capacity in the assembly line, as well as in the machine shop and the refrigeration unit departments.

Planners have estimated that the air conditioners could use the same facilities in the production of parts in the machine shop and refrigeration unit departments and that the only manufacturing facility that would need to be added is a final assembly line for the air conditioners. The capacity estimates and worker-hour requirements for both products are summarized in Table 2-4. The planners also estimate that the contributions to profit and overhead for the two products will be \$50 for the air conditioners and \$38 for the refrigerators. Note that any additional labor required by the air conditioners has been deducted as a variable cost from the gross revenue per air conditioner to obtain the \$50 contribution.

The planners then solve a simple linear programming problem for the optimum product mix and obtain the schedule of 1250 refrigerators and 2000 air conditioners per month.

At the executive committee meeting where the planners make a presentation of their analysis and recommendations, strong opposition is encountered from the vice-president in charge of marketing. She objects to the fact that the proposal calls for a reduction in the output of refrigerators from 2000 to only 1250 per month. "Why should we give away part of our market? Why not produce for the full market demand of 2000 refrigerators per month, we have the capacity?"

TABLE 2-4 **Worker-Hour Requirements and Capacities for Refrigerators and Air Conditioners**

	Worker-Hours per Unit		Capacities (Worker-Hours per Month)
	Refrigerators	Air Conditioners	
Air conditioner Assembly line	0	3	6000
Refrigerator Assembly line	2.9	0	8000
Machine shop	2.0	2.5	7500
Unit department	1.5	1.3	5000

a. Examine the profitability of alternate proposals for including the air conditioners in the product line.
b. In the linear programming solutions schedule, the machine shop is fully utilized. How much additional machine shop capacity would be required if refrigerator output were maintained at 2000 per month? Would the capacities of any other departments be affected?
c. What should the company do?

19. Tables 2-5 and 2-6 summarize extensive results of a long-range prediction of air travel and aircraft technology, using the Delphi method.

Study the results in Tables 2-5 and 2-6 carefully, and discuss the following:

a. What future do you see for supersonic passenger travel? If you were a manager of a major U.S. airline company, what plans would you make? What kind of commitment of resources would you make at this time?
b. If you were Secretary of Transportation, what plans would you be making for supersonic passenger travel?
c. What future do you see for subsonic passenger travel? If you were a manager of a major U.S. airline company, what plans would you make? What kind of commitment of resources would you make at this time?
d. Based on the results of the traffic projection in Table 2-6, what capacity increase would you plan for the years 1990 and 2000? How are your judgments affected by the environmental impact results reported in Table 2-6?

TABLE 2-5 **Results of Aircraft Technology Questionnaire, (1) Supersonic Passenger Transport, and (2) Subsonic Passenger Transport**

Final Results, May 1974

		Mean	Standard Deviation (consensus)
1.	SUPERSONIC PASSENGER TRANSPORT		
1.1	ANGLO FRENCH CONCORDE		
1.1.1	Do you believe the Concorde eventually will be allowed to overfly the U.S. at supersonic speeds?	No	(11 out of 12)
1.1.2	Do you believe the Concorde will be allowed to overfly European countries at supersonic speeds?	No	(7 out of 12)
1.1.3	What number of the present version of the Concorde do you estimate eventually will be sold?	28	14
1.1.4	When do you estimate production of the present version will be discontinued?	1979	1.3 years
1.1.5	How many of the present version do you estimate will be flying in 1990?	17	17
1.1.7	An advanced second generation version of the Concorde might capture a wider market:		

TABLE 2-5 (continued) **Results of Aircraft Technology Questionnaire, (1) Supersonic Passenger Transport, and (2) Subsonic Passenger Transport**

	Final Results, May 1974	Mean	Standard Deviation (consensus)
	(i) When do you estimate such an aircraft might be introduced into service?	1985	3.5
	(ii) Do you believe that the sonic boom can be reduced or eliminated so as to make overland flight possible?	No	(8 out of 12)
	(iii) If the answer to (ii) is yes, when do you estimate this will be achieved?	2000	—
	(iv) What do you expect the characteristics of the advanced Concorde to be?		
	(a) Passenger capacity	180	50
	(b) Range (kms)	7500	1000
	(c) Cruising speed (mach no.)	2.4	0.5
	(d) Direct operating cost c/passenger km	1.2	0.3
1.1.8	How many of the advanced version do you estimate will be flying in 1990?	60	34
1.1.9	How many of the advanced version do you believe will be flying in 2000?	85	97
1.2	TU 144 (USSR SUPERSONIC)		
1.2.1	Do you believe the TU 144 will be allowed to fly over the USSR at supersonic speed?	yes	(11 out of 11)
1.2.2	What do you expect will be the maximum number produced?	67	54
1.2.3	How many TU 144's do you estimate will be flying in 1990?	75	54
1.3	U.S. SUPERSONIC TRANSPORT		
1.3.1	Should a development program for a U.S. supersonic transport be undertaken; when do you estimate it might go into commercial service?	1986	2.5 years
1.3.2	What do you expect will be the characteristics of this transport?		
	(a) Passenger capacity	350	60
	(b) Range (kms)	7500	2000
	(c) Cruising speed (mach no.)	2.9	0.4
	(d) Service altitude (kms)	21	3
	(e) Direct operating costs c/pass. km	1.0	145
1.34	How many U.S. SST's might be flying in 1990?	165	145
1.3.5	How many of this type might be flying in 2000?	225	130
2.	SUBSONIC PASSENGER AIRCRAFT		
2.1	What number of the following aircraft types, which are now or will shortly be in service, do you expect to be flying in 1990?		
	(a) B747	500	190
	(b) DC10 and L 1011	860	250
	(c) 600–2000 passenger subsonic	330	220
	(d) B727	610	315
	(e) U.S. and European Twin Air Buses	980	450
	(f) USSR (medium to long range) subsonics	625	210
2.2	What do you expect the characteristics of the next generation of medium to long range subsonic transport type will be?		

TABLE 2-5 (continued) **Results of Aircraft Technology Questionnaire, (1) Supersonic Passenger Transport, and (2) Subsonic Passenger Transport**

	Final Results, May 1974	Mean	Standard Deviation (consensus)
	(a) Passenger capacity	600–800	(6 out of 12)
	(b) All cargo payload, tons	50–100	
	(c) Range (kms)	8,000–10,000 km	
2.3	When do you estimate this type might be introduced into service?	1983	3 years
2.4	How many of this new type might be flying in 2000?	600	250
2.5	Looking beyond the year 2000, what will be the characteristics of the next advanced type of medium to long range subsonic jet?		
	(a) Passenger capacity	800–1000	
	(b) All cargo payload-metric tons	200	
	(c) Range (kms)	8,000–10,000 km	
	(d) Type of fuel	No consensus	
2.6	When do you estimate this type will be introduced into service?	2000	
2.7	How many of this type might be flying in 2000?	70	50
3.	ALL CARGO TRANSPORT		
3.1	How many all cargo or convertible versions of the following subsonic types might be flying in 1990?		
	(a) B747	90	40
	(b) DC20 and L 1011	80	45
	(c) U.S. and European Twin Air Buses	40	30
	(d) USSR (medium to long range subsonic transports	80	35
	(e) Aircraft, type described in 2.2	150	110
3.2	Do you anticipate that by 1990 there will be an all cargo version of the supersonic transport flying? If so, how many of the following types might be flying in 1990?		
	(i) U.S. Supersonic	0	0
	(ii) Advanced Anglo French Concorde	0	0
	(iii) TU 144	5	12
3.3	How many all cargo versions of the following types might be flying in 2000?		
	(a) B747	70	50
	(b) DC10 and L 1011	70	60
	(c) Advanced Anglo French Concorde	0	0
	(d) USSR Supersonic	0	0
	(e) Aircraft type described in 2.2	250	140
4.	ADVANCED FLIGHT SYSTEMS		
4.1	A hypersonic transport has been proposed that will have an estimated speed of 8,000 kms per hour and fly in the upper stratosphere. This type might eliminate the sonic boom associated with the present supersonic types. When do you estimate this type might be introduced into service?	2020	8
4.2	What do you expect the characteristics of this type to be?		

TABLE 2-5 (continued) **Results of Aircraft Technology Questionnaire, (1) Supersonic Passenger Transport, and (2) Subsonic Passenger Transport**

Final Results, May 1974

		Mean	Standard Deviation (consensus)
	(a) Passenger capacity	320	60
	(b) Payload (metric tons)	100	40
	(c) Type of propulsion system	Rocket	
	(d) Fuel	LH_2	
	(e) Direct operating cost c/pass. km	1.5	0.5
4.3	How many of this type do you estimate will be flying in 2025?	90	80
4.4	The Ballistic-Glide Transport is also a possible future flight system. This will have an estimated transit time of 45 min. to any place on the globe and will be rocket propelled with turbo jets for maneuvering and landing. This transport will be subject to high takeoff acceleration and to zero gravity during flight. A passenger capacity of 200 has been proposd. When do you estimate that this type will go into service?	2030+	
4.5	How many of this type will be flying in the year 2025?	3	7

SOURCE: J. M. English, and G. L. Kernan, "The Prediction of Air Travel and Aircraft Technology to the Year 2000 Using the Delphi Method," *Transportation Research, 10*, 1976, pp. 1–8.

TABLE 2.6 **Results of Traffic Projection Questionnaire**

		Mean	Standard Deviation (consensus)
1.	ENVIRONMENT		
1.1	Do you believe that SST flights over populated areas in the U.S. will be permitted?	No	(14 out of 16)
1.2	What effect will airport saturation have on the growth of air traffic over the next two decades?	Slight	(7 out of 16)
1.3	Do you believe that SST flights over populated areas in Europe will be permitted?	No	(11 out of 16)
1.4	What effect will noise control at airports have on the growth of air traffic over the next two decades?	Slight	(8 out of 16)
1.5	Do you anticipate that real fuel costs in the year 1990 will be the same as today, higher or lower?	Significantly higher	(15 out of 16)
1.6	Do you believe that the relative price of air travel to other commodities will change significantly over the next two decades?	No	(10 out of 14)

TABLE 2-6 **Results of Traffic Projection Questionnaire**
(continued)

		Mean	Standard Deviation (consensus)
1.7	What effect will this price change have on the present growth of air travel?	Slight decrease (6 out of 16)	
1.8	Do you foresee any future developments that will significantly change future predictions of 1990 air traffic? If so, could you please elaborate.	The general consensus is that substitute transport modes and higher fuel prices will reduce the growth rate of air travel.	
1.9	Do you believe that the annual growth rate in world air travel over the next two decades will be: (a) Much the same as it has been (b) Will increase slightly (c) Will increase significantly (d) Will decrease slightly	Will decrease slightly (7 out of 16)	
2.	ROUTE PATTERN		
2.1	The breakdown of total passenger kilometers by percentage on principal routes is given below. Do you anticipate that there will be any significant changes in this pattern 20 years from now?		
	Routes		
	U.S. Domestic 1970...36.3	1990...31.0	4.5
	U.S.S.R. Domestic 17.1	17.5	2.0
	U.S. to Europe 10.5	11.0	3.5
2.	(i) World Billion Passenger Kms in 1900	2,668	1,400
	(ii) World Billion Passenger Kms in 2000	4,965	5,400
	(iii) World Billion Passenger Kms in 2020	13,800	23,000
2.3	The average world passenger load factor for 1972 was 0.52. What do you estimate the average load factor will be in 1990 for:		
	(i) Subsonic	0.65	0.07
	(ii) Supersonic	0.65	0.12
2.4	Estimate of 1990 World Fleet		

Subsonic or Supersonic	No. of Engines	Passenger Capacity	Mean	Standard Deviation (consensus)
Sub	4	120– 200	1100	500
Sub	3	90– 160	1500	250
Sub	4	350– 500	1000	550
Sub	3	250– 300	1000	—
Sub	2	80– 140	700 km	750
Sub	2	200– 300	320	300
Sub	4	500–1000	120	350
Super	4	110	40	32
Super	4	200	63	40
Super	4	350	70	50

		Mean	Standard Deviation (consensus)
3.1	World Air Freight		
	Billion Ton Kilometers in 1990	126	61
	Billion Ton Kilometers in 2000	280	235
	Billion Ton Kilometers in 2020	958	1200

SOURCE: J. M. English and G. L. Kernan, "The Prediction of Air Travel and Aircraft Technology to the Year 2000 Using the Delphi Method," *Transportation Research, 10,* 1976, pp.1–8.

20. The following is an article, "Incorporating Judgments in Sales Forecasts; Application of the Delphi Method at American Hoist & Derrick," by Shankar Basu* and Roger G. Schroeder,† reprinted from a 1977 issue of *Interfaces*. Read the article with the following questions in mind: The forecasts for the American Hoist & Derrick Company were for five years in advance. How much faith do you think should be put in these forecasts? Would you be willing to make plans that involved large investments in plant and equipment based on the forecasts? How do you evaluate the forecast methodology used for longer-term forecasts?

INCORPORATING JUDGMENTS IN SALES FORECASTS: APPLICATION OF THE DELPHI METHOD AT AMERICAN HOIST & DERRICK‡

Shankar Basu
and
*Roger G. Schroeder***

ABSTRACT

In many organizations complete reliance on historical data is not an adequate basis for forecasting future sales. Since underlying conditions or assumptions may be changing, a means of incorporating management judgment in sales forecasts is needed. This paper reports on the development and application of a Delphi method ("opinion methodology") for sales forecasting at the American Hoist & Derrick Company. Although the Delphi method has only been in use for one year, the sales forecast error for 1975 was reduced to less than 1 percent, whereas sales forecast errors for the previous ten years were significantly higher.

Introduction

American Hoist & Derrick is a well-known manufacturer of construction equipment, with annual sales of several hundred million dollars. Their sales forecast is an actual planning figure—not merely a goal—since it is used to develop the master production schedule, cash flow projections, and work force plans. Consequently, the top management personnel at American Hoist & Derrick are extremely concerned with predicting sales as accurately as possible. Due to this concern, management is reluctant to rely on any single forecasting method. Thus, while the Delphi method of sales

* Manager, Marketing Analysis, American Hoist & Derrick Co., St. Paul, Minnesota.
† Professor of Management Science, Graduate School of Business Administration, University of Minnesota, Minneapolis, Minnesota.
‡ Reprinted with permission from *Interfaces*, 7(3), May 1977.

forecasting is emphasized in this paper, the 1975 sales forecast was developed using a number of methods [Chambers et al., 1971].

In the past, American Hoist & Derrick sold everything they could make. Sales forecasts were derived by a few key individuals, who utilized various methods to analyze the data from plant managers and sales personnel, but relied principally upon selective judgment. Over the past ten years, these subjective forecasts have been in significant error and this has caused a great deal of concern in top management.

Beginning with the 1975 sales forecast, top management wanted to assess the sales potential accurately, in order to determine just how fast the production capacity should be expanded. Such an estimate could not be based solely upon historical sales, since these only reflected previous production constraints. Additionally, rapidly changing economic conditions made the past a relatively poor predictor of the future—for production costs as well as expected sales. Consequently, the managers decided to temper historical data with informed judgment by utilizing the Delphi method to develop a five-year sales forecast. However, alternate forecasts were also prepared, using regression analysis and exponential smoothing, in order to avoid undue reliance on any single method and to provide a bench mark with which the other results could be compared.

Due to space limitations, only the Delphi method is discussed in this paper. The next section describes the actual use of the Delphi method at American Hoist & Derrick; the following section summarizes the results and conclusions of its use. The last section, an Appendix, details the general Delphi technique with particular emphasis on developing sales forecasts, for those unfamiliar with the basic method. [The Appendix is not included in this reprint.]

Delphi Method Use at American Hoist & Derrick

Formulating the Delphi Study. *The well-known Delphi method consists of: (1) a panel of experts, (2) a series of rounds, and (3) a questionnaire for each round. Each member of the panel responds anonymously to the questionnaire on each round and the summarized responses of the panel are fed into the next round. At AH&D it was decided to use three rounds for the Delphi study.*

In constructing the panel a total of 23 key individuals were selected. The panel selection was based on the following criteria:

- Personnel who had been doing these forecasts intuitively,
- Personnel who were responsible for using these sales forecasts,
- Personnel whose activities were affected by these forecasts,
- Personnel who had a strong knowledge of market place and corporate sales.

Care was taken to include knowledgeable personnel from different functional areas of the corporation.

In the case of sales forecasting the Delphi questionnaire should request not only sales estimates of interest, but also such information as an industry projection and a business indicator. This additional information will provide a check against the sales figure through correlation and it also helps the respondents develop their process of estimation. A close look at the corporate revenue for American Hoist and Derrick over the last five years revealed that the construction equipment group generated approximately 60 percent of the total revenue. The most logical industry that total company sales would correlate with is, therefore, the construction equipment industry. A graphical and analytical check demonstrated that these two time series were indeed correlated. Next, leading, roughly coincident, and lagging indicators were tried for best fit with the construction equipment industry. The closest fit was obtained with GNP in current dollar series and GNP was therefore determined to be the business indicator of interest.

The questionnaire then requested the following four estimates on each of the three rounds:

- Gross National Product, current dollars;
- Construction equipment industry shipments, current dollars;
- American Hoist and Derrick construction equipment group shipments, dollars;
- American Hoist and Derrick, corporate value of shipments, current dollars.

The First Round. *To help obtain a realistic median on the first round, all four estimates which were being requested had data input. The data input were actual figures for the past five to seven years. Compound growth rate and graphical representation of each series were included. Opinion on percent increase expected for the next five years was requested. [Figure 2-6 is a part of the first round questionnaire for the construction industry estimate. A similar page was included for each of the other three estimates requested.]*

The responses on the first round were collected and summarized. A standard statistical analysis of the responses was conducted which generated for each of the four series—number of observations, largest observation, smallest observation, range of observations, mean, confidence limits, standard deviation and median.

The Second Round. *The second round questionnaire included feedback on actual responses together with the above standard statistical analysis. Each panel member's response was indicated and his revised estimated was requested in the second round. [See Table 2-7 for a sample of the data feedback for the construction industry estimate.] On the second round all members of the panel were asked for explanations of their responses, irrespective of whether their responses fell in the top, bottom or interquartile range. The logic behind including the entire population was that the*

Construction Industry Estimate

The following figures and graph show historical sales figures for power cranes, shovels, walking draglines and walking cranes for the last seven years. Please indicate your estimate for each of the next six years. (% increase expected and graphically).

Year	% Increase	Industry Sales Mil $	Calendar Year	% Increase Expected
1967		511	1974	-----
1968	2.5	524	1975	-----
1969	9.7	575	1976	-----
1970	0.4	598	1977	-----
1971	6.5	637	1978	-----
1972	12.8	719	1979	-----
1973	35.4	974		

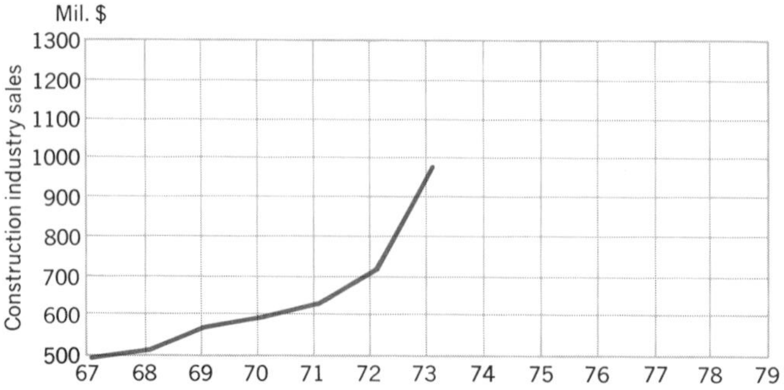

FIGURE 2-6
Sample: Partial First Round Request for Information.

explanations of any person might be important and meaningful for the total population.

The response to the second round was of a "brain storming" nature. There was an outpouring of explanations from the panel members. The individual reasons for estimates were collected, summarized and listed under separate headings of GNP, CE Industry, AH&D Construction Equipment group and AH&D total shipments. All these explanations were then categorized into positive and negative factors for each of the four estimates. Relevant published data and opinion of well-known economists were also collected and prepared for input into the third round.

The Third Round. *The third and final questionnaire was designed with extreme care. The format for this round was:*

- Restatement of the initial question,
- Feedback of a respondent's first and second estimates,
- Second round response, standard statistical analysis,

- Factors that were considered important by the respondents in reaching their estimate in the second round,
- Related facts, figures and views of external experts,
- A request for the respondent's revised estimate,
- A request for comments or opinions.

TABLE 2-7 **Sample: Partial Second Round Feedback of Estimates**

		Construction Equipment Industry Annual % Increase in Shipments					
You Are		1974	1975	1976	1977	1978	1979
	1	5.0	7.0	7.0	6.0	6.0	6.0
	2	25.0	20.0	15.0	20.0	20.0	15.0
	3	6.5	5.4	7.6	8.2	10.0	10.0
	4	15.0	10.0	8.0	0.0	2.0	5.0
	5	20.0	20.0	8.0	8.0	6.0	7.0
	6						
	7						
	8	15.0	3.0	7.0	6.0	6.5	4.0
	9						
	10	20.0	10.0	5.0	10.0	10.0	10.0
	11	7.0	8.0	5.0	5.0	6.0	
	12	20.0	4.0	5.0	16.0	2.2	4.0
	13	13.0	5.0	4.5	8.5	8.8	10.2
	14	25.0	2.0	3.0	6.0	7.0	8.0
	15	20.0	10.0	5.0	5.0	5.0	7.0
	16	35.0	0.0	7.0	9.0	0.0	(10.0)
	17						
	18	30.0	5.0	6.0	6.0	6.0	6.0
	19	(20.0)	5.0	9.0	3.0	3.0	3.0
	20	10.0	5.0	7.0	10.0	8.0	10.0
	21						
	22	20.0	10.0	20.0	35.0	25.0	15.0
	23						
High		35.0	20.0	20.0	35.0	25.0	15.0
Low		(20.0)	0.0	3.0	0.0	0.0	(10.0)
Range		55.0	20.0	17.0	35.0	25.0	25.0
Mean		15.7	7.6	7.6	9.5	7.7	7.1
Std. Dev.		12.3	5.5	4.1	8.0	6.3	5.7
95% Confidence		9.3/22.0	4.7/10.4	5.4/9.7	5.4/13.6	4.5/10.0	4.0/10.2
Your Revised Estimate Explanation		....		....		....	

Table [2-8] shows that portion of the third round questionnaire applicable to sales in the entire construction equipment industry; similar formats were used for Gross National Product (GNP) sales in the AH&D construction equipment group, and total sales for American Hoist & Derrick.

Statistical analysis on the third round was considered as the final forecast by the panel. The standard deviation and 95 percent confidence band had narrowed sufficiently to consider this forecast a reasonable consensus of the panel.

TABLE 2-8 **Sample: Partial Third Round Input of Data and Request for Final Estimate**

Construction Equipment Industry

1. *Question: Restated*	Estimate percent increase in CE industry shipments in current dollars for calendar years.				
	1975	*1976*	*1977*	*1978*	*1979*
2. *Your first est.:*					
Your second est.:					
3. Second Round Response:					
High	10.0	10.0	10.0	10.0	10.0
Low	4.0	5.0	6.0	5.0	6.0
Range	6.0	5.0	4.0	5.0	4.0
Mean	7.14	7.56	7.92	6.7	7.75
Std. Dev.	2.29	1.34	1.79	1.51	1.51
95% Conf.	5.5/8.8	6.6/8.5	6.6/9.2	5.6/7.8	6.6/8.8

4. *Factors that were considered by the respondents:*

Negative
- Long delivery time
- Slow recovery in industry from recession
- Interest rate, financial control over expansion, repair and maintenance of current equipment
- Continued slowdown in housing
- Environmental factors causing project delays
- Drying up of federal funds
- Downtrend in backlogs

Positive
- Price increases
- Accelerated export business, continued worldwide demand of machinery for energy related projects

5. *Related figures for your consideration:*
(Source: F. W. Dodge)

Construction contract awards in April and May were 32% above the depressed 1st quarter average. First quarter capital appropriation jumped 66% to 16.9 billion. This follows a 104% increase between the third and fourth quarters of 1974.

6. *Your revised estimate:*

	1975	*1976*	*1977*	*1978*	*1979*
Percent increase expected:					

7. *Comments, other facts and figures you would like to see or show:*

TABLE 2-9

Actual Versus Forecast Sales (American Hoist & Derrick)

	1975	1976
Actual sales (millions)	$359.1	$410
Forecast sales (millions) Using Delphi method	$360.2	$397
Forecast error	+0.3%	−3.3%

Results and Conclusions

Sales Forecast Accuracy. *The use of the Delphi method had a direct and definite impact on the American Hoist & Derrick sales projection. Top management was presented three forecasts: one developed using the Delphi method; another used regression analysis; and a third forecast developed using exponential smoothing. The Delphi forecast was considered most credible by top management because it incorporated the experienced judgment of 23 key corporate individuals. This confidence was subsequently justified, as indicated in Table [2-9]. The 1975 Delphi sales forecast was within $1.1 million of the actual 1975 sales, that is,* the sales forecast error was reduced to less than one-third of one percent. *Due to an extensive management reorganization and reassignment of key individuals in 1976, the Delphi method was not used to develop a new five-year sales forecast or update the previous one. In spite of this, the 1976 sales forecast—corresponding to the second year of the original five-year projection—was within $13 million of actual 1976 sales, resulting in a forecast error of less than four percent.* This was considerable improvement over the previous forecast errors of plus or minus 20 percent. *Additionally, the Delphi forecasts were more accurate than the forecasts developed using regression analysis or exponential smoothing. The forecasts developed using these latter techniques were incorrect by $30–50 million, indicating a forecast error on the order of 10–15 percent.*

One negative aspect of the study is the time period required to develop the forecast: approximately three months to complete all three rounds, evaluate the results, and produce a final consensus. However, the authors estimate that the next Delphi sales forecast, to be developed in late 1977, will be completed in only four to six weeks. This improved response time is expected to result from a Delphi education-training program and the use of precoded input forms in conjunction with a General Electric Statistical analysis system for data analysis.

Uniform Sales Estimates. *Apart from the improvement in forecast accuracy, the next most beneficial aspect of the Delphi study was that it provided a reasonably uniform estimate of sales among different managers. The managers on the panel were presented with accurate past and present economic conditions and data on the business environment. Through successive rounds of Delphi, they developed a congruence in outlook on business conditions and corporate sales. Table [2-10] indicates*

TABLE 2-10 **Convergence of Some Estimates**

	GNP Estimates 1978 % Growth				
	High	Low	Range	Standard Deviation	Median
Round 1	12.0	0.0	12.0	2.9	6.5
2	10.0	0.0	10.0	2.6	6.0
3	8.5	5.0	3.5	1.5	5.9
	Construction Equipment Industry Estimates 1978 % Growth				
	High	Low	Range	Standard Deviation	Median
Round 1	25.0	0.0	25.0	6.3	5.9
2	15.0	0.0	15.0	3.9	6.0
3	13.0	5.0	8.0	2.4	6.3

the median, standard deviation, and range of the responses for each of the three rounds for estimates of the GNP and total sales in the construction equipment industry. The tendency toward consensus is indicated by the reduction in the range, standard deviation, and in the 95% confidence interval from one round to the next. All numerical estimates exhibited this type of reduction in variance.

This tendency toward congruence resulted in a relatively uniform base for future decision making among the participants in the study. Managers reacted very favorably to involvement in the forecast, and exhibited interest by requesting the results of the study. Particular enthusiasm was shown in the second round, when each manager was asked for the reasons behind his/her forecasts. A great many corporations lack this uniformity of outlook, and quite divergent assumptions are made about sales potential. Even if there is a published forecast for sales, it probably will not have been internalized by managers to the extent that occurs from participation in a Delphi study. At American Hoist & Derrick there was a very marked divergence in sales estimates among managers in the first round; this divergence was greatly reduced upon completion of the third round.

Conclusions

The following conclusions were arrived at by the authors in conducting the Delphi study for sales forecasting:

1. Delphi has definite utility as an analytical tool for predicting sales of a corporation. As the forecasts incorporate anticipation of the future by experienced and qualified individuals, the results seem to be meaningful and realistic. Though the

results can be used singularly, it is suggested that they be used in conjunction with other quantitative approaches to sales forecasting.

2. Apart from its analytical value, a Delphi study has an inherent educational value. The corporate officers are presented with past and present economic conditions and status. As a result, there is the development of a congruence in outlook on business conditions and corporate sales volume. This provides for a more uniform singular base for decision making by the different managers involved in the study.

REFERENCES

Abernathy, W. J., and P. L. Townsend. "Technology, Productivity and Process Change," *Technological Forecasting and Social Change,* 7(4), 1975.

Abernathy, W. J., and K. P. Wayne. "Limits of the Learning Curve," *Harvard Business Review,* September-October 1974, pp. 109–119.

Ahl, D. H. "New Product Forecasting Using Consumer Panels," *Journal of Marketing Research,* 7(2), May 1970, pp. 159–167.

Armstrong, J. S., and M. C. Grohman. "A Comparative Study of Methods for Long-range Market Forecasting," *Management Science, 16*(5), January 1969.

Bass, F. M. "A New Product Growth Model for Consumer Durables," *Management Science, 16*(5), January 1969.

Bass, F. M., C. W. King, and E. A. Pessemeier. *Applications of the Sciences in Marketing Management.* John Wiley & Sons, New York, 1968.

Basu, S., and R. G. Schroeder. "Incorporating Judgments in Sales Forecasts: Application of the Delphi Method at American Hoist & Derrick," *Interfaces,* 7(3), May 1977, pp. 18–27.

Boeing Commercial Aircraft Company. "Studies of the Impact of Advanced Technologies Applied to Supersonic Transport Aircraft," Interim Summary Report 3, Market Analysis (TASK II), DG-22529-2, 1973.

Boyer, C. H. "Lockheed Links Design and Manufacturing," *Industrial Engineering, 9*(1) January 1977, pp. 14–21.

Box, G. E. P., and G. M. Jenkins. *Time Series Analysis, Forecasting, and Control.* Holden-Day, San Francisco, 1970.

Bright, J. R. *Automation and Management.* Graduate School of Business Administration, Harvard University, Boston, 1958.

Buffa, E. S., and J. G. Miller. *Production-Inventory Systems: Planning and Control.* (3rd ed.). Richard D. Irwin, Homewood, Ill., 1979.

Campbell, R., "A Methodological Study of the Use of Experts in Business Forecasting," Unpublished Ph.D. dissertation, UCLA, 1966.

Chambers, J. C., S. K. Mullick, and D. D. Smith. "How to Choose the Right Forecasting Technique," *Harvard Business Review,* July-August 1971, pp. 45–74.

Chase, R. B., and N. J. Aquilano. *Production and Operations Management* (rev. ed.). Richard D. Irwin, Homewood, Ill., 1977.

Claycamp, H. J., and L. E. Liddy. "Prediction of New Product Performance: An Analytical Approach," *Journal of Marketing Research, 6*(4), November 1969, pp. 414–421.

Dalkey, N. C., and O. Helmer. "An Experimental Application of the Delphi Method to the Use of Experts," *Management Science, 9* (6), April 1963.

Dalkey, N. C. et al. *Studies in Quality of Life,* Lexington Books, Lexington, Mass. 1972.

Douglas Aircraft Company. "Studies of the Impact of Advanced Technologies Applied to Supersonic Transport Aircraft (TASK II)," Market Analysis and Economic Ground Rules, April 1973.

English, J. M., and G. L. Kernan. "The Prediction of Air Travel and Aircraft Technology to the Year 2000 Using the Delphi Method," *Transportation Research, 10,* 1976, pp. 1–8.

Gerstenfeld, A. "Technological Forecasting," *Journal of Business, 44*(1), January 1971, pp. 10–18.

Hahir, J. P. "A Case Study on the Relationship Between Design Engineering and Production Engineering," *Proceedings, Fifth Annual Industrial Engineering Institute.* University of California, Berkeley-Los Angeles, 1953.

Hudson, R. G., J. C. Chambers, and R. G. Johnston. "New Product Planning Decisions Under Uncertainty," *Interfaces, 8*(1), Part 2, November 1977, pp. 82–96.

Makridakis, S., and S. C. Wheelwright. *Forecasting: Methods and Applications,* John Wiley & Sons, New York, 1978.

Meyers, S., and D. Marquis. *Successful Industrial Innovations.* National Science Foundation, Washington, D.C., 1969, pp. 69–70.

Milkovich, G. T. et al. "The Use of the Delphi Procedures in Manpower Forecasting," *Management Science,19* (3), October 1972, pp. 211–221.

North, H. Q., and D. L. Pyke. "Probes of the Technological Future," *Harvard Business Review,* May-June 1969.

Papen, G. W. "Minimizing Manufacturing Costs Through Effective Design," *Proceedings, Sixth Annual Industrial Engineering Institute.* University of California, Berkeley-Los Angeles, 1954.

Parker, G. G. C., and E. L. Segura. "How to Get a Better Forecast," *Harvard Business Review,* March-April 1971, pp. 99–109.

Starr, M. K. *Product Design and Decision Theory.* Prentice-Hall, Englewood Cliffs, N.J., 1963.

U.S. Department of Transportation. "Propulsion Effluents in the Stratosphere," Climatic Assessment Program, Monograph 2(2nd ed.). U.S. Government Printing Office, Washington, D.C., 1974.

Utterback, J. M., and W. J. Abernathy. "A Dynamic Model of Process and Product Innovation by Firms," *Omega,* 1975.

Wheelwright, S. C., and S. Makridakis. *Forecasting Methods for Management.* John Wiley & Sons, New York, 1977.

Yourdon, E. "CALL/360 Costs," *Datamation,* November 1970, pp. 22–28.

CHAPTER 3 Capacity Planning

THE LONG-RANGE OPERATIONS STRATEGY OF AN ORGANIZATION is expressed, to a considerable extent, by capacity plans. It is in connection with capacity planning that the following issues must be considered. What are the market trends, both in terms of market size and location, and technological innovations? How accurately can these factors be predicted? Is a technological innovation on the horizon that will have an impact on product or service designs? How will capacity needs be affected by the new products? Are there process innovations on the horizon that may affect production methods? Is a more continuous productive system justified in the near future? How are capacity needs affected by process innovations? Will it be profitable to integrate vertically during the planning horizon? In planning new capacity, should existing policies for using overtime and multiple shifts be reviewed? In planning new capacity, should we expand existing facilities or build new plants? What is the optimal plant size? Should a series of smaller units be added as needed, or should larger capacity units be added periodically? Should the policy be to provide capacity so that some lost sales may be incurred, or is demand to be met?

The foregoing strategic issues must be resolved as a part of capacity planning. In assessing alternatives, the revenues, capital costs, and operating costs may be compared, but managers may need to trade off the possible effects of the strategic issues against economic advantages. The process of capacity planning may be summarized as follows:

1. Predict future demands, including the possible impact of technology, competition, and other events.

2. Translate predictions into physical capacity requirements.

3. Generate alternate capacity plans related to requirements.

4. Analyze economic effects of alternate plans.

5. Identify risks and strategic effects of alternate plans.

6. Decide on a plan for implementation.

PREDICTING FUTURE CAPACITY REQUIREMENTS

Long-range forecasts of demand are difficult. There are always contingencies that can have important effects, such as recessions, wars, oil embargos, or sweeping technological innovations. Therefore, predicting demand also requires an assessment of contingencies. The contingencies are apt to be rather different, depending on the situation. Mature products are more likely to have stable and predictable growth, while the markets for new products may be quite uncertain.

Mature Products with Stable Demand Growth

Many products and services enjoy mature, stable markets. Examples of products and commodities are steel, aluminum, fertilizer, cement, and automobiles. Examples of services are airline travel and health care. In the previous chapter, Table 2-2 summarized prediction and forecasting methods that have value for the longer term. Recall that the predictive methods were Delphi, market surveys, and historical and life cycle analyses. The causal forecasting methods were regression and econometric models. In addition to these formal predictive methods, executive opinion and extrapolation are common methods for estimating future demand.

For mature, stable products, causal models are often appropriate. In a comparative study, Armstrong and Grohman [1972] showed that an econometric model was superior to subjective methods in forecasting the U.S. air travel market. Also, Basu and Schroeder [1977] showed that long-range forecasts based on executive opinion could be improved by incorporating the Delphi method at the American Hoist & Derrick Company.

Given long-range predictions of demand, we must generate capacity requirements. It is unlikely that these capacity needs will be uniform throughout the productive system. A balance of capacities of subunits exists that reflects the discrete nature of capacity. For example, the existing receiving, shipping, and factory warehouse area may accommodate a 50 percent increase in output, but the assembly line may already be operating at full capacity and the machine shop at 90 percent of capacity. The capacity gaps can then be related to future capacity requirements, as in Table 3-1.

In Table 3-1, predicted capacity requirements are shown for an enterprise through 1990. The presumption is that these predicted requirements are the expected values that take into account contingencies for the situation. Optimistic and pessimistic predictions of requirements could also be made.

In Table 3-1 the projected gaps are shown in parentheses. Currently there is slack capacity in the machine shop and in the receiving, shipping, and warehouse areas.

TABLE 3-1 **Predicted Requirements, Current Capacities, and Projected Capacity Differences**

	Capacity, Units per Year			
	Current, 1980	1982	1985	1990
Predicted capacity requirements	10,000	12,000	15,000	20,000
Machine shop capacity	11,000	—	—	—
Capacity (gap) or slack	1,000	(1,000)	(4,000)	(9,000)
Assembly capacity	10,000	—	—	—
Capacity (gap) or slack	—	(2,000)	(5,000)	(10,000)
Receiving, shipping, and factory warehouse capacity	15,000	—	—	—
Capacity (gap) or slack	5,000	3,000	—	(5,000)

TABLE 3-2 **Expected, Optimistic, and Pessimistic Predictions of Requirements**

	Capacity Units per Year			
	Current, 1980	1982	1985	1990
Expected capacity requirements[a]	10,000	12,000	15,000	20,000
Optimistic requirements	10,000	14,500	25,000	62,000
Pessimistic requirements	10,000	11,000	12,800	16,000

[a] From Table 3-1

In two years, however, both the machine shop and assembly line will need additional capacity. These capacity gaps will grow as shown for 1985 and 1990. On the other hand, the receiving, shipping, and factory warehouse capacities will be adequate through 1985.

Identifying the size and timing of projected capacity gaps provides an input to the generation of alternate plans. We may plan to meet demand either by providing the expected required capacity or partially by utilizing alternate sources, or we may absorb some lost sales. We can provide the needed capacity in smaller increments as needed or in larger increments that may involve initial slack capacity. We may enlarge existing facilities, establish new producing locations for the additional capacity, or relocate the entire operation.

New Products and Risky Situations

It is difficult to predict capacity requirements for new products initially or in the rapid developmental phase of product life cycles. There are also situations involving mature, stable products, such as oil, where the capacity planning environment is risky owing to unstable political factors. The prediction of capacity requirements in these kinds of situations needs to place greater emphasis on the distribution of expected demand. Optimistic and pessimistic predictions can have a profound effect on capacity requirements.

For example, suppose that the product represented in Table 3-1 is in the rapid developmental stage of its life cycle. There may be considerable uncertainty about the future market because of economic factors and developing competition.

Table 3-2 includes optimistic and pessimistic capacity predictions that affect capacity requirements drastically. The optimistic requirement schedule assumes approximately a 20 percent growth rate in demand, and the pessimistic schedule, only a 5 percent growth rate. A 20 percent growth rate might be justified given favorable economic conditions and a slight gain in market share in spite of competition. A 5 percent growth rate might be justified given the success of foreign competition and a smaller market share even though the market as a whole is assumed to expand.

TABLE 3-3 **Predicted Requirements, Current Capacities, and Projected Capacity Differences for the Optimistic Prediction**

	Capacity, Units per Year			
	Current, 1980	1982	1985	1990
Predicted optimistic capacity requirements	10,000	14,500	25,000	62,000
Machine shop capacity	11,000	—	—	—
Capacity (gap) or slack	1,000	(3,500)	(14,000)	(51,000)
Assembly capacity	10,000	—	—	—
Capacity (gap) or slack	—	(4,500)	(15,000)	(52,000)
Receiving, shipping, and factory warehouse capacity	15,000	—	—	—
Capacity (gap) or slack	5,000	500	(10,000)	(47,000)

Tables 3-3 and 3-4 show the widely differing capacity needs for the optimistic and pessimistic predictions. If we assume the optimistic schedule we need large capacity additions quickly and huge capacity additions within 10 years. If we fail to provide the capacity, we may miss the market, and lost sales could be an important opportunity cost. On the other hand, if we assume the pessimistic schedule, we need only modest amounts of capacity within five years that might be provided by multiple shifts and overtime. Even the 10-year capacity gaps seem relatively modest. How do we make capacity plans under such uncertain conditions?

The capacity planning problem may be set in either of the two demand prediction situations just discussed, or between them. In either the stable or uncertain demand situations, we need to consider the effect of contingencies. A formal methodology for

TABLE 3-4 **Predicted Requirements, Current Capacities, and Projected Capacity Differences for the Pessimistic Prediction**

	Capacity, Units per Year			
	Current, 1980	1982	1985	1990
Predicted pessimistic capacity requirements	10,000	11,000	12,800	16,000
Machine shop capacity	11,000	—	—	—
Capacity (gap) or slack	1,000	—	(1,800)	(5,000)
Assembly capacity	10,000	—	—	—
Capacity (gap) or slack	—	(1,000)	(2,800)	(6,000)
Receiving, shipping, and factory warehouse capacity	15,000	—	—	—
Capacity (gap) or slack	5,000	4,000	2,200	(1,000)

considering these contingencies is the Delphi method. The dominant informal methodology is executive opinion of what will happen.

GENERATION OF ALTERNATE CAPACITY PLANS

When capacity gaps have been identified, alternate plans can be considered. These alternatives may involve the size and timing of added capacity, the use of overtime and multiple shifts, outside capacity sources, absorption of lost sales, and the location of new capacity.

Large or Small Capacity Increments

When an enterprise enjoys stable demand growth, the issues are centered on how and when to provide the capacity, rather than *if* capacity should be added. Taking the data for expected capacity requirements from Table 3-1, there is a linear growth in capacity requirements of 1000 units per year. One issue is whether capacity should be added more often in smaller increments to keep up with demand (Figure 3-1*a*) or in larger increments by a slower schedule (Figure 3-1*b*).

Both Figures 3-1*a* and 3-1*b* assume that demand will be met through production, so that there will be slack capacity immediately after an addition. The slack capacity declines as requirements increase and falls to zero when the next increment to capacity

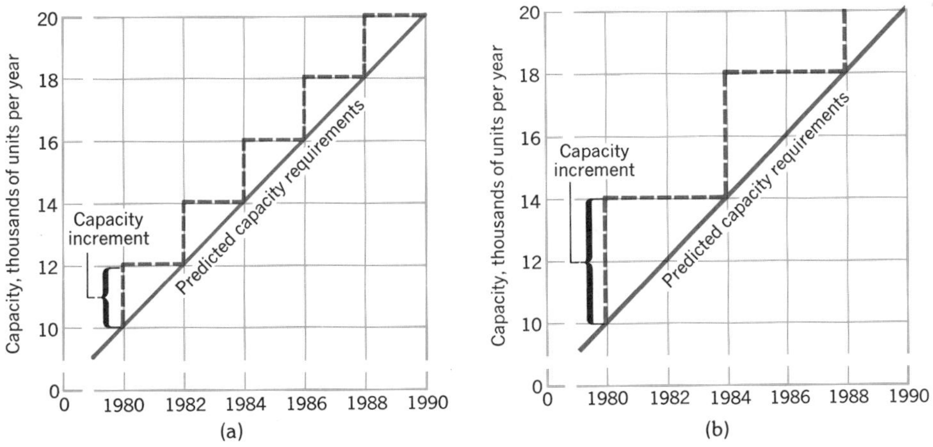

FIGURE 3-1
Capacity increments to meet requirements, *(a)* through 2000-unit increments each two years, and *(b)* through 4000-unit increments each four years.

is installed, if timing is perfect. Whether smaller or larger increments of capacity will be more economical depends on the balance of incremental capital and operating costs for a particular organization and whether or not economies of scale exist. A unit of capacity added now may cost less than a unit added later, and yet the slack capacity must be carried as additional overhead until it is actually productive.

Alternate Sources of Capacity

Another issue in generating capacity plans is whether or not alternate capacity sources can be used near a capacity limit. Figure 3-1 assumes that demand is met through regular productive capacity. Figure 3-2 assumes that the timing of increments to capacity makes it necessary to use overtime, multiple shifts, and subcontracting where feasible. The cost effects of using alternate sources of capacity are to trade off some of the costs of carrying slack capacity against the costs of overtime and multiple shift premium, productivity losses resulting from pushing capacity beyond normal limits, and the extra costs of subcontracting units of output. Again, whether or not the use of alternate sources of capacity will be more economical for a particular organization depends on the balance of incremental capital and operating costs.

Lost Sales

Another alternative to meeting demand through regular productive capacity, or alternate capacity sources, is to absorb some lost sales. This is a risky strategy, since it is possible that market share could be lost. On the other hand, near capacity limits,

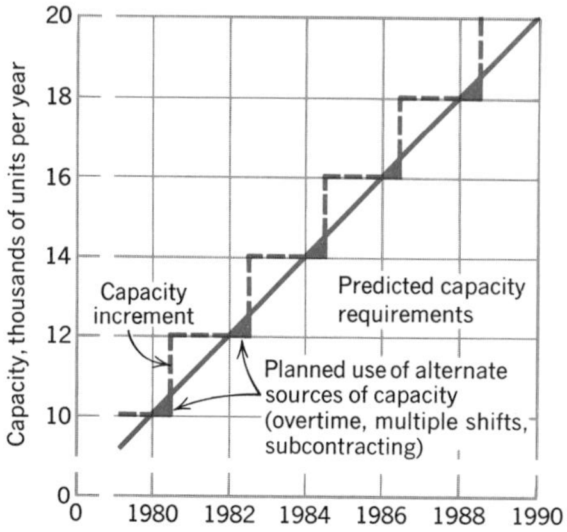

FIGURE 3-2
Capacity increments timed to use alternate sources of capacity to meet requirements.

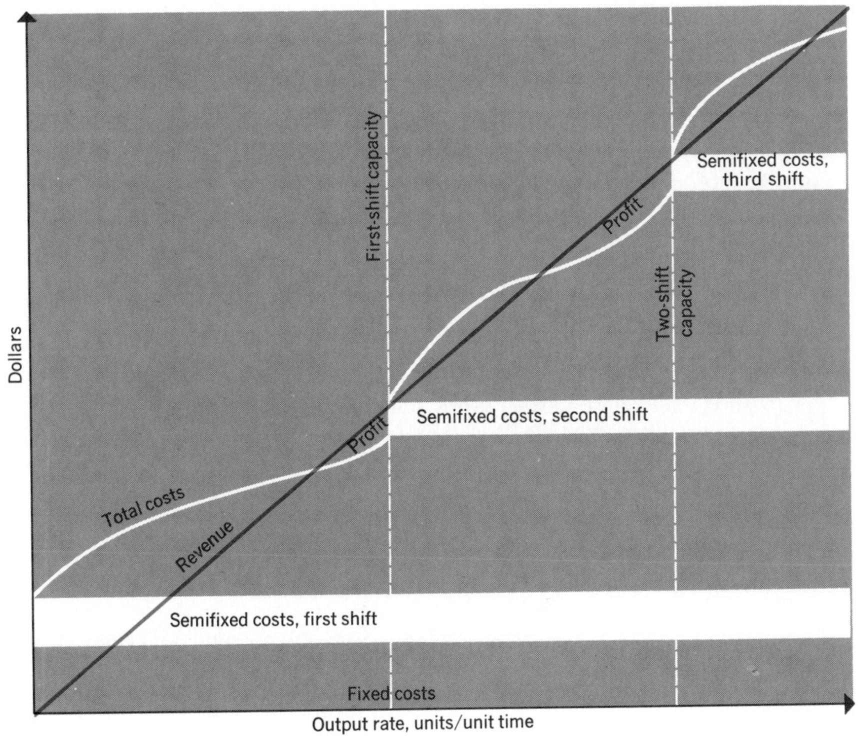

FIGURE 3-3
General structure of costs over a wide range of volume.
SOURCE: *E. S. Buffa and W. J. Taubert, "Evaluation of Direct Computer Search Methods for the Aggregate Planning Problem,"* Industrial Management Review, *Fall 1967.*

contributions decline because of overtime and shift premium and productivity losses. Thus, absorbing lost sales could be more economical in some situations. Yet managers hesitate to take the risk of losing market share. They may be forced into absorbing lost sales at capacity limits but would resist the idea of planning to absorb lost sales as a part of a capacity planning strategy.

The question of the location of new capacity is strategically important, involving an assessment of market location, system distribution costs, and other factors. We shall defer discussion of location until the next chapter.

Cost Behavior in Relation to Volume

Figure 3-3 shows a general picture of what happens to costs as volume increases. We are particularly interested in cost behavior at capacity limits of first and second shifts, since these are the conditions that prevail when capacity is added. Near capacity limits, variable costs increase as a result of increased use of overtime and subcontracting and because of congestion when facilities are maximally utilized.

On the other hand, when new capacity is first installed, it is not fully utilized unless the expansion is long overdue. Therefore, variable costs for the new capacity are likely to be relatively high, reflecting poor utilization of labor and other resources. But the new capacity relieves the stress on existing facilities, making it possible to eliminate overtime and/or multiple shifts.

The combination of new and existing capacity then reflects a variable cost structure that will improve as the new capacity becomes loaded. The fixed costs of existing capacity are spread over a larger and larger number of units as the volume is driven through capacity limits to second and even third shift levels. Thus the fixed cost per unit declines as existing facilities become more fully utilized. New capacity that is relatively poorly utilized will have high fixed costs per unit.

Economies of Scale

The nature of cost structures just discussed suggests that for a given facility there should be an optimum output that minimizes fixed plus variable costs. Figure 3-4 shows unit variable and fixed cost data for a simulated manufacturing enterprise that was driven through an operating range that included the use of a second shift. Variable costs per unit were computed and are plotted as data points. Unit costs vary, depending on the amount of overtime used for the direct and indirect work force and the costs incurred by management decisions to expand or contract the workforce. In general, overtime was used increasingly as production exceeded 450 units per month, and a second shift was required above 550 units per month. The fixed cost per unit curve in Figure 3-4 is simply the $30,000 fixed cost divided by the number of units produced. The total unit cost curve is the sum of the variable plus fixed costs and exhibits an optimum unit cost at about 525 units per month. In this case the second shift may be economical, depending on demand, expansion possibilities, and so on.

We can characterize the plant represented by the costs of Figure 3-4 as a 525 units per month plant; that is its minimum unit cost capacity, given the first and second shift capacities. Another way to characterize that plant is in terms of its normal (no overtime) capacity of 450 units per month. Since optimal plant operating points are usually not known, it is common to state normal capacities. In these terms, the minimum unit cost capacity for the plant of Figure 3-4 is $(525/450) \times 100 = 154$ percent of normal.

Usually there are economies of scale that may come from two basic sources: lower fixed cost per unit and/or lower variable costs per unit. The lower fixed costs accrue because plant and equipment costs of larger plants are less than proportional to capacity. Larger plants are likely to have better balance of subunits with less slack capacity in subunits. Lower variable costs may also accrue to the larger plant because

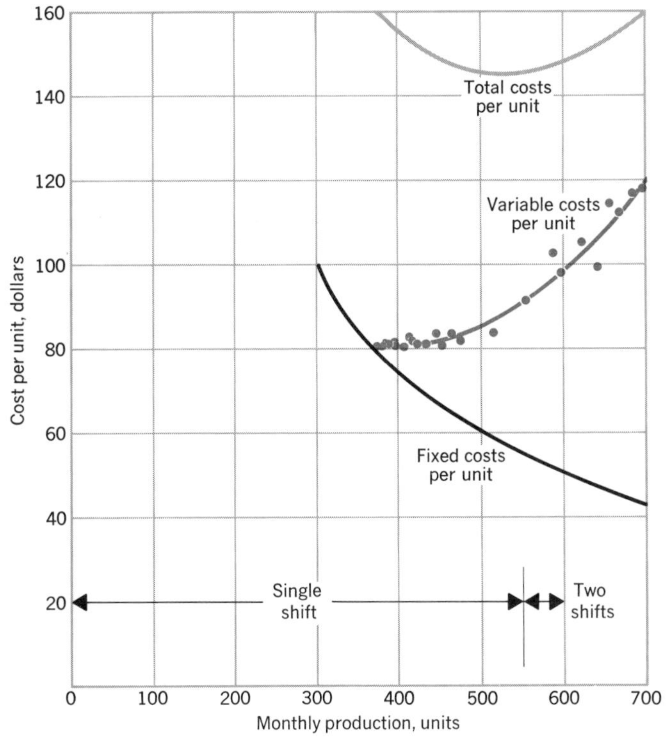

FIGURE 3-4
Unit variable, fixed, and total costs for a plant in relation to volume. Normal capacity without overtime is 450 units per month, and *minimum unit cost* capacity is 525 units per month.

larger volume may justify more mechanization and automation. The result is that minimum unit costs could be substantially less for larger plants, as shown in Figure 3-5.

ECONOMIC EVALUATION OF CAPACITY PLANS

Capacity plan alternatives may involve different size units that may differ in their productivity, reflecting economies of scale. As shown in Figure 3-1 and 3-2, the timing of investments in new capacity can be quite different for alternate plans. Timing of investments depends on the choice of the size of capacity increments and on the use of alternate sources of capacity. If plans involve lost sales, then lost contribution becomes a future opportunity cost. Since all of these costs are future costs, and the

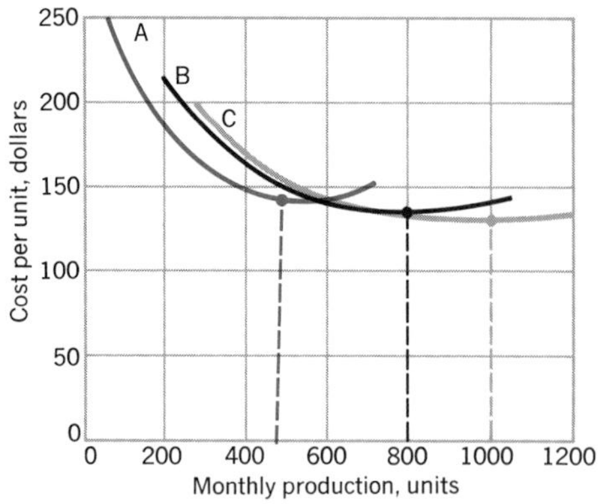

FIGURE 3-5
Economies of scale illustrated by three successively larger plants. *Minimum unit cost* plant sizes are A = 525 units per month, B = 800 units per month, and C = 1000 units per month.

time spans may be long, a discounted cash flow analysis is appropriate for comparing alternatives. In the following two examples we shall compute net present values as a criterion.

Mature Products with Stable Demand Growth

An Example. As an example, assume the situation described by Figures 3-1 and 3-2. In order to simplify the analysis, we shall consider only the first four years. We shall compute for four basic altenatives:

1. Capacity added January 1, 1980 and 1982 in increments of 2000 units.
2. Capacity of 4000 units added January 1, 1980.
3. Capacity added as of July 1, 1980 and 1982 in increments of 2000 units, depending on overtime and multiple shifts to meet requirements during the first six months of 1980 and 1982.
4. Capacity of 4000 units added January 1, 1981, depending on overtime and multiple shifts to meet requirements during 1980.

Costs. There are economies of scale in the larger plant of 4000 units that are reflected both in the original investment and in the operating costs. Also, units produced using overtime and multiple shifts cost an additional $1.00 per unit. The investment and operating costs are summarized in Table 3-5. In the analysis we shall assume that the cost of capital for the enterprise is 15 percent.

TABLE 3-5 **Original Investment Requirements and Operating Costs**

Plant Size, Units per Year	Original Investment	Operating Costs per Unit	Operating Costs per Unit When Using Alternate Sources of Capacity
2000	$1,000,000	$10	$11
4000	1,800,000	9	10

Alternative 1. Figure 3-6 summarizes the structure of cash flows that must be considered in determining the net present value for Alternative 1. The first investment of $1 million is already at present value. The second investment must be discounted to present value by the present value factor of a single payment two years hence at 15 percent, $PV_{sp} = 0.756$. The present value factors are available from Table H-1 in Appendix H or may be computed from $PV_{sp} = 1/(1 + i)^n$, where i = annual interest rate in decimals, and n = number of years. For this example, $1/(1 + 0.15)^2 = 0.756$.

The incremental operating costs are related to the actual product produced using

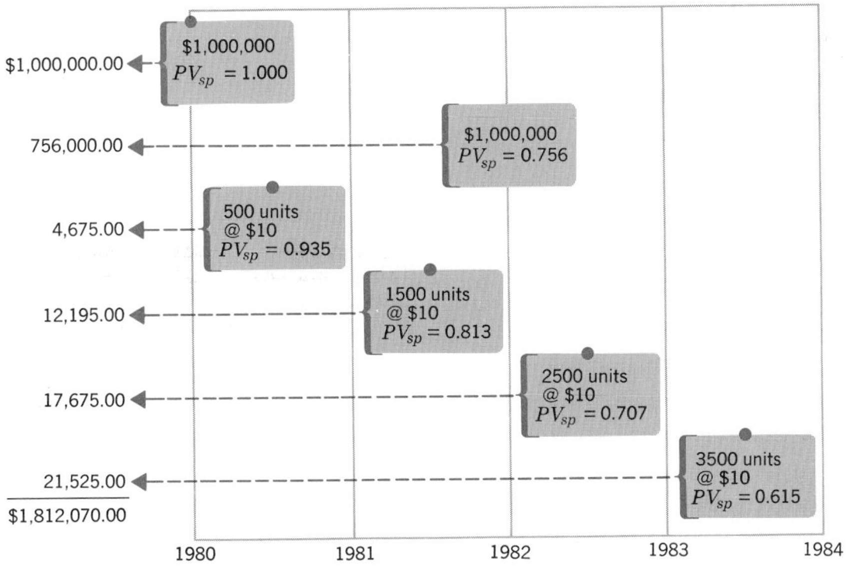

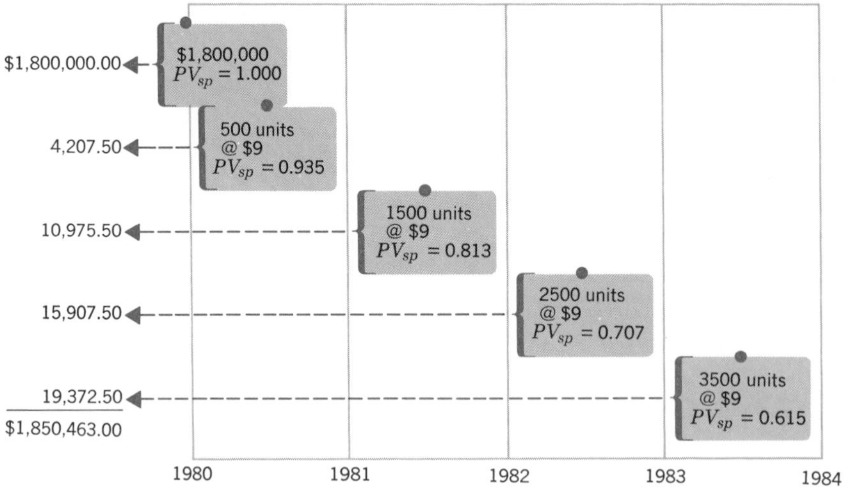

FIGURE 3-7
Present values for Alternative 2.

the new capacity. When the new capacity comes "on-stream" at the beginning of 1980, it will be entirely slack capacity. Requirements increase linearly as in Figure 3-1*a*, and during the first year, 500 units will be produced by the new capacity. During the second year, 1500 units will be produced, and at the end of the second year, the first expansion will be fully utilized.

In 1982, the second 2000 units of capacity will be added to the first expansion, and the pattern repeats. The 500 units produced by the second expansion are added to the 2000 units produced by the first expansion. Then, during 1982 the units produced by new capacity are 2000 on number 1 plus 500 units on number 2. In 1983, 2000 units will be produced by the first expansion plus 1500 units by the second expansion. At the end of 1983 the second expansion will be fully utilized.

In reducing the operating costs to present values we assume that the costs for the 500 units produced by the new plant during 1980 are centered in the middle of the year. The value, $PV_{sp} = 0.935$ was obtained from Table H-1 by linear interpolation. The interpolated values vary slightly from the formula which yields $PV_{sp} = 0.933$. The present value for the production costs in 1980 is, then, $500 \times 10 \times 0.935 =$ \$4675.00. Similarly, the production costs from added capacity of succeeding years is assumed to be centered during the year. The present value factors used are the interpolated values.

Figure 3-6 shows the total present value of the investment and operating costs for two cycles of capacity additions to be \$1,812,070. This value will be compared with the values derived by similar methods for the other three alternatives.

Alternative 2. Figure 3-7 summarizes the cash flows for Alternative 2. The only investment required is in 1980. The production costs from the new capacity are less than those for Alternative 1 because of the economy of scale effect. The economies of scale in capital and operating costs are not sufficient to counterbalance the advantage of Alternative 1 which delays part of the investment in capacity until 1982. Still, Alternative 2 has a strategic advantage of greater slack capacity during the four-year period.

Alternative 3. Alternative 3 is similar to 1 except that each investment in capacity can be delayed six months by depending on overtime and multiple shifts to meet requirements during that time interval. Recall that using alternate capacity sources results in an incremental cost of $1.00 per unit. See Figure 3-8 for the structure of present values.

The production cost must now reflect the timing and extra cost of the output produced by alternate sources. In 1980, 125 units are produced during the first six months at $11 per unit. We have centered these costs at the end of the third month in 1980, and the present value factor of $PV_{sp} = 0.968$ is interpolated from Table H-1. Production costs for the balance of 1980 are centered at the end of the ninth month. Production costs in 1982 are handled in a similar way. The total present value

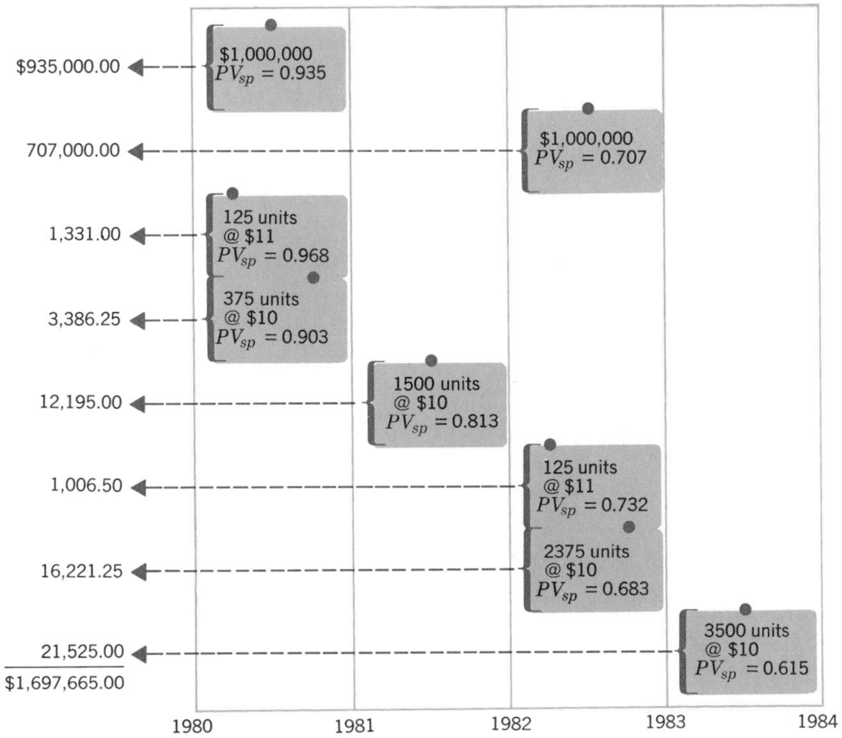

FIGURE 3-8
Present values for Alternative 3.

figure for Alternative 3 shows a net advantage from delaying the investments and obtaining fuller utilization of the new capacity.

Alternative 4. This alternative is similar to Alternative 2 but uses the investment delay concept of Alternative 3. The large plant is installed in 1981, and the 500-unit capacity requirement in 1980 is met through overtime and multiple shifts (see Figure 3-9).

The present values for the four alternatives are:

1. $1,812,070
2. $1,850,463
3. $1,697,665
4. $1,616,931

The cost structures in this example favor the use of alternate capacity sources in order to delay capital investments for either the small or large plants. Delaying the investment has the advantage of obtaining better utilization of the new capacity when it is installed. Alternative 4, involving the larger plant with economies of scale, has the minimum present value cost. Alternative 4 also has the strategic advantage over Alternative 3 of providing greater slack capacity during four years. This slack capacity could be put to use at no incremental investment cost if demand should be greater than expected.

Note that without the use of alternate capacity sources in Alternatives 1 and 2, the

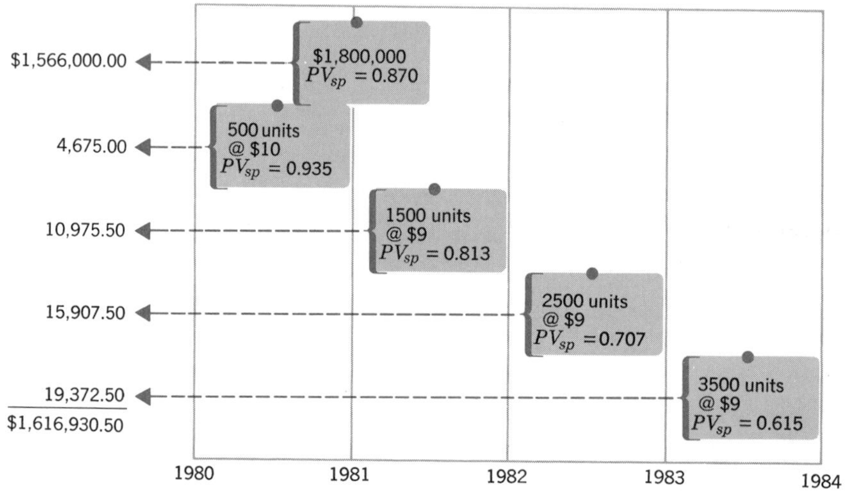

FIGURE 3-9
Present values for Alternative 4.

economy of scale effect in Alternative 2 is not great enough to counterbalance the large initial investment. Another way of looking at the cost difference between Alternatives 1 and 2 is that Alternative 2 results in poor utilization of the larger plant. The poor utilization results in relatively large capital costs per unit which counterbalance the economy of scale advantage.

Variations. The basic methodology illustrated by the example can be applied to a host of variations in capacity expansion strategy by expanding the alternatives list. For example, added alternatives could involve different locations coupled with expansion, answering the question of whether or not relocation is justified at this time. Erlenkotter's [1973] dynamic programming models of expansion consider expansion and location as joint decisions. Vertical integration proposals can be included as alternatives, if equivalent raw materials and other affected costs are accounted for in all alternatives. Other possibilities include the capacity effects of new products or process innovations.

Inflation in both investment requirements and operating costs can be taken into account by the present value methodology without difficulty. Also, assumptions regarding the centering of operating expenses made in the example can be relaxed by constructing a more detailed model.

New Products and Risky Situations

When the color TV market began to develop rapidly, the need for new capacity was apparent, but how fast would the market develop? Could existing facilities for black and white TV be converted? Would market shares remain stable? Would color TV repeat the growth pattern of black and white? Similar uncertainties occurred in other new products, such as pocket calculators and microwave ovens.

Market uncertainties might also occur in mature products that have enjoyed stable growth, because of impending technological innovations or political uncertainties. Imagine the market uncertainties in the mechanical calculator field when solid-state electronics became practical. The design of an electronic desk calculator became an obvious objective. What would an electronic desk calculator cost? Would the mechanical calculator become completely obsolete or would the electronic model be expensive, leaving a market for mechanicals? Suppose a shipping company regularly did a substantial business from the United States West Coast to the East Coast and Europe. What effect would the closing of the Panama Canal have on operations and capacity needs?

If demands are uncertain, lead times can be important. It may take considerable time for planning, for obtaining government permits that currently involve environmental impact studies, and for construction. The length of these lead times becomes of even greater importance when planning for products with uncertain demand. Events can happen within the lead times that change the logical alternatives.

Capacity planning in these situations requires an assessment of the risks. The effect of the probability that risky events will occur needs to be accounted for. If the market is uncertain, a probabilistic prediction of the market provides basic data.

An Example. Suppose that we are planning future capacity for a product that is in the rapid developmental phase. Present annual capacity is 20,000 units. New competition is becoming very aggressive, but the enterprise expects to retain its market share. The sales department even feels that market share could be increased with aggressive promotion. Estimates of the total market vary, some feeling that growth might be explosive in the next four to five years. On the other hand, there is the additional uncertainty concerning continuing technological innovation that could stunt the growth of the current line. Thus, expected, optimistic, and pessimistic market predictions are made and assigned probabilities that each might occur. The predictions are converted to capacity requirements per year as follows:

	1980, Current	1981	1982	1983	1984
Optimistic ($p = 0.25$)	17,000	24,000	34,000	48,000	66,000
Expected ($p = 0.50$)	17,000	20,000	24,000	29,000	35,000
Pessimistic ($p = 0.25$)	17,000	19,000	21,000	23,000	25,000

The optimistic requirements are based on the assumption of a 40 percent annual growth, the expected, a 20 percent annual growth; and the pessimistic, only a 10 percent annual growth (capacities rounded).

Strategies. Three alternate strategies are developed, each designed with the three market assumptions in mind:

1. Install new capacity in 1982, 1983, and 1984 in increments of 15,000 units.

2. Install new capacity in 1982, 1983, and 1984 in increments of 5000 units.

3. No capacity additions.

The 15,000-unit capacity additions require an investment of $800,000 each, and the 5000-unit additions require an investment of $300,000 each, reflecting an investment economy for the larger units. The operating costs per unit are the same for both sizes of capacity additions. Given each of the three strategies, the outcomes will depend on which requirements schedules actually occur. When requirements exceed capacity, sales are lost, so the cost of lost contribution of $50 per unit must be taken into account in evaluating the alternatives.

For each of the three strategies, any of the three market assumptions could occur with the stated probabilities. Figure 3-10 is a decision tree representing the strategies and events. In order to evaluate the three alternate strategies, we must compute the present value of each of the nine possible outcomes.

Present Values of Outcomes. The present values for each of the three outcomes given each of the three basic strategies are shown in Tables 3-6, 3-7, and 3-8. The calculations for Alternative 1 in Table 3-6 are typical. In Alternative 1a of Table 3-6, the optimistic requirements in relation to the proposed capacity with additions provides the basis for computing lost sales. Since there are no operating cost economies

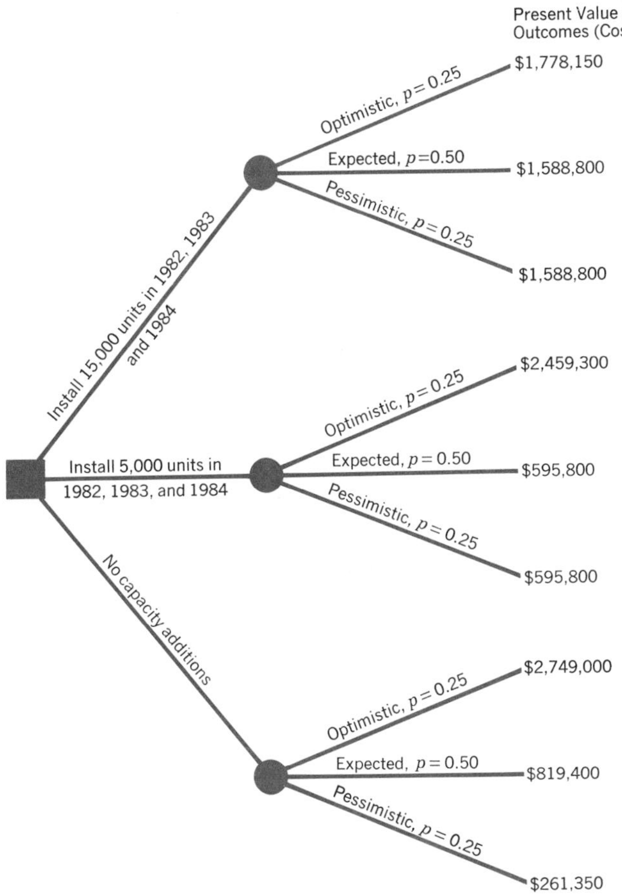

FIGURE 3-10
Decision tree for the risky market example.

of scale between the large- and small-capacity additions, the operating cost differences between alternatives are measured by the lost sales.

The present value calculations reduce all future net cash flow in the four-year planning horizon to the planning base period of 1980, at the interest rate of 15 percent. The capacity additions occur two, three, and four years hence, and each are related to present value. The lost sales are assumed to be centered within the year that they occur, and the present value factors are interpolated from Table H-1. The present value of the investments and lost sales costs of $1,778,150 for Alternative 1 is the outcome for that alternative shown in the decision tree of Figure 3-10. All the cost outcomes shown in Figure 3-10 are computed similarly in Tables 3-6, 3-7, and 3-8.

TABLE 3-6 **Alternative 1: New Capacity in 1982, 1983, and 1984 in Increments of 15,000 Units at a Cost of $800,000 Each (interest rate = 15 %)**

		1980, Current	1981	1982	1983	1984
1a.	Optimistic Requirements	—	24,000	34,000	48,000	66,000
	Proposed additions to capacity	—	—	15,000	15,000	15,000
	Proposed capacity	20,000	20,000	35,000	50,000	65,000
	Lost sales	—	4,000	0	0	1,000
	Present values:					
	New capacity, 1982, 800,000 × 0.756 =				$604,800	
	New capacity, 1983, 800,000 × 0.658 =				526,400	
	New capacity, 1984, 800,000 × 0.572 =				457,600	
	Lost sales, 1981, 4,000 × 50 × 0.813 =				162,600	
	Lost sales, 1984, 1,000 × 50 × 0.535 =				26,750	
	Total				$1,778,150	
1b.	Expected requirements	—	20,000	24,000	29,000	35,000
	Proposed additions to capacity	—	—	15,000	15,000	15,000
	Proposed capacity	20,000	20,000	35,000	50,000	65,000
	Lost sales	—	0	0	0	0
	Present values:					
	New capacity, 1982, 800,000 × 0.756 =				$604,800	
	New capacity, 1983, 800,000 × 0.658 =				526,400	
	New capacity, 1984, 800,000 × 0.572 =				457,600	
	Total				$1,588,800	
1c.	Pessimistic requirements	—	19,000	21,000	23,000	25,000
	Proposed additions to capacity	—	—	15,000	15,000	15,000
	Proposed capacity	20,000	20,000	35,000	50,000	65,000
	Lost sales	—	0	0	0	0
	Present values:					
	Same as for 1b.				$1,588,800	

TABLE 3-7 **Alternative 2: New Capacity in 1982, 1983, and 1984 in Increments of 5000 Units at a Cost of $300,000 Each (interest rate = 15 %)**

		1980, Current	1981	1982	1983	1984
2a.	Optimistic requirements	—	24,000	34,000	48,000	66,000
	Proposed additions to capacity	—	—	5,000	5,000	5,000
	Proposed capacity	20,000	20,000	25,000	30,000	35,000
	Lost sales	—	4,000	9,000	18,000	31,000
	Present values:					
	New capacity, 1982, 300,000 × 0.756 =				$226,800	
	New capacity, 1983, 300,000 × 0.658 =				197,400	
	New capacity, 1984 300,000 × 0.572 =				171,600	
	Lost sales, 1981, 4,000 × 50 × 0.813 =				162,600	
	Lost sales, 1982, 9,000 × 50 × 0.707 =				318,150	
	Lost sales, 1983, 18,000 × 50 × 0.615 =				553,500	
	Lost sales, 1984, 31,000 × 50 × 0.535 =				829,250	
	Total				$2,459,300	

TABLE 3-7 (continued) **Alternative 2: New Capacity in 1982, 1983, and 1984 in Increments of 5000 Units at a Cost of $300,000 Each (interest rate = 15 %)**

	1980, Current	1981	1982	1983	1984
2b. Expected requirements	—	20,000	24,000	29,000	35,000
Proposed additions to capacity	—	—	5,000	5,000	5,000
Proposed capacity	20,000	20,000	25,000	30,000	35,000
Lost sales	—	0	0	0	0
Present values:					
New capacity, 1982, 300,000 × 0.756 =				$226,800	
New capacity, 1983, 300,000 × 0.658 =				197,400	
New capacity, 1984 300,000 × 0.572 =				171,600	
Total				$595,800	
3b. Pessimistic requirements	—	19,000	21,000	23,000	25,000
Proposed additions to capacity	—	—	5,000	5,000	5,000
Proposed capacity	20,000	20,000	25,000	30,000	35,000
Lost sales	—	0	0	0	0
Present values:					
Same as for 2b.				$595,800	

TABLE 3-8 **Alternative 3: No Capacity Additions (interest rate = 15%)**

	1980, Current	1981	1982	1983	1984
3a. Optimistic requirements	—	24,000	34,000	48,000	66,000
Proposed capacity	20,000	20,000	20,000	20,000	20,000
Lost sales	—	4,000	14,000	28,000	46,000
Present values:					
Lost sales, 1981, 4,000 × 50 × 0.813 =				$162,600	
Lost sales, 1982, 14,000 × 50 × 0.707 =				494,900	
Lost sales, 1983, 28,000 × 50 × 0.615 =				861,000	
Lost sales, 1984, 46,000 × 50 × 0.535 =				1,230,500	
Total				$2,749,000	
3b. Expected requirements	—	20,000	24,000	29,000	35,000
Proposed capacity	20,000	20,000	20,000	20,000	20,000
Lost sales	—	0	4,000	9,000	15,000
Present values:					
Lost sales, 1982, 4,000 × 50 × 0.707 =				$141,400	
Lost sales, 1983, 9,000 × 50 × 0.615 =				276,750	
Lost sales, 1984, 15,000 × 50 × 0.535 =				401,250	
Total				$819,400	
3c. Pessimistic requirements	—	19,000	21,000	23,000	25,000
Proposed capacity	20,000	20,000	20,000	20,000	20,000
Lost sales	—	0	1,000	3,000	5,000
Present values:					
Lost sales, 1982, 1,000 × 50 × 0.707 =				$35,350	
Lost sales, 1983, 3,000 × 50 × 0.615 =				92,250	
Lost sales, 1984, 5,000 × 50 × 0.535 =				133,750	
Total				$261,350	

Expected Values of Alternate Strategies. In order to evaluate the three strategies, we compute their expected values by "rolling back" the decision tree. We multiply the present value of the cost outcome by the probability of its occurrence and add the three probability-weighted values to obtain the expected value for the strategy. This is done for all three strategies in Table 3-9.

Alternative 2, involving the installation of the three 5000-unit-capacity additions, produces the lowest expected present value costs. Note that Alternative 2 incurs lost sales only if the optimistic prediction actually materializes, but it requires much smaller capital investment than Alternative 1. Alternative 3 requires no capital investments but incurs large costs of lost sales even for the pessimistic requirements prediction. On balance, Alternative 2 is the most economical capacity plan, given the probabilities of obtaining each of the requirement's predictions.

Recall, however, that it is not the expected value of costs that will actually materialize. The actual costs will depend on the actual capacity requirements. If, in fact, the optimistic prediction materializes, the present value of costs would be $2,459,300, which is $681,150 greater than if the 15,000-unit additions had been installed. How would managers respond to the situation if by mid-1981 (with the first 5000-unit-capacity addition under construction) it appeared that the optimistic prediction was valid? Of course, they would adjust plans, presumably through a new set of predictions and alternatives. There would be no logic in following through with the original decision by installing the two remaining small-capacity additions.

The example is relatively simple in the alternatives selected and in the cost structure represented. With large future costs being involved, managers would probably consider more alternatives and perhaps more than three probabilistic predictions. Also, a more complex cost structure might represent the situation more accurately. But these are straightforward extensions of the same methodology.

RISKS AND STRATEGIC EFFECTS

Many important effects of alternate capacity plans can be represented in an expanded decision tree. For example, if a technological breakthrough can be assigned a probability of perhaps $p = 0.30$, then a branch in the tree can be structured for "no breakthrough" ($p = 0.70$) and "breakthrough" ($p = 0.30$). Other risks can be handled similarly in the formal decision tree analysis.

TABLE 3-9 **Expected Values of Three Alternate Strategies**

Alternative 1:	0.25 × 1,778,150	+ 0.50 × 1,558,800	+ 0.25 × 1,588,800	= $1,636,137
Alternative 2:	0.25 × 2,459,300	+ 0.50 × 595,800	+ 0.25 × 595,800	= $1,061,675
Alternative 3:	0.25 × 2,749,000	+ 0.50 × 819,400	+ 0.25 × 261,350	= $1,162,288

Many strategic effects, however, must be evaluated through managerial trade-off. If the most economical plan involves lost sales and possibly a decline in market share, the manager must weigh the lower expected cost against the loss of market share. Specific capacity plans might have an impact on competition, flexibility of operations, market locations, labor policies, market share, and so on.

IMPLICATIONS FOR THE MANAGER

Capacity planning involves top managers in decisions that have important strategic implications. These decisions often have an impact on and are affected by other enterprise functions in addition to operations. They interface with the marketing function in terms of product strategies and market predictions. They interface with engineering and technology in terms of both product and process innovations. Capacity decisions may commit resources that commonly represent the major assets of a firm, requiring financing by debt or equity instruments. Finally, capacity decisions may also set directions for the philosophy governing the productive system design. Will the design emphasize cost and availability, suggesting a continuous system, or flexibility and quality, suggesting an intermittent system?

Predicting long-range market trends is difficult because events that may cause major demand shifts are unpredictable. Nevertheless, there are many commodities and mature products whose demand seems caused by basic trends in population, industrialization, urbanization, and so forth. Causal forecasting models have been shown to be useful for these types of products. Even in these stable product situations, however, economic and political events can have an impact on capacity needs to an extent not foreseen by historical data-based models. Executive opinion, possibly formalized through the Delphi method, may be the basis for these long-range predictions. New product predictions are most often based on these latter methods, often in combination with market surveys and reference to product life cycle analyses.

Given market predictions, the capacity planning options available to managers involve frequent smaller capacity additions versus less frequent larger units and the use of other sources of capacity, either marginally or as a part of an ongoing capacity philosophy. There is no single right answer to these options. Rather, the answer is in the appropriate analysis that provides logical alternatives and evaluation methods. Each situation has its unique growth and cost structure characteristics. In general, deterministic methods are adequate in mature product situations, but probabilistic methods are really needed when market predictions are risky.

While formal economic evaluation is extremely important because of the capital commitments, the results must be regarded as one important input to the decision process. Strategic factors must be weighed and may be the basis for a final choice. With the economic analysis of alternatives available, and an identification of the strategic effects of each alternative, managers can make trade-offs. In effect, the trade-off process allows a pricing of strategic effects.

REVIEW QUESTIONS AND PROBLEMS

1. Review the six-step process for capacity planning. Now think in terms of proposals that you know about or have heard about for expanding electric utility capacity through nuclear power plants.
   Do you think all the steps were followed? Which steps involved the most time? Identify the risks and strategic effects to the company.

2. What are the appropriate methods for predicting future requirements for:
   a. Electric power
   b. Oil tankers
   c. Hula-hoops

3. Why might there be imbalances in the existing capacities of departments in a hospital such as x-ray, laboratory, intensive care, and the like?

4. A fertilizer company is reviewing its future capacity needs. The company has concluded that its share of the market demand will expand at the rate of 20 percent per year for the next five years. Production is highly capital intensive, and transportation costs for distribution are also important.
   What kinds of alternatives do you think should be considered?

5. Using the cost-volume relationship shown in Figure 3-3 as a background:
   a. Under what conditions would you expect minimum unit cost plant size (output) to include a second shift?
   b. What are the factors that might result in a larger plant being more cost effective than a smaller one?

6. Define the following terms: regular capacity, optimal capacity, and maximum capacity.

7. Home computers are now coming on the market. They are quite powerful, involving keyboard input, video output, and very substantial memory capacities, and are designed to use rather capable languages such as BASIC. Prices range from $600 to $1500 and are declining.
   If you were tooling up to produce such a product, how would you go about assessing the market for the next five years? What kinds of capacity strategies would you generate? What decision methodology would you use?

8. Referring to question 7, what strategic factors should be weighed and traded off against objective cost-profit results in your strategies?

9. New alternatives are made available for the example of economic analysis of a mature product given in the text. The present value calculations for the text example are given in Figures 3-6, 3-7, 3-8, and 3-9.

A radically new process technology has been developed by the company engineers that has generated tremendous enthusiasm. The process has been automated so that capital and material costs dominate; labor costs have been almost eliminated. Therefore, operating costs have been drastically reduced. The result is that two new alternatives have been added to the four given in the text. The two new alternatives involve the possibility of expansion, including the automated processes and costs as follows:

a. Capacity added January 1, 1980 and 1982 in increments of 2000 units.
b. Capacity of 4000 units added January 1, 1980.

These plants would be designed for 24-hour operation because of their capital intensive nature. Therefore, other plans involving alternate sources of capacity are not feasible.

There is an economy of scale in the investment cost of the larger plant. The operating costs of the two plant sizes are the same, however, by the nature of the new process. The investment and operating costs are shown in Table 3-10.

Compute present values for the two alternatives and compare the results with the four alternatives computed in the text. What decisions should be made? Why?

10. This problem is an extension of the text example for the economic evaluation of a product in its rapid development phase. The decision tree structure is given in Figure 3-10.

A technological breakthrough in an automated production process is now available that can be incorporated with new capital additions. The result would be a dramatic cost decrease of $30 per unit. On the other hand, it is estimated that plant investment costs would increase to $900,000 each for the 15,000-unit plant, and $375,000 each for the 5000-unit plant.

The optimistic, expected, and pessimistic requirements schedules, as well as the probability estimates, remain the same as in the text example. The internal rate of return is 15 percent.

a. What are the present values of each of the nine possible outcomes?
b. What are the expected values of the three alternatives?
c. Which alternative would you choose? Why?

TABLE 3-10

Investment and Operating Costs for Problem 9

Plant Size, Units per Year	Original Investment	Operating Costs per Unit
2000	$1,300,000	$1
4000	2,000,000	1

SITUATIONS

11. CHEMCO is a chemical manufacturing company that has been successful in research and development. It has built its reputation by exploiting its excellent research staff's ability to develop new and useful products. The firm has been able to capitalize on being an innovator, reaping the high profits that result from being first with a product and facing little initial competition. The company has promoted new products strongly, obtaining an indentification with them that has carried over into longer-term market dominance in many cases. A recent new product PRIMEBEEF seemed to have remarkable effects as a cattle feed additive. It resulted in a higher proportion of high-grade beef.

Having been involved in many new product introductions, CHEMCO has learned to deal with the market uncertainties of new products. Therefore, when initial market tests for PRIMEBEEF were successful, capacity planning became an issue. CHEMCO developed flexible capacity plans that took account of contingencies. The potential market was large but not certain. It was known that competitors were already attempting imitations. Therefore, part of the strategy was to expand output as soon as possible to establish their market position. The capacity planning issues were centered in plant size and expansibility of a small plant, should that be the decision. After making market estimates, CHEMCO decided on a 10-year planning horizon.

Market Scenarios

Marketing predictions could be framed in several scenarios that were structured as follows with probability estimates:

a. Demand would be high initially, product identification successful, and demand would remain high. Probability = 0.60.
b. Demand would be high in the initial two years, but competition would be so keen that demand would be low thereafter (third through tenth years). Probability = 0.10.
c. Demand would be initially low and remain low. The product would never really be successful. Probability = 0.30.

From the above three scenarios, they noted that the probability of an initial (first two years) high demand was $p = 0.70$, which formed the basis for a capacity strategy that would start with a small plant that could be expanded after two years if, in fact, demand was high during the initial period.

Alternatives

Two basic capacity strategies were based on the market predictions: Alternative 1—build a large plant, cost, \$3,000,000; Alternative 2—build a small but expansible plant initially, cost \$1,300,000. If initial demand is high, decide within two years whether or not to expand it at a cost of \$2,200,000. This decision involved risks also, since even if initial demand were high ($p = 0.70$), the probability was only $p = 0.60$ that it would be high thereafter. Therefore, the conditional probability that demand would be high following a decision to expand was $p = 0.60/0.70 = 0.86$.

Revenue Patterns

Estimates of annual income were made under the assumptions of each alternate demand pattern as follows:

1. A large plant with high volume would yield \$1,000,000 annually in cash flow.

2. A large plant with low volume would yield only \$100,000 annually because of high fixed costs and inefficiencies.

3. A small plant with low demand would be economical and would yield annual cash income of \$400,000.

4. A small plant, during an initial period of high demand, would yield \$450,000 per year, but this yield would drop to \$300,000 yearly in the long run because of competition. (The market would be larger than under number 3 but would be divided up among competitors.)

5. If the small plant were expanded to meet sustained high demand, it would yield \$700,000 cash flow annually and would be less efficient than a large plant built initially.

6. If the small plant were expanded but high demand were not sustained, estimated annual cash flow would be \$50,000.

Analysis and Decision

CHEMCO decided that the capacity planning program was definitely risky and that a careful analysis should be made before attempting a decision. The president of the company stated that he wished to preserve CHEMCO's historical record of maintaining an 18 percent before tax return on investments. He also raised the questions of whether or not they had all the information they needed and whether there were other strategies that should be considered.

What analysis should be made? What should CHEMCO do?

REFERENCES

Armstrong, J. S., and M. C. Grohman. "A Comparative Study of Methods for Long-range Market Forecasting," *Management Science, 19* (2), October 1972, pp. 211–221.

Basu, S., and R. G. Schroeder. "Incorporating Judgments in Sales Forecasts: Application of the Delphi Method at American Hoist & Derrick," *Interfaces, 7* (3), May 1977, pp. 18–27.

Buffa, E. S., and J. S. Dyer. *Management Science/Operations Research: Model Formulation and Solution Methods.* John Wiley & Sons, New York, 1977, chapter 2.

Buffa, E. S., and W. H. Taubert. "Evaluation of Direct Computer Search Methods for the Aggregate Planning Problem," *Industrial Management Review,* Fall 1967, pp. 19–36.

Culliton, J. W. *Make or Buy.* Division of Research, Harvard Business School, Boston, 1942.

Erlenkotter, D. "Sequencing Expansion Projects," *Operations Research, 21,* 1973, pp. 542–553.

Huettner, D. *Plant Size, Technological Change, and Investment Requirements,* Praeger Publishers, New York, 1974.

Magee, J. F. "Decision Trees for Decision Making," *Harvard Business Review,* July-August, 1964.

Manne, A. S. (editor). *Investments for Capacity Expansion: Size, Location, and Time Phasing.* MIT Press, Cambridge, Mass., 1967.

Manne, A. S. "Waiting for the Breeder," *Review of Economic Studies, 41* (Supplement), 1974, pp. 47–65.

Marshall, P. W. et al. *Operations Management: Text and Cases.* Richard D. Irwin, Homewood, Ill., 1975, pp. 312–322.

Morris, W. T. *The Capacity Decision System.* Richard D. Irwin, Homewood, Ill., 1967.

Nord, O. C. *Growth of a New Product—Effects of Capacity Expansion Policies.* MIT Press, Cambridge, Mass., 1963.

Petersen, E. R. "A Dynamic Programming Model for the Expansion of Electric Power Systems," *Management Science, 20* (4) Part II, December 1973, pp. 656–664. pp. 656–664.

Rose, L. M. *Engineering Investment Decisions: Planning Under Uncertainty.* Elsevier Scientific Publishing Company, Amsterdam, 1976.

Skinner, W. *Manufacturing in the Corporate Strategy.* John Wiley & Sons, New York, 1978.

Spetzler, C. S., and Zamora, R. M. "Decision Analysis of a Facilities Investment and Expansion Problem," in *Decision and Risk Analysis: Powerful New Tools for Management,* edited by A. Lesser, Jr. *Proceedings of the Sixth Triennial Symposium, The Engineering Economist,* Hoboken, New Jersey, 1972, pp. 27–51.

CHAPTER 4 Location and Distribution

THE LOCATION OF FACILITIES INVOLVES A COMMITMENT OF RE-sources to a long-range plan. Thus the predictions of the size and location of markets are of great significance. Given these predictions we establish facilities for production and distribution that require large financial outlays. In manufacturing organizations, these capital assets have enormous value, and even in service organizations, the commitment of resources may be very large. Location and distribution is perhaps even more important, because these plans represent the basic strategy for accessing markets and may have significant impact on revenue, costs, and service levels to customers and clients.

It is not immediately obvious that location is a dominant factor in the success or failure of an enterprise. Indeed, it is not uniformly important for all kinds of enterprises. Decentralization within industries must mean that many good locations exist or that the location methods used could not discriminate between alternate locations.

General technological constraints will commonly eliminate most of the possible locations. Or, to take the opposite point of view, a technological requirement may dominate and that activity is then oriented toward the technical requirement. For example, mining is raw material oriented, beer is water oriented, aluminum reduction is energy oriented, and service activities including sales are consumer or client oriented. If some technological requirement, such as raw material location, water, or energy, does not dominate, then manufacturing industries are often transportation oriented.

The criterion for choice of location is intended to be profit maximization for economic activities. If the prices of products are uniform in all locations, then the criterion becomes one of minimizing relevant costs.

If the costs of all inputs are independent of location but product prices vary, then the criterion for locational choice becomes maximum revenue. In such instances, locations will gravitate to the location of consumers, and the general effect will be to disperse or decentralize facilities.

If all prices and costs are independent of location, then choice will be guided by proximity to potential customers or clients, to similar and competing organizations, and to centers of economic activity in general.

INDUSTRIAL PLANT LOCATIONS

In most plant location models, the objective is to minimize the sum of all costs affected by location. Some items of cost, such as freight, may be higher for city A and lower for city B, but power costs, for example, may have the reverse pattern. We are seeking the location that minimizes costs on balance.

In attempting to minimize costs, however, we are thinking not only of today's costs but of long-run costs as well. Therefore, we must be interested in predicting the influence of some of the intangible factors that may affect future costs. Thus, factors such as the attitude of city officials and townspeople toward a new factory site in their city may be an indication of future tax assessments. Poor local transportation facilities may mean future company expenditures to counterbalance this disadvantage. A short labor supply may cause labor rates to be bid up beyond rates measured during a location survey. The type of labor available may indicate future training expenditures. Thus, while a comparative cost analysis of various locations may point toward one community, an appraisal of intangible factors may be the basis of a decision to select another. The result is an excellent example of a managerial decision with multiple criteria, where trade-offs must be made between the various values and criteria.

The general problem indicating the nautre of trade-offs required is illustrated by Table 4-1. In Table 4-1a we see the results of comparative cost analyses for six alternate sites, and site 1 has the lowest projected monthly cost of $28,237. In Table 4-1b, however, there are listed 14 subjective factors that management felt were important in this particular location study. The relative importance of these factors is not immediately obvious, nor is it obvious how they should be related to the objective costs. A formal methodology is needed to provide the basis for managerial decisions.

TABLE 4-1 **(a) Objective Factor Costs and (b) Subjective Factors for Six Sites**

(a)

Site	Material	Marketing	Utilities	Labor	Building	Taxes	Total Objective Factor Cost (OFC)
1	$1079	$1316	$ 9,460	$12,773	$514	$3095	$28,237
2	945	1485	11,563	11,249	563	3470	29,275
3	490	1467	12,768	10,422	539	3580	29,266
4	979	1600	10,548	12,159	490	3755	29,531
5	925	1263	10,898	12,333	612	3701	29,732
6	1507	1950	11,628	12,244	612	3393	31,334

(b)

Availability of transportation	Union activities	Community services	Competition
Industrial sites	Recreation facilities	Employee transportation facilities	Complementary industries
Climate	Housing	Cost of living	Availability of labor
Educational facilities	Future growth		

SOURCE. Adapted from P. A. Brown, and D. F. Gibson, "A Quantified Model for Facility Site Selection–Application to a Multiplant Location Problem," *AIIE Transactions*, 4(1), March 1972, pp. 1–10. Copyright American Institute of Industrial Engineers; reprinted with permission.

A PLANT LOCATION MODEL

A model that attempts to deal with the multidimensional location problem was developed by Brown and Gibson [1972]. This model classifies criteria affecting location according to the model structure, quantifies the criteria, and achieves the balancing or trade-off among criteria.

Classification of Criteria

The model deals with any list of criteria set by management but classifies them as follows:

1. *Critical*—criteria are critical if their nature may preclude the location of a plant at a particular site, regardless of other conditions that might exist. For example, a water-oriented enterprise, such as a brewery, would not consider a site where a water shortage was a possibility. An energy-oriented enterprise, such as an aluminum smelting plant, would not consider sites where low-cost and plentiful electrical energy was not available. Critical factors have the effect of eliminating sites from consideration.

2. *Objective*—criteria that can be evaluated in monetary terms, such as labor, raw material, utilities, and taxes, are considered objective. A factor can be both objective and critical; for example, the adequacy of labor would be a critical factor, while labor cost would be an objective factor.

3. *Subjective*—criteria characterized by a qualitative type of measurement. For example, the nature of union relationships and activity may be evaluated, but its monetary equivalent cannot be established. Again criteria can be classified as both critical and subjective.

Model Structure

For each site i, a location measure LM_i is defined that reflects the relative values for each criterion.

$$LM_i = CFM_i \times [X \times OFM_i + (1 - X) \times SFM_i] \tag{1}$$

where CFM_i = the critical factor measure for site i ($CFM_i = 0$ or 1).

OFM_i = the objective factor measure for site i ($0 \le OFM_i \le 1$, and $\sum_i OFM_i = 1$)

SFM_i = the subjective factor measure for site i ($0 \le SFM_i \le 1$, and $\sum_i SFM_i = 1$)

X = the objective factor decision weight ($0 \le X \le 1$)

The critical factor measure CFM_i is the sum of the products of the individual critical factor indexes for site i with respect to critical factor j. The critical factor index for each site is either 0 or 1, depending on whether the site has an adequacy of the factor or not. If any critical factor index is 0, then CFM_i and the overall location measure LM_i are also 0. Site i would therefore be eliminated from consideration.

The objective criteria are converted to dimensionless indices in order to establish comparability between objective and subjective criteria. The objective factor measure for site i, OFM_i, in terms of the objective factor costs, OFC_i, is defined as follows:

$$OFM_i = \left[OFC_i \times \sum_i (1/OFC_i)\right]^{-1} \tag{2}$$

The effect of equation 2 is that the site with the minimum cost will have the largest OFM_i, the relationships of total costs between sites are retained, and the sum of the objective factor measures is 1.

The subjective factor measure for each site is influenced by the relative weight of each subjective factor and the weight of site i relative to all other sites for each of the subjective factors. This results in the following statement:

$$SFM_i = \sum_k (SFW_k \times SW_{ik}) \tag{3}$$

where SFW_k = the weight of subjective factor k relative to all subjective factors, and
SW_{ik} = the weight of site i relative to all potential sites for subjective factor k

Preference theory is used to assign weights to subjective factors in a consistent and systematic manner. The procedure involves comparing subjective factors two at a time. If the first factor is preferred over the second, then the numerical value of 1 is assigned to the first factor and 0 to the second, and vice versa for the opposite result. If one is indifferent regarding the two factors, a rating of 1 is given both factors. Procedures are also included for higher-order rankings. As with objective factors, the ratings are normalized, so that the sum of subjective weightings for a given site adds to 1.

Finally, the objective factor decision weight, X, must be determined. This factor establishes the relative importance of the objective and subjective factors in the overall location problem. The decision is commonly based on action by a mangement committee, reflecting policies, past data, and an integration of a wide variety of subjective factors. The determination of X could logically be subjected to a Delphi process (see Chapter 2).

With all the data inputs, equation 1 can be used to compute the location measure LM_i, for each site, and the site that receives the largest LM_i is selected. Brown and Gibson extend the model to multiplant location and present a computed example of the evaluation of six sites, involving capacity constraints. Sensitivity analyses are shown to indicate how decisions would change when the objective factor decision weight, X, is varied from 0 to 1.0. The entire procedure has been programmed for electronic computing using a 0-1 programming algorithm capable of treating problems as large as 150 variables and 50 constraints.

TABLE 4-2

Objective and Subjective Factor Measures and Location Measures for Six Sites ($X = 0.8$)

Site	Objective Factor Measure	Subjective Factor Measure	Location Measure
1	0.17433	0.22839	0.18514
2	0.16814	0.13638	0.16179
3	0.16819	0.15257	0.16507
4	0.16669	0.24880	0.18311
5	0.16556	0.23386	0.17922
6	0.15709	0.00000	0.12568

SOURCE. P. A. Brown, and D. F. Gibson, "A Quantified Model for Facility Site Selection–Application to a Multiplant Location Problem," *AIIE Transactions*, 4(1), March 1972, pp. 1–10. Copyright American Institute of Industrial Engineers; reprinted with permission.

Table 4-1 supplies the general data on objective costs and subjective factors for the Brown and Gibson example. Table 4-2 summarizes the objective, subjective, and overall location measures for the six sites for $X = 0.8$. Site 1 produces the largest overall location measure. Note, however, the sensitivity analysis shown in Figure 4-1. Site 1 is indicated for values of X of 0.73 or more, while site 4 would be the choice for values of X less than 0.73. Management is therefore in a position to evaluate the limited range of choices that have finally resulted from the study.

CAPITAL EXPENDITURE—VOLUME EFFECTS

Another important variable between alternate locations is the relationship between fixed and variable expenses. The fixed investment costs can differ considerably, depending on local construction and land costs and variations in the particular site selected. Therefore, the concept of break-even analysis could be used to contrast the objective cost factors of individual locations. Such a break-even analysis is valid only for volumes near the design or break-even capacity, since large changes in volume would entail differences in capital investment as well as possible differences in variable costs.

MULTIPLANT LOCATION

Multiplant location is influenced by existing locations as well as the kinds of economic factors that we have already discussed. Each location considered must be placed in

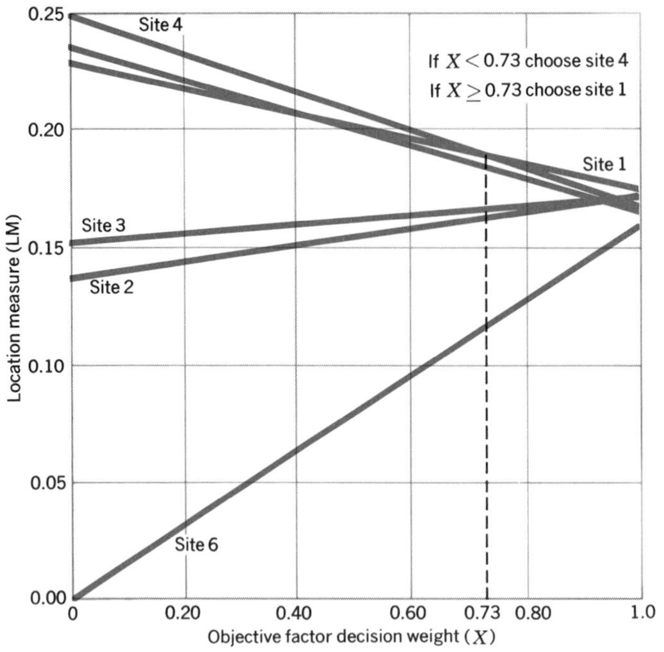

FIGURE 4-1
Sensitivity of location measures to changes in objective factor decision weight.
SOURCE: *P. A. Brown, and D. F. Gibson, "A Quantified Model for Facility Site Selection—Application to a Multiplant Location Problem,"* AIIE Transactions, *4 (1), March 1972, pp. 1-10. Copyright © American Institute of Industrial Engineers; reprinted with permission.*

economic perspective with the existing plants and market areas. The objective factor measures focus on the minimizing of total production-distribution costs. This aim is somewhat different from the location analysis for a single plant, because each alternate location requires a different allocation of capacity to markets in order to minimize overall costs. The formal problem can be placed in a linear programming framework and solved in a distribution table. See Appendix C for a review of linear programming-distribution methods.

An Example

The Pet Food Company has experienced increasing demand for its dog food and cat food products, particularly in the South and Southwest. (The Pet Food Company is used as an example in Appendix C in discussing linear programming concepts in distribution). As a result of this market expansion, the company is now considering the construction of a new plant with a capacity of 20,000 cases per week.

Surveys have narrowed the choice to three general locations: Denver, Los Angeles, and Salt Lake City. Because of differences in local wage rates and other costs, the

TABLE 4-3 **Production and Distribution Costs, Plant Capacities, and Market Demands for the Pet Food Company, Existing and Proposed Plants**

	Distribution Costs per 1000 Cases to These Markets Centers					Normal Plant Capacity, 1000 Cases per Week	Production Cost per 1000 Cases
	Atlanta (V)	Buffalo (W)	Cleveland (X)	Denver (Y)	Los Angeles (Z)		
Existing plants							
Chicago (A)	18	16	12	28	54	46	270
Houston (B)	24	40	36	30	42	20	265
New York (C)	22	12	16	48	44	34	275
Proposed plants							
Denver (D)	40	40	35	2	31	20	262
Los Angeles (E)	57	70	64	31	3	20	270
Salt Lake (F)	50	50	46	14	19	20	260
Market demand, 1000 cases per week	30	18	20	15	37		

estimated production costs per thousand cases are different, as indicated in Table 4-3. The distribution costs from each of the proposed plants to the five distribution points and the new estimated market demands in each area are also shown in Table 4-3. The production costs in the existing plants at Chicago, Houston, and New York are $270, $265, and $275 per thousand cases, respectively.

These data are summarized in the three distribution tables shown in Figure 4-2, one for each of the three possible configurations that include the new plant at Denver, Los Angeles, or Salt Lake City. Since production costs vary in the alternate locations, we must minimize the sum of the production plus distribution costs. For example, the typical item of cost represented by the Chicago-Atlanta square in Figure 4-2 is made up of the production cost at Chicago of $270 plus the distribution cost from Chicago to Atlanta of $18 from Table 4-3, or $270 + $18 = $288 per week. Similarly, for the proposed plant at Salt Lake City, the cost shown in square Salt Lake City-Buffalo in Figure 4-2 is the production cost at Salt Lake City of $260 plus the Salt Lake City-to-Buffalo distribution costs of $50, or $260 + $50 = $310 per week.

The important question now is: Which location will yield the lowest production plus distribution costs for the system of plants and distribution centers?

To answer this question, we solve the three linear programming distribution problems one for each combination. Figure 4-2 shows the optimum solutions for each configuration. In this instance, the objective cost factors favor the Los Angeles location. The Los Angeles location results in the lowest production plus distribution cost of $34,411 per week. The Los Angeles location is $439 per week (almost $23,000 per year) less costly than the Salt Lake City location and $464 per week (more than $24,000 per year) less costly than the Denver location. The reason that the Los Angeles location is less costly is that the lower distribution costs more than compensate for the higher production costs, as compared to the other two alternatives.

From Factories \ To Distr. Points	Atlanta (V)	Buffalo (W)	Cleveland (X)	Denver (Y)	Los Angeles (Z)	Available from factories, 1000's
Chicago (A)	288 (30)	286	282 (16)	298	324	46
Houston (B)	289	305	301	295	307 (20)	20
New York (C)	297	287 (18)	291 (4)	323	319 (12)	34
New Plant at Denver	302	302	297	264 (15)	293 (5)	20
Required at distribution points, 1000's	30	18	20	15	37	120

Production cost = $32,310
Distribution cost = 2,565
Total $34,875
(a)

FIGURE 4-2
Optimum production-distribution solutions for three proposed locations for the additional Pet Food Company plant. (*a*) New plant at Denver, (*b*) new plant at Los Angeles, and (*c*) new plant at Salt Lake City.

The combined production-distribution analysis then provides input concerning the objective factor costs (OFC) in the Brown-Gibson location model discussed previously. The subjective factors would be evaluated as before and a final decision would be based on both objective and subjective factors and the relative weights placed on them.

LOCATIONAL DYNAMICS FOR MULTIPLANTS

Suppose that the Pet Food Company decides to build the Los Angeles plant. The balance of cost factors that produced the solution shown in Figure 4-2*b* could change, however. Then, the allocation of capacity to markets should also change in order to minimize relevant costs. Thus, location analysis is a continuous consideration rather than a one-shot analysis performed only at the time of expansion.

Let us assume that after the Los Angeles plant was built, the Pet Food Company experienced a net decline in demand because of the entry in the market of aggressive new competitors. Instead of a total demand of 120,000 cases per week as projected in the original location analysis, only 105,000 cases are required.

The result is that any three of the plants can meet the demand by using overtime capacity. The company is now faced with comparing the objective and subjective factors of five production-location alternatives. The five alternatives are: operate all plants at partial capacity, plus four additional alternatives that involve shutting down one of the plants and meeting requirements using the other three plants operating on

To Distr. Points / From Factories	Atlanta (V)	Buffalo (W)	Cleveland (X)	Denver (Y)	Los Angeles (Z)	Available from factories, 1000's
Chicago (A)	288 (26)	286	282 (20)	298	324	46
Houston (B)	289	305	301	295 (15)	307 (5)	20
New York (C)	297 (4)	287 (18)	291	323	319 (11)	34
New Plant at Los Angeles	327	340	334	301	273 (20)	20
Required at distribution points, 1000's	30	18	20	15	37	120

Production cost = $32,470
Distribution cost = 1,941
Total $34,411
(b)

To Distr. Points / From Factories	Atlanta (V)	Buffalo (W)	Cleveland (X)	Denver (Y)	Los Angeles (Z)	Available from factories, 1000's
Chicago (A)	288 (30)	286	282 (16)	298	324	46
Houston (B)	289	305	301	295 (15)	307 (5)	20
New York (C)	297	287 (18)	291 (4)	323	319 (12)	34
New Plant at Salt Lake City	310	310	306	274	279 (20)	20
Required at distribution points, 1000's	30	18	20	15	37	120

Production cost = $32,270
Distribution cost = 2,580
Total $34,850
(c)

overtime schedules.

In order to compare the alternatives, five different linear programming distribution tables would be developed. In order to keep the alternatives involving overtime capacity within the linear programming framework, the overtime capacity would be regarded as a separate source of supply. In actual shipment, units produced on overtime would not be segregated; overtime capacity would simply reflect higher costs of production. Five optimal production-distribution tables would be generated and the variable plus fixed costs of operation compared for the five alternatives. The alternative with the lowest cost would be the one favored on the basis of objective factor costs. The final decision would necessarily be influenced by both objective and subjective factors, since shutting down a plant has a number of important effects on employee and community relationships.

WAREHOUSE LOCATION

Whereas industrial plant location is often oriented toward dominant factors, such as raw material sources or even personal preferences of owners, warehouse location is definitely distribution oriented. Although the particular site choice will be affected by subjective factors, such as those included in the Brown and Gibson model, the focus of interest in the warehouse location problem is on minimizing distribution cost. One of the reasons that warehouse location is interesting is that the problem occurs more frequently than plant location and can be evaluated by objective criteria.

Earlier efforts in logistics and distribution management attempted to define the most appropriate customer zones for existing warehouses. These graphical approaches centered on the determination of lines of constant delivery costs. Since the 1950s, however, there has been a variety of attempts to deal with warehouse location as a variable to be determined, using mathematical programming, heuristic and simulation approaches, and branch and bound methods.

The general nature of the problem is to determine warehouse location within the constraints of demand in customer zones in such a way that distribution cost is minimized for a given customer service level. Warehouse capacity is determined as a part of the solution. Customer service is defined in terms of delivery days, thus limiting the number of warehouses that can service a given zone. Distribution costs are the sum of transportation cost, customer service cost, and warehouse operating costs. The warehouse operating costs break down into costs that vary with volume, fixed costs of leasing or depreciation, fixed payroll, and fixed indirect costs.

First, let us dispose of the possibility of calculating the distribution costs of all warehouse-customer zone combinations and simply selecting the combination that has the minimum cost property. The impracticality of this enumeration approach is discussed by Khumawala and Whybark [1971]. *Consider the following characteristics of a medium-sized manufacturing firm that distributes only in the United States:*

1. *5000 customers or demand centers*
2. *100 potential warehouse locations*
3. *5 producing plants*
4. *15 products*
5. *4 shipping classes*
6. *100 transportation rate variables involving direction of shipment, product, geographical area, minimum costs, rate breaks, and so forth*

One single evaluation can be made by making an assignment of customers to warehouses for each of the product lines and then using the computer to search the minimum freight rates for that assignment. That would determine the total cost of that particular warehouse location alternative. Each other assignment would be evaluated the same way until all were complete and the least cost alternative found. Although this looks feasible, the company described above has over 12 million alternate distribution systems. Even if the evaluation of each alternative could be performed in just three seconds, the evaluation of all alternatives would take over one year of computer time at 24 hours per day.

Since the "brute force" approach is impractical, we must consider techniques that make some trade-off with reality through simplifying assumptions. We will discuss three practical approaches that have been used in large-scale applications.

Esso—The Branch and Bound Technique

Effroymson and Ray [1966] developed a model, later improved by Khumawala [1971], that involves a procedure using branch and bound methods and linear programming to produce optimal solutions with reasonable computing time. The location system was applied in the Esso Company to several location problems involving 4 plants, 50 warehouses, and 200 customer zones. The procedure involves the application of rules for including or excluding warehouse locations, depending on whether or not their competitive savings cover their fixed costs of operation. Linear programming is used at points in the procedure to compute lower bound costs. By following out the branches, computing upper and lower bound costs, warehouse locations can be either definitely included in the optimum solution or excluded, leading to the final optimum solution. An example of the procedure together with a case history is given in Atkins and Shriver [1968].

The nature of the procedure is best explained in the context of an example. Table 4-4 summarizes the fixed and variable distribution costs for supplying four customer zones centered at Los Angeles, Detroit, Dallas, and New York from four *possible*

TABLE 4-4 **Fixed and Variable Distribution Costs from Four Possible Warehouse Locations to Four Customer Zones**

		Variable Distribution Costs From Warehouses to Customer Zones Below			
Warehouse	Annual Warehouse Fixed Cost	C_1 Los Angeles	C_2 Detroit	C_3 Dallas	C_4 New York
W_1 Denver	$4500	$ 3000	$10,000	$ 8000	$18,000
W_2 Chicago	6000	9000	4000	6000	5000
W_3 Philadelphia	6000	12,000	6000	10,000	4000
W_4 Houston	8000	8000	6000	5000	12,000

warehouse locations at Denver, Chicago, Philadelphia, and Houston. The question is, which warehouse locations should be selected in order that total distribution cost is minimized?

Step 1

Determine whether there are any warehouse locations that *must* be in the optimum solution, because the *minimum* variable distribution cost saving that results from opening them is greater than the fixed costs required to open them. To do this for each warehouse, we compare the smallest variable cost for each customer zone with the next most costly warehouse to supply that zone. If the saving is greater than the fixed cost for the warehouse, then the warehouse must be used in the optimum solution.

1. W_1 (Denver) saves a minimum of $8000 − $3000 = $5000 by supplying C_1 (customer zone 1) centered at Los Angeles, since supplying from Houston would cost $8000. The saving of $5000 is greater than the $4500 fixed cost at W_1, and therefore, W_1 must be in any final solution.

2. W_2 (Chicago) saves only $6000 − $4000 = $2000 for C_2, but this is less than the $6000 fixed cost for W_2.

3. W_3 (Philadelphia) saves only $5000 − $4000 = $1000 for C_4, but this is less than the $6000 fixed cost for W_3.

4. W_4 (Houston) saves only $6000 − $5000 = $1000 for C_3, but this is less than the $8000 fixed cost for W_4.

The result of step 1 is that W_1 must be included in the optimum solution and that W_2, W_3, and W_4 may or may not be included, depending on further computations.

Step 2

Determine if any warehouse location can be eliminated from the solution by the following rule: given that W_1 is fixed open, if for any other warehouse location the *maximum* saving achieved by opening it is *less* than the fixed cost, then that warehouse can be eliminated from the solution; it cannot possibly compete with other warehouse locations.

1. If W_2 is opened, given that W_1 is already open, then the maximum saving is ($10,000 − $4000 for C_2) + ($8000 − $6000 for C_3) + ($18,000 − $5000 for C_4) = $21,000. Since this maximum saving is greater than the $6000 fixed cost, W_2 cannot be eliminated at this point.

2. The maximum saving if W_3 is opened, given W_1 open, is ($10,000 − 6000 for C_2) + (18,000 − $4000 for C_4) = $18,000. Since this maximum saving is greater than the $6000 fixed cost, W_3 cannot be eliminated at this point.

3. The maximum saving if W_4 is opened, given that W_1 is open, is (\$10,000 − \$6000 for C_2) + (\$8000 − \$5000 for C_3) + (\$18,000 − \$12,000 for C_4) = \$13,000. Since this maximum saving is greater than the \$8000 fixed cost, W_4 cannot be eliminated at this point.

The result of step 2 is that the final status of W_2, W_3, and W_4 is still undetermined, since they are neither definitely in the solution from step 1 nor definitely eliminated in step 2. W_1 is definitely in the solution from step 1.

Step 3

From step 2, we see that the largest maximum saving occurs by opening W_2. Therefore, we open W_2 in addition to W_1 and repeat step 2, calculating the maximum savings by opening each of the other two warehouses to see if they can be eliminated with W_1 and W_2 open.

1. The maximum saving that results if W_3 is opened, given that W_1 and W_2 are open, is \$5000 − \$4000 = \$1000 for C_4. Since this maximum saving is less than the \$6000 fixed cost, W_3 is eliminated.

2. The maximum saving if W_4 is opened, given that W_1 and W_2 are open, is \$6000 − \$5000 = \$1000 for C_3. Since this maximum saving is less than the \$8000 fixed cost, W_4 is eliminated.

Step 4

Assuming W_1 and W_2 in the solution, determine the minimum distribution cost for this branch, which has now been completed, since each warehouse has been evaluated both opened and closed. Costs are \$10,500 (fixed) + (\$3000 for C_1 + \$4000 for C_2 + \$6000 for C_3 + \$5000 for C_4) (variable) = \$28,500.

Step 5

Now calculate for the alternate branch, where W_2 is assumed closed. Under these conditions, W_3 must be open, because the maximum saving exceeds its fixed cost. The total distribution cost is \$10,500 (fixed) + (\$3000 for C_1 + \$6000 for C_2 + \$8000 for C_3 + \$4000 for C_4) (variable) = \$31,500. Calculating for the other combinations of W_3 and W_4 either open or closed can only increase distribution cost. The result at the end of step 4, with the W_1 and W_2 locations selected, is the optimum solution.

In the actual program, linear programming is used to compute the lower bound cost at the end of step 1 with W_1 fixed open and again at the end of step 2 to determine which branch to follow. In our example, the branch involving W_1 and W_2 would have been chosen and the upper bound of \$28,500 determined. The other branch involving W_2 being closed would need to be evaluated, however, because the linear programming lower bound was less than the current upper bound. On following that branch, however, the result was inferior to the solution with W_1 and W_2 open.

Hunt-Wesson Foods, Inc.—An Application of Mathematical Programming

Hunt-Wesson Foods, Inc., produces several hundred distinguishable commodities at 14 locations and distributes nationally through 12 distribution centers. The company decided to undertake a study of its distribution system, particularly the location of distribution warehouses. Five changes in location were indicated involving the movement of existing distribution centers as well as the opening of new ones. The cost reductions resulting from the study were estimated to be in the low seven figures.

Geoffrion and Graves [1974] formulated the Hunt-Wesson distribution problem in such a way that the multicommodity linear programming subproblem decomposes into as many independent classical transportation problems as there are commodities. The resulting problem structure included 17 commodity classes, 14 plants, 45 possible distribution center sites, and 121 customer zones. Thus, three levels of distribution were accounted for—plants, distribution centers or warehouses, and customer zones.

Demand for each commodity at each customer zone was known. Demand was satisfied by shipping via regional distribution centers (warehouses), with each customer zone being assigned exclusively to a warehouse. Upper and lower limits were set on the annual capacity of each warehouse. Warehouse location sites were selected to minimize total distribution costs, which were composed of fixed plus variable cost components.

The Hunt-Wesson warehouse location study indicates that the cost reductions possible are of great significance and justify careful study. Geoffrion and Graves state that similar results were obtained in an application for a major manufacturer of hospital supplies with 5 commodity classes, 3 plants, 67 possible warehouse locations, and 127 customer zones.

Ralston Purina—An Application of Simulation

Markland [1973] applied a computer simulation methodology in evaluating field warehouse location configurations and inventory levels for the Ralston Purina Company.

The basic structure of product flow is a multilevel, multiproduct distribution system involving plant warehouses, field warehouses, wholesalers, and finally retail grocers. Note that shipments from the five warehouses may go to other plant warehouses, to field warehouses, or to wholesalers. Inventories are maintained at the plant warehouse, field warehouse, and wholesaler levels. Also, shipments may go from any of the five field warehouses to any of the 29 demand analyses areas representing the wholesale level or to any of the other field warehouses.

Model Structure. Markland modeled the distribution system in the basic format of system dynamics as a dynamic feedback control system where product flow is the main control variable. Product flow and inventory level equations were written to represent all flow combinations and inventory levels at plant and field warehouses. Constraints on maximum inventory levels at plant and field warehouses were established, as well as constraints on maximum plant production capacity.

Sample Results. The system was simulated for six different field warehouse configurations and for different inventory service levels. The number of field warehouses was varied from zero to the existing five. Thirty-two field warehouse location patterns were tested involving combinations of warehouses, using a procedure of dropping warehouses from the existing pattern. In the example given, Ralston Purina saved \$132,000 per year by consolidating field warehouses from five to three. The proposed elimination of intermediate warehouses as a policy would have increased costs by \$240,000 per year, as compared to the optimal policy of using only three field warehouses. In addition to the preceding result, it was found that an 85 percent inventory service level minimized distribution cost.

The three warehouse location methodologies discussed are apparently all quite powerful and capable of dealing with problems of practical size. The real advantage of the simulation methodology is that nonlinear costs can be represented easily and that the other features, such as inventory policy, can be included. The advantage of the Hunt-Wesson application is in its capability for handling extremely large-scale systems and in representing the storage in transit costs that are important for firms such as Hunt-Wesson and Ralston Purina. Finally, the branch and bound procedure is efficient in computing time.

LOCATION OF REGIONAL HEALTH SERVICES

Whereas industrial plant and warehouse location is influenced strongly by distribution costs, the location of services is oriented toward the location of users. Retail outlets will seek out locations that can maximize their revenue. Medical facilities need to be placed within the reach of those who are ill. Fire stations and ambulance services need to be located to provide a certain minimum response time. Thus the services location problem, as well as the techniques used, is somewhat different from the industrial plant and warehouse location problems.

The objective of health care planners is often stated as allocating facilities to locations such that primary health care demanded by the population is maximized. Implementing such an objective, however, depends on how one weights and trades off several criteria. In order to make rational location plans, demographic data are needed that characterize aspects of user behavior. Abernathy and Hershey [1972] developed a location model that provides optimum locations within a region for different criteria so that decision makers have a basis for trade-off analysis in determining actual locations.

Model Description

The model assumes a defined medical service area, such as that shown in Figure 4-3, where the numbered geographic areas represent census blocks. The *xy* axes provide a grid for specifying the location of populations and health care facilities. Three cities are located within the area of Figure 4-3, and demographic data are available

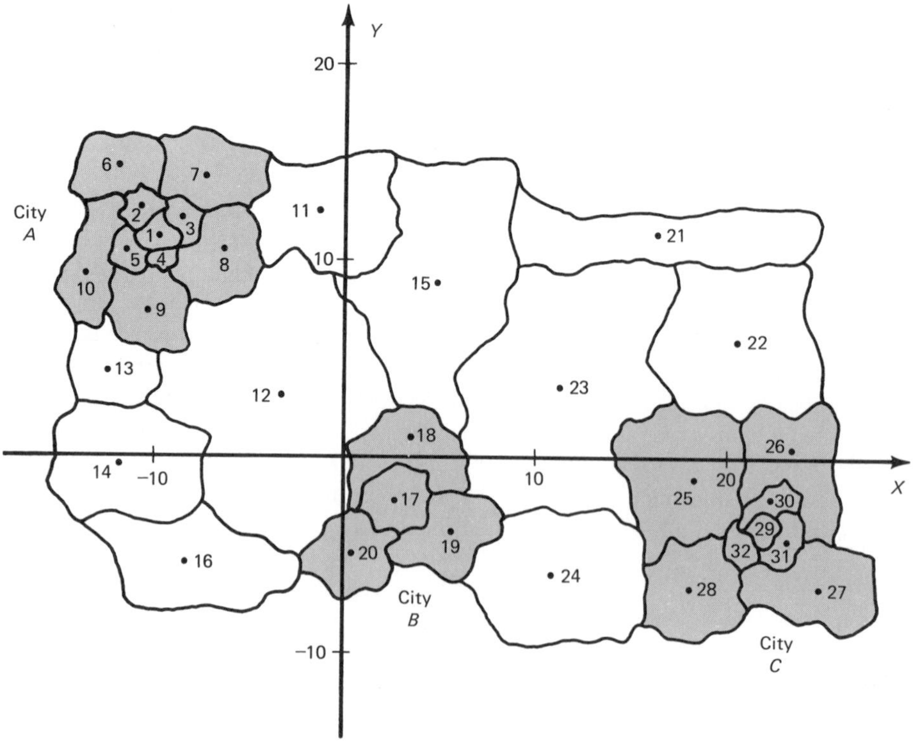

FIGURE 4-3
A hypothetical medical service area with 32 census blocks and three cities. City populations are (approximately) $A = 17{,}000$, $B = 9000$, and $C = 13{,}000$. Distances on xy axes are in miles.
SOURCE: *W. J. Abernathy and J. C. Hershey, "A Spatial-Allocation Model for Regional Health-Services Planning,"* Operations Research, 20, *May-June 1972, pp. 629–642.*

concerning the behavior of the population in regard to the use of medical facilities. The demographic data are divided into strata that exhibit relatively homogeneous patterns of medical facility use. The key variables measure the extent to which the propensity to seek care decreases as distance to a health facility increases.

Based on the stratified data, equations are developed that describe the utilization of a facility in relation to distance and the probability of choosing a particular center relative to the distance to the nearest center. Total demand is determined by summing the demand over all centers, strata, and census blocks. The model determines the (xy) coordinates for health centers in such a way that some criterion is optimized. In the example problem, four criteria were used:

1. *Maximum utilization*: maximize the total number of visits to centers.

2. *Minimum distance per capita*: minimize the average distance per capita to the closest center.

3. *Minimum distance per visit*: minimize the average per visit travel distance to the nearest center.

4. *Minimum percent degradation in utilization*: minimize the average reduction in the number of visits individuals in the community make as a percentage of the visits made if they were in immediate proximity to a center.

The system was optimized by a computer search methodology that searches a multidimensional criterion surface to find the optimum point. Any of the preceding criteria could be used, as well as others.

An Example

Abernathy and Hershey provide the results of a study for the medical service region of Figure 4-3. Data were given for the key variables in four strata, together with the coordinate locations of population centers by stratum. Optimum locations were generated using Search Decision Rule (SDR) for one, two, three, and four centers for each of three crtieria independently. Table 4-5 shows the coordinate locations of centers for the three different criteria. Figure 4-4 shows the locations of one center in relation to the grid and the three cities for the three criteria.

If there were only one health center for the region, its optimum location would depend on which criterion is used, because different geographic areas have different behavior patterns, as shown in Figure 4-4. When criterion 1 (maximize utilization) is used, the center is located near the center of city C. This is because this area contains a large number of individuals for whom distance is a strong barrier. But this location choice tends to increase overall distance per capita. Individuals whose use is curtailed most by long travel distance are favored by this criterion; however, it shifts the transportation costs to those who are less sensitive to distance. Criterion 2 (minimum distance per capita) results in a location near the population–distance centroid for the region, near city B. Criterion 3 (minimum distance per visit) results in a location near the center of the largest city, city A. This is regarded as a local criterion, since it does not represent needs throughout the region in a consistent way and so results in the lowest utilization.

Figure 4-5 provides the decision maker with information for use in deciding on trade-offs among criteria. Each of the three graphs presents the value of each criterion when it serves as the basis for optimizing the location. Each graph also provides a comparison of the criterion with the degradation in the same criterion measure when each of the other two criteria is optimized. In general, for each criterion, there is a significant improvement as the number of centers is increased above one. Note, however, that the rate of improvement declines rapidly as more centers are added. The decision maker can compare the cost implications of suggested numbers, sizes, and locations of centers with the marginal benefits of a given level of service. These benefits are measured in terms of the three different criteria.

TABLE 4-5 **Location Coordinates in Miles for Three Criteria and Different Numbers of Centers** (see Figures 4-4 and 4-5 for locations of coordinates)

Center Number	Criterion: (1) Maximize Utilization		(2) Minimize Distance per Capita		(3) Minimize Distance per Encounter	
	x	y	x	y	x	y
I With 1 center						
1	21.00	−3.00	0.64	1.20	−8.70	10.00
II With 2 centers						
1	21.4	−3.7	17.6	−3.30	18.50	−3.30
2	−9.89	10.4	9.89	10.4	−9.90	10.40
III With 3 centers						
1	22.40	−3.1	21.52	−2.78	22.30	−3.20
2	−10.16	10.40	−10.20	10.40	−10.20	10.40
3	3.63	−2.75	3.60	−2.80	3.60	−2.80
IV With 4 centers						
1	22.40	−3.14	22.00	−3.50	21.23	−3.08
2	−10.20	10.40	−10.10	10.30	−9.80	10.40
3	3.59	−2.78	2.69	−4.80	3.61	−2.70
4	11.32	−2.25	3.76	3.04	−11.35	3.00
V With 5 centers[a]						
1	22.40	−3.10				
2	−9.72	10.61				
3	3.24	−3.19				
4	−11.62	3.24				
5	11.04	−2.00				

[a] Determined only for the first criterion.
SOURCE. W. J. Abernathy, and J. C. Hershey, "A Spatial-Allocation Model for Regional Health-Services Planning," *Operations Research*, *20* (3), May–June 1972, pp. 629–642.

LOCATION OF EMERGENCY UNITS

The location of emergency units, such as ambulances, fire stations, and police services, has a dynamic character. The nature of demand for service and response time requirements create the need for adaptation as units are called or dispatched to perform service, leaving their previous locations unprotected. Since personnel costs constitute as much as 90 to 98 percent for emergency services, fixed locations are less justified.

In general, fire stations are usually at fixed locations when they are dispatched, while ambulances and police cars are more mobile and may be dispatched from any location. The distinction between fixed and mobile locations breaks down in periods of high demand, since the units may be dispatched directly from one incident to the next.

The location problem for emergency units might be more correctly termed an "allocation" or "deployment" problem. It is the strategy by which the services meet

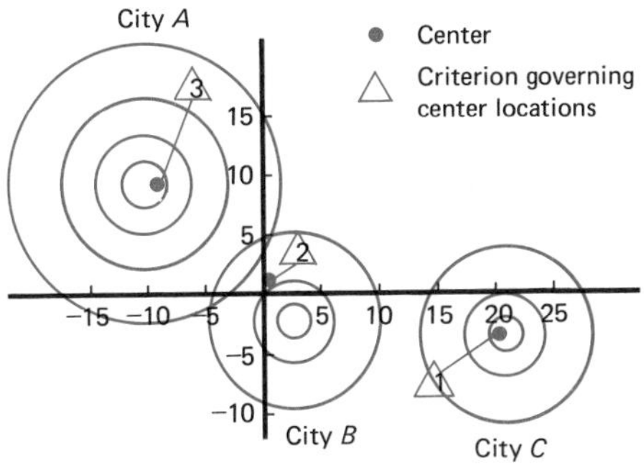

FIGURE 4-4
Location of one center based on three different criteria.
SOURCE: *W. J. Abernathy and J. C. Hershey, "A Spatial-Allocation Model for Regional Health-Services Planning,"* Operations Research, 20, *May-June 1972, pp. 629–642.*

some standard of performance, such as reaction time. Therefore, bound up within the location problem we have: the total capacity of the system, the location or patrol patterns, priorities for different kinds of calls, which units are actually dispatched in response to a call, and the circumstances under which relocation or redeployment takes place. The major objective of the design of an emergency system is to reduce to a low level the possibility that a true emergency call will need to be backlogged.

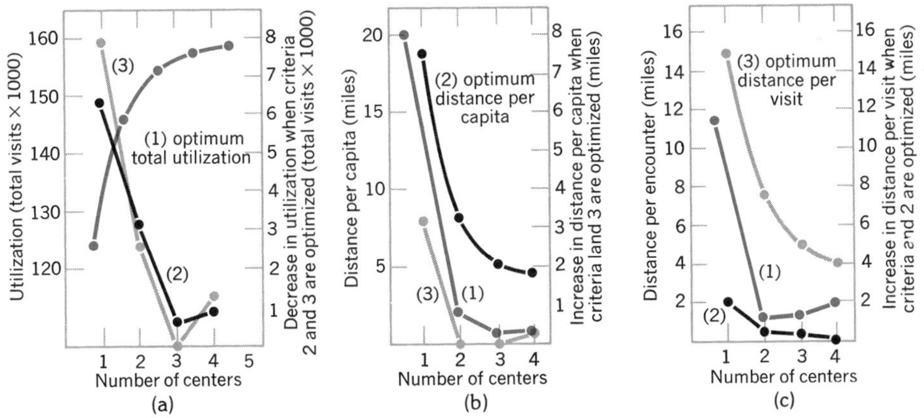

FIGURE 4-5
Differential effects of three criteria for one to four centers. *(a)* Total Utilization (utilization when optimized for utilization and decrease in utilization with optimization for (2) distance per capita (3) distance per visit). *(b)* Distance per Capita (distance per capita when optimized for distance per capita, and increase in distance per capita with optimization for (1) utilization (3) distance per visit). *(c)* Distance per Visit (distance per visit when optimized for distance per visit, and increase in distance per visit with optimization for (1) utilization (2) distance per capita).
SOURCE: *W. J. Abernathy and J. C. Hershey, "A Spatial-Allocation Model for Regional Health-Services Planning,"* Operations Research, 20 *(3), May-June 1972, pp. 629–642.*

Design of Response Areas

The location problem is closely tied to the area for which the emergency unit has primary responsibility. Several objectives may be involved in the actual design of these response areas: minimum response time, work load balancing, demographic homogeneity, and administrative requirements. Rules of thumb have used square or circular patterns to minimize travel time to the scene of an emergency. Speed may depend on specific streets used. If travel time rather than distance is taken as the criterion, a "square" travel time response area would have its longer dimension correspond with the higher-speed streets. Closely related to the design of response areas are site selection for new and changed facilities and the dynamic problems of prepositioning and repositioning to change locations as load develops.

Location of Fire Units

In the case of fire units, prepositioning means determining the location of the fixed-position fire stations themselves. The fire station location problem can be divided into two main classes. First are those where the call rate is high relative to fire protection capacity, such as in New York City or Chicago. In these situations, queuing analysis and stochastic simulation are appropriate, and the concepts of prepositioning and repositioning represent important strategies. Where the call-capacity ratio is relatively low, as in Denver, a static approach may be justified.

Relocation of Fire Units

Kolesar and Walker [1974] designed a fire company relocation algorithm for use in New York City. The New York problem is typical of large cities, where more than one serious fire may be in progress simultaneously. They state that, on the average, 10 such problems occur daily in New York. A five-alarm fire in Manhattan could deplete the area of half of its fire-fighting units, resulting in a sharp degradation of fire protection. Under such circumstances, it is common practice to relocate the remaining units in selected "empty" firehouses to anticipate further alarms. They developed a computer-based system to relocate in such a way that minimum total expected response times result.

Referring to Figure 4-6, suppose that two fires have simultaneously required the services of the seven fire companies shown; one region is uncovered. The problem is how to redeploy in order to minimize expected response time. The model assumes that the call rate will be random. The "square root law" is applied to approximate the expected response distance. This law states that the expected response distance is proportional to the square root of the area served by a fire company.

The algorithm for relocation performs with the following main steps:

1. Determine the need for relocation. An uncovered response area is detected by

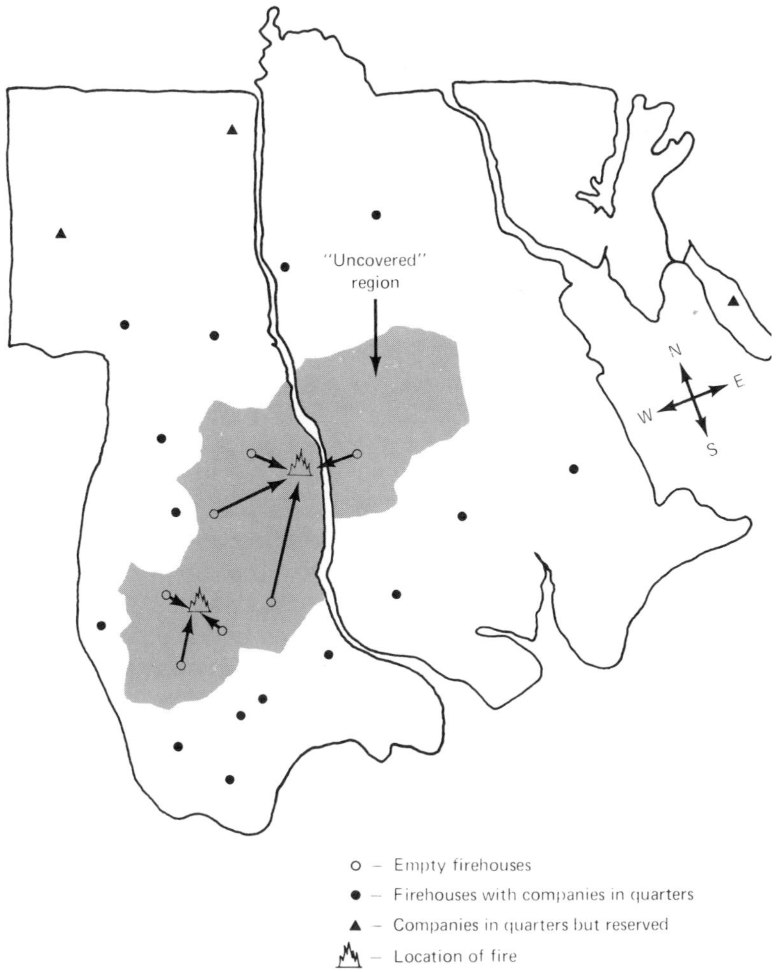

FIGURE 4-6
A sample relocation problem.
SOURCE: *P. Kolesar and W. E. Walker, "An Algorithm for the Dynamic Relocation of Fire Companies,"* Operations Research, 22, *March-April 1974, pp. 249–274.*

a program called "trigger." This program comes into play whenever a fire requires the use of at least three engines and two ladders. An uncovered area is one that falls below the minimum standard of having at least one company within x minutes of every alarm box.

2. Determine which of the empty firehouses should be filled. A heuristic rule is used to select the firehouse associated with the largest number of uncovered response areas.

3. Determine the available companies to relocate. A heuristic rule chooses available companies with the lowest relocation costs. A feasibility check is made to be sure that no area becomes uncovered as a result of a potential relocation.

4. Solve a linear programming assignment problem to minimize total relocation distance.

Now we return to Figure 4-6, which shows the locations of two fires, the locations of the seven ladder companies working at the fires, and the region left uncovered. In step 1, it was found that there were nine uncovered response areas.

Step 2 indicated that there were two solutions, each considering four of the seven empty firehouses that provide a minimum coverage.

In step 3, the program selects the companies to relocate. The least cost assignment is indicated by the colored arrows in Figure 4-7.

The least travel distance solution produced by step 4 is shown by the solid arrows. The solution produced by step 4 results in a 22 percent reduction in travel distance and only a 9 percent increase in cost.

Location of Ambulances

The ambulance location problem is similar in general terms to the fire station location problem. The response time depends on the state of the system at the time a call for service occurs. However, an ambulance normally responds with only one unit and is usually occupied with that call for only a relatively short service time. The number and location of hospitals will have an effect on the time that an ambulance is tied up. The longer the ambulance is tied up giving service, the more likely another call will be received while the unit is busy. The mean call rate is typically a function of the time of day, so that the size of the ambulance fleet in service also varies during the day. In Los Angeles, typical response times are 3.5 minutes, and ambulance service times are 17 minutes.

Fitzsimmons [1973] developed a simulator for an emergency medical system and validated it both in comparison to analytical models and data from the Los Angeles Ambulance System. In the model, travel time is computed as the sum of x and y distance-times to correspond to the usual urban rectangular layout plan. While each vehicle had a home base, the simulator was designed to include a mobile system (where ambulances can be dispatched en route). The simulator included capability for both ambulances and helicopters.

Given the validated simulation model, we have an effective vehicle for the evaluation of alternatives in terms of numbers and locations of both ambulances and hospitals and for the evaluation of alternate policies. The alternatives were evaluated mainly in terms of response time.

Number and Location of Ambulances. The effect on response time of having 1 to 10 ambulances in the system was evaluated for single and dispersed home stations. The single home stations were at a hospital. In general, the response time for a single

FIGURE 4-7
Solutions to the sample relocation problem.
SOURCE: *P. Kolesar and W. E. Walker, "An Algorithm for the Dynamic Relocation of Fire Companies,"* Operations Research, 22, *March-April 1974, pp. 249–274.*

station levels off at about three ambulances in the system, as shown in Figure 4-8; however, response time for dispersed deployment continues to decline. Waiting time for the single station falls to near zero with three ambulances in the system.

A series of 20 runs evaluated hypotheses concerning location deployment patterns, and dispersed deployment dominated the single station alternative for all criteria. Furthermore, optimal locations improved mean travel time to the scene by about 12

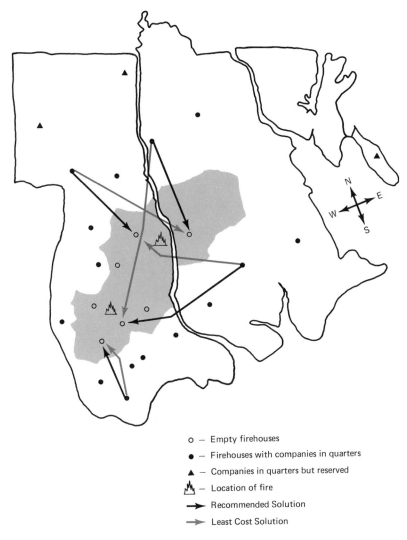

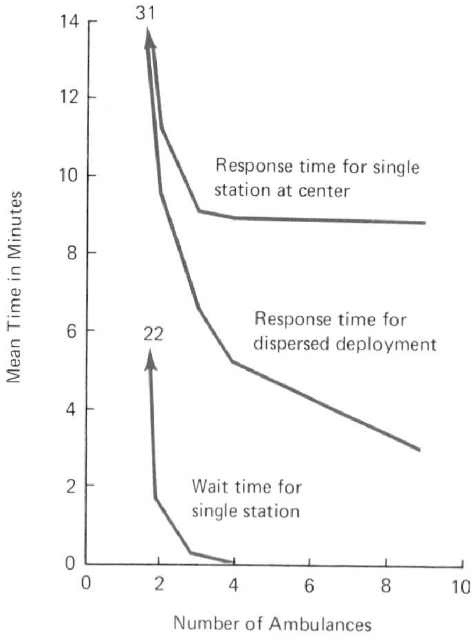

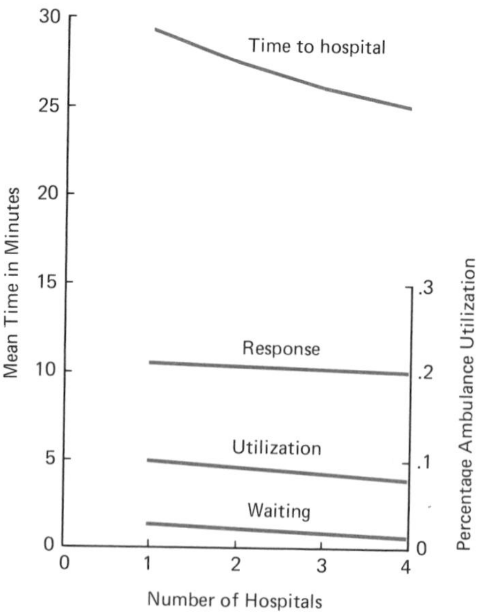

FIGURE 4-8
Response times for single ambulance station and dispersed locations for uniform distribution of calls within the geographical area.
SOURCE: *J. A. Fitzsimmons, "Emergency Medical Systems: A Simulation Study and Computerized Method for Deployment of Ambulance," Ph.D. dissertation, University of California, Los Angeles, 1970.*

FIGURE 4-9
Effect of the number of hospitals in the system on response time.
SOURCE: *J. A. Fitzsimmons, "Emergency Medical Systems: A Simulation Study and Computerized Method for Deployment of Ambulances," Ph.D. dissertation, University of California, Los Angeles, 1970.*

percent, as compared to existing locations. Finally, it was found that optimal deployment was a function of load.

Number and Location of Hospitals. A series of simulation runs was made in which the number of hospitals was varied from one to four, using a single ambulance located centrally. Mean time to the hospital was reduced with the addition of hospitals to the system, but response time and waiting time were only slightly reduced, and ambulance utilization declined about 2 percent, as shown in Figure 4-9. Mean travel time to the scene actually increased slightly. Locating hospitals optimally reduced mean waiting time only slightly because of reduced ambulance utilization.

Control Policies. Alternate dispatch policies were evaluated involving no radio communication, radio dispatch but without mobile transmitters on ambulances, and two-way radio communication. These alternate policies were evaluated under loads varying from call rates of 15 to 45 per day. Simulated response time was reduced by 7 to 8 percent with radio dispatch; however, the two-way radio dispatch system was very little better than the basic system. Nevertheless, the two-way system was recommended, since it effected some improvement.

Even though ambulance systems may be physically dispersed, simulation experiments showed a definite advantage of pooling ambulances into one central dispatch command. Of course, this result would be predicted from queuing theory (see Appendix D).

Alternate ambulance deployment policies were also evaluated, indicating an advantage for an adaptive system. The adaptive system allows repositioning of vehicles as load builds up, instead of having each vehicle return to its home base at the end of an incident.

Fitzsimmons [1970] developed a computer program named "Computerized Ambulance Location Logic" (CALL) to determine optimum ambulance deployment. Based on evaluations for particular ambulance locations, the SDR routine directs changes in the ambulances' locations to decrease the system's mean response time. The SDR program as an optimizing vehicle will be described in Chapter 6 and was also applied in the regional health facility location model described earlier in this chapter.

Results. The CALL model was applied to the problem of establishing locations for 14 ambulances in Los Angeles. The ambulance system in Los Angeles is under the jurisdiction of the fire department, and the problem was to select 14 out of 34 sites that could house an ambulance unit. The fire department had made preliminary decisions on location with the object of locating an ambulance within 2.5 miles of every point in the city, or a five-minute response time at 30 miles per hour. Application of CALL resulted in the relocation of nine ambulances. The relocations produced an almost 9 percent reduction in mean response time and a 33 percent improvement in the probability of a response time greater than six minutes. An important side effect of the new deployment was a more balanced work load as a result of shorter response times.

It was found that ambulances had a 10 percent average utilization rate. As in all productive systems having large demand variation, good service is achieved with a large fraction of server idle time, so that an ambulance unit is likely to be available when an emergency occurs.

LOCATION OF RETAIL OUTLETS

Early models of retail outlet location were based on the gravity model. The models involved the hypothesis that two cities attract retail trade from an intermediate town approximately in direct proportion to the populations of the two cities and in inverse proportion to the square of the distances from the two cities to the intermediate town. While the gravity model had severe limitations, the gravity, or attraction concepts have carried over into models that are nearly as simple but inject some aspects of consumer behavior.

Huff [1962] developed a model in which the utility of a given shopping center is directly proportional to the ratio S/T^{λ}, the ratio of the size of the shopping center to the travel time. Size is measured in terms of square footage of selling space. The travel time is modified by λ, which is estimated empirically to reflect the effect of travel time on various kinds of shopping trips. Huff cities empirical evidence that these two factors exert an influence on a consumer's choice of a shopping center and may be the only variables needed to predict such behavior.

Based on the utilities calculated for all centers, one computes the probability of a consumer at a given origin traveling to a given shopping center. When this probability is multiplied by the number of consumers in a defined area, we have an approximation of the expected number of consumers who are likely to travel to a given shopping center for a given type of shopping trip. The basic statement of the model is then

$$E_{ij} = P_{ij}C_i = \frac{\dfrac{S_j}{T_{ij}^{\lambda}}}{\displaystyle\sum_{j=1}^{n}\frac{S_j}{T_{ij}^{\lambda}}} \cdot C_i$$

where E_{ij} = the expected number of consumers at i that are likely to travel to shopping center j

C_i = the number of consumers at i

P_{ij} = the probability of a consumer at a given point of origin i traveling to a given shopping center j

S_j = the size of a shopping center j

T_{ij} = the travel time involved in getting from a consumer's travel base i to shopping center u

λ = a parameter which is to be estimated empirically to reflect the effect of travel time on various kinds of shopping trips.

EMPIRICAL VALIDATION

A suburban community was selected within the metropolitan Los Angeles area that had within it three rather distinct neighborhoods. These neighborhoods were the points of consumer origin. Data were gathered by questionnaire from householders in the three neighborhoods regarding which shopping center they had last patronized, which center they normally patronized for several kinds of purchases, and information on family income. Information was also gathered on 14 shopping centers within a 20-mile radius from each of the neighborhoods. Table 4-6 indicates the data on gross size of the 14 shopping centers and the travel times to each of the shopping centers from each of the three neighborhoods, respectively. Estimates of the value of λ were also obtained for each of the different kinds of shopping trips.

Comparisons of observed and expected numbers of consumers going to each of the 14 shopping centers were then generated. Table 4-7 is an example of one of

TABLE 4-6 **Sizes and Proximity of Neighborhoods to Selected Shopping Centers**

Shopping Center (j)	Gross Size [Sq. Ft. of Selling Area] (S_j)	Travel Time [Minutes] from Neighborhood 1 (T_{1j})	Travel Time [Minutes] from Neighborhood 2 (T_{2j})	Travel Time [Minutes] from Neighborhood 3 (T_{3j})
J_1	239,000	2.8	3.6	4.2
J_2	236,000	11.1	6.8	8.6
J_3	326,000	13.3	17.0	14.3
J_4	97,000	15.0	14.7	8.6
J_5	1,250,000	15.4	16.1	20.5
J_6	281,000	14.0	17.8	15.2
J_7	228,000	17.8	17.4	11.5
J_8	326,000	21.7	21.4	15.5
J_9	203,000	22.9	22.7	27.0
J_{10}	222,000	20.0	17.8	27.1
J_{11}	502,000	15.7	19.2	17.4
J_{12}	425,000	27.6	27.7	25.8
J_{13}	134,000	10.8	7.7	5.2
J_{14}	121,000	10.4	9.4	16.1

SOURCE. D. L. Huff, *Determination of Intra-Urban Retail Trade Areas*, Graduate School of Management, UCLA, 1962.

these involving furniture purchases. The correlation coefficient between observed and expected values for each of the three neighborhoods is quite high.

USE OF THE MODEL

Based on the model, one can estimate the number of consumers coming to a given shopping area for a given type of purchase. Then, by determining from survey data average household income in the area and the average family budget figures for various kinds of purchases, the annual sales potential for a shopping center is determined (multiply each of the budget figures by the expected number of consumers for that type of purchase). Thus the model produces data on alternate locations concerning the sales potential. One would choose a location that would maximize potential revenue. Given the location in a shopping center, the problem then becomes one of finding a suitable site, based on a host of intangible criteria and values.

IMPLICATIONS FOR THE MANAGER

The emphasis in industrial plant location is to minimize costs; however, we are speaking of longer-run costs, and many intangible factors may influence future costs. Thus,

TABLE 4-7 **Comparison of Observed and Expected Number of Consumers from Each of the Three Neighborhoods Who Last Made a Furniture Purchase at One of the Specified Shopping Centers,[a] Undifferentiated for Age or Income[b] (Distance measured in travel time minutes)**

Shopping Center	Neighborhood 1		Neighborhood 2		Neighborhood 3	
	Obs.	Exp.	Obs.	Exp.	Obs.	Exp.
J_1	51	51.66	68	65.83	80	78.21
J_2	0	1.50	4	16.72	1	7.43
J_3	0	1.30	0	3.27	1	1.95
J_4	0	0.00	0	1.33	0	3.05
J_5	3	3.43	24	14.06	11	2.30
J_6	6	0.98	6	2.55	12	1.38
J_7	0	0.00	3	2.17	3	2.78
J_8	0	0.00	1	2.00	0	15.0
J_9	0	0.00	0	1.10	0	0.00
J_{10}	2	0.00	16	2.02	4	0.00
J_{11}	0	0.00	0	3.88	1	1.58
J_{12}	0	1.31	0	1.50	1	0.00
J_{13}	0	0.91	0	7.28	8	21.82
J_{14}	0	0.91	6	4.29	0	0.00
Total	62	62.00	128	128.00	122	122.00

	Neighborhood 1	Neighborhood 2	Neighborhood 3
[a]r =	.99	.94	.96
λ =	2.542	2.115	3.247
[b]Avg. adult age	37.15	35.52	38.10
Avg. stated income	6130	7246	6611

SOURCE. D. L. Huff, *Determination of Intra-Urban Retail Trade Areas,* Graduate School of Management, UCLA, 1962.

a manager is faced with making trade-offs between tangible costs in the present and a myriad of subjective factors that may influence future costs. The Brown-Gibson model provides a framework for the integration of objective and subjective factors using preference theory to assign weights to factors in a consistent manner.

Multiplant location is influenced by existing locations, since each location considered must be placed in economic perspective with the existing plants and market areas. Each alternate location considered results in a different allocation of capacity to markets, and the manager's objective is to minimize costs for the system as a whole. The concepts of locational dynamics provides the manager with a basis for balancing costs among producing plants as demand shifts in different market areas. In some instances, an appropriate economic decision might be to close a plant, enlarging capacity in the remaining plants through the use of overtime capacity.

While it is often true that plant location (particularly the location of individual versus multiple plants) is dominated by the owner's or manager's personal preference for location, warehouse location is dominated by the cost criterion. Here, the managerial objective of minimizing distribution cost can be enhanced considerably through the use of well-developed computer-assisted distribution planning systems.

The very nature of service operations requires a manager to seek locations based on an analysis of the location of users. Thus, while warehouses and many industrial plant locations are distribution oriented, the location of service operations is user oriented. Service facilities must be decentralized and relatively local in nature in order to bring the service rendered to the users. The basis on which locations in relation to users is chosen depends strongly on the nature of the system.

The managers of health service facilities are faced with complex criteria in attempting to provide locations that give the best public service. The Abernathy-Hershey study indicated the need for trading off criteria of maximum utilization, minimum distance per capita, and minimum distance per encounter.

The managers of emergency systems such as ambulances, fire protection, and police protection are faced with the need for rapid response. The random nature of the timing and frequency of calls for service, in combination with varying time for response and service, place the problem within the general framework of a waiting line problem. But the managers of such systems would have to provide extremely large and expensive capacity in order to meet the response standards imposed if facilities were provided for peak demand. These facilities would necessarily be idle most of the time. Since queuing of calls to any great extent is not acceptable, deployment strategies have been used to provide service at reasonable cost. Thus, locations become mobile for ambulances and police protection through two-way communication systems. Instead of always returning to a home base, ambulances and patrol cars may be redeployed in transit. In fire protection systems, units may be relocated (redeployed) when a major fire occurs in order to minimize expected response time if another major fire were to break out.

Managers of retail stores focus on maximizing revenue in determining locations. The basic model developed by Huff involves the hypothesis that two cities attract retail trade from an intermediate town approximately in direct proportion to the population of the two cities and in inverse proportion to the square of the distances of the two cities to the intermediate town, modified by some aspects of consumer behavior.

REVIEW QUESTIONS AND PROBLEMS

1. In the concepts of locational choice, what is meant by a location oriented toward a technical requirement? Give examples.

2. In locational choice, under what conditions must the criterion be maximum profit? Maximum revenue? Minimum relevant costs?

3. In the Brown-Gibson location model, what is the rationale for weighting subjective factors?

5. In the Brown-Gibson location model, how are the relative weights between objective and subjective factors determined in the overall location problem? Are location choices sensitive to this relative weighting?

6. How can differences in capital expenditure requirements between locations be traded off against differences in variable costs?

7. How is the problem of locating a single plant different from that of locating an additional plant in an existing system of producing facilities?

8. Define the warehouse location problem.

9. Table 4-8 gives fixed and variable costs for three possible warehouse locations at Detroit, Philadelphia, and Dallas to four customer zones located in San Francisco, Chicago, Houston, and New York. Of the proposed warehouse locations, are there any that must be in any optimum solution by virtue of the fact that the smallest variable distribution cost saving that results from opening them is greater than the fixed costs required to open them?

10. Using the data in Table 4-8, can any of the three warehouse locations be eliminated from the solution by virtue of the fact that a warehouse may have been fixed open in any optimum solution from problem 9 and that the maximum saving by opening an additional warehouse is less than the fixed cost?

11. Suppose that we decide to open W_1 and W_3 for the Table 4-8 data. Can we now eliminate W_2 from the solution?

12. If W_1 and W_3 are assumed in the solution for the Table 4-8 data, what is the minimum total distribution cost?

13. If W_2 and W_3 are assumed in the solution for the Table 4-8 data, what is the minimum total distribution cost? Which warehouse locations represent an optimum solution?

TABLE 4-8 **Fixed and Variable Distribution Costs from Three Possible Warehouse Locations to Four Customer Zones**

Warehouse	Annual Warehouse Fixed Cost	Variable Distribution Costs from Warehouses to Customer Zones			
		C_1 San Francisco	C_2 Chicago	C_3 Houston	C_4 New York
W_1 Detroit	3.0	9	4	6	5
W_2 Philadelphia	3.5	12	6	10	4
W_3 Dallas	2.5	8	6	3	12

14. In the location of regional health services, how could the decision maker use the data provided by Figure 4-5?

15. Under what conditions might the location of emergency units, such as fire station, be regarded as a dynamic relocation and redeployment problem? A static location problem?

16. Kolesar and Blum developed a relocation algorithm for fire units. Suppose two fires break out simultaneously. How would the algorithm function in relocating the remaining units?

17. In the model for locating retail outlets, what factors are used to predict consumer behavior?

18. In the model used for locating retail outlets, how is the probability that a given customer will travel to a given shopping center calculated? What aspects of consumer behavior are implied by this calculation?

19. Assume the data given in the example for the location of regional health services. Figure 4-4 shows the location of one center based on the *independent* optimization of each of the three criteria: (1) maximum utilization, (2) minimum distance per capita, and (3) minimum distance per visit. Criterion (1) favors a location near the center of city C, criterion (2) favors a location near city B, and criterion (3) a location near the center of city A.

    Assume that only one center will be financed, so that we must decide on a single location that best satisfies our trade-off among the three criteria. We know, however, that the three criteria are not independent of each other.

    How much weight in the final location of the health center do you think should be given each of the three criteria? Where do you think the center should be located in Figure 4-4?

SITUATIONS

20. A company supplies its products from three factories and five distribution warehouses. The company has been expanding its sales efforts westward and in the South and Southwest. It has been supplying these markets from existing distribution centers, but current volume in the new locations has raised the question of the advisability of a new warehouse location.

    Three possible locations are suggested because of market concentrations: Denver, Houston, and New Orleans. Data concerning capacities, demands, and costs are given in Table 4-9.

    Based on these data, which warehouse location should be chosen? What additional criteria might be invoked to help make a choice? How should management decide whether or not to build the new warehouse or continue to

TABLE 4-9 **Production Costs, Distribution Costs, Plant Capacities, and Market Demands for Situation 20**

From plants \ To Distribution Center	Distribution Costs per Pair, Handling, Warehousing, and Freight								Normal Weekly Capacity Pairs	Unit Production Cost
	Existing Warehouses					Proposed New Warehouses				
	Atlanta	Buffalo	Cincinnati	Cleveland	Milwaukee	Denver	Houston	New Orleans		
Atlanta	$0.27	$0.46	$0.43	$0.45	$0.48	$0.65	$0.58	$0.55	25,000	$2.62
Chicago	0.49	0.48	0.42	0.44	0.32	0.50	0.54	0.60	20,000	2.68
Detroit	0.50	0.38	0.41	0.36	0.42	0.55	0.60	0.65	27,000	2.70
Forecast, weekly Market demand, Pairs	8000	15,000	11,000	13,000	9000	16,000	16,000	16,000		

TABLE 4-10 **Distribution Costs, Production Costs, Fixed Costs, Plant Capacities, and Market Demands for Situation 21**

From Plants	Distribution Costs per Unit					Normal Weekly Capacity, Units	Unit Production Cost	Fixed Costs	
	Milwaukee	Cleveland	Cincinnati	Erie	Mobile			When Operating	When Shut Down
Detroit—Reg.	$0.50	$0.44	$0.49	$0.46	$0.56	27,000	$2.80	$14,000	$6,000
Detroit—OT						7,000	3.52		
Hammond—Reg.	0.40	0.52	0.50	0.56	0.57	20,000	2.78	12,000	5,000
Hammond—OT						5,000	3.48		
Mobile—Reg.	0.56	0.53	0.51	0.54	0.35	25,000	2.72	15,000	7,500
Mobile—OT						6,000	3.42		
Forecast, Weekly Market Demand, Units	9,000	13,000	11,000	15,000	8,000				

supply from the existing warehouses? *Hint*: Three linear programming distribution tables must be formulated, solved, and compared.

21. A company has three plants located in Detroit; Hammond, Indiana; and Mobile, Alabama that distribute to five distribution centers located at Milwaukee; Cleveland; Cincinnati; Erie, Pennsylvania; and Mobile, Alabama. Table 4-10 shows distribution costs per unit, together with data on production costs at regular and overtime, fixed costs when the plants are operating and when shut down, plant capacities at regular time and the additional output available at overtime, and market demands.

The problem facing the company is that the demand for their products has fallen somewhat owing to an economic recession. The forecasts indicate that the company cannot expect rapid recovery from the demand levels shown in Table 4-10. They are considering the possibility of closing a plant but are also quite sensitive to the impact of such an action on employee, union, and community relationships. The nature of their products requires semiskilled personnel for a substantial fraction of the labor force in each plant.

a. What configuration of production plants operating and product distribution system would be most economical?
b. How do you, as the manager of this operation, trade off the economic advantages of minimum cost operation for the recession period against the subjective factors? What decision should you make, as manager?

Hint: Four linear programming distribution tables must be formulated, solved, and compared, one with all plants operating, and one each with one of the plants shut down and capacity being met with the other two through overtime, if necessary.

22. The Woods Furniture Company of Grand Rapids is an old-line producer of quality wood furniture. They are still located at the original plant site in a complex of buildings and additions to buildings. The plant is bursting at the seams again and is in need of further expansion. There is no room left for expansion on the old site, so a variety of alternatives has been considered in and around Grand Rapids. The final alternatives include the removal of all rough mill operations to a second site in town and the building of an entirely new plant on available property about five miles outside of town.

At that point, the treasurer comes to the president with a deal she has uncovered through a contact with a North Carolina wood supplier. The town of Lancaster, North Carolina, is anxious to bring in new industry and is willing to make location there very attractive. After considerable discussion, it is agreed that Woods should at least listen to a proposal. It turns out that the inducements are substantial. They include a free site large enough for Woods' operations, plus options to buy adjacent land for expansion at an attractive price within five years, no city taxes for five years, the building of an access road to the property, and the provision of all utilities to the site at no cost.

After hearing the proposal, Woods' officers retire to an executive committee meeting. In a burst of enthusiasm, the treasurer moves that the proposal be accepted.

a. What additional information should Woods obtain in order to evaluate the proposal?
b. How should the Woods Company make the location decision?
c. Should Woods accept the proposal from the town of Lancaster, North Carolina?

23. The Pumpo Pump Company is currently located in the industrial district of a large city, and several problems have led to the decision to relocate. First and most important, the present site is old and inefficient, and there is no room left for easy expansion. A contributing factor is high labor rates, which translate almost directly into high labor costs in the products.

The company has been able to operate with an independent union, and the president feels that relationships have been excellent. Recently, there has been a great deal of pressure for the union to affiliate with the Teamsters. Although this fact is not the major reason for moving, the president hopes that a carefully planned move may avoid future labor problems, which he feels would have an adverse affect on the company. As the search for a new site continues, the president seems to place more emphasis on this labor relations factor.

The search has finally narrowed to four cities, and comparative data are shown in Table 4-11. Objective factor costs that can be measured and seem to be affected by alternate locations are shown in Table 4-11a. The president is surprised by the range in these costs, city D having costs 1.9 times those of city A.

Seven subjective factors have been isolated as having importance, and the search staff has made a preliminary rating of each factor (see Table 4-11b). The first six factors were rated on a scale of "excellent, plentiful, very good, good, adequate, or fair." The seventh factor, union activity, was rated "active, significant, moderate, or negligible."

When the summary data were presented to the president, he was impressed, but expressed concern about how to equate the various objective and subjective factors and requested further study. The search staff has decided to use the Brown-Gibson location model.

a. Compute the objective factor measure (OFM) using equation 2.
b. Compute the subjective factor decision weights (SFW_k), the *relative weights* to be assigned to each subjective factor. In order to do this, compare factors two at a time and conclude for each paired comparison which factor, in your judgment, is more important. Assign the more important factor a value of 1 and the less important a value of 0. If you feel that the two factors are of equal value, assign a 1 to both.

   If the process is carried out by a group, develop a table on a chalkboard with the factors in a column at the left and the comparisons to be made across the top. When all combinations of comparisons have been made,

TABLE 4-11 **(a) Objective Factor Costs, Millions of Dollars per Year, and (b) Subjective Factors, for Four Sites, for the Pumpo Pump Company**

	Labor	Trans-portation	Real Estate Taxes	Fuel	State Taxes	Electric Power	Water	Total Objective Factor Costs
City A	1.50	0.60	0.03	0.04	0.02	0.04	0.02	2.25
City B	1.60	0.70	0.05	0.06	0.04	0.07	0.03	2.55
City C	1.85	0.60	0.06	0.06	0.05	0.05	0.03	2.70
City D	3.45	0.50	0.10	0.06	0.08	0.06	0.05	4.30

(a)

	Subjective Factor						
	Labor Supply	Type of Labor	Attitude	Appear-ance	Transpor-tation	Recreation	Union Activity
City A	Adequate	Good	Good	Fair	Good	Good	Significant
City B	Plentiful	Excellent	Very good	Good	Very good	Very good	Negligible
City C	Plentiful	Excellent	Very good	Good	Good	Very good	Negligible
City D	Plentiful	Excellent	Good	Good	Very good	Very good	Active

(b)

total the 1's in each row representing the sum of the preference values for that factor. The factor weight is then the factor sum divided by the total preference values for all factors. As a check on work, the sum of all factor weights, should equal 1.0.

If the process is done individually, place the name of each factor on a separate piece of paper. Make the two-by-two comparisons, tallying the values on the backs of the pieces of paper that identify the factors. Sum and compute weights as indicated previously.

c. Determine site weights (SW) for each factor. The determination of site weights for each factor follows a similar procedure. Comparisons of each site for each subjective factor must be made, one factor at a time. The data rating each factor for each site given in Table 4-11b serve as a guide for the weighting process.

A separate table of comparisons is required for each factor. For example, using the factor of labor supply, develop a comparison table with the four sites in a column at the left and the six comparisons to be made across the top. Insert 1's and 0's in the table, representing the results of comparison. Total the 1's in each row representing the sum of preference values for that site, and compute site weights. The site weights for the factor, labor supply, are $A = 0.0$, $B = 0.3333$, $C = 0.3333$, $D = 0.3333$, as a checkpoint for this development.

The result of this step is a table that gives the site weights for each subjective factor; that is, a 7×4 table of 28 site weights.

d. Compute the subjective factor measure (SFM) for each of the four sites, using equation 3. For each site, SFM is the sum of the successive multiplication of the factor weights determined previously by the site weights, for each factor. For each site, there are seven such multiplications that produce the SFM for that site. As a check, the sum of the SFMs for the four sites should be 1.0.
e. Compute the final location measures (LM) for each site. In order to do this, you must decide on the proportion of the decision weight that you wish to place on objective factors. This is a judgmental process, but you should be able to justify why you have chosen a given objective factor decision weight, X. Given the selection of a value of X, the final location measures are calculated using equation 1. As a final check, the total of the location measures for all four sites should be 1.0.
f. How sensitive is the final decision as indicated by the LMs to variation in the objective factor decision weight, X? If the value you have selected for X were to change slightly, would the location selected by the model change?
g. Given the results of the Brown-Gibson model for Pumpo Pump Company, what decision do you think should be made? Are you satisfied that the model has allowed you to make the necessary trade-offs between objective and subjective factors and that your values (the decision maker's) have been properly and effectively represented in the trade-off process?

REFERENCES

Abernathy, W. J., and J. C. Hershey. "A Spatial-Allocation Model for Regional Health-Services Planning," *Operations Research, 20,* May-June 1972, pp. 629–642.

Alcouffe, A., and G. Muratet. "Optimal Plant Location," *Management Science, 23* (3), November 1976, pp. 267–274.

Atkins, R. J., and R. H. Shriver. "New Approach to Facilities Location," *Harvard Business Review,* May-June 1968, pp. 70–79.

Beckman, M. *Location Theory.* Random House, New York, 1968.

Brown, P. A., and D. F. Gibson. "A Quantified Model for Facility Site Selection—Application to a Multiplant Location Problem," *AIIE Transactions, 4,* March 1972, pp. 1–10.

Buffa, E. S., and A. E. Bogardy. "When Should a Company Manufacture Abroad?" *California Management Review, 2,* 1960.

Buffa, E. S., and J. S. Dyer. *Management Science/Operations Research: Model Formulation and Solution Methods.* John Wiley & Sons, New York, 1977.

Chaiken, J. M., and R. C. Larson. "Methods for Allocation Urban Emergency Units: A Survey," *Management Science, 19,* Part 2, December 1972, pp. 110–130.

Easton, A. *Complex Managerial Decisions Involving Multiple Objectives.* John Wiley & Sons, New York, 1973, plant-location example, p. 288.

Effroymson, M. A., and T. A. Ray. "A Branch-Bound Algorithm for Plant Location," *Operations Research, 14,* May-June 1966, pp. 361–368.

Erlenkotter, D. "Facility Location with Price-Sensitive Demands: Private, Public, and Quasi-Public," *Management Science, 24*(4), December 1977, pp. 378–386.

Fitzsimmons, J. A. "Emergency Medical Systems: A Simulation Study and Computerized Method for Deployment of Ambulances." Ph.D. dissertation, UCLA, 1970.

Fitzsimmons, J. A. "A Methodology for Emergency Ambulance Deployment," *Management Science, 19,* February 1973, pp. 627–636.

Geoffrion, A. M. "A Guide to Computer-assisted Methods for Distribution Systems Planning," *Sloan Management Review, 16,* Winter 1975, pp. 17–41.

Geoffrion, A. M., and G. W. Graves "Multicommodity Distribution System Design by Benders Decomposition," *Management Science, 20*(5), January 1974, pp. 822–844.

Geoffrion, A. M. "Better Distribution Planning with Computer Models," *Harvard Business Review,* July-August 1976, pp. 92–99.

Gibson, D. F., and J. Rodenberg. "Location Models for the Forest Products Industry and Other Applications," *AIIE Transactions, 7,* June 1975, pp. 143–152.

Huff, D. L. "Determination of Intra-Urban Retail Trade Areas," University of California-Los Angeles, Graduate School of Management, 1962.

Huff, D. L. "A Programmed Solution for Approximating an Optimum Retail Location," *Land Economics, 42,* August 1966, pp. 293–303.

Keuhn, A. A., and M. J. Hamburger. "A Heuristic Program for Locating Warehouses," *Management Science, 19,* July 1963, pp. 643–666.

Khumawala, B. M. "An Efficient Branch and Bound Algorithm for the Warehouse Location Techniques," *The Logistics Review, 7,* 1971.

Khumawala, B. M., and D. C. Whybark. "A Comparison of Some Recent Warehouse Location Techniques," *The Logistics Review, 7,* (1971).

Kolesar, P., and E. H. Blum. "Square Root Laws for Fire Engine Response Distances," *Management Science, 19,* August 1973, pp. 1368–1378.

Kolesar, P., and W. E. Walker. "An Algorithm for the Dynamic Relocation of Fire Companies," *Operations Research, 22,* March-April 1974, pp. 249–274.

Larson, R. C. *Urban Police Patrol Analysis.* MIT Press, Cambridge, Mass., 1972.

Lee, S. M., and L. J. Moore. "Optimizing Transportation Problems with Multiple Objectives," *AIIE Transactions, 5,* December 1973, pp. 333–338.

Magee, J. F. *Industrial Logistics.* McGraw-Hill, New York, 1968.

Markland, R. E. "Analyzing Geographically Discrete Warehouse Networks by Computer Simulation," *Decision Sciences, 4,* April 1973, pp. 216–236.

Markland, R. E. "Analyzing Multi-Commodity Distribution Networks Having Milling-in-Transit Features," *Management Science, 21,* August 1975, pp. 1405–1416.

Markland, R. E., and R. J. Newett. "Production-Distribution Planning in a Large-Scale Commodity Processing Network," *Decision Sciences, 7,* October 1976, pp. 579–594.

McAllister, D. M. "Local Recreation Facility Location," in *Regional Environmental Management: Selected Proceedings of the National Conference,* edited by L. E. Coate and P. A. Bonner. John Wiley & Sons, New York, 1975.

Reed, R., Jr. *Plant Location, Layout, and Maintenance.* Richard D. Irwin, Homewood, Ill., 1967.

Schollhammer, H. *Locational Strategies of Multinational Firms.* Center for International Business, Los Angeles World Trade Center, Los Angeles, 1974.

Shycon, H. N., and R. B. Maffei. "Simulation Tool for Better Distribution," *Harvard Business Review,* November-December 1960.

Swoveland, C., D. Ueno, I. Vertinsky, and R. Vickson. "Ambulance Location: A Probablistic Enumeration Approach," *Management Science, 20,* Part 2, December 1973, pp. 686–698.

Whitman, E. S., and W. J. Schmidt. *Plant Relocation: A Case History of a Move.* American Management Association, New York, 1966.

PART

THREE

OPERATIONS PLANNING AND CONTROL

WE NOW TURN OUR ATTENTION TO THE SHORTER-TERM, DAY-to-day, month-to-month kinds of decisions that bear on operations planning and control. Given the physical system, what kinds of plans are necessary for effective operation, what kinds of controls are necessary, and what are the criteria by which we select from among alternate plans and controls?

Part Three is concerned with such questions as: What policies and procedures will guide us in setting basic activity rates? Should we hire or lay off personnel, and in what numbers? When is using overtime justified instead of increasing the size of the work force? When should we take the risk of accumulating seasonal inventories to stabilize employment? How big should inventories be to sustain the process? What policies and procedures are appropriate for controlling inventories and reordering One effective and convenient method for accomplishing differential weighting and forecasts for the first several months of the computer usage data.
be maximized, or is there a value to idleness? How do we maintain the reliability of the productive system so that specified quantity and quality are produced? When is a preventative maintenance policy justified? What policies and procedures can be effective in controlling labor and other costs? Is it possible to control *costs,* or is it really the activities that must be controlled?

CLASSIFICATION OF SYSTEMS

It is useful to review general classifications of productive systems, because the nature of the most important operating problems is quite different for different systems. Recall that we established two bases of classification: (1) continuous versus intermittent systems, and (2) the output of inventoriable versus noninventoriable items.

Continuous Versus Intermittent Systems

Continuous systems are typified by production lines, continuous chemical processes, and in general, production systems of enterprises that produce the high-volume standardized products for which our society is noted. There are also some nonmanufacturing systems that follow the continuous model, such as the cafeteria line and some large-scale office operations. The basis for the term *continuous* is that the nature of demand results in continuous or near continuous use of facilities. The physical flow of whatever is being processed may also be either continuous or approach continuous movement.

On the other hand, intermittent systems are those where the physical facilities must be flexible enough to handle a wide variety of products and sizes, or where the basic nature of the activity imposes changes of the design of the system outputs from time to time. In such instances, no single sequence pattern of processes is appropriate; therefore, the relative location of the operations must be a compromise that is best for all products. The emphasis is on flexibility of product design, processes, flow paths through the system, and so on. The system is termed *intermittent* because the nature of product demand results in intermittent use of facilities. Intermittent systems are characterized by custom- or job-order-type machine shops, batch-type chemical operations, general offices, large-scale one-time projects, hospitals and other health care facilities, and so forth. The basis of the continuous-intermittent classification is also the general nature of the physical layout of facilities in the productive system.

Systems for Inventoriable Versus Noninventoriable Items

Figure III-1 is a diagram of a production-inventory system for an inventoriable product. It emphasizes the broad flow characteristics of the system as a whole. In Figure III-1, the manufacturing phase could be either continuous or intermittent. Of course, many intermittent manufacturing systems produce to inventory. Such systems may be arranged physically and scheduled internally for intermittent flow. Common examples are the machine shops of the large automotive companies. Such shops are closed to job orders from outside the organization and may produce a set of products repetitively, in cycles. Equipment is time-shared among many different products, and

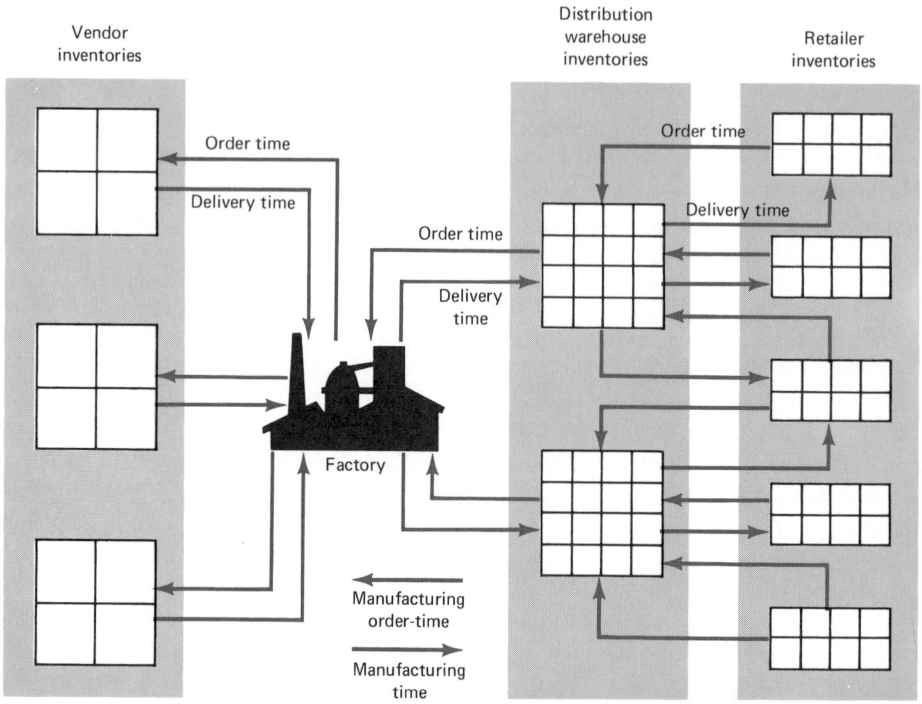

FIGURE III-1
Supply-production-distribution system for an inventoriable product. Times lags are characteristic between and within each stage, and inventories serve vital functions at each stage.

the products are produced to inventory because they are based on standard designs for which predictable markets exist. Such intermittent shops are called *closed job shops* because they are closed to outside job orders.

Thus, we may have production-inventory systems in the inventoriable classification whose physical configuration may be continuous as well as intermittent. If we refer again to Figure III-1, we find a commonly occurring enterprise specialization in the distribution of products. These enterprises specialize in the distribution of products. The portion of Figure III-1 under managerial control for distribution systems focuses mainly on the inventories. The operations phase of such enterprises concerns the replenishment of inventories, the control of inventory levels, and shipment. Service rendered may be judged in terms of frequency of stockout or inability to supply the demands for products from inventories. Distribution systems are nearly pure inventory systems from an operations management viewpoint and are of interest simply because they are common in society. Thus, production-inventory systems for inventoriable items include systems in which production is continuous, systems in which production may be intermittent, and systems involving only distribution. It is important to group systems involving inventoriable items together, because we can use inventories effectively in planning production, and work force requirements.

Custom products and services require production-inventory systems that have no finished goods inventories. Nevertheless, many such systems may face inventory problems for raw materials and materials in process and for supplies that may be used up as a part of the productive process. Open job shops and large-scale one-time projects are prominent examples. The open job shop produces to job order or contract. Since these orders or contracts may never be repeated, flexibility is a crucial requirement. Managerial attention is focused on scheduling and utilizing workers and machines to meet agreed due dates and quality standards.

The difference between the open job shop and the large-scale project in our classification is largely one of scale and complexity. There is no clear dividing line between the two systems, except that when a contract is large enough and complex enough to justify the special PERT-type planning and scheduling techniques it is called a large-scale project.

Nonmanufacturing and service systems are the other prominent types of systems that produce no inventoriable output. The demand for their services must be met from current capacity, and inventories cannot be used as a means of buffering the fluctuations in demand. This fact has far-reaching importance in planning and scheduling for such systems.

A Comparison

To summarize, the continuous-intermittent classification results in the following:

Continuous Systems	Intermittent Systems
Production-inventory systems for high-volume standardized products	Job shops (open and closed)

Distribution systems	Large-scale projects
Service and nonmanufacturing systems	Service and nonmanufacturing systems

This classification is most useful for designing the layout of physical facilities and for indicating the nature of detailed scheduling of both workers and machines.

The inventoriable-noninventoriable item classification results in the following:

Systems for Inventoriable Items	Systems for Noninventoriable Items
Continuous high-volume standardized product systems	Open job shop systems
Closed job shop systems	Large-scale projects
Distribution systems	Service and nonmanufacturing systems

This classification is most useful in determining the nature of appropriate planning for production and work force requirements. The production-inventory-system concept also is useful in helping us focus on the system as a whole. This should help us avoid the organization suboptimization that can result if the organization is segmented into producing and distributing functions.

OPERATIONS PLANNING AND CONTROL SYSTEMS

Figure III-2 presents the broad relationships of the plans and controls for the operations phase of a productive activity. Physically, the productive process takes as inputs labor, materials, equipment and physical facilities, and energy, and converts these inputs into useful outputs of goods and services. Above the productive process shown in Figure III-2, we have outlined the planning processes in block form. The basis for short-term plans centers in forecasts, and forecasting will be the subject of the next chapter.

Based on forecasts, we must make aggregate plans and schedules. These aggregate plans set the basic activity and personnel levels in the short term—they represent short-term capacity decisions. Such decisions involve determining whether to enlarge or contract the size of the labor force, how much to use overtime capacity, and if we are dealing with an inventoriable product, whether to build or draw down inventories. Aggregate planning methods (see Chapter 6) become the basis for other plans, such as raw material ordering and inventories, equipment, and detailed personnel schedules.

Below the productive process in Figure III-2 are the systems for controlling the quality, quantity, and costs for the plans made. As in all kinds of control processes, we need a way of monitoring the aspects over which we wish to establish control, and we need standards for comparison. The control system makes comparisons,

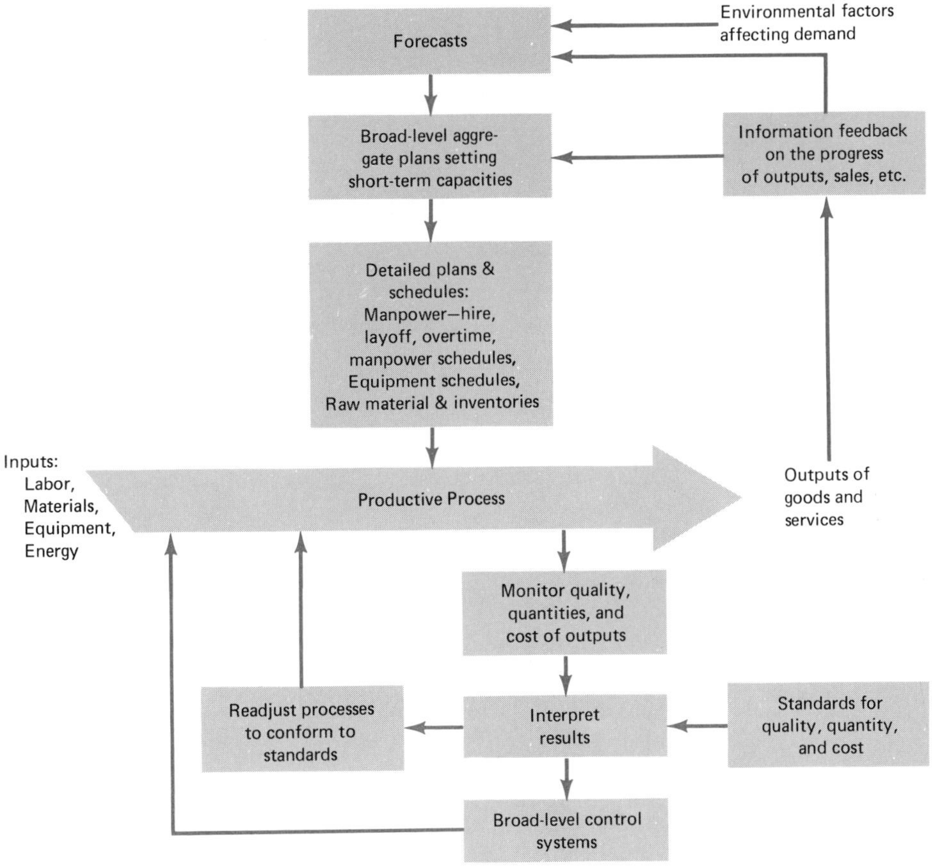

FIGURE III-2
Operations planning and control systems.

interprets results, and takes action to readjust processes to conform to standards. Management attempts to achieve a system optimum by means of various types of controls. Inventories are significant in many systems, and Chapter 7 considers the functions and control models for inventories. Industrial planning, scheduling, and control systems are the subject of Chapters 8 and 9, and Chapter 10 discusses project planning and control. Some of the special problems of service systems are covered in Chapter 11, and finally, the concepts and methods for maintaining system reliability are discussed in Chapter 12.

CHAPTER 5 Demand Forecasting for Operations

PLANNING AND CONTROL FOR OPERATIONS DEPENDS ON THE astute combination of intelligence about what is actually happening to demand and the forecasting of what we expect to happen. We need to make plans that range from a day-to-day to a year-to-year basis. The planning horizon depends on the particular application but is relatively short. We need forecasting methods that are relatively inexpensive to install and maintain and that can be adapted to situations involving a large number of items to be forecast. This means that the data input and storage requirement should be modest and that computerized methods are a likely mechanism to update forecast data as needed.

Table 5-1 summarizes some of the most pertinent and useful forecasting methods under three main headings. The time series models are applicable to shorter-range forecasts for operations and are of particular value for inventory and production control. The causal methods are most appropriate for short- and medium-range forecasting, particularly as a basis for the aggregate planning methods that we discuss in Chapter 6. The predictive methods are appropriate for longer-range forecasts and were discussed in connection with strategic decisions involving facilities planning, location, and product development.

Although good forecasting models have clearly shown their worth in business, industry, and government, there is an art to forecasting. The best results are seldom obtained through the mechanical application of a model. Subjective inputs from knowledgeable people can improve forecast accuracy. Two of the "situations" at the end of the chapter provide information concerning actual experience in combining the inputs from more than one source, including subjective inputs.

REQUIREMENTS OF FORECASTS FOR OPERATIONS

The demand forecasting function serves many broad managerial purposes in both profit and nonprofit organizations. To be useful for operations planning and control, it is important that demand forecast data be available in a form that can be translated into demands for material, time in specific equipment classifications, and demands for specific labor skills. Therefore, forecasts of gross dollar demand, demand by customer or client classification, or demand by broad product or service classifications are of limited value for operations.

Planning and control for operations must necessarily take place at several different levels. Therefore, it is unlikely that one kind of forecast can serve at all levels. We require forecasts of different time spans to serve as the basis for operating plans. These plans are (1) plans for current operations and for the immediate future; (2) intermediate-range plans to provide for the required capacities of personnel, materials, and equipment for the next 1 to 12 months; and (3) long-range plans for capacity, locations, changing product and service mix, and the exploitation of new products and services. We discussed longer range strategic plans and decisions in Part Two. Our present interest, therefore, is in forecasting needs to serve intermediate-range plans and plans for current operations.

TABLE 5-1 **Methods of Prediction and Forecasting**

Method	General Description	Applications	Relative Cost
Time series forecasting models:			
Moving averages	Forecast based on projection from time series data smoothed by a moving average, taking account of trends and seasonal variations. Requires at least two years of historical data.	Short-range forecasts for operations, such as inventory, scheduling, control, pricing, and timing special promotions.	Low
Exponential moving averages	Similar to moving averages, but averages weighted exponentially to give more recent data heavier weight. Well adapted to computer application and large numbers of items to be forecast. Requires at least two years of historical data.	Same as above.	Low
Fourier series least squares fit	Fits a finite Fourier series equation to empirical data, projecting trend and seasonal values. Requires at least two years of historical data. Computer required.	Same as above.	Medium
Causal forecasting methods:			
Regression analysis	Forecasts of demand related to economic and competitive factors that control or *cause* demand, through least squares regression equation.	Short- and medium-range forecasting of existing products and services. Marketing strategies, production and facility planning.	Medium
Econometric models	Based on a system of interdependent regression equations.	Same as above.	High
Predictive methods:			
Delphi	Expert panel answers a series of questionnaires where the answers of each questionnaire are summarized and made available to the panel to aid in answering the next questionnaire.	Long-range predictions, new products and product development, market strategies, pricing and facility planning.	Medium-high
Market surveys	Testing markets through questionnaires, panels, surveys, tests of trial products, analysis of time series.	Same as above.	High
Historical analogy and life cycle analysis	Prediction based on analysis of and comparison with growth and development of similar products. Forecasting new product growth based on the S-curve of introduction, growth, and market saturation.	Same as above.	Medium

The required range of the forecast must be matched with the decision that it will affect. If the decision will be required in three months, a one-month forecast is valueless. On the other hand, it is not wise to select a forecasting model that has an acceptable one-month error but has poor accuracy when projected ahead three months. Therefore, a major criterion for model selection is the match between decision time, forecast range, and forecasting accuracy.

COMPONENTS OF DEMAND

Figure 5-1 shows the three-year record of the demand for a computing service. The company involved derives income from the use of its proprietary programs, and the forecasting of usage can have an important bearing on many factors in planning operations. In general, Figure 5-1, shows an increasing demand for the service with considerable variation in demand. It is the types of variation that are of interest.

Some of the variation in demand can be explained by a statistical model and some cannot. The unexplainable variations we call "random" variations. In forecasting demand for operations, we would like not to respond to what we think might be simply random variations. Since many different customers are using the company's programs to suit their own schedules and needs, much of the seemingly odd increases or decreases from month to month come from this source. Company A using the programs may decide to initiate a new planning cycle that may have affected demand in August 1978, while company B may shut down for two weeks in August for vacations. The aggregation of all users' needs produces this kind of unexplained or random variation. Trying to find out why each month's demand was what it was is probably not worth the time and cost.

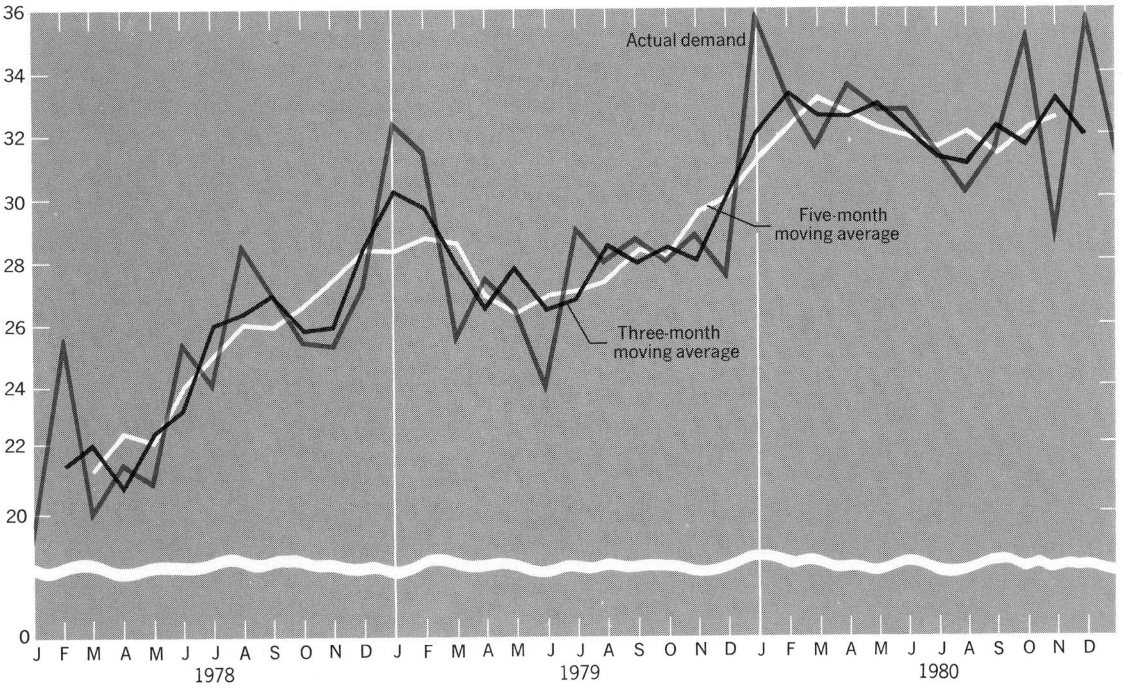

FIGURE 5-1
Monthly program usage with three- and five-month moving averages shown.

Although some of the variation in Figure 5-1 is not explainable, the aggregation of usage by all customers does produce a logical pattern. In spite of the random variations, there is some *average* demand for any particular period. For example, the average for the first quarter of 1978 is 21.51, for the first six months 22.03, and for the full year of 1978 it is 24.06. Each of these measures of average demand has some meaning, although in the form given they have limited usefulness.

It is obvious from Figure 5-1 that there is an upward *trend* in the demand for program usage, and the averages for the first quarter, first six months, and entire year verify an upward trend during 1978. The same conclusion is not quite so clear in 1979, however. The comparable figures for the first quarter, first six months, and full year of 1979 are: 29.72, 27.80, and 28.03. Each comparable figure indicates a growth in demand from year to year; however, both the graph and the 1979 figures indicate that demand may exhibit *seasonal* variation.

The components of demand that we need to take into account in forecasting models are then random, average levels, trend, and seasonal. Cyclic variations related to the business cycle are beyond our scope.

TIME SERIES FORECASTING METHODS

Moving Averages

The common way of smoothing the effects of random variations in demand is to estimate average demand by some kind of moving *average*. Table 5-2 gives some sample demand data taken from Figure 5-1 for the first six months.

A moving average is the average of *n* values centered on the period in question. For example, the first three months of actual demand in Table 5-2 are 19.36, 25.45,

TABLE 5-2

Actual Demand for Program Usage and Three- and Five-Month Moving Averages

Date	Actuals	Three-Month Moving Average	Five-Month Moving Average
1978, Jan.	19.36	—	—
Feb.	25.45	21.51	—
Mar.	19.73	22.22	21.36
Apr.	21.48	20.66	22.57
May	20.77	22.56	22.24
June	25.42	23.33	23.96
July	23.79	25.85	25.03
Aug.	28.35	26.31	—
Sept.	26.80	—	—

and 19.73, and the three-month moving average is computed as (19.36 + 25.45 + 19.73)/3 = 21.51. Therefore, the estimate of February demand with random variations discounted by the averaging process is 21.51. When we estimate for March, we drop the January figure of 19.36 and add the April figure of 21.48. The new moving average centered on March is then (25.45 + 19.73 + 21.48)/3 = 22.22. If we were computing a five-month moving average for March, the data would still be centered on that month. For example, the five-month moving average for the first five months' actual data from Table 5-2 is (19.36 + 25.45 + 19.73 + 21.48 + 20.77)/5 = 21.36.

Both Figure 5-1 and Table 5-2 show that actual demand is quite variable. The three-month moving average is much more stable, however, because the demand for any one month receives only one-third weight. Extreme values are discounted; if they are simply random variations in demand, we are not strongly influenced by them when we gauge demand by the three-month moving average.

Greater smoothing effect is obtained by averaging over a longer period, as is shown by the five-month moving average in Table 5-2. Extreme values are discounted even more, since each period demand carries only a one-fifth weight in the moving average. Figure 5-1 shows three- and five-month moving averages plotted in comparison with actual demand for monthly program usage.

Note the smoothing effects and how the moving average lines reveal the trend and seasonal components in the data. The five-month moving average produces greater smoothing effect, as we would expect. Here we note a conflict of objectives, however. The five-month average discounts random effects more effectively, but the three-month moving average gives more weight to the most recent data. Since the program usage data in Figure 5-1 show both trend and seasonal components, we have a keen interest in emphasizing the most current data in the moving average.

Exponentially Weighted Moving Averages

One effective and convenient method for accomplishing differential weighting and smoothing is by exponentially weighted moving averages. The simplest exponential smoothing model estimates average smoothed demand for the current period S_t by adding or subtracting a fraction, α (alpha), of the difference between actual current demand D_t and the last smoothed average S_{t-1}. The new smoothed average S_t is then

New smoothed average = old smoothed average + α(new demand − old smoothed average)

Or, stated symbolically,

$$S_t = S_{t-1} + \alpha(D_t - S_{t-1}) \qquad (1)$$

The smoothing constant, α, is between 0 and 1 with commonly used values of 0.01 to 0.30. Equation 1 can be rearranged as follows:

New smoothed average = α(new demand) + $(1 - \alpha)$(old smoothed average)

or

$$S_t = \alpha D_t + (1 - \alpha)S_{t-1} \tag{2}$$

If $\alpha = 0.10$, then Equation 2 says that the smoothed average in the current period S_t will be determined by adding 10 percent of the new actual demand information D_t and 90 percent of the last smoothed average S_{t-1}. For example, if $\alpha = 0.1$, $D_t = 19.36$, and $S_{t-1} = 23.00$, then the new smoothed average is

$$S_t = 0.1 \times 19.36 + 0.8 \times 23.00 = 1.94 + 18.40 = 20.34$$

Since the new demand figure D_t includes possible random variations, we are discounting 90 percent of those variations. Obviously, small values of α will have a stronger smoothing effect than large values. Conversely, large values of α will react more quickly to real changes in actual demand (as well as to random variations). For example, if $\alpha = 0.4$ and the other data remain the same, the new smoothed average would be

$$S_t = 0.4 \times 19.36 + 0.6 \times 23.00 = 7.74 + 13.80 = 21.54$$

The components of D_t and S_{t-1} are now weighted quite differently, giving considerably more weight to current actual demand, D_t. The choice of α is normally guided by judgment, although studies could produce optimal values that minimize forecast errors.

Equation 2 actually gives weight to all past actual demand data. This occurs through the chain of periodic calculations to produce smoothed averages for each period. In Equation 2, for example, the term S_{t-1} was computed from

$$S_{t-1} = \alpha D_{t-1} + (1 - \alpha)S_{t-2}$$

which includes the previous actual demand D_{t-1}. The S_{t-2} term was calculated in a similar manner that included D_{t-2}, and so on back to the beginning of the series. Therefore, the smoothed averages are based on a sequential process representing all previous actual demands.

Figure 5-2 shows comparative weightings given data by three- and five-period moving averages and by exponentially weighted moving averages with $\alpha = 0.1$ and 0.3. Note the effectiveness of the exponentially weighted averages in placing heavier weight on the most recent data. Another factor implicit in Figure 5-2 is that exponentially weighted data give a weight to all prior actual demand data, although the effect of old data will be small.

It is important to place the time periods for S_t, D_t, and S_{t-1} in perspective and to recognize that the new smoothed average is not an extrapolation beyond known demand data. Instead, it is the most current smoothed average. It is not a forecast, but a statement of current demand.

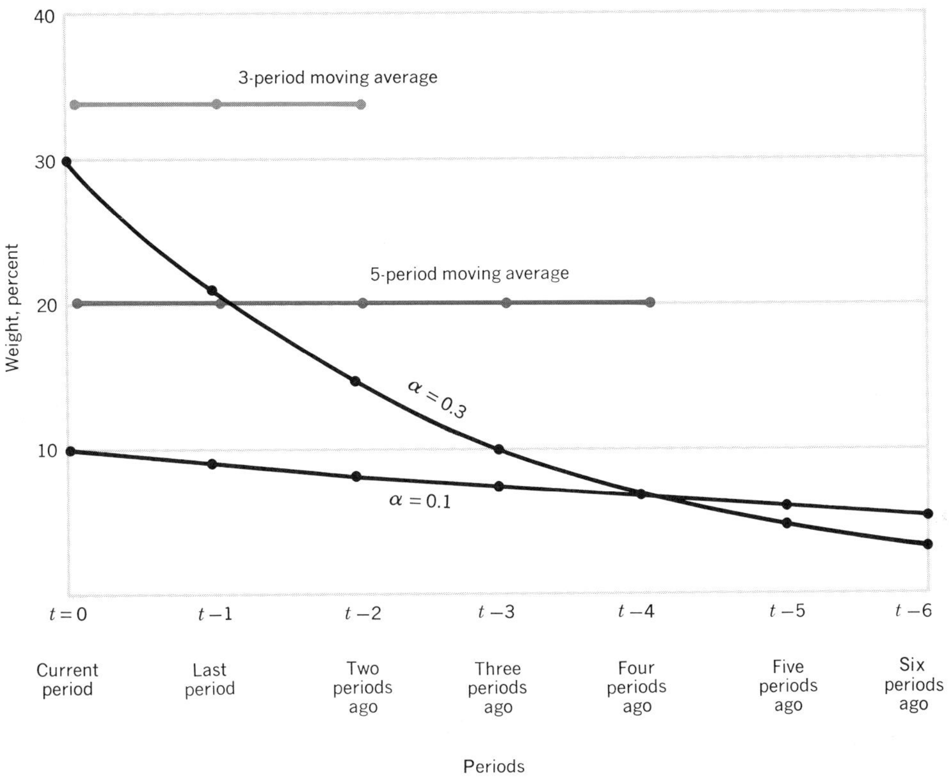

FIGURE 5-2
Comparative weightings given past data by three- and five-period moving averages and by exponentially weighted moving averages with $\alpha = 0.1$ and 0.3.

Extrapolation and Forecast

Since no trend or seasonality is included in the model, direct extrapolation from S_t to infer a forecast is justified. Therefore, the forecast for the upcoming period F_{t+1} is taken directly as the computed value of S_t. Table 5-3 shows computations and forecasts for the first several months of the computer usage data.

Forecast Errors

Forecast errors are defined as

$$\text{Forecast error} = D_t - F_t$$

Forecast errors provide a measure of accuracy and for comparing the performance of alternate models. Three error measures are commonly used:

TABLE 5-3

Sample Computations for S_t and the Forecast, F_{t+1}, for the Simple Exponential Smoothing Model. Data for Actuals from Table 5-2. $\alpha = 0.2$.

Date	Actuals	Smoothed Average, S_t	Forecast, F_{t+1}
Initial	—	23.0	—
1978, Jan.	19.36	22.272	—
Feb.	25.45	22.908	22.27
Mar.	19.73	22.272	22.91
Apr.	21.48	22.114	22.27
May	20.77	21.845	22.11
June	25.42	22.560	21.85
July	—	—	22.56

1. Average error (AE)
2. Mean absolute deviation (MAD)
3. Mean squared error (MSE)

The average error should be near zero for a larger sample; otherwise the model exhibits *bias*. Bias indicates a systematic tendency for overforecasting or underforecasting. But AE obscures variability, since positive and negative errors cancel out.

The mean absolute deviation (MAD) provides additional information useful in selecting a forecasting model and its parameters. MAD is simply the sum of errors without regard to algebraic sign, divided by the number of observations.

The mean squared error provides information similar to MAD but penalizes larger errors. MSE is computed by squaring the individual errors and dividing by the number of observations.

Many forecasting computer programs automatically report all three error measures.

Trend Model

The apparent trend in the exponential smoothed averages is the difference between the successive values, $S_t - S_{t-1}$. If we attempted to compensate for trend using this raw measure, we would have a rather unstable correction, since random effects are still present. Sometimes we might record a negative trend, when in fact the general trend was positive. To minimize these irregular effects, we can stabilize the raw trend measure in the same way we stabilized actual demands, by applying exponential smoothing.

We can smooth the $(S_t - S_{t-1})$ series with a smoothing constant, β (beta). The smoothing constant need not have the same value as α used in smoothing D_t.

The updated value of T_t, the smoothed trend, is

$$T_t = \beta(S_t - S_{t-1}) + (1 - \beta)T_{t-1} \quad (3)$$

Our computation of S_t must now reflect trend, so Equation 2 is modified by simply adding the old smoothed trend to the old smoothed average as follows:

$$S_t = \alpha D_t + (1 - \alpha)(S_{t-1} + T_{t-1}) \quad (4)$$

Equations 3 and 4 yield smoothed current values of demand and trend, so to forecast for the upcoming period, we extrapolate by adding T_t to the current smoothed average S_t as follows:

$$F_{t+1} = S_t + T_t \quad (5)$$

If we apply this trend model to the first several months of the computer usage data, we have Table 5-4, using $\alpha = 0.2$ and $\beta = 0.1$.

In applying the model, note that the current smoothed average S_t must be computed first by Equation 4, since this value is required to compute the current smoothed trend T_t by Equation 3. Using the data for March 1978 in Table 5-4, the forecast for April would be computed as follows:

$$S_{\text{Mar.}} = 0.2 \times 19.73 + 0.8\ (20.9773 + 0.0990)$$
$$= 3.9460 + 16.8610 = 20.8070$$

$$T_{\text{Mar.}} = 0.1\ (20.8070 - 20.9773) + 0.9 \times 0.0990$$
$$= -0.0170 + 0.0891 = 0.0721$$

$$F_{\text{Apr.}} = 20.8070 + 0.0720 = 20.8791$$

The model using a direct estimation of trend is valuable when trend and random variation is present in the data. In addition, the concept is useful in some of the more complex models that combine estimates of average demand, trend, and seasonal components.

The simple exponentially smoothed forecast and the trend model forecast are

TABLE 5-4 **Sample Computations for Model Using Direct Estimation of Trend**
$\alpha = 0.2$, $\beta = 0.1$

Date	Actuals, D_t	S_t, Equation 4	T_t, Equation 3	F_{t+1}, Equation 5
Initial	—	20.00	0	—
1978, Jan.	19.36	19.8720	− 0.0128	—
Feb.	25.45	20.9773	0.0990	19.86
Mar.	19.73	20.8070	0.0721	21.08
Apr.	21.48	20.9993	0.0841	20.88
May	20.77	21.0207	0.0778	21.08
June	25.42	21.9628	0.01642	21.10
July	23.79	—	—	22.13

plotted in relation to actual demand in Figure 5-3 for the program usage data. Since the raw data exhibits a trend, the simple model lags in its response. The trend model corrects for this lag and lies above the simple model forecast after initial conditions lose their effect.

For this model applied to the three-year record of the computer usage data, with $\alpha = 0.2$ and $\beta = 0.1$, AE = 0.33, MAD = 2.34, and MSE = 9.06. The choice of the two smoothing constants has an important influence on forecast errors. Holding $\alpha = 0.2$, the same computer usage data were run for values of $\beta = 0.01, 0.05, 0.1, 0.2, 0.3$. Subsamples yielded MAD = 2.23, 2.16, 2.34, 2.23, and 2.44, respectively. Since the objective in developing a forecasting system is to minimize forecast errors, testing the sensitivity of errors to parameter values is an important step in refining a system.

Model for Seasonals

The basis for taking direct account of seasonal variations is to construct a seasonal index using the preceding year's data. For example, if we take the 1978 computer usage actual demands and divide each monthly demand by the annual average, we have a set of indexes. The average demand during 1978 was 24.07, so the index for

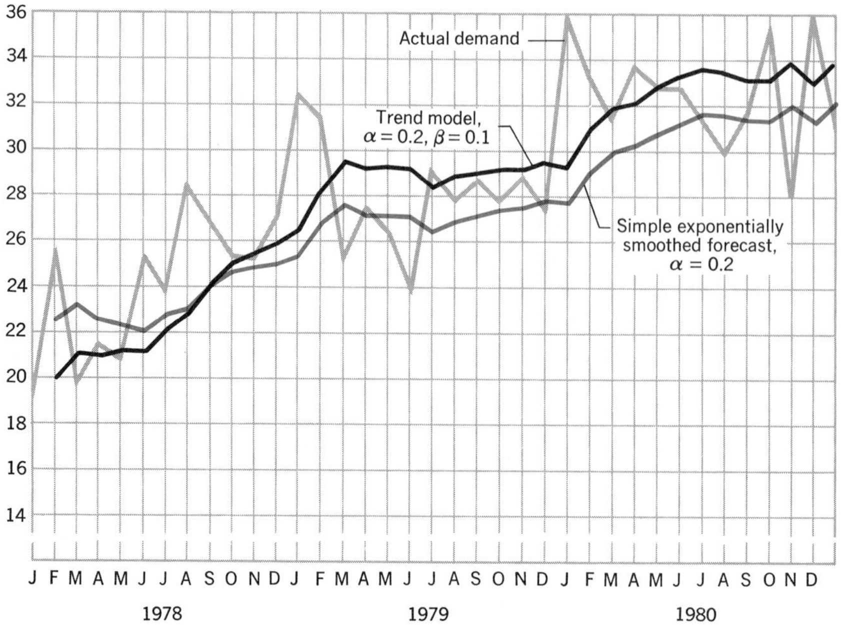

FIGURE 5-3
Forecasts of program usage by simple exponential smoothing and by the trend model.

TABLE 5-5 **Sample Computations for Seasonal Model**
$\alpha = 0.1$, $\gamma = 0.3$

Date	Actuals, D_t	S_t, Equation 6	I_t, Equation 7	F_{t+1}, Equation 8
1978, Jan.	19.36		0.804	
Feb.	25.45		1.057	
Mar.	19.73		0.819	
Apr.	21.48		0.892	
May	20.77		0.863	
June	25.42		1.056	
July	23.79		0.988	
Aug.	28.35		1.178	
Sep.	26.80		1.113	
Oct.	25.32		1.052	
Nov.	25.22		1.048	
Dec.	27.14	30.00 (assumed)	1.128	—
1979, Jan.	32.52	31.0447	0.877	—
Feb.	31.33	30.9043	1.044	32.81
Mar.	25.32	30.9054	0.819	25.31
Apr.	27.53	30.9012	0.892	27.57
May	26.38	30.8679	0.860	26.67
June	23.72	30.0273	0.976	32.60
July	29.14	—	—	29.67

January would be 19.36/24.07 = 0.804, and for February, 25.45/24.07 = 1.057. These initial indexes for 1978 are shown in Table 5-5.

Given the indexes, we can normalize actual demand figures by dividing by the previous year's index in that period, I_{t-L} (L is the number of periods in one cycle, 12 if data are by months, 4 if data are quarterly). Therefore, if actual demand for February 1979 is $D_t = 31.33$, we divide by the index for February in the previous year, 1978, $I_{t-12} = 1.057$ to obtain 31.33/1.057 = 29.64. The effect of this process is to deseasonalize by decreasing adjusted demand during high demand periods and increasing it during low demand periods. The deseasonalized smoothed average S_t is then

$$S_t = \alpha(D_t/I_{t-L}) + (1 - \alpha)S_{t-1} \tag{6}$$

However, the seasonal indexes for 1978 are reflective of only that year's experience. If the seasonal cycle repeated itself accurately each year, using the 1978 indexes each year would be appropriate. Since random variations are a component, however, a single year's history as a basis for seasonal indexes is normally replaced by some averaging process, such as exponentially weighted averaging. Therefore, we use the following equation to update the seasonal indexes, where γ (gamma) is the smoothing constant for seasonal indexes:

$$I_t = \gamma(D_t/S_t) + (1 - \gamma)I_{t-L} \tag{7}$$

The actual demand D_t is divided by the new smoothed average S_t, computed by Equation 6 to reflect the amount by which D_t exceeds or falls short of the deseasonalized average. This variation from the deseasonalized average is weighted by the smoothing constant. The old seasonal index is last year's index and is weighted by $1 - \gamma$. The new index I_t is stored, to be used in computations for S_t and I_t next year.

The result, after the process has been in operation for several years, is that each seasonal index is based on seasonal variation that occurred L, $2L$, $3L$, and so on, periods ago. The most recent data are weighted more heavily, depending on the value of the smoothing constant γ.

In order to forecast for the upcoming period, $t + 1$, we carry forward the most current smoothed average, S_t, but modify it by the seasonal index for the upcoming period, I_{t-L+1}. Equation 8 has the effect of reseasonalizing the formerly deseasonalized smoothed average.

$$F_{t+1} = S_t I_{t-L+1} \tag{8}$$

Table 5-5 applies the seasonal model to the first several months of the computer usage data, using $\alpha = 0.1$ and $\gamma = 0.3$. In applying the model, the current smoothed average S_t is computed first by Equation 6, since this value is required to update the seasonal indexes by Equation 7 and to compute the forecast by Equation 8.

Using the data for April 1979 in Table 5-5, the forecast for May would be computed as follows:

$$\begin{aligned} S_{\text{Apr.}} &= 0.1\left(\frac{27.53}{0.892}\right) + 0.9 \times 30.9054 \\ &= 3.0863 + 27.8149 = 30.9012 \\ I_{\text{Apr.}} &= 0.3\left(\frac{27.53}{30.9012}\right) + 0.7 \times 0.892 \\ &= 0.2673 + 0.6244 = 0.8917 \\ F_{\text{May}} &= 30.9012 \times 0.863 = 26.667 \end{aligned}$$

For the seasonal model applied to the three-year record of the computer usage data, AE $= -0.15$, MAD $= 3.38$, and MSE $= 20.53$ ($\alpha = 0.1$, $\gamma = 0.3$).

Forecasting accuracy is not as good with the seasonal model. Why? First, it does not account for trend, which is a factor in the data. Second, an entire year of the data is consumed in order to initialize the seasonal indexes. The remaining two-year sample may contain extreme values that carry larger weight in the smaller sample. Third, the seasonal cycles in the computer usage data may not be stable from year to year; the company was a new one with a short history.

As with the trend model, the choice of the smoothing constants affects forecasting errors. Subsamples used to compute MAD for several combinations of values of α and γ yield the following:

	γ				
α	0.01	0.05	0.10	0.20	0.30
0.1	3.67	3.60	3.54	3.44	3.38
0.2	3.65	3.75	3.73	3.60	3.37

Model for Trend and Seasonals

As might be expected, one can combine the trend model and the seasonal model. The equations to update smoothed trend and seasonals are the same, but the equation to compute the current value for the smoothed average, S_t, must reflect both the trend and seasonal variations.

$$S_t = \alpha(D_t/I_{t-L}) + (1 - \alpha)(S_{t-1} + T_{t-1}) \qquad (9)$$

The trend, and seasonal index equations are

$$T_t = \beta(S_t - S_{t-1}) + (1 - \beta)T_{t-1} \qquad (10)$$
$$I_t = \gamma(D_t/S_t) + (1 - \gamma)I_{t-L} \qquad (11)$$

Finally, to forecast for the upcoming period, $t + 1$, we combine the elements of Equations 5 and 8,

$$F_{t+1} = (S_t + T_t)I_{t-L+1} \qquad (12)$$

The value of S_t using Equation 9 must be computed first, since it is used in Equations 10, 11, and 12. As before, computing the updated seasonal index produces an index to be stored for use a year hence.

Table 5-6 applies the trend and seasonal model to the first several months of the computer usage data, using smoothing constants of $\alpha = 0.2$, $\beta = 0.3$, and $\gamma = 0.1$.

TABLE 5-6 **Sample Computations for the Trend and Seasonal Model**
$\alpha = 0.2$, $\beta = 0.3$, and $\gamma = 0.1$

Date	Actuals D_t	S_t, Equation 9	T_t, Equation 10	I_t, Equation 11	F_{t+1}, Equation 12
1978, Jan.	19.36			0.804	
Feb.	25.45			1.057	
Mar.	19.73			0.819	
Apr.	21.48			0.892	
May	20.77			0.863	
June	25.42			1.056	
July	23.79			0.988	
Aug.	28.35			1.178	
Sep.	26.80			1.113	
Oct.	25.32			1.052	
Nov.	25.22			1.048	
Dec.	27.14	30 (assumed)	1.0 (assumed)	1.128	—
1979, Jan.	32.52	32.8896	1.5669	0.823	—
Feb.	31.33	33.4932	1.2779	1.045	36.42
Mar.	25.32	34.0000	1.0465	0.812	28.48
Apr.	27.53	34.2099	0.7955	0.883	31.26
May	26.38	34.1179	0.5293	0.854	30.21
June	23.72	32.2102	−0.2017	1.024	36.59
July	29.14	—	—	0.982	31.62

The initial seasonal indexes are those used to intialize the seasonal model in Table 5-5.

Using the data for May 1979 in Table 5-6 to forecast for June, sample computations are

$$
\begin{aligned}
S_{\text{May}} &= 0.2(26.38/0.863) + 0.8(34.2099 + 0.7955) \\
&= 6.1136 + 28.0043 = 34.1179 \\
T_{\text{May}} &= 0.3(34.1179 - 34.2099) + 0.7 \times 0.7955 \\
&= -0.0276 + 0.5569 = 0.5293 \\
I_{\text{May}} &= 0.1(26.38/34.1179) + 0.9 \times 0.863 \\
&= 0.0773 + 0.7767 = 0.8540 \\
F_{\text{June}} &= (34.1179 + 0.5293)1.056 = 36.5874
\end{aligned}
$$

With three smoothing constants, the number of possible combinations increases substantially. Berry and Bliemel [1974] show how computer search methods can be used in selecting optimal combinations of the smoothing constants.

Adaptive Methods

As we have noted, it is common to use fairly small values of α in exponential smoothing systems in order to filter out random variations in demand. When actual demand rates increase or decrease gradually, such forecasting systems can track the changes rather well. If demand changes suddenly, however, a forecasting system using a small value of α will lag substantially behind the actual change. Thus, adaptive response systems have been proposed.

The basic idea of adaptive smoothing systems is to monitor the forecast error and, based on preset rules, to react to large errors by increasing the value of α. The effect of increasing α is to change the weights assigned to past data, increasing the weight placed on the most recent data. The forecast is then computed using the new weights. For example, if a step change in demand were to occur because of a radical change in the market, a large error would result. The large error signals that α should increase, giving greater weight to current demand. The forecast would then reflect the change in actual demand. When actual demand stabilizes at the new level, adaptive systems would reset the value of α to a lower level that filters out random variations effectively. A number of approaches to adaptive response systems have been proposed.

Although adaptive systems have an intuitively appealing rationale, at this writing their performance as compared to standard exponential smoothing forecast systems is in doubt. Two studies have been made that have produced conflicting results. Whybark [1972] tested six different forecasting models, four of which were adaptive. In general, the adaptive systems performed better than the simple exponential smoothing system.

A later study by Dancer and Gray [1977] compared the performance of a single exponential smoothing model, a double exponential smoothing model,* and one

* Double exponential smoothing is an alternate model that takes account of trend. It functions by processing the value of S_t computed by Equation 2 by smoothing it a second time, using Equation 2.

taking account of seasonal demand patterns, with the two adaptive systems that performed best in Whybark's study. Dancer and Gray found no significant differences in the performance of the adaptive systems compared to the standard models. These results are consistent with the results obtained by Church [1973] in developing a short-term forecasting system in connection with staff scheduling in a telephone business office. Additional research will be required to clarify the relative performance of adaptive models.

FOURIER SERIES FORECASTING METHODS

The speed, economy, and storage capacity of present-day computers have made it feasible to use fairly sophisticated mathematical models for forecasting. One such model is the least squares Fourier series forecasting model.

The mathematical background for this methodology was established by Joseph Fourier, a French physicist and mathematician. Fourier demonstrated that *any* periodic (i.e., seasonal) function that is finite, single-valued, and continuous over the period (season) may be represented by a mathematical series consisting of a constant term plus the sum of several sine and cosine terms.* The series is expressed as an infinite series, because in theory an infinite number of terms is required to duplicate mathematically the desired function with complete accuracy. As a practical matter, a finite number of terms in the range of 4 to 14 seems to produce excellent results in seasonal forecasting situations.

Figure 5-4 shows the result of applying the Fourier series forecasting model to the data for computer program usage, using six terms in the model. Note that the forecast generally follows the actual data; however, the peaks and valleys in actual demand are smoothed. For Figure 5-4, MAD = 1.86. (Note that the decimal point has not been printed for the actual demand, forecast demand, and forecast error. The standard output for the program includes the histogram of errors, as well as the summary statistics on forecast errors.)

Increasing the number of terms in the model will normally improve the fit to historical data; however, computer costs will increase. As a general rule the minimum number of model terms is equal to two times the number of peaks in a seasonal cycle plus 2. Looking at Figure 5-1, we would assume that the minimum number of terms for the computer usage example would be four. Figure 5-5 shows the output of the program when the number of terms is increased to 12. The fit has been improved and the variability of the forecast errors reduced. For the 12-term model, MAD = 1.76.

* The equation for the Fourier series is

$$F_t = a_1 + a_2 \sin wt + a_3 \cos wt + a_4 \sin wt + a_5 \cos wt + a_6 \sin wt + \ldots$$

where F_t = the numerical value of the series computed at time t
a_1 = a constant term
$a_2, a_3, \ldots$ = coefficients defining the amplitude of the harmonics
$w = 2\pi/T = 6.2832/T$
T = the length of the period (i.e., the number of forecast intervals per year)

FORECAST MODEL RESULTS

FCST NO.	ACTUAL DEMAND	FORECAST DEMAND	FORECAST ERROR
1	2425	2512	-87
2	1973	2408	-435
3	2148	2283	-135
4	2077	2259	-182
5	2542	2359	183
6	2379	2488	-109
7	2835	2540	295
8	2680	2501	179
9	2532	2463	69
10	2522	2528	-6
11	2714	2694	20
12	3252	2848	404
13	3133	2877	256
14	2532	2773	-241
15	2753	2648	105
16	2638	2624	14
17	2372	2724	-352
18	2914	2854	60
19	2775	2905	-130
20	2870	2866	4
21	2783	2828	-45
22	2885	2894	-9
23	2739	3059	-320
24	3580	3213	367
25	3310	3242	68
26	3139	3139	0
27	3362	3013	349
28	3266	2989	277
29	3265	3089	176
30	3137	3219	-82
31	2995	3270	-275
32	3156	3231	-75
33	3520	3194	326
34	2836	3259	-423
35	3604	3424	180
36	3152	3578	-426
37		3607	
38		3504	
39		3378	
40		3354	
41		3454	
42		3584	
43		3635	
44		3596	
45		3559	
46		3624	
47		3789	
48		3944	

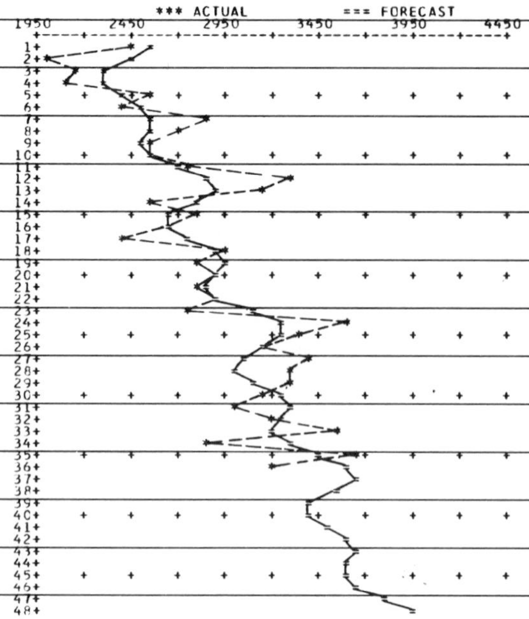

HISTOGRAM OF FORECAST ERRORS

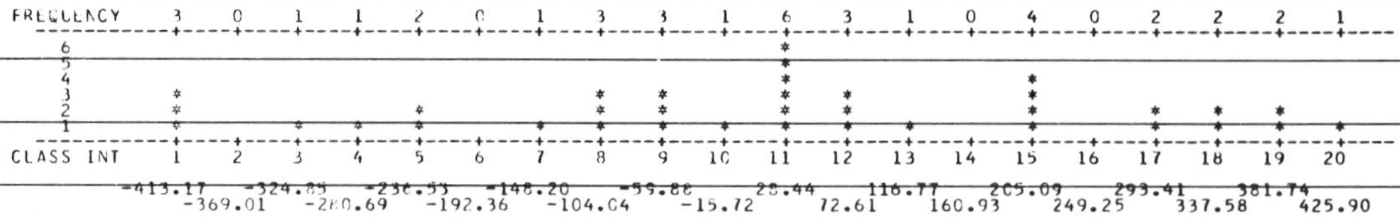

CLASS INTERVAL WIDTH	=	44.161682
MAXIMUM VALUE	=	403.817139
MINIMUM VALUE	=	-435.254883
MEAN VALUE	=	0.007270
STANDARD DEVIATION	=	234.283096

DISTRIBUTION OF FORECAST ERRORS

	NO. VALUES	PERCENT
MEAN -4 STD DEV	0	0.0
MEAN -3 STD DEV	0	0.0
MEAN -2 STD DEV	7	19.4
MEAN -1 STD DEV	10	27.8
MEAN 1 STD DEV	12	33.3
MEAN 2 STD DEV	7	19.4
MEAN 3 STD DEV	0	0.0
MEAN 4 STD DEV	0	0.0
TOTAL	36	100.0

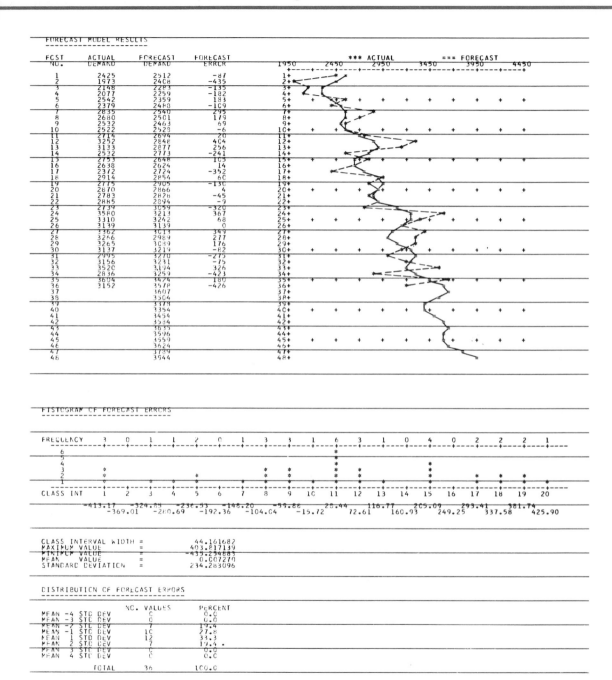

FIGURE 5-4
Computer output from six-term Fourier series forecasting model for the program usage data of Figure 5-1.

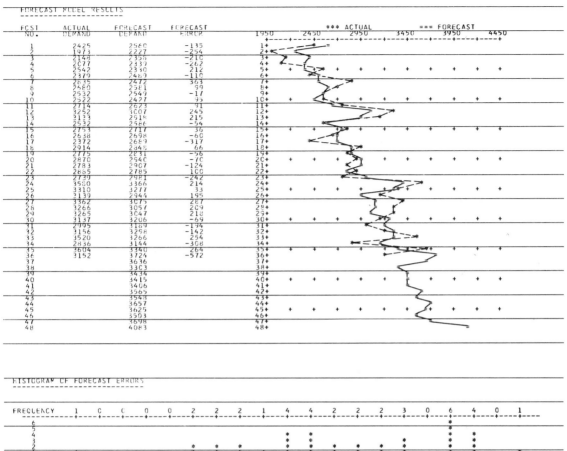

FORECAST MODEL RESULTS

FCST NO.	ACTUAL DEMAND	FORECAST DEMAND	FORECAST ERROR
1	2425	2560	-135
2	1973	2227	-254
3	2148	2358	-210
4	2077	2339	-262
5	2542	2330	212
6	2379	2489	-110
7	2835	2472	363
8	2680	2581	99
9	2532	2549	-17
10	2522	2427	95
11	2714	2623	91
12	3252	3007	245
13	3133	2918	215
14	2532	2586	-54
15	2753	2717	36
16	2638	2698	-60
17	2372	2689	-317
18	2914	2848	66
19	2775	2831	-56
20	2870	2940	-70
21	2783	2907	-124
22	2885	2785	100
23	2739	2981	-242
24	3580	3366	214
25	3310	3277	33
26	3139	2944	195
27	3362	3075	287
28	3266	3057	209
29	3265	3047	218
30	3137	3206	-69
31	2995	3189	-194
32	3156	3298	-142
33	3520	3266	254
34	2836	3144	-308
35	3604	3340	264
36	3152	3724	-572
37		3636	
38		3303	
39		3434	
40		3415	
41		3406	
42		3565	
43		3548	
44		3657	
45		3625	
46		3503	
47		3698	
48		4083	

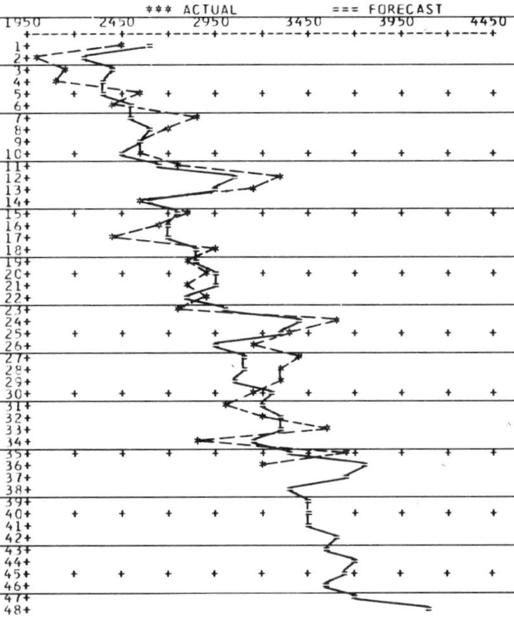

HISTOGRAM OF FORECAST ERRORS

```
FREQUENCY    1    0    0    0    0    2    2    2    1    4    4    2    2    2    3    0    6    4    0    1
-----------+----+----+----+----+----+----+----+----+----+----+----+----+----+----+----+----+----+----+----+----
    6                                                                                        *
    5                                                                                        *
    4                                                       *    *                           *    *
    3                                                       *    *                   *       *    *
    2                                  *    *    *          *    *    *    *    *    *       *    *
    1        *                         *    *    *    *     *    *    *    *    *    *       *    *         *
-----------+----+----+----+----+----+----+----+----+----+----+----+----+----+----+----+----+----+----+----+----
CLASS INT    1    2    3    4    5    6    7    8    9   10   11   12   13   14   15   16   17   18   19   20

     -547.67   -449.22   -350.77   -252.33   -153.88    -55.44     43.01    141.46    239.90    338.35
         -498.44   -400.00   -301.55   -203.10   -104.66     -6.21     92.23    190.68    289.13    387.57
```

CLASS INTERVAL WIDTH	=	49.223114
MAXIMUM VALUE	=	362.960938
MINIMUM VALUE	=	-572.278320
MEAN VALUE	=	0.004557
STANDARD DEVIATION	=	213.835175

DISTRIBUTION OF FORECAST ERRORS

	NO. VALUES	PERCENT
MEAN -4 STD DEV	0	0.0
MEAN -3 STD DEV	1	2.8
MEAN -2 STD DEV	5	13.9
MEAN -1 STD DEV	12	33.3
MEAN 1 STD DEV	10	27.8
MEAN 2 STD DEV	8	22.2
MEAN 3 STD DEV	0	0.0
MEAN 4 STD DEV	0	0.0
TOTAL	36	100.0

FIGURE 5-5
Computer output from twelve-term Fourier series forecasting model.

CAUSAL FORECASTING METHODS

When we have enough historical data and experience, it may be possible to relate forecasts to the factors in the economy that *cause* the trends, seasonals, and fluctuations. Thus, if we can *measure* the causal factors and we have determined their *relationships* to the product or service of interest, then we can compute forecasts of considerable accuracy.

The factors that enter causal models are of several types: disposable income, new marriages, housing starts, inventories, cost-of-living indexes, as well as predictions of dynamic factors and/or disturbances, such as strikes, actions of competitors, and sales promotion campaigns. The causal forecasting model expresses mathematical relationships between the causal factors and the demand for the item being forecast. As indicated in Table 5-1, there are two general types of causal models, and the costs range from medium to high for installation and operation.

Regression Analysis

Forecasting based on regression methods establishes a forecasting function called a "regression equation." The equation expresses the series to be forecast in terms of other series that presumably control or cause the sales to increase or decrease. The rationale can be general or specific. For example, in furniture sales we might postulate that sales are related to disposable personal income: if disposable income is up, sales will increase, and if people have less money to spend, sales will be down. Establishing the empirical relationship is accomplished through the regression equation. To take additional factors, we might postulate that furniture sales are controlled to some extent by the number of new marriages and the number of new housing starts. These are both specific indicators of possible demand for furniture.

Table 5-7 gives data on these three independent variables—housing starts, disposable income, and new marriages—and on sales of a hypothetical furniture company called the Cherryoak Company. We propose to build a relationship between the observed variables and company sales, where sales are dependent on, or caused by, the observed variables. Therefore, sales is termed the "dependent" variable, and the observed variables are called the "independent" variables. The correlation coefficients between sales (S) and each of the independent variables are:

1.	Disposable personal income (I)	0.805
2.	Housing starts (H)	0.435
3.	New marriages (M)	0.416

Since disposable income (I) correlates most strongly with company sales, let us start with it as an example. Using regression analysis we can determine the straight line that best fits the data expressing the relationship between sales (S) and disposable

TABLE 5-7 **Data for 24 Years (1947-1970) Used in Performing Regression Analysis to Forecast 1971 Sales of Cherryoak Company**

Year	Housing Starts (*H*) (thousands)	Disposable Personal Income (*I*) ($ billions)	New Marriages (*M*) (thousands)	Company Sales (*S*) ($ millions)	Time (*T*)
1947	744	158.9	2291	92.920	1
1948	942	169.5	1991	122.440	2
1949	1033	188.3	1811	125.570	3
1950	1138	187.2	1580	110.460	4
1951	1549	205.8	1667	139.400	5
1952	1211	224.9	1595	154.020	6
1953	1251	235.0	1539	157.590	7
1954	1225	247.9	1546	152.230	8
1955	1354	254.4	1490	139.130	9
1956	1475	274.4	1531	156.330	10
1957	1240	292.9	1585	140.470	11
1958	1157	308.5	1518	128.240	12
1959	1341	318.8	1451	117.450	13
1960	1531	337.7	1494	132.640	14
1961	1274	350.0	1527	126.160	15
1962	1327	364.4	1547	116.990	16
1963	1469	385.3	1580	123.900	17
1964	1615	404.6	1654	141.320	18
1965	1538	436.6	1719	156.710	19
1966	1488	469.1	1789	171.930	20
1967	1173	505.3	1844	184.790	21
1968	1299	546.3	1913	102.700	22
1969	1524	590.0	2059	237.340	23
1970	1479	629.6	2132	254.930	24

Note. Company sales and disposable per-capita income have been adjusted for the effect of inflation and appear in constant 1959 dollars.

SOURCE. G. C. Parker and E. L. Segura, "How to Get a Better Forecast," *Harvard Business Review*, March–April 1971, based on data from *Statistical Abstract of the United States*, Bureau of the Census, Washington, D.C.

income (I). From statistics we know that the regression equation represents a straight line that minimizes the square of the deviations from it and sets the sum of the simple deviations to zero. The regression equation for the data of company sales (S) versus disposable income (I) is

$$S = 72.5 + 0.23I \tag{13}$$

where the coefficient, 72.5, is the y axis intercept, and the slope of the straight line is 0.23. The form of the equation is the standard format of the equation of a straight line, $y = a + bx$, where y is the dependent variable, x the independent variable, a the y intercept, and b the slope. In regression analysis a and b are termed the "regression coefficients" and are the parameters that specify the equation.

The regression line is plotted in Figure 5-6, showing some specific points for selected years. These points illustrate the kinds of forecast errors that would have resulted if one had used this equation to forecast Cherryoak furniture sales. To use the regression equation to forecast sales, one simply inserts the value of I and computes sales S. For example, if $I = 700$, then the forecaster could compute the value of S as $S = 72.5 + 0.23 \times 700 = \$233.5$ (million).

Reliability of Forecast. There are a number of statistical tests that can be performed to help determine how good the regression equation is as a forecasting device. Data resulting from these statistical tests are commonly generated automatically in standard regression analysis computer programs. Our interest is particularly in the coefficients of determination and the standard error of estimate. In addition, there are important statistical tests concerning the significance of the regression coefficients that we will not discuss.

The coefficient of determination is simply r^2, the correlation coefficient squared.

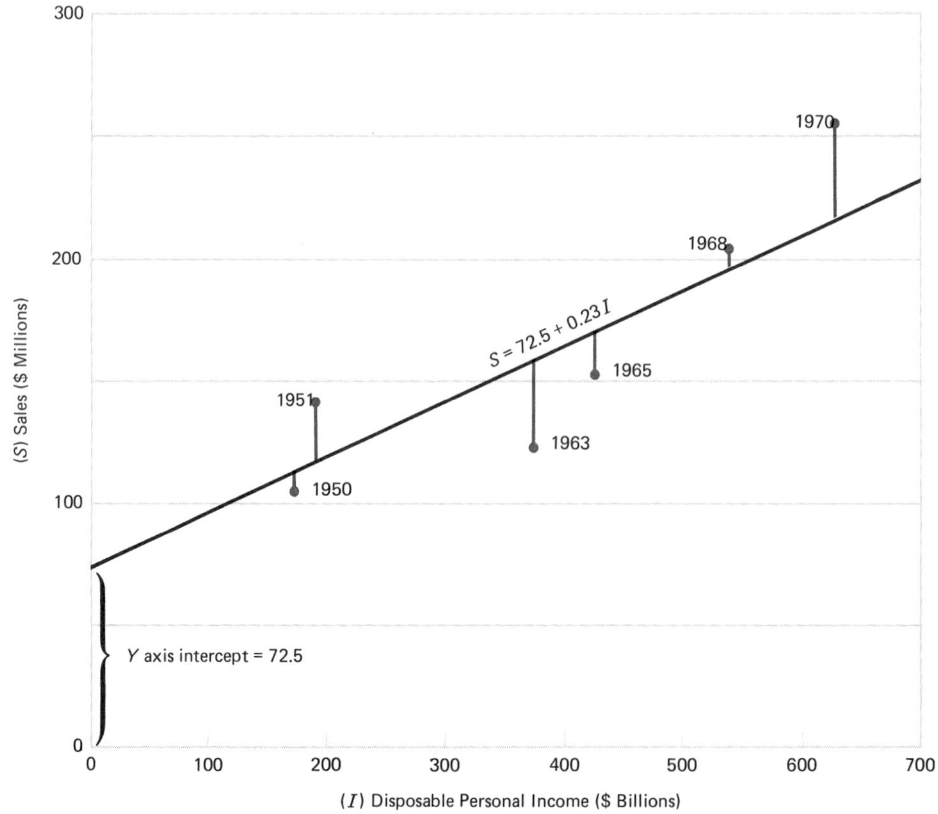

FIGURE 5-6
Simple regression line for sales dependent on disposable income. Data from Table 5-7.

For our example, $r^2 = 0.805^2 = 0.65$. The coefficient of determination states the proportion of the variation in the regression equation that is explained by the independent variable. For our equation, 65 percent of the variation in sales is controlled by variation in I, and 35 percent is unexplained. Thus, we can expect large forecast errors if we use Equation 13. Apparently, other variables account for a substantial fraction of the changes in S that actually occur.

The standard error of estimate indicates the expected range of variation from the regression line of any forecast made. For example, the standard error of estimate for our data and Equation 13 is 38.7. Since we assume a normal distribution of sales for each value of I, this means that we can expect with some confidence that two-thirds of the time our estimate of S will be in the range of $\pm$ 38.7. Therefore, if $I = 700$, then $S = 233.5$, as computed previously. However, with a standard error of estimate of 38.7, we are actually stating that two-thirds of the time we would expect the actual value of S to be in the range of 194.8 to 272.2, a rather broad range.

Obviously, we need to improve the forecasting ability of Equation 13, and we can accomplish this by including other causal factors in the regression equation.

Multiple Regression

The general concepts of simple regression analysis can be extended to include the effects of several causal factors through multiple regression analysis. For the data of Table 5-7, Parker and Segura [1971] show that the regression equation would be

$$S = 49.85 - 0.068M + 0.036H + 1.22I - 19.54T \qquad (14)$$

where S = gross sales per year
49.85 = base sales, or starting point from which other factors have an influence
M = new marriages during the year
H = new housing starts during the year
I = disposable personal income during the year
T = time trend ($T = 1, 2, 3, \ldots, n$)

and the coefficients that precede each of the causal factors represent the amount of influence on sales of each of the causal factors M, H, I, and T.

The coefficient of determination for Equation 14 is 0.92, and the standard error of estimate is 11.9, indicating that the value of the equation as a forecasting mechanism has been increased substantially over Equation 13.

Equation 14 is then improved as a forecasting mechanism by making additional changes. The factor of new marriages is dropped and last year's sales (S_{t-1} is substituted in order to improve the overall forecasting accuracy. Also, last year's housing starts (H_{t-1}) is substituted for H, since this allows for the lag we would expect between construction time and the time home furnishing expenditures might be made. The revised equation is

$$S = -33.51 + 0.373S_{t-1} + 0.033H_{t-1} + 0.672I_t - 11.03T \qquad (15)$$

Forecasting accuracy has improved again with $r^2 = 0.95$, and the standard error of estimate = 9.7. Table 5-8 summarizes the record of actual versus forecasted sales and forecast errors for the entire 24-year period, and Figure 5-7 shows a comparative graph of actual versus forecasted sales.

When forecasts must be made for longer terms, as when new products and services are contemplated, or when new facility locations and capacities are being considered, multiple regression is a logical forecasting method. It requires considerable time and cost, since various hypotheses regarding the effect of variables may need to be tested. However, standard computing programs for multiple regression are now widely available that ease the burden and reduce the cost of application.

A considerable historical record is necessary for regression analysis to have validity. As a rule of thumb, Parker and Segura state that a five-year record is needed for one

TABLE 5-8

Differences in Actual Sales of Cherryoak Company, 1947-1970, and Sales Forecasted by Multiple Regression
(In millions of dollars)

Year	Actual Sales	Predicted Sales	Difference	Ratio of Actual Sales to Predicted Sales
1947	92.29	93.04	−0.75	0.99
1948	122.44	117.72	4.72	1.04
1949	125.57	136.91	−11.34	0.92
1950	110.46	129.33	−18.87	0.85
1951	139.40	128.65	10.75	1.08
1952	154.02	154.90	−0.88	0.99
1953	157.59	144.88	12.71	1.09
1954	152.23	145.17	7.06	1.05
1955	139.13	135.64	3.49	1.02
1956	156.33	137.44	18.89	1.13
1957	140.47	149.27	−8.80	0.94
1958	128.24	134.99	−6.75	0.95
1959	117.45	123.56	−6.11	0.95
1960	132.64	127.31	5.33	1.04
1961	126.16	136.52	−10.36	0.92
1962	116.99	124.20	−7.21	0.94
1963	123.90	125.56	−1.66	0.99
1964	141.32	134.79	6.53	1.05
1965	156.71	156.61	0.10	1.00
1966	171.93	170.59	1.34	1.01
1967	184.79	187.90	−3.11	0.98
1968	202.70	198.75	3.95	1.02
1969	237.34	227.95	9.39	1.04
1970	254.93	263.92	−8.99	0.96

SOURCE. G. C. Parker and E. L. Segura, "How to Get a Better Forecast," *Harvard Business Review,* March–April 1971.

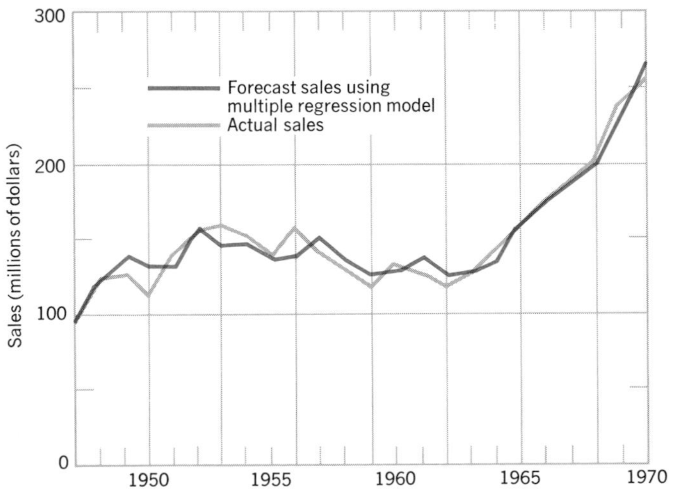

FIGURE 5-7
Forecast versus actual sales. Data plotted from Table 5-8.

independent variable, eight years for two independent variables, and a longer history for three or more independent variables. These data requirements are often severe limitations to application.

Furthermore, there are four very important assumptions made in regression analysis that should be met. First, there is the assumption of linearity that states that the dependent variable is linearly related to the independent variables. Where the linear relationship does not hold, transformations can often be made that make it possible to meet the requirements of this assumption. The second basic assumption in regression analysis is that the variance of errors is constant. The third assumption is that errors from period to period are independent of each other, or not autocorrelated. Finally, regression analysis assumes that the errors are normally distributed. The nature and importance of these assumptions is covered in greater detail in Benton [1972], and Makridakis and Wheelwright [1978]. Obviously, considerable knowledge of statistical methods is required for the appropriate application of regression analysis.

Beyond possibly ignoring one or more of the important assumptions, one of the great dangers in misapplying regression analysis is in assuming that a good fit to historical data guarantees that the regression equation will be a good forecasting device. The regression equation itself should be an expression of a good causal theory relating to factors in the regression model. In addition, we also need an understanding of the potential importance of factors that are not included in the model.

One of the differences, then, between time series forecasting models and causal methods is that time series accept increases or decreases in demand in an unbiased way, being rather mindless about the reasons for the increase or decrease. On the other hand, causal methods demand an explanation within the rationale of the forecasting system for demand changes that occur.

Econometric Forecasting Methods

In simplest terms, econometric forecasting methods are an extension of regression analysis and include a system of simultaneous regression equations. If, for example, in Equation 15 we attempted to include the effect of price and advertising, then there is the possibility of an interdependence, where our own sales can have an effect on these factors as well as vice versa.

For example, assume that sales is a function of GNP, price, and advertising. In regression terms we would assume that all three independent variables are exogenous to the system and thus are not influenced by the level of sales itself or by one another. This is a fair assumption as far as GNP is concerned. But if we consider price and advertising, the same assumption may not be valid. For example, if the per unit cost is of some quadratic form, a different level of sales will result in a different level of cost. Furthermore, advertising expenditures will influence the price of the product, since production and selling costs influence the per unit price. The price, in turn, is influenced by the magnitude of sales, which can also influence the level of advertising. All of this points to the interdependence of all four of the variables in our equation. When this interdependence is at all strong, regression analysis cannot be used. If we want to be accurate, we must express this sales relationship by developing a system of four simultaneous equations that can deal with the interdependence directly.

Thus in econometric form, we have

Sales	= f(GNP, price, advertising)
Cost	= f(production and inventory levels)
Selling expenses	= f(advertising, and other selling expenses)
Price	= f(cost + selling expenses)

Instead of one relationship, we now have four. "As in regression analysis, we must (1) determine the functional form of each of the equations, (2) estimate in a simultaneous manner the values of their parameters, and (3) test for the statistical significance of the results and the validity of the assumptions."

Econometric models have been used to date largely in connection with relatively mature products where a considerable historical record is available and in industry and broad economic forecasts. For example, the Corning Glass Works developed econometric models to forecast television tube sales [Chambers et al., 1971]. These models were used to forecast sales six months to two years in the future to help spot turning points enough in advance to assist decisions for production and employment planning.

Industry econometric models have been developed to forecast activity in the forest products industry. Also, the economic forecasting models developed at UCLA and the Wharton School are econometric models.

IMPLICATIONS FOR THE MANAGER

If there is a single, most important set of data for managers, it is forecast data. Virtually every really important decision in operations depends in some measure on a forecast of demand. For example, the broad aggregate decisions concerning hiring and layoff of personnel, use of facilities, overtime, and the accumulation of seasonal inventories are developed in relation to a forecast of expected demand. In service industries, such as hospitals or telephone exchange offices, it is necessary to provide service 24 hours per day, seven days per week. Scheduling personnel to perform these services, who normally work only 40 hours per week, depends on the demand for service. Inventories of raw materials, work in process, and finished goods are of great importance to manufacturing systems. Inventories of supply items are important to the smooth functioning of most service systems. Constructing rational inventory policies depends on forecasts of usage.

Managers are keenly aware of their dependence on forecasts. Indeed, a great deal of executive time is spent in concern over trends in economic and political affairs and how events will affect demand for their products or services. An issue is the relative value of executive opinion versus quantitative forecasting methods.

Executives are perhaps most sensitive to events that may have a significant impact on demand. Quantitative techniques are probably least effective in "calling the turn" that may result in sharply higher or lower demand, since quantitative methods depend on historical data.

Management science has attempted to automate sensitivity to sharp changes in demand through adaptive systems and through causal models. As we noted in our discussion of adaptive forecasting systems, success has been illusive, and it is not clear that the performance of adaptive systems is superior. Causal models suffer least from sluggishness, since they are constructed to reflect the effects on demand that are caused by indicators in an industry or in the economy.

The manager's choice of forecasting models needs to take account of the forecasting range required, accuracy requirements, and cost. The exponential smoothing methods are particularly adaptable to short-range forecasting and are inexpensive to install. The annual cost of maintaining such systems increases as the forecasting range is shortened, simply because the system must be updated more often.

Exponential smoothing forecasts are particularly adaptable to computer systems, since a minimum number of data items need to be stored for each item to be forecast. The data storage requirements increase as one moves from the simple to the more complex models that take account of trend and seasonal variations. Whether or not the additional complexity is justified depends on the balance between the forecasting accuracy needed and cost. The fact that exponential smoothing systems are adaptable to electronic computing makes them particularly attractive for forecasting large numbers of items as often is required for inventory controls.

When the forecasting horizon requirement is somewhat longer, and the need for accuracy is greater, causal methods should be favored by management. These conditions are common when managers must make broad level plans for the allocation of capacity sources that may involve the expansion or contraction of short-term capacity. These aggregate plans are discussed in the next chapter and may involve hiring or layoff of personnel, use of overtime, the accumulation of seasonal inventories, the use of back ordering or lost sales, or the use of subcontracting. These are important managerial decisions, and forecasting accuracy requirements may justify the costs of causal methods.

Box and Jenkins [1970] have developed a highly sophisticated forecasting methodology in which a general class of forecasting methods is postulated for a particular situation. In stage one, a specific model is tentatively identified as the forecasting method best suited to that situation. The forecasting models may range from moving averages through exponential and adaptive methods to regression analysis. In stage two, the postulated model is fit to the available historical data, and a check is run to determine whether or not the postulated model is adequate. Various statistical tests are involved. If the postulated model is rejected, stage two is used to identify an alternate model, which is then tested. The process is repeated until a satisfactory model has been identified.

Makridakis, Hodgsdon, and Wheelwright [1974] have developed a generalized interactive forecasting system within the framework of the Box-Jenkins philosophy. The interactive computing system is divided into two main segments. The first segment allows the user to perform preliminary analyses in order to identify two or three forecasting models as the most likely candidates. The user can then select any of the candidate methods and have that model used in the second segment of the system.

These more sophisticated forecasting systems are costly, but may be justified to reduce the risk involved in important managerial decisions.

REVIEW QUESTIONS AND PROBLEMS

1. What common components of demand do we wish to take into account in a forecasting system for operations?

2. What are the general smoothing effects of a small number of periods in a moving average (perhaps three to five periods)? What are the effects on the ability of a moving average to track a rapidly changing demand?

3. What are the general smoothing effects of values of α in the 0.01 to 0.20 range in exponential forecasting systems? How do systems with low values of α track rapidly changing demand?

4. Show that exponential forecasting systems actually give weight to all past data.

5. How can one rationalize the fact that to extrapolate or forecast using the basic model of Equation 2, we simply use the most current smoothed average, S_t, and phase it forward by one period?

6. If one used a no-trend model to forecast demand where a trend actually exists, what is the nature of the error that results?

7. Figure 5–8 is a graph of 5.5 years of the monthly average number of patients in a large hospital. Table 5–9 gives the corresponding data.
   a. Compute three- and five-month moving averages for the first year and plot the results as forecasts of F_{t+1} on a graph in comparison with the actual values. How do you evaluate the results as forecasts of inpatient census?
   b. Compute exponentially smoothed forecasts for the first two years of data using Equation 2, assuming a smoothing constant of $\alpha = 0.10$ (remember that the forecasts to be plotted are F_{t+1}). How do you evaluate the results as forecasts?
   c. If the Fourier series program is available, compute forecasts using 6 and 14 terms.

8. What is the raw measure of trend used in the trend model? Why do we smooth it?

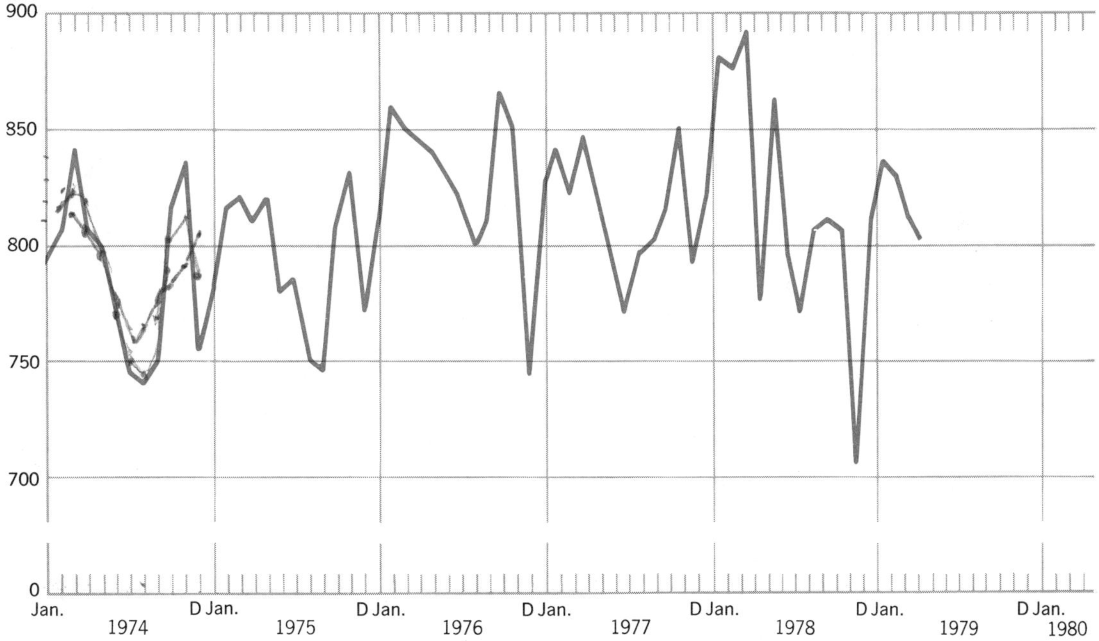

FIGURE 5-8
Monthly average inpatients at a large hospital.

9. Using the data in Table 5–9, compute forecasts for February and March, 1974. Assume that initial values of S_{t-1} and T_{t-1} are 800 and 0.0, respectively, and $\alpha = 0.2$, $\beta = 0.1$.

10. Recognizing that for the trend model, $F_{t+1} = S_t + T_t$, how can Equation 4 be simplified?

11. Using the data for 1974 in Table 5–9, construct an *initial* set of monthly seasonal indexes. Why do we smooth these indexes for subsequent years rather than simply using the initial values?

12. Using the data for 1975 in Table 5–9, and the initial monthly seasonal indexes generated in the preceding exercise, compute the forecasts for February and March, 1975, using the seasonal model. $S_{t-1} = 820$, $\alpha = 0.2$, and $\gamma = 0.3$

13. Using the data for 1975 in Table 5-9, and the 1974 initial seasonal indexes generated previously, compute the forecasts for February and March, 1975, using the trend *and* seasonal model. $S_{t-1} = 820$, $T_{t-1} = -1.0$, $\alpha = 0.2$, $\beta = 0.1$, and $\gamma = 0.3$.

14. What is the general structure of adaptive forecasting systems?

15. What is the rationale behind the Fourier series forecasting methodology?

16. Distinguish the statistical methodology of causal methods of forecasting from time series methods.

TABLE 5-9 **Monthly Average Number of Inpatients at a Large Hospital**

Month	Year 1974	1975	1976	1977	1978	1979
January	795	780	815	830	820	820
February	810	820	865	840	880	835
March	840	825	850	825	875	830
April	820	815	845	845	890	815
May	800	825	840	830	775	800
June	765	780	825	810	865	
July	745	785	820	770	795	
August	740	750	800	795	770	
September	750	745	810	805	805	
October	820	830	870	815	810	
November	840	810	850	850	805	
December	755	770	745	790	705	

17. As in the Cherryoak Company example used in the text, if we find a regression equation that fits historical data accurately, why not assume that it will be a good forecasting device? Why have a theory to explain why the equation fits the forecasts?

18. Define the coefficient of determination and the standard error of estimate as measures of forecast reliability in regression analysis.

19. What are the assumptions made in regression analysis?

20. How is econometric forecasting different from regression analysis?

21. Using the final regression model of Equation 15 developed by Parker and Segura for the Cherryoak Company, and the data in Table 5–7, compute the sales forecast for 1970. Check your answer with the forecasts given in Table 5–8.

22. The standard error of estimate for Equation 15 was given as 9.7. Interpreting the significance of the forecast of 227.95 for 1969 given in Table 5–8, what is the probability of actual sales as low as 227.95 − 9.7 = $218.25 million?

SITUATIONS

23. The hospital for which data are given in Table 5–9 and plotted in Figure 5–8 is attempting to improve its nurse scheduling system. Initially, it is concentrating on broad planning for the aggregate levels of nursing personnel needed.

In the past, the number of nursing personnel had been determined from the peak demand expected through the year. With the heavy pressure to reduce costs, the hospital administrator was now considering alternatives that took account of shorter-term variations in the patient load. Therefore, the initial effort was to improve forecasting. An exponential smoothing model was applied to the 5.5-year historical record, using a smoothing constant of $\alpha = 0.2$. These studies of the historical record resulted in forecasting error measurements of MAD = 31.9 for the simple exponential smoothing model. The hospital administrator felt that the errors were too large, stating that he could do almost as well based on his knowledge of fluctuations. For example, he expected a relatively small number of patients in December because those with "elective" medical problems avoided the holiday season. These elective cases usually came in the spring, in his experience.

The administrator was contemplating the use of more sophisticated forecasting models and wondered whether a model that accounted for trends and seasonals would result in greater forecasting accuracy. He also wanted advice concerning the possible use of a regression model.

Finally, the hospital administrator was deeply concerned about the 1978 data. The hospital had a record average of 890 patients in April of that year but a record low of only 705 in December. He felt that the situation was explainable by the fact that the load had built up because of the increase in population in the community and that the low in December reflected the opening of enlarged facilities by the nearby county hospital. He wondered, however, what plans he should make for the balance of 1979 and for 1980.

What recommendations would you submit to the hospital administrator?

24. Berry, Mabert, and Marcus [1975] report a study of forecasting teller window demand at the Purdue National Bank. Exponential smoothing using the equivalent of seasonal factors that reflected special day effects were the unique part of the forecasting model. Trend was not included in the model, since no appreciable trend was observed in the historical data.

Causes of Demand Variation

Analysis of historical data included substantial variation in the number of "cash slips" processed per day by tellers. There were significant differences in cash slip volume depending on the day of the week, as follows:

Day	Daily Average	Standard Deviation
Monday	422	113
Tuesday	293	113
Wednesday	385	100
Thursday	305	135
Friday	508	169

The average load on Tuesday, for example, was only 58 percent of the average Friday load. Part of this variation was random or unpredictable as a result of individuals' particular reasons for going to the bank on a particular day, to the weather, and so forth. In addition to the random variation, however, a substantial amount of the variation could be attributed to weekly, biweekly, and monthly payments. Taking Monday as an example, when the Mondays that involved known reasons for demand variation were removed from the historical data, average demand dropped from 422 to 407 cash slips, but even more significantly the standard deviation dropped from 113 to only 65. In other words, Monday's demand becomes somewhat more predictable when special days are taken into account.

Eight different kinds of special days were isolated that had a significant effect on the teller window demand,

1. Regular days

2. Academic paydays

3. Fiscal year paydays (monthly)
4. Days on which both an academic and fiscal year paydays occurred
5. Days following an academic payday
6. Days following a fiscal year payday
7. Days that followed a fiscal year payday and which occurred on an academic payday
8. Biweekly paydays (Wednesday)

Forecasting model

From the data regarding the effect of special days on demand, a set of seasonal type indexes were constructed and taken into account in computations by the following three equations:

$$S_t = \alpha(D_t/F_m) + (1 - \alpha)S_{t-1} \quad (16)$$
$$F_m = \gamma(D_t/S_t) + (1 - \gamma)F_m \quad (17)$$
$$F_{t+1} = S_t \times F'_m \quad (18)$$

where F_m = special day factor for day m

Note that the equations are very nearly the same as our seasonal model represented by Equations 6, 7, and 8. The differences are in the seasonal indexes (special day factors). The special day factor for day m, F_m, is constructed from the historical data depending on the code of eight types of special days and combinations of days listed previously. F_m is used in Equation 16 to compute the smoothed average S_t at the end of each day, based on the cash slip count for the day, D_t. These factors deflate actual demand in computing S_t. The special day factor F_m is then updated by current data in Equation 17 and used to reflate S_t for the forecast computation in Equation 18. Because the bank's special day schedule is dominated by a single large employer (Purdue University), special day factors could be computed for long periods in advance, making long-range forecasts possible. Equation 18 makes it possible to forecast n periods in advance by inserting the undated special day factor for the future day.

Comparative Models

A remaining question was, what is the incremental gain in adding successive refinement to the computation of special day factors in the model. Five models were constructed that progressively took account of additional causes of demand variation, and were tested on the historical data. In each instance, experiments were run to determine the parameter values, α and γ, that minimized forecast error. The results for the optimal values of the parameters are shown in Table 5-10.

How do you evaluate the results? What effect do you think results from the fact that the special day factors are static, that is, not time-dependent as in the seasonal model represented by Equations 6, 7, and 8? Since the model isolates causes of variation, is it a causal model? If you were the manager of the Purdue National Bank, which of the forecasting models, if any, would you install?

One of the results of the model indicates that the teller staffing requirements for Mondays, for example, varies between four and nine tellers, depending on the particular special day factor. How could you staff these extreme variations in load?

25. A study by Reisman et al. [1976] was conducted in a manufacturing company, after experience with an existing forecasting system during the 1974 recession. The company produced residential and light commercial air conditioners and heating units. The recession had resulted in a 25 to 30 percent decline in housing starts, a major factor in the company's sales. The company forecasting system was producing grossly overoptimistic forecasts. A revised forecasting

TABLE 5-10 **Purdue National Bank Forecasting Study. Forecast Error Standard Deviation for Optimal Parameters α and γ**

Model Accounts for Special Day Variations Listed	Monday	Tuesday	Wednesday	Thursday	Friday
Basic model, no special days	102.2	144.8	160.0	76.0	166.2
Paydays	93.0	97.9	94.1	71.5	83.3
Paydays, and day following	83.3	60.7	79.2	66.1	81.8
Academic paydays, fiscal year paydays, days on which both types of paydays occurred, day following any payday, academic payday that follows a fiscal year payday	65.3	57.5	79.3	61.2	61.1
All of the above plus biweekly paydays (Wed. only)	—	—	67.3	—	—

SOURCE. W. L. Berry, V. A. Mabert, and M. Marcus, "Forecasting Teller Window Demand with Exponential Smoothing," *Institute for Research in the Behavioral, Economic, and Management Sciences*, Purdue University, Paper No. 536, November 1975.

system was established that combined two kinds of objective forecasts plus a subjective forecast from the field, as shown in Figure 5-9.

Objective Forecasts

Two different objective forecasts were used, adaptive exponential smoothing and regression analysis. Since the company had been caught in a downturn, both objective forecasts were designed to react to rapid changes in market indicators.

The adaptive exponential smoothing model used historical data for computing projections and was regarded as "backward looking." Adaptive smoothing techniques incorporate mechanisms for sensing rapid changes in demand, even though the basic parameter value may be set to a relatively small value in order to filter out random fluctuations in demand. The model used also incorporated adjustments for seasonality. The system was tested against a moving average by subjecting both to a 50 percent increase in demand over a six-month period. The adaptive smoothing model sensed the change of trend after about two to three months and very quickly adapted

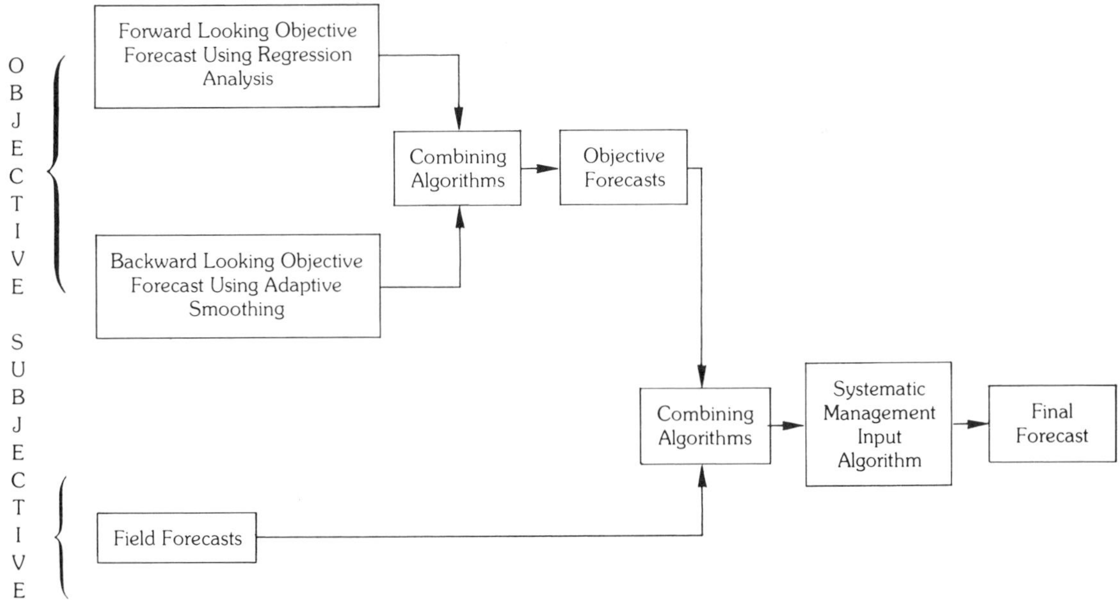

FIGURE 5-9
Forecasting system combining objective and subjective forecasts.
SOURCE: *A. Reisman, K. Gudapati, R. Chandrasekaran, P. Darukhanavala, and D. Morrison, "Forecasting Short-Term Demand,* Industrial Engineering, *8 (5), May 1976, pp. 38–45.*

itself to the new level. On the other hand, the moving average forecast was quite sluggish in responding to the rapid increase in demand.

The forecast based on regression analysis was regarded as "forward looking" because it involved projections of variables known to be associated with the company's demand through previous regression analyses. For example, the regression equation for cooling units involved the variables gross private domestic investment (PDIC), private housing starts (HST), and government purchases of goods and services (GVTC). The resulting regression equation ($r^2 = 0.762$) for cooling units was

$$\text{Deseasonalized cooling units} = -57{,}725.5 + 292.058(\text{PDIC}) + 7822.676(\text{HST}) + 216.536(\text{GVTC})$$

The result of the regression equation was reseasonalized in quoting forecasts. For the forecast to be forward looking, it must be remembered that forecast values of the three variables must be inserted in the regression equation in order to obtain an estimate of future sales units. The real value of the regression forecast was in its behavior with respect to sharp upturns and downturns. The response of the regression model was very similar to that of the adaptive smoothing model, the main difference being that there was no lag, since the projections of regression variables that cause demand produced immediate forecast changes in response to changes in the causal variables.

The two objective forecasts were combined into a final objective forecast in proportion to their error contributions. This was accomplished by taking the weighted sum of the two forecasts, with weights that were inversely proportional to the MAD of the two forecasting techniques.

Field Forecasts

The field forecasts were subjective and involved the sensitivity of field managers to what was happening in their regions. The field forecasts were submitted in a systematic way so that the results from 28 districts could be combined and merged with the objective forecasts. Systematic procedures were used for tracking and modifying field forecasts in order to improve their accuracy and determine their contributions in the final forecasts. Correction factors were computed for each region and each month by dividing actual demand by the field forecast, resulting in a correction ratio. The corrections ratios were then smoothed by taking a three-month smoothing average. The correction factors were applied to field forecasts, and the final field forecast was then the sum of corrected field forecasts.

Combining Objective and Subjective Forecasts

The final objective forecast was then combined with the final field forecast by weighting the two in proportion to their error contributions, following the same procedure for combining the two objective forecasts.

How do you evaluate the new forecasting system? Of what value are the field forecasts in relation to the objective forecasts? Will the field forecasts become more accurate by virtue of the information feedback concerning their errors? Since the forward-looking forecast is only as accurate as the projections of economic indicators used in the regression equation, do we have a forecast or "crystal ball gazing?" If you were president of the company, would you install this forecasting system? Would you modify the system? If so, how?

26. Mabert [1976] reports a comparative study of the performance of a sales force-executive opinion approach to forecasting single-item demand over a five-year period with three statistical techniques. The study was carried out in the Duncan Manufacturing Company and dealt with one of their main products.

The company used a judgmental approach in preparing annual forecasts of demand. Each fall, sales personnel submitted estimates of customer needs for the coming year. The process involved managerial review of these estimates and the determination of the final demand forecast by period. The results of the judgmental approach were compared with three statistical forecasting methods: an exponential smoothing model taking account of seasonal factors, similar to the one presented in the chapter; a harmonic model similar to the Fourier series methodology presented in the chapter; and the Box-Jenkins methodology.

Twleve years of historical data on actual demand and company forecasts were available for analysis. The first seven years of data were used to identify and estimate the appropriate statistical models and their parameters used for comparison. The models were then used to forecast the last five years with comparative performance data, as shown in Table 5-11.

Statistical analysis showed that the Box-Jenkins and Fourier series model

TABLE 5-11 **Comparative Results of Company Forecasts and Three Models**

Technique	MAD (percent)	Hours Required for Model Development	Hours Required for Forecasting	Hours Required for Total	Compute Time Required for development and run (seconds)	Time Elapsed Before Forecast Produced (days)
Company	15.9	—	103.0	103.0	—	27
Exponential smoothing	15.1	7.1	4.9	12.0	2.19	2
Harmanic	14.1	6.9	4.3	11.2	1.21	2
Box-Jenkins	14.0	11.6	5.9	17.5	2.31	2.5

SOURCE. V. A. Mabert, "Statistical Versus Sales Force-Executive Opinion: A Time Series Analysis Case Study," *Decision Sciences*, 7(2), April 1976, pp. 310–318.

forecast errors were significantly different from the company's errors but showed no significant difference between the three statistical procedures. The company forecasts required at least six times the hours of the next lowest technique. The 103 hours for the company forecasts included the time of the sales staff, district managers, and the vice-presidents involved.

Based on the comparative data, how do you evaluate statistical forecasting versus sales force-executive opinion? Wherein lies the advantage of each? Based on the comparative studies, if you were president of the Duncan Company, would you replace the company procedures with one of the statistical forecasting techniques? Why?

REFERENCES

Benton, W. K. *Forecasting for Management.* Addison-Wesley, Reading, Mass., 1972.

Berry, W. L., and F. W. Bliemel. "Selecting Exponential Smoothing Constants: An Application of Pattern Search," *International Journal of Production Research, 12*(4), July 1974, pp. 483–500.

Berry, W. L., V. A. Mabert, and M. Marcus. "Forecasting Teller Window Demand With Exponential Smoothing," *Institute for Research in the Behavioral, Economic, and Management Sciences,* Purdue University Paper No. 536, November 1975.

Box, G. E. P., and G. M. Jenkins. *Time Series Analysis, Forecasting, and Control.* Holden-Day, San Francisco, 1970.

Brown, R. G. *Smoothing, Forecasting and Prediction.* Prentice-Hall, Englewood Cliffs, N.J., 1963.

Brown, R. G. *Statistical Forecasting for Inventory Control.* McGraw-Hill, New York, 1959.

Buffa, E. S., and J. G. Miller. *Production-Inventory Systems: Planning and Control* (3rd ed.). Richard D. Irwin, Homewood, Ill., 1979.

Buffa, F. P. "The Application of a Dynamic Forecasting Model with Inventory Control Properties," *Decision Sciences, 6*(2), April 1975, pp. 298–306.

Chambers, J. C., S. K. Mullick, and D. D. Smith. *An Executive's Guide to Forecasting.* John Wiley, New York, 1974.

Chambers, J. C., S. K. Mullick, and D. D. Smith. "How to Choose the Right Forecasting Technique," *Harvard Business Review,* July–August 1971, pp. 45–74.

Church, J. G. "Sure Staf: A Computerized Scheduling System for Telephone Business Offices," *Management Science, 20*(4), December 1973, Part II, pp. 708–720.

Dancer, R., and C. Gray. "An Empirical Evaluation of Constant and Adaptive Computer Forecasting Models for Inventory Control," *Decision Sciences, 8*(1), January 1977, pp. 228–238.

Groff, G. K. "Empirical Comparison for Models for Short Range Forecasting," *Management Science, 20*(1), September 1973, pp. 22–31.

Mabert, V. A. "Forecast Modification Based Upon Residual Analysis: A Case Study of

Check Volume Estimation," *Decision Sciences,* 9(2), April 1978, pp. 285-296.

Mabert, V. A. "Statistical Versus Sales Force-Executive Opinion Short Range Forecasts: A Time Series Analysis Case Study," *Decision Sciences,* 7(2), April 1976, pp. 310–318.

Makridakis, S., A. Hodgsdon, and S. C. Wheelwright. "An Interactive Forecasting System," *The American Statistician, 28*(4), November 1974, pp. 153–158.

Makridakis, S., and S. C. Wheelwright. *Forecasting Methods and Applications,* John Wiley, New York, 1978.

Miller, J. G., W. L. Berry, and C. F. Lai. "A Comparison of Alternative Forecasting Strategies for Multi-Stage Production-Inventory Systems," Krannert Graduate School of Industrial Administration, Purdue University, Paper No. 441, April 1974.

Reisman, A., K. Gudapati, R. Chandrasekaran, P. Darukhanavala, and D. Morrison. "Forecasting Short-term Demand," *Industrial Engineering, 8*(5), May 1976, pp. 38–45.

Trigg, D. W., and A. G. Leach. "Exponential Smoothing with an Adaptive Response Rate," *Operational Research Quarterly, 18*(1), March 1967, pp. 53–59.

Wheelwright, S. C., and S. Makridakis. *Forecasting Methods for Management* (2nd ed). John Wiley, New York, 1977.

Whybark, D. C., "A Comparison of Adaptive Forecasting Techniques," *The Logistics and Transportation Review, 8*(3), 1972, pp. 13–26.

Winters, P. R., "Forecasting Sales by Exponentially Weighted Moving Averages," *Management Science, 6*(3), April 1960, pp. 324–342.

CHAPTER 6 Aggregate Planning—Short-term Capacity Planning

MOST MANAGERS WANT TO PLAN AND CONTROL OPERATIONS at the broadest level through some kind of aggregate planning that bypasses details of individual products and the detailed scheduling of facilities and personnel. This fact is a good illustration of how managerial behavior actually employs system concepts by starting with the whole. Management would rather deal with the basic relevant decisions of programming the use of resources. This is accomplished by reviewing projected employment levels and by setting activity rates that can be varied within a given employment level by varying hours worked (working overtime or undertime). Once these basic decisions have been made for the upcoming period, detailed scheduling can proceed at a lower level within the constraints of the broad plan. Finally, last-minute changes in activity levels need to be made with the realization of their possible effects on the cost of changing production levels and on inventory costs if they are a part of the system.

What is needed first for aggregate plans is to develop some logical overall unit of measuring output, for example, gallons of paint in the paint industry, cases in the beer industry, perhaps equivalent machine-hours in mechanical industries, beds occupied in hospitals, or pieces of mail in a post office.

Second, management must be able to forecast for some reasonable planning period, perhaps up to a year, in these aggregate terms. Finally, management must be able to isolate and measure the relevant costs. These costs may be reconstructed in the form of a model that will permit near optimal decisions for the sequence of planning periods in the planning horizon.

The sequential nature of the decisions should be kept in mind. A decision on employment levels and activity rates made for an upcoming period cannot be termed either right or wrong, good or bad. Decisions will also be made two periods hence based on the decisions just made, on new information about the actual progress of sales, and on the forecasts for the balance of the planning horizon. The result is that all decisions are right or wrong only in terms of the sequence of decisions over a period of time.

Most of our discussion deals with systems that produce inventoriable items, where the existence of finished goods inventories can make possible a trade-off for the costs of changes in employment level. However, systems that cannot store their output are discussed in Chapter 10, "Large-scale Projects," and Chapter 11, "Service Systems."

NATURE OF AGGREGATE PLANNING

Aggregate planning increases the range of alternatives for capacity use that can be considered by management. The concepts raise such broad basic questions as: To what extent should inventory be used to absorb the fluctuations in demand that will occur over the next 6 to 12 months? Why not absorb these fluctuations by simply varying the size of the work force? Why not maintain a fairly stable work force size and absorb fluctuations by changing activity rates through varying work hours? Why not maintain a fairly stable work force and activity rate and let subcontractors wrestle with the problem of fluctuating order rates? Should the firm purposely not meet all demands?

In most instances, it is probably true that any one of these pure strategies would not be as effective as a balance among them. Each strategy has associated costs, and we seek an astute combination of the alternatives.

COSTS

The basic picture of costs is shown in Figure 6-1. The same basic structure is appropriate for manufacturing concerns, hospitals, the post office, and many other service organizations. For some kinds of governmental operations, the revenue line could be substantially different, depending on the basis for obtaining budget to cover expenses.

What happens to total costs as volume increases? This cost behavior is of the

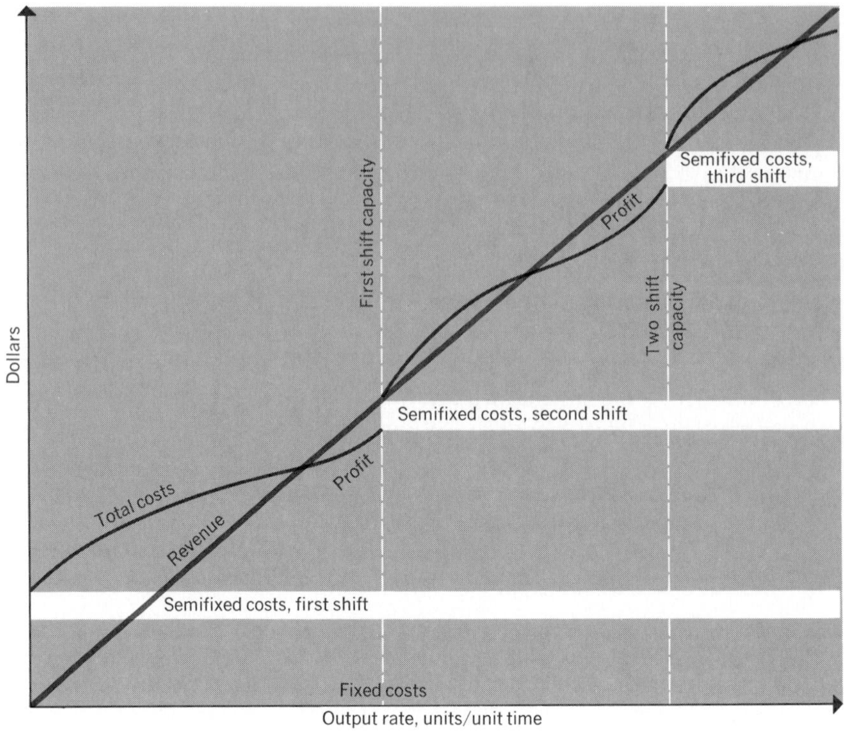

FIGURE 6-1
General structure of costs over a wide range of volume.
SOURCE: *E. S. Buffa, and W. J. Taubert, "Evaluation of Direct Computer Search Methods for the Aggregate Planning Problem,"* Industrial Management Review, *Fall 1967.*

greatest interest, especially near or at the sharp changes in the organization of resources that occur with the addition or deletion of second and third shifts. Multiple shifts are common in manufacturing, hospitals, and the post office; however, a single-shift operation would have a structure similar to that shown for the first shift in Figure 6-1. Following the conventional accounting definition of fixed, semifixed, and variable costs, the interesting factors that dominate the shape taken by the total cost curve are in the semifixed and variable cost categories.

To put on a second shift would require a reorganization of the supervisory structure and other managerial functions, causing a quantum step up in costs. In addition, there may be required some physical changes in the reorganization of facilities to maintain a two-shift operation. Thus, we observe the discontinuities in the total cost curve at the limits of one- and two-shift capacity.

But why should we not expect the variable costs to be linear? There are a variety of reasons, and we illustrate some of them in Figure 6-2. First, it is important to recognize that we must look at the system as something dynamic, changing in response to changes in demand, product or service mix, technology, and changing organizational and social patterns. To regard the enterprise as being static is to miss the entire flavor and character of the problems of operations management.

In Figure 6-2*a* the payroll costs may be close to being linear, although even this assumption could be attacked on the basis of the supply and demand for labor as indicated by the dashed line in Figure 6-2*a*. In any case, the productivity of labor in relation to volume of activities shown in Figure 6-2*b* makes the labor cost per unit a nonlinear function. Viewing labor in the aggregate and not considering substantial changes in basic technology, we would expect the toe of the productivity curve to exhibit start-up difficulties that would be reflected in low output per worker-hour. In the middle range of one-shift operations, the curve may be approximately linear. But as we approach the limits of one-shift capacity, productivity falls off because of increased congestion, cramped quarters, interference, and delays. There is a logical tendency to try to achieve the increased output near the limits of one-shift capacity through the use of overtime with the existing work force and its higher cost of marginal productivity. Changes in technology might change the level and contour of the curve, but the general nonlinearities would remain.

In Figure 6-2*c* we see two cost components related to material. We assume that the cost of materials could be linear with volume (dashed line), but quantity discounts and an aggressive procurement staff should produce economies at higher volumes (solid line). As volume goes up, the size of the aggregate inventory necessary to maintain the production-distribution process increases also, but not in direct proportion.

In Figure 6-2*d* we examine the relative cost of holding either too much or too little inventory. This presupposes that there exists, for each level of operation, some ideal aggregate inventory to sustain the productive process. Inventories, however, might vary from this ideal amount for two possible reasons. First, they might be different from the ideal levels because of the capriciousness of consumer demand. If demand were less than expected, we would incur extra inventories and the cost of holding them. If demand were greater than expected, we would incur back-order or shortage

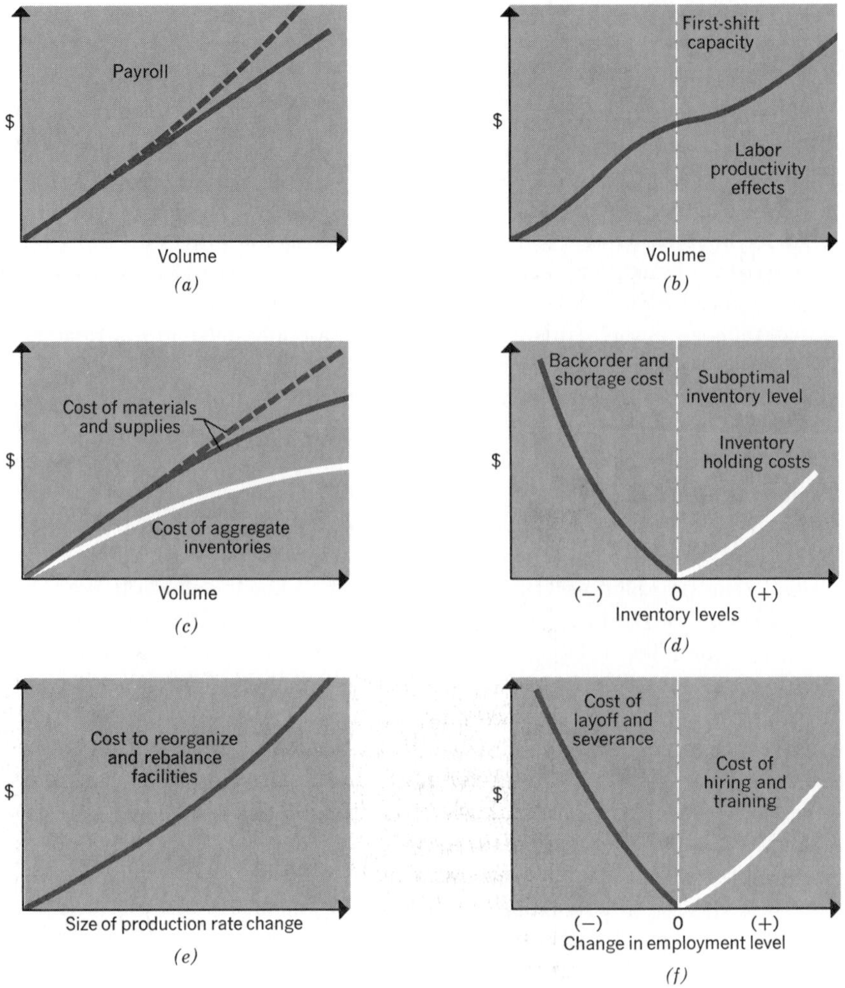

FIGURE 6-2
Selected cost behavior patterns.

costs and the possible opportunity costs of lost sales or volume of activity. A second basic reason why aggregate inventories might differ from ideal levels is as a result of conscious managerial design. Management might consciously accumulate extra inventories in a slack demand season in order to reduce costs of hiring, layoff, and overtime. These costs of holding too much or too little inventory are probably not linear for larger ranges.

Finally, in Figures 6-2*e* and 6-2*f*, we have included two items of cost associated with changes in output and employment levels. When output rates are changed, there are some costs in reorganizing and replanning for the new level and in rebalancing crews and facilities.

Now let us return to Figure 6-1. Cost-volume patterns will not be linear for many of the reasons discussed, and therefore, profit appears in a "pocket" beyond a break-even point. If the total cost curve lies entirely above the revenue line, then losses will be less in the upper region. Profit increases to a point where the diseconomies discussed become effective. At that point, profit is reduced as volume continues to increase and may disappear as we approach the limits of one-shift capacity.

The reasons for the relative increases in costs and decline in profits are related to the behavior of some of the cost components, but the profit decline generally is associated with the decline of labor productivity, use of overtime, and the relative inefficiency of newly hired labor. These cost increases normally are somewhat greater than the savings in material and inventory costs, when operating near a capacity limit. When the decision to add a second shift is made, there will be increased semifixed costs in order to put on the second shift. As we progress into the second shift volume range, labor productivity will be relatively low, reflecting start-up conditions, and a pattern similar to the first shift will be repeated.

In some of the analyses of operations management problems, we shall assume linear approximations to cost functions in order to simplify and conceptualize problems. In many instances, the assumption of linearity is an excellent one, but near a capacity limit it may be dangerous.

Many costs affected by aggregate planning and scheduling decisions are difficult to measure and are not segregated in accounting records. Some, such as interest costs on inventory investment, are opportunity costs. Other costs are not measureable, such as those associated with public relations and public image. However, all the costs are real and bear on aggregate planning decisions.

PROBLEM STRUCTURE

The simplest structure of the aggregate planning problem is represented by the single-stage system shown in Figure 6-3. In Figure 6-3 the planning horizon is only one period ahead; therefore, we call Figure 6-3 a "single-stage" system. The state of the system at the end of the last period is defined by W_0, P_0, and I_0, the aggregate work force size, production or activity rate, and inventory level, respectively. The ending state conditions become the initial conditions for the upcoming period. We have a forecast of the requirements for the upcoming period; through some process, decisions are made that set the size of the work force and production rate for the upcoming period. Projected ending inventory is then, $I_1 = I_0 + P_1 - F_1$ where F_1 is forecasted sales.

The decisions made may call for hiring or laying off personnel, thus expanding or contracting the effective capacity of the productive system. The work force size, together with the decision on activity rate during the period, then determines the re-

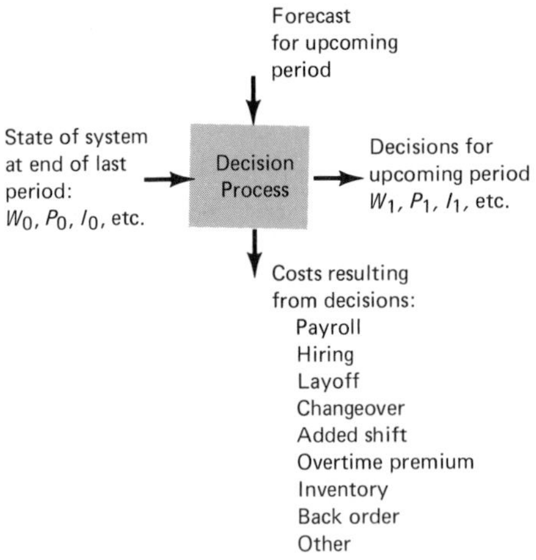

FIGURE 6-3
Single-stage aggregate planning decision system where planning horizon is only one period. W = size of work force, P = production or activity rate, and I = inventory level.

quired amount of overtime, inventory levels or back ordering, whether or not a shift must be added or deleted, and other possible changes in operating procedure. The comparative costs that result from alternate decisions on work force size and production rate are of great interest in judging the effectiveness of the decisions made and the decision process used. The comparative cost of a sequence of such alternate decisions is also of interest in judging the applicability of the single-stage model.

Let us suppose that we make a sequence of decisions by the structure of the single-stage model of Figure 6-3. If the forecasts for each period for the first four periods are progressively decreasing, our decision process responds by decreasing both the work force size and the activity rate in some combination, incurring layoff and changeover costs. Then, for the fifth through tenth periods, we find the period forecasts are progressively increasing, and each period the decision process calls for hiring personnel and increased activity rates, incurring more hiring and changeover costs. The single-period planning horizon has made each independent decision seem internally logical but has resulted in the laying off of workers only to hire them back again.

Had we been able to look ahead for several periods with an appropriate decision process, we might have decided to stabilize the work force size, at least to some extent, and absorb the fluctuations in demand in some other way. For example, we could have changed activity rate through the use of overtime and undertime or by carrying extra inventories through the trough in the demand curve. It appears that broadening the planning horizon can improve the effectiveness of the aggregate planning system.

Figure 6-4 shows a multistage aggregate planning system where the horizon has been expanded with forecasts for each period. Our objective is the same as before: to make decisions concerning the work force size and production rate for the upcoming period. In doing so, however, we consider the sequence of projected decisions in relation to forecasts and their cost effects. The decision for the upcoming period is to be affected by the future period forecasts, and the decision process must consider the cost effects of the sequence of decisions. The connecting links between the several stages are the *W*, *P*, and *I* values that are at the end of one period and the beginning of the next. The feedback loop from the decision process may involve some iterative or trial-and-error procedure to obtain a solution.

DECISION PROCESSES FOR AGGREGATE PLANNING

Given the structure of Figure 6-4, our interest is drawn to the decision processes. There are several, classified as graphic, mathematical, heuristic, and computer search methods. We classify them further as static versus dynamic, as well as single-stage versus multistage models. We discuss the graphic, mathematical, and computer search methods.

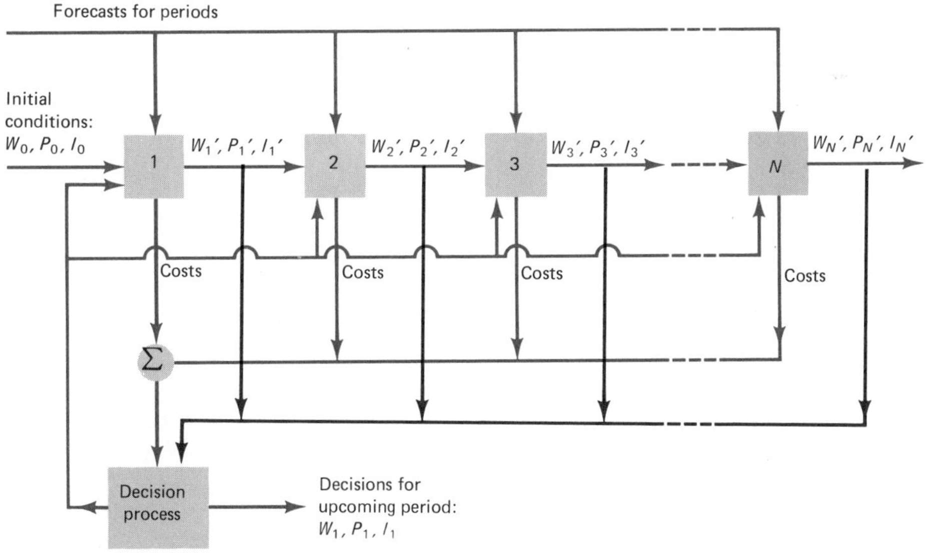

FIGURE 6-4
Multistage aggregate planning decision system for planning horizon of *N* periods.

Graphic Methods

Table 6-1 shows a forecast of expected production requirements in column 4, by months; these data are cumulated by months in column 5. The ratio of peak to valley in the requirements schedule is \$11,000 in July to \$4,000 in February and March, or 11,000/4,000 = 2.75.

Note, however, that the number of working days per month varies considerably. Column 2 shows that the number of working days varies from 23 in March and August to only 11 in September. (The plant closes for two weeks in September owing to vacations scheduled to coincide with required plant maintenance.) Therefore, the swing in production requirements per production day (see column 9 of Table 6-1) varies from 6500/11 = 591 in September to only 4000/23 = 174 in March, or a ratio of 591/174 = 3.40. This substantial variation in daily production requirements is shown by the graph of requirements in Figure 6-5. Developing aggregate schedules that meet these seasonal requirements is part of the problem. Developing schedules that minimize the incremental costs associated with meeting requirements is the challenge.

Assume that normal plant capacity is 350 units per day. Additional capacity can be obtained by employing labor at overtime to a maximum capacity of 410 units per day. The *additional* cost per unit is \$10 for units produced during overtime hours.

Buffer Inventories and Maximum Requirements. Column 6 of Table 6-1 shows

TABLE 6-1

Forecast of Production Requirements and Buffer Inventories: Cumulative Requirements, Average Buffer Inventories,[a] and Cumulative Maximum Production Requirements

(1) Month	(2) Production Days	(3) Cumulative Production Days	(4) Expected Production Requirements	(5) Cumulative Production Requirements Col. 4 Cumulated	(6) Required Buffer Inventories	(7) Cumulative Maximum Production Requirements. Col. 5 + Col. 6	(8) Col. 2 × Col. 6	(9) Production Requirements per Production Day. Col. 4 ÷ Col. 2
January	22	22	5,000	5,000	2,800	7,800	61,600	227.3
February	20	42	4,000	9,000	2,500	11,500	50,000	200.0
March	23	65	4,000	13,000	2,500	15,500	57,500	173.9
April	19	84	5,000	18,000	2,800	20,800	53,200	263.2
May	22	106	7,000	25,000	3,200	28,200	70,400	318.2
June	22	128	9,000	34,000	3,500	37,500	77,000	409.1
July	20	148	11,000	45,000	4,100	49,100	82,000	550.0
August	23	171	9,000	54,000	3,500	57,500	80,500	391.3
September	11	182	6,500	60,500	3,000	63,500	33,000	590.9
October	22	204	6,000	66,500	3,000	69,500	66,000	272.7
November	22	226	5,000	71,500	2,800	74,300	61,600	227.3
December	18	244	5,000	76,500	2,800	79,300	50,400	277.8
							743,200	

[a] Average buffer inventory = 743,200/244 = 3045.9 units.

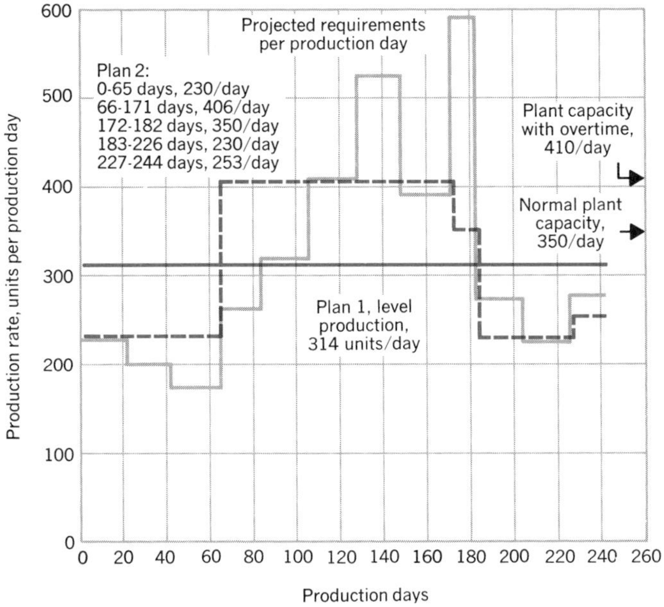

FIGURE 6-5
Comparison of two production plans that meet requirements.

buffer inventories stated as minimum stocks. Their purpose is to provide for the possibility that market requirements could be greater than expected. When we add these buffer inventories in each month to the cumulative production requirements in column 5, we have cumulative maximum requirements, as shown in column 7. Column 8 provides the basis for weighting the buffer inventory by production days and for computing the average buffer inventory of 3045.9 units, at the bottom of the table.

Plan 1, Level Production. The simplest production plan is to establish an average output level that meets annual requirements. The total annual requirements are shown as the last figure in the cumulated requirements schedule in column 5 of Table 6-1, 76,500 units. Since there are 244 working days, an average daily output of 76,500/244 = 314 units should cover requirements. We may find problems with such a schedule because of timing, but we shall see what to do about these problems later. Our strategy is simple: accumulate seasonal inventory during the slack requirements months to be used during peak requirements months. The level production plan is shown in relation to the production requirements per day in Figure 6-5 as Plan 1.

The inventory requirements for Plan 1 are calculated in Table 6-2. The production in each month is computed in column 3 and accumulated in column 4 to produce a schedule of units available each month, starting with a beginning inventory of 2800

TABLE 6-2 **Calculation of Seasonal Inventory Requirements for Plan 1**

(1) Production Days	(2) Production Rate, Units, per Day	(3) Production in Month, Units, Col. 1 × Col. 2	(4) Cumulative Units Available. Cumulative Production + Beginning Inv. (2800)	(5) Cumulative Maximum Requirements, from Col. 7 of Table 6-1	(6) Seasonal Inventory[a] = Col. 4 − Col. 5	(7) Col. 1 × Col. 6
22	314	6,908	9,708	7,800	1,908	41,976
20	314	6,280	15,988	11,500	4,488	89,760
23	314	7,222	23,210	15,500	7,710	177,330
19	314	5,966	29,176	20,800	8,376	159,144
22	314	6,908	36,084	28,200	7,884	173,448
22	314	6,908	42,992	37,500	5,492	120,824
20	314	6,280	49,272	49,100	172	3,440
23	314	7,222	56,494	57,500	−1,006	−23,138
11	314	3,454	59,948	63,500	−3,552	−39,072
22	314	6,908	66,856	69,500	−2,644	−58,168
22	314	6,908	73,764	74,300	− 536	−11,792
18	314	5,652	79,416	79,300	116	2,088
244						635,840

[a] Average seasonal inventory = 635,840/244 = 2605.9 units.

units. Then, by comparing the units available to column 4 with the cumulative maximum requirements schedule in column 5, we generate the schedule of seasonal inventories in column 6.

Looking at the seasonal inventories for Plan 1 in column 6, they vary from a maximum of 8476 units in April to a minimum of −3552 in September. The significance of the negative seasonal inventories is that the plan calls for dipping into buffer stocks. In August, we propose to use 1006 out of the planned buffer of 3500 units, but in September, we actually exceed the planned buffer by 552 units. In other words, the plan would actually require back ordering or the possible loss of the sale of 552 units in September. The plan recovers in subsequent months and meets aggregate requirements, but incurs total shortages for the year of 7738 units.

Now, a managerial judgment must be made. Should we make plans that *depend* on the use of buffer inventories? Or, should we make plans that expect shortages resulting in back ordering or even lost sales? These are not easy questions to answer. In some instances, it might be more economical to take the risk and plan to use the buffer stock or even to incur shortages. It depends on the balance of cost of inventories, the cost to produce the extra units, the cost of incurring shortages, and managerial policies concerning delivery service.

Suppose we decide not to plan to use the buffer inventory, since the buffer was designed to absorb unexpected sales increases. (If we plan to use them, they lose their buffering function.) How do we adjust the plan to take the negative seasonal inventories into account? All we need to do is increase the beginning inventory by the most negative seasonal inventory balance, −3552 units in September. This new beginning inventory level has the effect of increasing the entire schedule of cumulative units available in column 4 of Table 6-2 by 3552 units. Then, average seasonal inventories will also be increased by 3552 units.

The seasonal inventories for Plan 1 are calculated in Table 6-2 as 2605.9 units, weighted by production days, assuming that we use buffer stocks and record shortages as indicated in column 6. If we revise the plan so that the buffer inventories are not used, the average seasonal inventory would be 2605.9 + 3552 = 6157.9 units.

Assuming that inventory holding costs are $50 per unit per year and that shortage costs are $25 per unit short, we can now compute the relative cost of the variants of Plan 1. If beginning inventories are only 2800 units, the annual inventory cost is 50 × 2605.9 = $130,295, and the shortage cost is 25 × 7738 = $193,450. The total incremental cost is then $323,745.

By comparison, if we decide not to use up buffer inventory, the average seasonal inventories are 6157.9 units at a cost of 50 × 6157.9 = $307,895, the total incremental cost for comparison. Obviously, for these costs of holding inventories and incurring shortages, it is more economical to plan on larger inventories. In other situations, the reverse might be true. If the cost of shortages is only $20 per unit, it would be slightly more economical to take the risk. Or alternately, if the cost of holding inventories were $55 per unit per year (shortage cost $25), then the balance of costs would also favor taking the risk of incurring shortages.

Plan 1 has significant advantages. First, it does not require the hiring or layoff of personnel. It provides stable employment to workers and would be favored by organized labor. Also, scheduling is simple—314 units per day. From an incremental production cost viewpoint, however, it fails to consider whether or not there is an economic advantage in trading off the large seasonal inventory and shortage costs for overtime costs and/or costs incurred in hiring or laying off personnel to meet seasonal variations in requirements.

Plan 2, Using Hiring, Layoff, and Overtime. Note from Figure 6-5 that normal plant capacity allows an output of 350 units per day and that an additional 60 units per day can be obtained through overtime work. Units produced on overtime cost an additional $10 per unit in this example.

Up to the normal capacity of 350 units per day, we can increase or decrease output by hiring or laying off labor. A worker hired or laid off affects the net output rate by 1 unit per day. The cost of changing output levels in this way is $200 per worker hired or laid off, owing to hiring, training, severance, and other associated costs.

Plan 2 takes advantage of the additional options of changing basic output rates and using overtime for peak requirements. Plan 2 is shown in Figure 6-5 and involves two basic employment levels: labor to produce at normal output rates of 230 and 350 units per day. Additional variations are achieved through the use of overtime when needed. The plan has the following schedule:

0 to 65 days—produce at 230 units per day.

66 to 171 days—produce at 406 units per day (66 units per day at overtime rates; hire 120 workers to increase basic rate without overtime from 230 to 350 units per day).

172 to 182 days—produce at 350 units per day (no overtime).

183 to 226 days—produce at 230 units per day (lay off 120 workers to reduce basic rate from 350 to 230 units per day again).

227 to 244 days—produce at 253 units per day (23 units per day at overtime rates).

The calculations of the seasonal inventory requirements for Plan 2 are similar to those for Plan 1. The seasonal inventory for Plan 2 has been reduced to only 2356 units, 90 percent of Plan 1 with shortages and 38 percent of Plan 1 without shortages.

To counterbalance the inventory reduction, we must hire 120 workers in April and lay off an equal number in October. Also, we have produced a significant number of units at overtime rates from April to September and in December. The costs of Plan 2 are summarized in Table 6-3. The total incremental costs of Plan 2 are $229,320, 71 percent of Plan 1 with shortages and 75 percent of Plan 1 without shortages.

The trade-offs produce a somewhat more economical plan but require substantial fluctuations in the size of the work force. Whether the social and employee relations' consequences are to be tolerated is a matter for managerial judgment. If the employment fluctuation is greater than management feels can be tolerated, other alternatives involving a smaller employment fluctuation can be computed. Perhaps some of the variations can be absorbed by more overtime work, and in some kinds of industries, subcontracting can be used to meet the most severe peak requirements.

Plan 3, Adding Subcontracting as a Source. Assume that management wishes to consider a third alternative that involves smaller work force fluctuation, using overtime, seasonal inventories, and subcontracting to absorb the balance of requirements fluctuations. Plan 3 has the following schedule:

0 to 84 days—produce at 250 units per day.

85 to 128 days—produce at 350 units per day (hire 100 workers to increase the basic rate from 250 to 350 units per day).

129 to 148 days—produce at 410 units per day (60 units per day produced at overtime, plus 1700 units subcontracted).

149 to 171 days—produce at 370 units per day (20 units per day produced on overtime).

TABLE 6-3

Calculation of Incremental Costs for Plan 2

(1) Production Days	(2) Production Rate/Day	(3) Units of Production Rate Change	(4) Units Produced at Overtime Production Rates Greater Than 350 or 230 × Col. 1
22	230	–0–	–0–
20	230	–0–	–0–
23	230	–0–	–0–
19	406	120	1,064
22	406	–0–	1,232
22	406	–0–	1,232
20	406	–0–	1,120
23	406	–0–	1,288
11	350	–0–	–0–
22	230	120	–0–
22	230	–0–	–0–
18	253	–0–	414
		240	6,350

Production rate change costs = 240 × 200 = (A change in the basic rate of one unit requires the hiring or layoff of one worker at $200 each)	$48,000
Overtime costs at $10 extra per unit = 10 × 6,350 =	63,500
Seasonal inventory cost (2356.4 units at $50 per unit per year) = 50 × 2356.4 =	117,820
Total incremental cost =	$229,320

172 to 182 days—produce at 410 units per day (60 units per day produced on overtime, plus 1380 units subcontracted).

183 to 204 days—produce at 273 units per day (23 units per day produced on overtime; lay off 100 workers to reduce employment level from basic rate of 350 to 250 units per day).

205 to 244 days—produce at 250 units per day.

Plan 3 reduces seasonal inventories still further to an average of only 1301 units. Employment fluctuation is more modest, involving the hiring and laying off of 100 workers. Only 4066 units are produced at overtime rates, but a total of 3080 units are subcontracted at an additional cost of $15 per unit.

Table 6-4 summarizes the costs for all three plans. For the particular example, Plan 3 is the most economical, being 83.7 percent as costly as Plan 2 and only 59.3 percent as costly as Plan 1 with shortages. The buffer inventories are nearly the same for all plans, so their costs are not included as incremental costs.

TABLE 6-4 **Comparison of Costs of Alternate Production Plans**

Costs	Plan 1 With Shortages	Plan 1 Without Shortages	Plan 2	Plan 3
Shortages[a]	$193,450	—	—	—
Seasonal Inventory[b]	130,295	$307,895	$117,820	$ 65,070
Labor Turnover[c]	—	—	48,000	40,000
Overtime[d]	—	—	63,500	40,660
Subcontracting[e]	—	—	—	46,200
Totals	$323,745	$307,895	$229,320	$191,930

[a] Shortages cost $25 per unit.

[b] Inventory carrying costs are $50 per unit per year.

[c] An increase or decrease in the basic production rate of one unit requires the hiring or layoff of one employee at a hiring and training, or severance, cost of $200 each.

[d] Units produced at overtime rates cost an additional $10 per unit.

[e] Units subcontracted cost an additional $15 per unit.

Even though Plan 3 involves less employment fluctuation than Plan 2, it may be felt to be too severe. Other plans involving less fluctuation could be developed and their incremental costs determined in the same way.

Cumulative Graphs. Although Figure 6-5 shows the effects of the production rate changes quite clearly, it is actually somewhat easier to work with cumulative curves, as shown in Figure 6-6. The procedure is to plot first the cumulative production requirements. The cumulative maximum requirements curve is then simply the former curve with the required buffer inventories added for each period. The cumulative graph of maximum requirements can then be used as a basis for generating alternate program proposals. Any production program that is feasible, in the sense that it meets requirements while providing the desired buffer stock protection, must fall entirely above the cumulative maximum requirements line. The vertical distances between the program proposal curves and the cumulative maximum requirements curve represent seasonal inventory accumulation for each plan.

Graphic methods are simple and have the advantage of visualizing alternate programs over a broad planning horizon. The difficulties with graphic methods, however, are the static nature of the graphic model and the fact that the process is in no sense cost or profit optimizing. In addition, the process does not generate good programs itself but simply compares proposals made.

The alternate plan proposals used in connection with the graphic methods indicate the sensitivity in alternate plans to the use of various sources of short-term capacity: seasonal inventories, shortages, use of overtime capacity, and subcontracting. Mathematical and computer search models attempt to find optimal combinations of these sources of short-term capacity.

Mathematical Optimization Methods

We will discuss three mathematical optimization methods: the Linear Decision Rule, linear programming, and goal programming. All three have a basis for optimizing the model developed, so our interest is in appraising how closely the three models represent reality.

The Linear Decision Rule. The Linear Decision Rule (LDR) was developed in 1955 by Holt, Modigliani, Muth, and Simon [1955, 1956, 1960] as a quadratic programming approach for making aggregate employment and production rate decisions. The LDR is based on the development of a quadratic cost function for the company in question with cost components of (1) regular payroll, (2) hiring and layoff, (3) overtime, and (4) inventory holding, back ordering, and machine setup costs. The quadratic cost function is then used to derive two linear decision rules for computing work force level and production rate for the upcoming period based on forecasts of aggre-

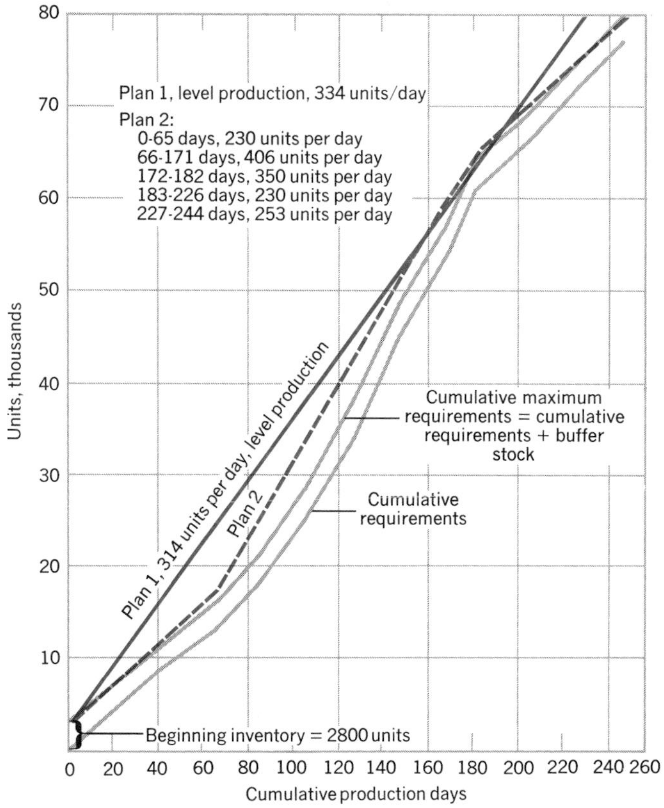

FIGURE 6-6
Cumulative graphs of requirements and alternate production programs.

gate sales for a preset planning horizon. The two linear decision rules are optimum for the model.

Figure 6-7 shows the form of the four components of the cost function. The work force size is adjusted in the model once per period with the implied commitment to pay employees at least their regular-time wages for that period. This is indicated in Figure 6-7*a*. Hiring and layoff costs are shown in Figure 6-7*b*; the LDR model approximates these costs with a quadratic function as shown. If the work force size is held constant for the period in question, then changes in production rate can be absorbed by the use of overtime and undertime. Undertime is the cost of idle labor at regular payroll rates. The overtime cost depends on the size of the work force, W, and on the aggregate production rate, P.

The form of the overtime-undertime cost function in relation to production rate is shown in Figure 6-7*c*, being approximated by a quadratic function. Whether overtime or undertime costs will occur for a given decision depends on the balance of costs defined by the horizon time. For example, in responding to the need for increased output, the costs of hiring and training must be balanced against the overtime costs. Or conversely, the response to a decreased production rate would require the balancing of layoff costs against the costs of undertime.

The general shape of the net inventory cost curve is shown in Figure 6-7*d*. When inventories deviate from ideal levels, either extra inventory costs must be absorbed if inventory levels are too high or costs of back ordering or lost sales will occur if inventory levels are too low. Again, these costs are approximated by a quadratic function in the LDR model.

The total incremental cost function is then the sum of the four component cost functions for a particular example. The mathematical problem is to minimize the sum of the monthly combined cost function over the planning horizon time of N periods. The result of this mathematical development is the specification of two linear decision rules to be used to compute the aggregate size of the work force and the production rate for the upcoming period. These two rules require as inputs the forecast for each period of the planning horizon in aggregate terms, the ending size of work force, and inventory level in the last period. Once the two rules have been developed for a specific application, the computations needed to produce the decisions recommended by the model require only 10 to 15 minutes by manual methods.*

An Example

An LDR model was developed for a paint company and applied to a six-year record of known decisions in the company. Two kinds of forecasts were used as inputs: a perfect forecast and a moving average forecast. The actual order pattern was extremely variable, involving both the 1949 recession and the Korean War. The graphical record of actual factory performance compared with the simulated performance of the LDR is shown in Figures 6-8 and 6-9 for production rates and work force levels. Costs were reconstructed for the six-year period of actual operation and pro-

* The equations for the Linear Decision Rule model are given in Appendix G.

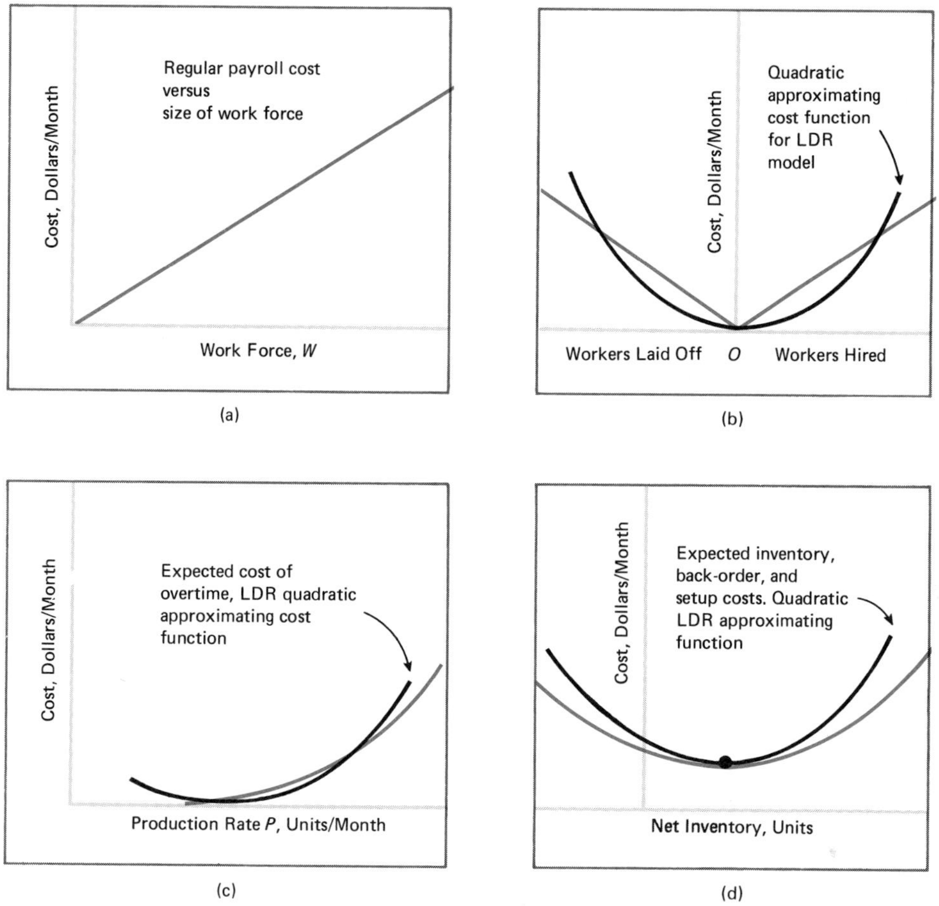

FIGURE 6-7
Approximating linear and quadratic cost functions used by the Linear Decision Rule (LDR) model. Colored lines represent presumed actual cost functions, and black lines represent the LDR approximating functions.

jected for the decision rules based on the nonquadratic cost structure originally estimated from paint company data. The cost difference between the actual company performance and performance with the LDR with the moving average forecast was $173,000 per year in favor of the LDR.

LDR has many important advantages. First, the model is optimizing, and the two decision rules, once derived, are simple to apply. In addition the model is dynamic and representative of the multistage kind of system that we discussed in connection with Figure 6-4. On the other hand, the quadratic cost structure may have severe limitations and probably does not adequately represent the cost structure of any organization.

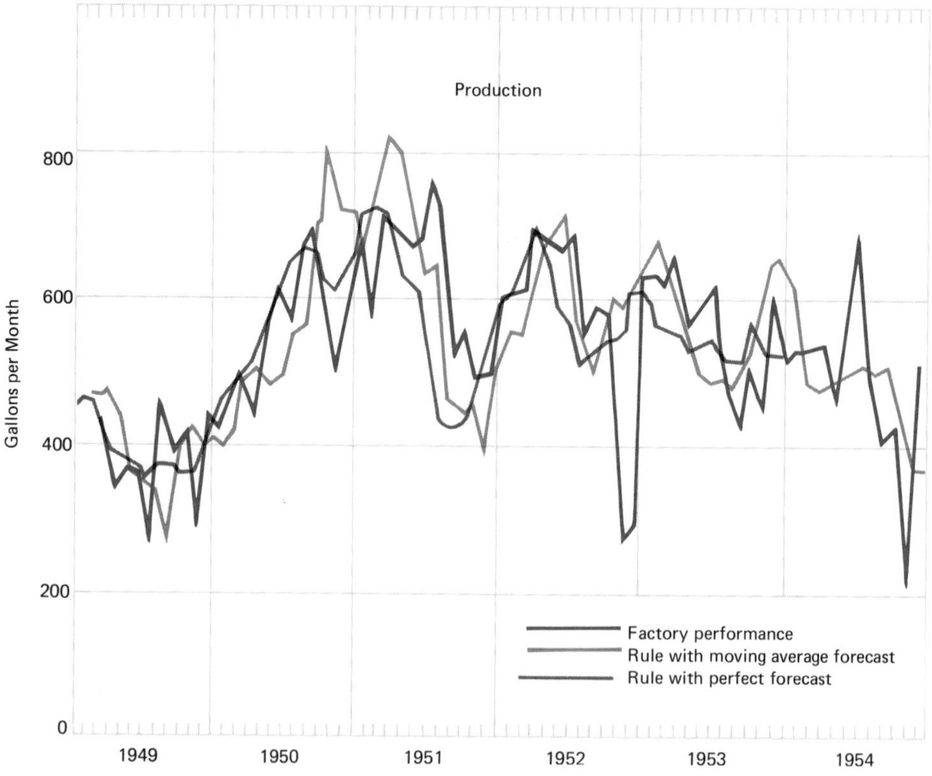

FIGURE 6-8
Comparative performance of the Linear Decision Rule (LDR) with actual factory performance for production rates.
SOURCE: *C. C. Holt, F. Modiglinai, and H. A. Simon, "A Linear Decision Rule for Production and Employment Scheduling,"* Management Science, 2 *(2), October 1955.*

Linear Programming Methods. The aggregate planning problem has been developed in the context of both simplex and distribution models of linear programming. Bowman [1956] proposed the *distribution model* of linear programming for aggregate planning. This model focused on the objective of assigning units of productive capacity, so that production plus storage costs were minimized and sales demand was met within the constraints of available capacity. The rim conditions in the distribution table form the constraints that sales requirements must be met on the one hand and that capacity limitations in the form of initial inventory, regular-time production capacity, and overtime production capacity must be met on the other hand. Both beginning and ending inventories must be specified for the program developed over the N periods in the planning horizon. The matrix elements are costs.

The criterion function to be minimized is the combined regular production, overtime production, and inventory cost. The output of the process is a program that specifies the amount of regular and overtime production in each period of the planning horizon.

The basic matrix can be extended to more than one product by establishing a separate column in each period for each product.

Distribution methods of linear programming have serious limitations when applied to the aggregate planning problem, and our interest in them is now mainly historical. First, the distribution model does not account for production change costs, such as hiring and layoff of personnel, and there is no cost penalty for back ordering or lost sales. Thus, resulting programs may call for changes in production levels in one period, requiring an expanded work force, only to call for the layoff of these workers in the future periods. Also, the linearity requirement often is too severe.

The *simplex method* of linear programming makes it possible to include production level change costs and inventory shortage costs in the model. Hanssman and Hess [1960] developed a simplex model that is entirely parallel with the Linear Decision Rule in terms of using work force and production rate as independent decision variables and in terms of the components of the cost model. The main difference between the Hanssmann-Hess and LDR models is that all cost functions must be linear and that linear programming is the solution technique in the H-H models. One's preference between the two models would depend on a preference for either the linear or quadratic cost model in a given application.

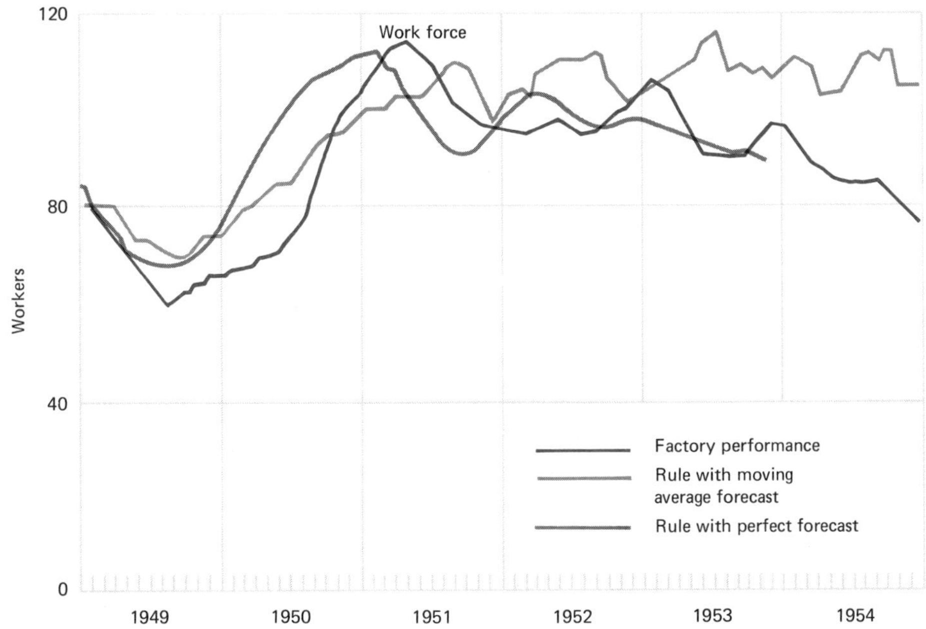

FIGURE 6-9
Comparative performance of the Linear Decision Rule (LDR) with actual factory performance for work force size.
SOURCE: *C. C. Holt, F. Modigliani, and H. A. Simon, "A Linear Decision Rule for Production and Employment Scheduling,"* Management Science, *2 (2), October 1955.*

Industrial applications of linear programming to aggregate planning problems are reported by Eisemann and Young [1960] in a study of a textile mill, by Fabian [1967] in a study of blast furnace production, and by Greene et al. [1959] in a study of the packing industry. Krajewski and Thompson [1975] applied linear programming for aggregate planning in public utilities.

Goal Programming. Lee and Moore [1974] have applied the general goal programming format to the aggregate planning problem. Goal programming is an extension of linear programming that uses the slack variables to free the user from the usual unidimensional objective function. Managerial goals are identified and placed in priority order. The following is a list of goals that pertain to aggregate planning in a specific example:

P_1 Operate within the limits of productive capacity.

P_2 Meet the contracted delivery schedule.

P_3 Operate at a minimum level of 80 percent of regular time capacity.

P_4 Keep inventory to a maximum of three units.

P_5 Minimize total production and inventory costs.

P_6 Hold overtime production down to a minimum.

The solution procedure successively seeks the achievement of goals in priority order, where higher priority goals are constraints that cannot be violated. Higher priority goals, then, can be achieved at the expense of lower priority goals.

In an example that used the preceding list of goals, Lee and Moore generated a solution in which, in terms of goal attainment, the results were as follows:

Production capacity (P_1)	Achieved
Product delivery (P_2)	Achieved
Minimum utilization of production capacity (P_3)	Achieved
Inventory capacity (P_4)	Achieved
Minimization of production and inventory cost (P_5)	Not achieved
Minimization of overtime production (P_6)	Not achieved

Therefore, a trade-off was made between (1) production, inventory, and overtime costs, and (2) the higher priority goals of production capacity, product delivery, minimum capacity utilization, and inventory capacity.

Search Decision Rule (SDR)

A computer optimum-seeking procedure may be used to evaluate systematically a cost or profit criterion function at trial points. In the procedure it is hoped that an optimum value will be eventually found, but there is no guarantee. In direct search methods of computer optimum seeking, the cost criterion function is evaluated at a point, the result compared to previous trial results, and a move determined on the basis of a set of heuristics. The new point is then evaluated, and the procedure repeated until either a better value of the function cannot be found or the predetermined computer time limit is exceeded. Taubert [1968, 1968a] experimented with computer search methods, using the paint company data as a vehicle.

The costs to be minimized are expressed as a function of production rates and work force levels in each period of the planning horizon. Therefore, each period included in the planning horizon requires the addition of two dimensions to the criterion function, one for production rate and one for work force size. The particular pattern search program used was written to handle a maximum of 20 independent variables, and therefore, the planning horizon time was limited to 10 months in Taubert's analysis of the paint company. The search program was set to end whenever the decrease in the objective cost function found by SDR's exploration of the response surface was less than $\$0.5 \times 10^{-6}$.

Table 6-5 shows a sample of the computer output for the first month of factory operation of the paint company. The computer output gives the first month's decision as well as an entire program for the planning horizon of 10 months. In the lower half of the table the program prints out for the entire planning horizon the component costs of payroll, hiring and layoff, overtime, inventory, and the total of these costs. Thus, a manager is provided not only with the immediate decisions for the upcoming month but also with the projected decisions based on monthly forecasts for the planning horizon time and the economic consequences of each month's decisions. Of course, the more distant projections lose much of their significance. The projections are updated with each monthly decision based on the most recent forecast and cost inputs.

Table 6-6 gives a month-by-month comparison for the first 24 months of the results obtained by the SDR program and those obtained by the two optimum decision rules for the LDR. The month-by-month decisions are not identical, but they are very close to each other; the 24-month production totals differ by only two gallons. The total cost of the SDR program exceeds the LDR total by only \$806 or 0.11 percent. This difference may be accounted for by the fact that the SDR used a planning horizon of only 10 months as compared to the 12-month horizon used by the LDR.

With the encouraging results in virtually duplicating the performance of LDR for the paint company, it was decided to test SDR in more demanding situations. Thus, models were developed for three rather different applications: (1) the Search Company [Buffa and Taubert, 1967, 1972], a hypothetical organization involving a much more complex cost model, including the possibility of using a second shift when needed; (2) the Search Mill [Redwine, 1971], based on disguised data obtained from a major American integrated steel mill; and (3) the Search Laboratory [Taubert,

TABLE 6-5 **SDR Output for the First Month of Factory Operation (Perfect Forecast)**

A. SDR Decisions and Projections

Month	Sales (gallons)	Production (gallons)	Inventory (gallons)	Work Force (men)
0			263.00	81.00
1	430	471.89	304.89	77.60
2	447	444.85	302.74	74.10
3	440	416.79	279.54	70.60
4	316	380.90	344.44	67.32
5	397	374.64	322.08	64.51
6	375	363.67	310.75	62.07
7	292	348.79	367.54	60.22
8	458	345.63	268.17	58.68
9	400	329.83	198.00	57.05
10	350	270.60	118.60	55.75

B. Cost Analysis of Decisions and Projections (dollars)

Month	Payroll	Hiring and Layoff	Overtime	Inventory	Total
1	26,384.04	743.25	2558.82	18.33	29,704.94
2	25,195.60	785.62	2074.76	24.57	28,080.54
3	24,004.00	789.79	1555.68	135.06	26,484.53
4	22,888.86	691.69	585.21	49.27	24,215.03
5	21,932.79	508.43	1070.48	0.36	23,512.06
6	21,102.86	383.13	1206.90	7.06	22,699.93
7	20,473.22	220.51	948.13	186.43	21,828.29
8	19,950.99	151.70	2007.33	221.64	22,331.66
9	19,395.30	171.76	865.74	1227.99	21,660.79
10	18,954.76	107.95	−1395.80	3346.46	21,012.37
					241,530.14

SOURCE. W. H. Taubert, "Search Decision Rule for the Aggregate Scheduling Problem," *Management Science, 14* (6), February 1968, pp. 343–359.

1968a], a fictitious name for a division of a large aerospace research and development laboratory. All three situations represent significant extensions beyond the paint company application with highly complex cost models and other factors to challenge the SDR methodology.

Flowers and Preston [1977] applied the SDR methodology to work force planning in one department of a tank trailer manufacturer. The department had a work force ranging from 14 to 23 employees, and the company employed a total of 250 to 350 persons. Because the department worked almost exclusively to order, the assumption of perfect forecasts was logical and was accepted by management as representing reality. The simplicity of the situation made it possible to restrict decisions to the single variable of work force size. SDR decisions were about 3 percent less costly than actual decisions. This saving was regarded as significant by management and led to their desire to extend the model to include the entire work force.

The preceding application of SDR restricted to a single department with a single decision variable made it possible to construct a dynamic programming model. Flowers, Dukes, and Curtiss [1977] report optimal costs for the tank trailer manufacturer. The SDR costs were within 0.3 percent of optimal.

Sectioning Search

Goodman [1973] proposed applying a modeling technique called sectioning search to the aggregate planning problem. The mathematical procedure is to vary each decision variable, one at a time, to find a preliminary value that optimizes the objective

TABLE 6-6 **A Comparison of Linear Decision Rule and Search Decision Rule for the First 24 Months of Operation with Paint Company Data (Perfect Forecast)**

		Production (gallons)		Work Force (workers)[a]		Inventory (gallons)		Monthly Cost (dollars)	
Month	Monthly Sales (gallons)	LDR	SDR	LDR	SDR	LDR	SDR	LDR	SDR
0				81	81	263	263		
1	430	468	472	78	78	301	305	29,348	29,705
2	447	442	443	75	74	296	301	27,797	27,930
3	440	416	418	72	71	272	279	26,294	26,460
4	316	382	385	69	68	337	348	24,094	24,415
5	397	377	376	67	66	317	327	23,504	23,436
6	375	368	366	66	64	311	318	22,879	22,672
7	292	360	360	65	63	379	386	22,614	22,539
8	458	382	382	65	63	303	309	23,485	23,322
9	400	377	379	66	64	280	288	23,367	23,331
10	350	366	366	67	64	296	304	22,846	22,569
11	284	365	359	69	67	377	379	23,408	23,004
12	400	404	401	72	70	381	380	25,750	25,654
13	483	447	447	75	74	345	344	28,266	28,367
14	509	477	479	79	78	313	314	30,180	30,408
15	500	495	498	83	81	307	312	31,310	31,479
16	475	511	510	87	86	343	348	32,422	32,481
17	500	543	547	91	90	386	394	34,858	35,074
18	600	595	592	96	94	380	387	38,118	38,216
19	700	641	642	100	98	321	328	40,849	41,110
20	700	661	659	103	101	282	287	41,848	41,898
21	725	659	658	105	103	216	220	41,945	41,981
22	600	627	624	106	105	244	245	39,074	38,940
23	432	605	601	107	106	417	413	38,134	37,928
24	615	653	655	109	108	455	454	41,785	42,003
Totals		11,621	11,619	1972	1936	7859	7970	734,176	734,982

[a] Rounded to the next larger number of workers.

SOURCE: W. H. Taubert, "Search Decision Rule for the Aggregate Scheduling Problem," *Management Science, 14* (6), February 1968, pp. 343–349.

function while holding the other decision variables fixed. When the preliminary optimum of the first variable is found, it is fixed, and the procedure is repeated, varying the second decision variable, and so on. The procedure is continued until the objective function cannot be improved by altering any single variable.

The advantages of sectioning search are as follows:

1. The method does not require a particular mathematical structure, as does LDR and linear programming.

2. Dimensionality is not a serious problem. Computational requirements increase only linearly as the problem size increases. It is also computationally fast, requiring only a few seconds of computer time for reasonably sized problems.

3. The method supplies an integer solution; this is particularly important for work force size.

4. The methodology is simple and can be readily understood by nontechnical personnel.

The main disadvantage of sectioning search is shared with all search techniques: the solution for complex problems it not a provable optimum.

Goodman applied the method to the paint company model used to develop and test LDR and duplicated cost results within 0.2 percent of the optimal cost.

Because of the advantages cited previously, the method appears to hold considerable promise and deserves further research.

COMPARATIVE PERFORMANCE OF AGGREGATE PLANNING DECISION PROCESSES

Because the LDR is optimal for the model, it has commonly been used as a standard for comparison by proposers of new decision processes for aggregate planning. The availability of a standard for comparison has been particularly valuable for decision processes that are not mathematically optimal. By substantially duplicating LDR performance on standard problems, such as the paint company, the general validity of such methods has been established. Such comparisons, however, fall short of validating the performance of any decision process in real environments. LDR assumes that cost structures are quadratic in form, when in fact cost functions may take a variety of mathematical forms. Thus the best evaluation of the comparative performance of alternate decision processes is in the real world, where we are attempting to optimize the costs (or profits) actually found, instead of a restrictive model of cost behavior.

Lee and Khumawala [1974] report a comparative study carried out in the environment of a firm in the capital goods industry having annual sales of approximately

$11 million. The plant was a typical closed job-shop manufacturing facility in which parts were produced for inventory and then assembled into finished products either for inventory or for customer order. A model of the firm was developed that simulated the aggregate operation of the firm. Demand forecasting in the model provided an option to use either a perfect or imperfect forecast, and four alternate aggregate planning decision processes were used to plan production and work force size.

The four decision processes were LDR, SDR, Parametric Production Planning, and the Management Coefficients model. Parametric Production Planning is a decision process proposed by Jones [1967] that uses a coarse grid search procedure to evaluate four possible parameters associated with minimum cost in the firm's cost structure. The cost structure is developed for the particular firm and is free of constraints on mathematical form. The four parameters are then inserted in decision rules for work force size and production rate. There is no guarantee of optimality.

The Management Coefficients model was proposed by Bowman [1963] and established the *form* of decision rules through rigorous analysis but determines the *coefficients* for the decision rules through statistical analysis of management's own past decisions. The theory behind Bowman's rules is rooted in the assumption that management is actually sensitive to the same behavior used in analytical models and that management behavior tends to be highly variable rather than consistently above or below optimum performance.

The assumption in the Management Coefficients model, then, is that management's performance using the decision rules can be improved considerably simply by applying the rules more consistently. In terms of the usual dish-shaped criterion function, variability in applying decision rules is more costly than being slightly above or below optimum decisions, but being consistent, since such functions are commonly quite flat near the optimum.

Results

Table 6-7 summarizes comparative profit performance for actual company decisions and the four test decision models. When the imperfect forecast available to the management of the firm is used, all four decision models result in increased profits. The minimum mean profit increase compared to company decisions is $187,000 (4 percent) using the Management Coefficients model. The maximum mean increase is $601,000 (14 percent) using SDR. The contrast between the profit figures for the perfect and imperfect forecast gives a measure of the value of forecast information. While perfect information has value ($119,000 for SDR), it is less significant than the decision process ($601,000 difference between SDR and company decisions, and $414,000 difference between SDR and Management Coefficients).

JOINT DECISIONS AMONG OPERATIONS, MARKETING, AND FINANCE

The entire thrust of aggregate planning and scheduling methods is to employ systems concepts in making the key decisions for operations. The results of coordinating

TABLE 6-7 **Comparative Profit Performance**

	Imperfect Forecast	Perfect Forecast
Company decisions	$4,420,000	—
Linear Decision Rule	$4,821,000	$5,078,000
Management Coefficients model	$4,607,000	$5,000,000
Parametric Production Planning	$4,900,000	$4,989,000
Search Decision Rule	$5,021,000	$5,140,000

SOURCE: W. B. Lee and B. M. Khumawala, "Simulation Testing of Aggregate Production Planning Models in an Implementation Methodology," *Management Science, 20* (6), February 1974, pp. 903–911.

decisions concerning activity levels, work force size, use of overtime, and inventory levels amply illustrate that these kinds of decisions should be made jointly rather than independently. To make them independently is to suboptimize. But why stop with the operations function? Would even better results be obtained if some of the key operational decisions were made jointly with other key decisions in marketing and finance?

Tuite [1968] proposed merging marketing strategy selection and production scheduling. Holloway [1969] proposed price as an independent variable coupled with allocations of compensatory promotion budgets. Bergstrom and Smith [1970] proposed estimating revenue versus sales curves for each product in each time period, the amount to be sold considered as a decision variable dependent upon price and possibly other parameters. Finally, Damon and Schramm [1972] proposed to make decisions jointly in production, marketing, and finance. In their model, marketing sector decisions are made with respect to price and promotion expenditures, and the finance sector decisions are made with respect to investment in marketable securities and short-term debt incurred or retired. The solution technique was a computer search methodology similar to SDR.

AGGREGATE PLANNING FOR NONMANUFACTURING SYSTEMS

The general nature of aggregate planning and scheduling problems in nonmanufacturing settings is basically similar in that we are attempting to build a cost or profit model in terms of the key decision variables for short-term capacity. The degrees of freedom in adjusting short-term capacity in nonmanufacturing settings are likely to be fewer, however, because of the absence of inventories and subcontractors as sources of capacity. The result is that the manager is more likely to absorb fluctuations in demand rather directly by varying work force size, hours worked, and overtime.

We will not attempt any detailed discussion of aggregate planning in nonmanufacturing here, since separate chapters follow dealing with operations planning and control in large-scale projects and in service systems. These chapters will include the special problems of both aggregate and detailed schedules in such systems.

THE MANAGER AND COST-REDUCING RESPONSES

Aggregate scheduling is a formal approach to smoothing that in its most developed state extends into the marketing and financial functions. In its broadest conception, decisions are made jointly concerning production, employment, prices, compensatory marketing and promotion expenditures, and short-term investments. Still, there are smoothing strategies used by managers that are not included in any formal model. Galbraith [1969] included among these strategies attempts to influence and manipulate the demand function, adaptation of the organization in various ways, and coordination with other organizations.

Influencing Demand

As we noted, some of the formal models include attempts to shift demand from sales peaks to valleys through counterseasonal pricing and promotion of their products and services. These models include decision variables regarding price and promotion funding through known relationships regarding price elasticity of demand and response to promotion. The models are attempts to formalize well-known managerial behavior illustrated by post-Christmas sales and the airlines' offerings of special prices during the off-season and during the night, when equipment would otherwise get poor utilization. Galbraith also gives examples in the post office and the highly seasonal flower industry.

In addition, however, managers look at their demand in the aggregate and seek to expand product lines with products that have demand that is counterseasonal to their existing lines but that use the same basic production technology. Vergin [1966] found in his analysis of eight manufacturing organizations with seasonal demand patterns that the counterseasonal product was the dominant managerial strategy, almost to the exclusion of other very attractive alternatives, such as aggregate scheduling.

Organization Adaptations

Mangers make fundamental adaptations of their organizations in an attempt to smooth demand in relation to often fixed resources. For example, in one of the early applications of the LDR in a chocolate factory, the firm changed its location to a rural area in order to take advantage of the farm labor available in the fall and winter season. Thus, a hiring and layoff smoothing strategy became compatible with seasonal production. The guaranteed annual wage in the meat-packing industry makes it possible to vary the length of the work week without substantial wage variations. Of course, the use of counterseasonal products is also an adaptive response.

Coordination with Other Organizations

One of the common managerial strategies has been to subcontract needs above certain capacity limits when possible. Such a managerial strategy involves coordination between two usually different firms in the marketplace. A larger integrated firm may make subcontracting arrangements with smaller, more flexible firms operating in the same field. Galbraith [1969] quotes examples in the coal and oil-refining industries.

In other situations, managerial strategy may involve coordination between producers and customers in which the producer may be able to increase lot sizes, split lots, or delay or speed up deliveries in order to smooth the work load. In return the customer receives preferential supply treatment.

Finally, it may be possible for some organizations to form a coalition that can have the effect of smoothing work loads. For example, electric utilities join together in networks of supply that make it possible to meet peak demands for the network system as a whole. If each individual utility had to meet its individual peak demands, higher plant investment would be required in the aggregate and for each organization. The airlines have found that by sharing equipment when route structures are noncompetitive and counterseasonal, the organizations involved can achieve somewhat better equipment utilization.

IMPLICATIONS FOR THE MANAGER

The aggregate planning problem is one of the most important to managers, since it is through these plans that major resources are deployed. Through the mechanisms of aggregate planning, management's interest is focused on the most important aspects of this deployment process; basic employment levels and activity rates are set, and where inventories are available as a part of the strategy, their levels are also set. Given managerial approval of these broad level plans, detailed planning and scheduling of operations can proceed within stated operating constraints.

We have discussed the structure of the aggregate planning problem and a number of alternate decision processes. At this point the graphic methods are probably used most frequently. Mathematical and computer search methods have been developed in an effort to improve on the traditional methods by making the process dynamic, optimum seeking, and representative of the multistage nature of the problem. Several models have been of value mainly as stepping-stones to more useful models that represent reality more accurately. The most important single stepping-stone has been the LDR; however, its original advantage in requiring only simply computations has been largely offset by the computer.

Presently, the computer search methods seem to offer the most promise because of their greater flexibility in representing costs that really occur in organizations. Although some of the analytical methods do produce optimum solutions, we must remember that it is the model that is being optimized. The real-world counterpart of the model is also optimized only if the mathematical model duplicates reality. The computer search methods are only optimum seeking by their nature but do not suffer from the need to adhere to strict mathematical forms in the model and can therefore more nearly duplicate reality in cost and profit models.

The extension of aggregate planning models to make joint decision among the production, marketing, and finance functions is most encouraging and demonstrates progress in our ability to employ systems concepts.

Perhaps the greatest single contribution of formal models to aggregate planning is that they provide insight for the manager into the nature of the resource problem faced. Managers need to understand that good solutions ordinarily involve a mixed strategy that uses more than one of the available options of hiring-layoff, overtime, inventories, outside processing, and so forth. The particular balance for a given organization will depend on the balance of costs in that organization. Even if formal models are not used for decision making, they may be useful as managerial learning devices concerning the short-term capacity economics of their enterprise. Their judgment about the most advantageous combination of strategies at any particular point in the seasonal cycle can be developed through a gaming process.

The manager must be concerned not only with the direct economic factors that enter the aggregate planning problem but also with the human and social effects of alternate plans. These kinds of variables are not included in formal models; however, by considering a range of alternate plans that meet the human and social requirements in varying degrees, managers can make trade-offs between costs and the subjective values. Thus the formal models can help managers generate aggregate plans that are acceptable on the basis of broadly based criteria.

Inventories are not available to absorb demand fluctuations in service and non-manufacturing situations. While this variable is not available as a managerial strategy, the aggregate planning problem in such organizations is conceptually similar. Thus, managers must focus their strategies on hiring and layoff, the astute use of normal labor turnover, and the allocation of overtime and undertime. In service-oriented situations, the use of part-time workers is often an effective strategy.

REVIEW QUESTIONS AND PROBLEMS

1. What is the meaning of the term "aggregate plan"? What are the objectives of aggregate plans? What are the inputs and the nature of outputs?

2. Place aggregate planning in context with the term "planning horizon." What is the appropriate horizon for aggregate planning?

3. Discuss the relevant cost components involved in aggregate planning decisions.

4. Under what conditions would a single-stage aggregate planning decision system be appropriate?

5. In what ways does a multistage aggregate planning decision system take account of realities that in fact would affect decisions for production rates and work force size?

6. Appraise graphic methods of aggregate planning.

7. Compare the Linear Decision Rule with the multistage aggregate planning decision system discussed and summarized by Figure 6-4. What compromises with reality, if any, have been made by the LDR model?

8. Compare the Hanssmann-Hess linear programming model of aggregate planning with the multistage decision process discussed and summarized by Figure 6-4. What compromises with reality have been made, if any, in the linear programming model?

9. As a decision system, contrast the SDR with the LDR. What are the advantages and disadvantages of each?

10. Discuss the unique methodology of sectioning search. How does it differ from the methodology of SDR?

11. Referring to Figure 6-7:
    a. Rationalize why the cost of overtime should increase at an increasing rate as production rate increases.
    b. If inventory varies from the optimal level (minimum cost level), why should the incremental costs increase at an increasing rate as represented in Figure 6-7*d*?

12. Criticize the usefulness and validity of the strict aggregate planning concept, that is, making decisions solely in aggregate terms of work force size and production rate.

13. What is the meaning of the term "capacity" in aggregate planning models? How does a decision to hire, lay off, or subcontract affect capacity? How does physical or limiting capacity affect these decisions?

14. Cost comparisons between the result of actual managerial decisions and those

produced by solving decision rule models are typically made by running both sets of decisions through the cost model and then comparing the results. Does this methodology seem valid? If not, what other approach might be followed?

15. Account for the difference in performance of the several aggregate planning decision systems in the study summarized by Table 6-7.

16. What values are gained by expanding aggregate planning decision systems to produce joint decisions among operations, marketing, and finance? Are there any disadvantages?

17. Table 6-8 gives data that show projected requirements for the production of a medium-priced camera, together with buffer stock requirements and available production days in each month. Develop a chart of cumulative requirements and cumulative maximum requirements for the year, plotting cumulative production days on the horizontal axis and cumulative requirements in units on the vertical axis.

18. Using the data of Table 6-8, compare the total incremental costs involved in a level production plan, in a plan that follows maximum requirements quite closely, and in some intermediate plan. Normal plant capacity is 400 units per working day. An additional 20 percent can be obtained through overtime but at an additional cost of $10 per unit. Inventory carrying cost is $30 per unit per year. Changes in production level cost $5000 per 10 units in production rate. Extra capacity may be obtained through subcontracting certain parts at an extra cost of $15 per unit. Beginning inventory is 600 units, or must be determined for some plans.

TABLE 6-8

Projected Production and Inventory Requirements

Month	Production Requirements	Required Buffer Stocks	Production Days
Jan.	3000	600	22
Feb.	2500	500	18
Mar.	4000	800	22
Apr.	6000	1200	21
May	8000	1600	22
June	12,000	2400	21
July	15,000	3000	21
Aug.	12,000	2400	13
Sept.	10,000	2000	20
Oct.	8000	1600	23
Nov.	4000	800	21
Dec.	3000	600	20
	87,500	17,500	244

19. Given the data in Table 6-4:
    a. What value of inventory carrying cost would make Plans 1 and 2 equally desirable?
    b. What hiring-layoff cost makes Plans 1 and 2 equally desirable?
    c. What subcontracting cost makes Plans 2 and 3 equal?

SITUATIONS

20. Schwarz and Johnson [1978] have reevaluated the empirical performance of the paint company application of the LDR. The results of the paint company applications of LDR compared to company performance are summarized in Table 6-9. The inventory related costs (inventory plus back order costs = 361 + 1566 = 1927) account for 1927 × 100/4085 = 47.2 percent of the total company performance costs. On the other hand, these costs account for only (451 + 616 = 1067) × 100/3222 = 33.1 percent of LDR total costs with the moving average forecast, and (454 + 400 = 854) × 100/2929 = 29.2 percent of LDR costs with the perfect forecast.

    Schwarz and Johnson focus attention on the source of cost reductions in Table 6-9. They point out that inventory-related cost reductions by themselves account for nearly all the cost reduction achieved, as shown in Table 6-10. They then state the following hypothesis:
    The LDR (substitute any other aggregate planning model you like) has not been implemented because, despite its conceptual elegance, most of the cost savings of the LDR may be achieved by improved aggregate inventory management alone.

TABLE 6-9 **Comparative Costs in Thousands for the Paint Company, 1949–1953.**

		Decision Rule	
Type of Cost	Company Performance	Moving Average Forecast	Perfect Forecast
Regular payroll	$1,940	$1,834	$1,888
Overtime	196	296	167
Inventory	361	451	454
Backorders	1,566	616	400
Hiring and layoffs	22	25	20
Total cost	$4,085(139%)	$3,222(110%)	$2,929(100%)

SOURCE: C. C. Holt, F. Modigliani, and H. A. Simon, "A Linear-Decision Rule for Production and Employment Scheduling," *Management Science*, 2(2), October 1955.

TABLE 6-10 **Source of Major Cost Reductions for the LDR Application in the Paint Company**

	LDR (Moving Average Forecast) versus Company Performance	LDR (Perfect Forecast) versus Company Performance
Total cost reduction	4085 − 3222 = 863	4085 − 2929 = 1156
Inventory related cost reduction	1927 − 1067 = 860	1927 − 854 = 1073
Percent, inventory to total cost reduction	99.7%	92.8%

The authors argue that if company management had begun with an initial gross inventory of 380 units (corresponding to the LDR's ideal *net* inventory of 320 units) as the LDR did, total company costs would have been only $3,228,000. This is approximately the same total cost achieved by LDR using the moving average forecast. In addition, they state, "More important, *any* initial gross inventory between 380 and 700 would have resulted in total company costs *lower* than the LDR with moving average forecasts." Schwarz and Johnson state further, "Please note that we do *not* claim that the paint company management was following the 'wrong' inventory policy, although given the firm's true backorder and holding costs it would appear that it was. We also do not claim that management *could be expected* to have made exactly the same sequence of monthly production and workforce decisions as they did before, given a higher buffer inventory. We *do* claim the following: virtually all of the LDR's cost savings *could* have been obtained without recourse to any aggregate planning model. All that was necessary was a different inventory management policy: in this case a significantly higher buffer inventory."

Given the empirical findings by Schwarz and Johnson, do you feel that a company should ignore aggregate planning concepts and simply focus on aggregate inventory management, as implied in the authors' hypothesis? Would the results have been the same in another organization where the inventory costs were twice what they were in the paint company? Ten times what they were in the paint company? In relation to the preceding two questions, how do you interpret the statement by Schwarz and Johnson: "Please note that we do *not* claim that the paint company management was following the 'wrong' inventory policy, although given the firm's true backorder and holding costs it would appear that it was"? If we accept the Schwarz-Johnson implied conclusion, that only inventory management is needed, what recommendations regarding aggregate planning would you make to organizations that produce no inventoriable product?

21. The hospital admissions system is used as an overall mechanism for planning and scheduling the use of hospital facilities. The basic unit of capacity is the hospital bed, and bed occupancy level is the variable under management control. Bed occupancy level, in turn, opens up revenue flow from the broadly based health care services of the hospital. The utilization and resulting revenue from these hospital services depend, in turn, on patient mix selection that affects length of stay of patients. The demand on other hospital services, such as x-ray and other laboratories, then flows from patient mix.

The problem of operations management is not simply to maximize bed occupancy level, since emergency demands must be met and since the cost of such services incurred when bed occupancy levels are too high increases rapidly. Thus, an aggregate planning and scheduling model must take account of scheduling the flow of elective arrivals and the patient mix therefrom, as shown in Figure 6-10. This flow can be regulated to some substantial degree, since the actual admission date for such cases can be scheduled, and this represents the hospital manager's buffer against the seeming capriciousness of demand. The flow of emergency cases, however, is on a random arrival basis and compounds the problem greatly. The cost-effectiveness of the entire hospital admissions system is dependent on variables that follow probabilistic distributions. Most costs are fixed in nature (rooms, beds, laboratory facilities, basic staff), comprising up to 75 percent of total hospital costs. Yet the demand on these services is highly variable in nature.

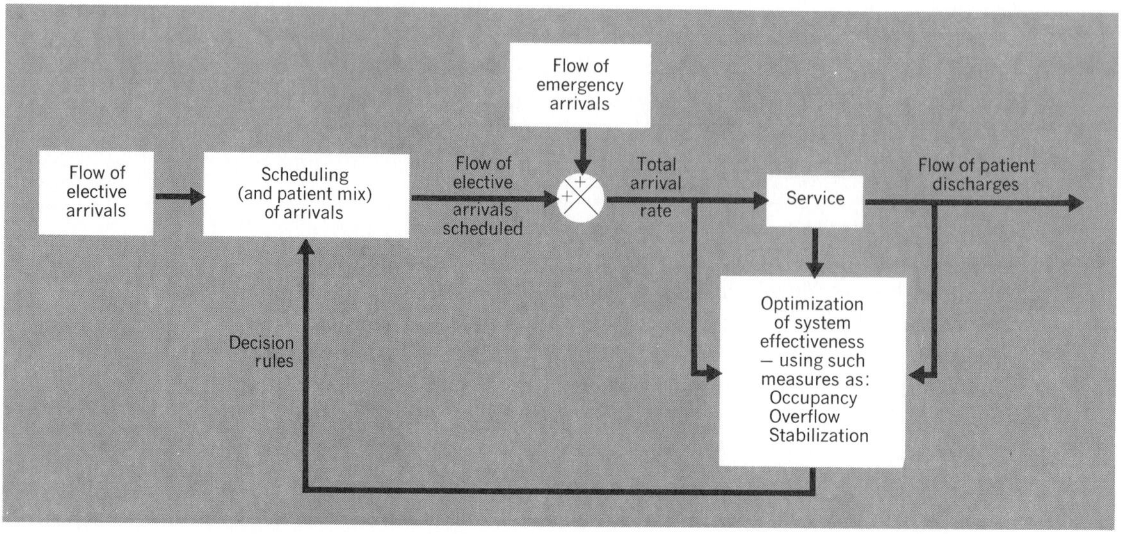

FIGURE 6-10
An overview of the hospital admissions process.
SOURCE: *J. H. Milsum, E. Turban, and I. Vertinsky, "Hospital Admissions Systems: Their Evaluation and Management,"* Management Science, 19 *(6), February 1973, p. 655.*

Do you feel that aggregate planning is a useful concept in hospitals? What form should it take, that is, should it be a cost-minimizing approach, an attempt merely to find feasable solutions, a methodology designed to maximize service to patients, or what? Do you feel that any of the formal models discussed in the chapter can be applied?

22. Taubert [1968] developed an aggregate planning cost model for the Search Laboratory, a hypothetical name for a real company. The laboratory is housed in a 100,000 square foot facility employing a staff of 400. Approximately 300 members of the staff are classified as direct technical employees, and the balance are indirect administrative support for the operations of the laboratory.

The laboratory offers a capability through its scientific staff and facilities, and widely fluctuating employment could severely impair this capability. The research programs of the laboratory are funded by both the government and the corporation, and an important part of the operating environment is wide fluctuations in government sales and rapid shifts in technology. Thus the operations planning problem is defined by the need for employment stability on the one hand and wide fluctuations in government sales on the other.

Specifically, the operations planning problem is centered in a monthly decision by the director to determine the size of the scientific staff and administrative staff as well as the allocation of the scientific staff to government contracts, corporate research programs, and overhead. Overhead charges arise when there are no contracts or corporate research programs available for scientists. This charge is in addition to the charges normally made to overhead for the usual indirect costs. In effect, then, overhead is used as a buffer to absorb fluctuations in the demand for scientific personnel. The four independent decision variables incorporated in the aggregate planning model are as follows:

1. The size of the scientific staff:
   WG_t, personnel allocated to government contracts
   WR_t, personnel allocated to corporate research programs
   WO_t, personnel allocated to overhead

2. WI_t, the size of the indirect administrative support staff

Cost Model

Figure 6-11 shows the 12 cost relations that form the components for the cost model of the Search Laboratory. Note that a variety of mathematical relationships are included, such as linear, piecewise linear, constraints, and nonlinear forms. Taubert also built into the model a complete set of equations representing the overhead cost structure used to compute the overhead rate for any given set of decision variables. The resulting overhead rate is then used to compute the monthly government sales volume, which in turn is compared to a cumulative sales target.

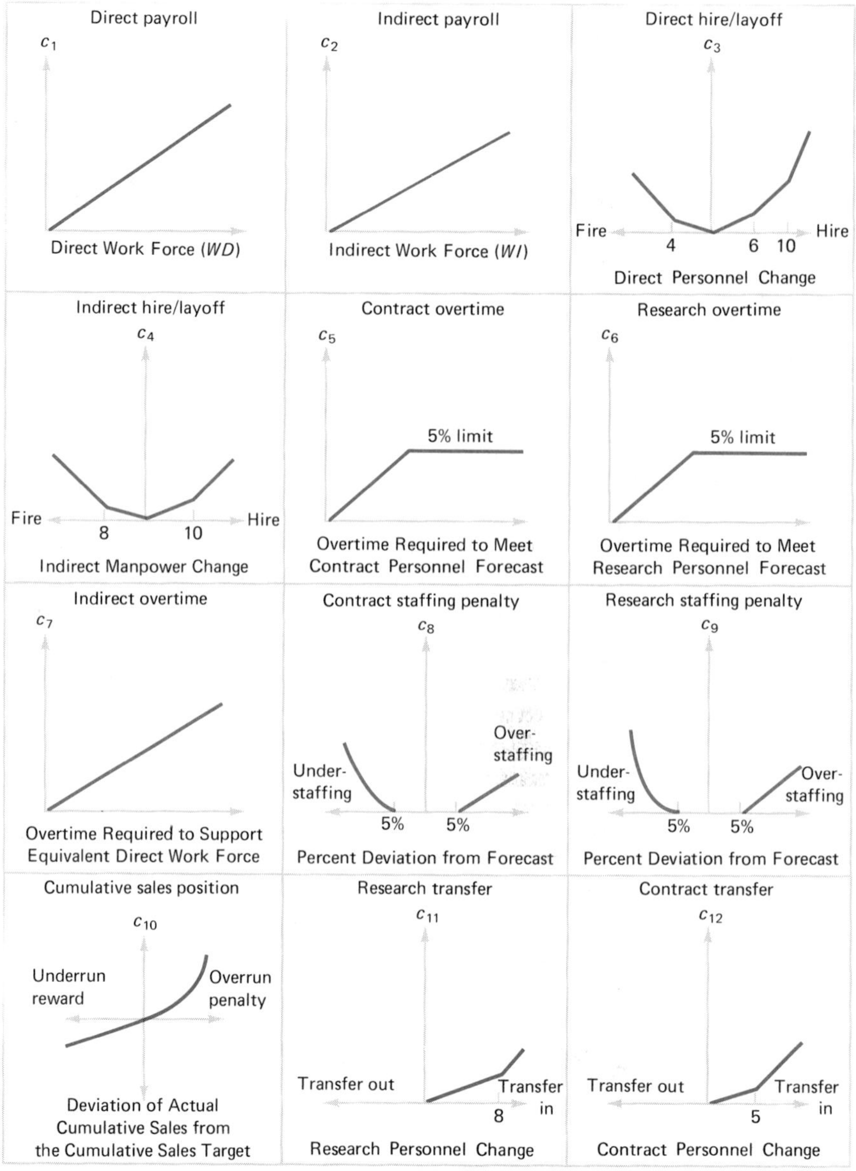

FIGURE 6-11
The 12 cost relationships for the Search Laboratory cost model.
SOURCE: *W. H. Taubert, "The Search Decision Rule Approach to Operations Planning." Unpublished PhD dissertation, UCLA, 1968.*

The inputs to the decision system are monthly forecasts of contract personnel, research personnel, overhead personnel, and a cumulative sales target that represents the financial plan of the laboratory. The total personnel forecast must be met, and it is a part of the director's operations planning problem to determine the best combination of decision variables that will accomplish the objective. Failure to meet the personnel requirements increases costs, and this effect is also implemented in the cost model.

Results

Taubert validated the cost model against the financial record of the laboratory over a five and one-half-year period. Following the validation, the decision system was operated for each month in the five and one-half-year test period. A six-month planning horizon was used that required SDR to optimize a 24-dimensional response surface (four decisions per month for a six-month planning horizon). Figure 6-12 summarizes the comparative results, Figure 6-12*a* showing the contrast between SDR decisions on contract and research personnel compared with forecasts, and Figure 6-12*b* showing a similar comparison of actual management decisions compared with forecasts. Note that the SDR decisions responded much more smoothly to fluctuating personnel forecasts than did actual management decisions.

The costs resulting from SDR decisions compared to actual management decisions indicated that SDR would have produced cost savings. Over the five and one-half-year test period, the SDR advantage ranged from a high of 19.7 percent to a low of 5.2 percent, averaging 11.9 percent over the entire test period. The SDR decisions produced lower overhead rates and significant cost reductions in direct payroll, research program staffing, sales target penalties and rewards, and direct hiring costs. It achieved these results largely through the more extensive use of overtime.

If you were the manager of the Search Laboratory, would the aggregate planning system developed be helpful in making decisions? If so, what kinds of decisions? The Search Laboratory is a system without inventories. With inventories not available as a trade-off, are the manager's hands tied? Is there any flexibility left for him?

23. As a follow-up on the Search Laboratory model discussed in situation 22, Taubert [1968] subsequently disaggregated the decision variables for the size of the scientific staff into six departments. In effect, each department was considered a miniature laboratory with its own contract and research manpower forecast as a sales target. Thus the allocation of research personnel had to be made in each of the six departments. Although some transferring of personnel was allowed, this practice was limited because of the fact that, in general, scientists are not interchangeable and cannot readily be shifted from one department or technical expertise to another simply to meet fluctuating personnel needs. Residual departmental personnel adjustments then had to be handled by hiring and layoff decisions.

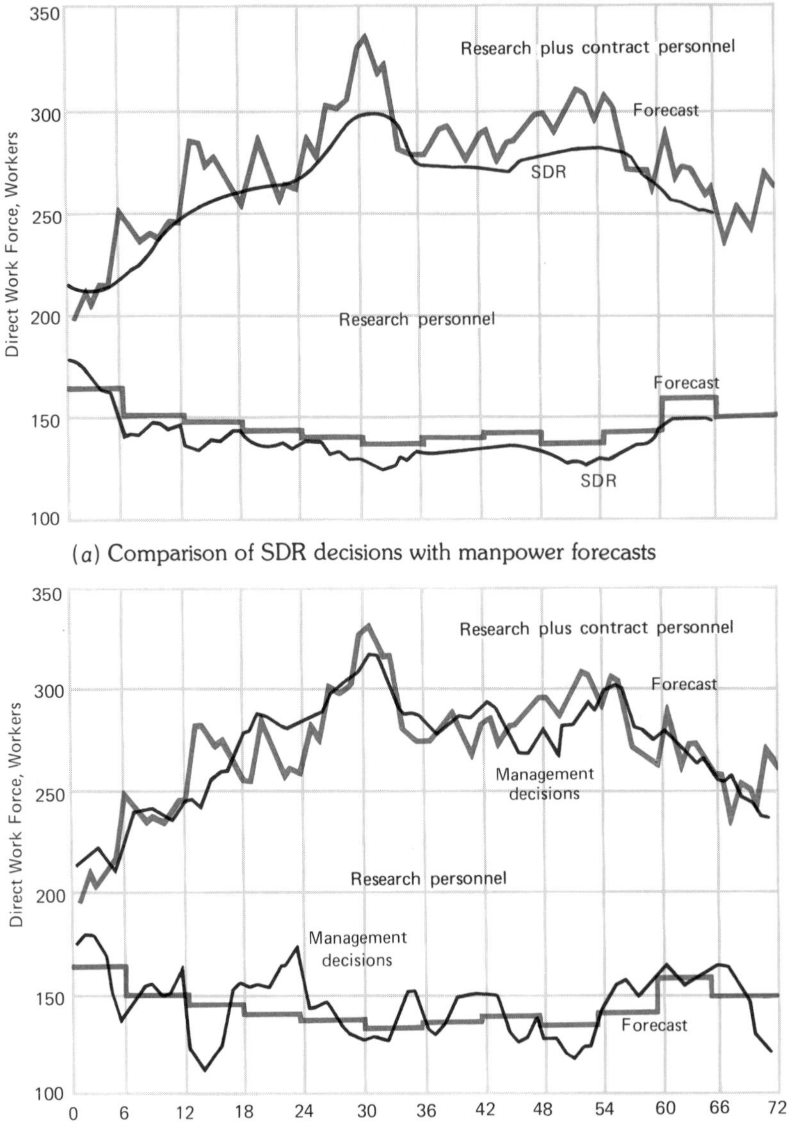

(b) Comparison of management decisions with manpower forecasts

FIGURE 6-12
Results of SDR decisions for Search Laboratory I compared to forecasts and actual management decisions.
SOURCE: *W. H. Taubert, "The Search Decision Rule Approach to Operations Planning." Unpublished PhD dissertation, UCLA, 1968.*

The SDR decisions followed the forecasts with a smooth response, carrying members of the scientific staff in overhead for short periods of time when faced with downturns in the personnel forecasts.

How do you evaluate the disaggregate planning system? Would the additional detail be useful, or simply excess baggage?

24. Feinberg [1972] and Geoffrion, Dyer, and Feinberg [1972] developed an aggregate planning model for a school of management and applied it at the Graduate School of Management at U.C.L.A. The faculty of the school may be viewed as engaging in three primary activities: formal teaching, school service (e.g., administration and curriculum development), and activities such as research and student counseling. The formal teaching occurs at three levels: graduate, lower-division undergraduate, and upper-division undergraduate. The basic planning unit of output is the equivalent course section and all activities are related to that equivalence. Thus, "course releases" are given for such items as administrative activities, curriculum development, and research, so that the overall allocation of faculty effort can be planned in terms of equivalent course section capacity.

Figure 6-13 summarizes an aggregate planning and scheduling model for such a school. The major features of the model are the identification of the variables under school control, those not controllable by the school (such as policies, procedures, and restrictions), and the multiple criteria and output by which resulting aggregate schedules are judged. The capacity to carry on the various activities is complicated by the differing roles of the various kinds of faculty. For example, lecturers provide only teaching service, being assigned nine quarter courses per academic year but no administrative, committee, or other assignments. Regular faculty teach five quarter courses per year but carry on extensive other activities, such as counseling, committee service, and research. Teaching assistants assist in courses and sometimes teach sections of courses in the ratio of one full-time assistant for six equivalent course sections. University rules permit only faculty with certain qualifications to teach graduate level courses, upper-division undergraduate courses, and so on.

Thus, administrators are forced to think in terms of various decision alternatives in filling vacancies in the faculty, as well as in deciding on the other variables under control. The number of students admitted can be controlled, the aggregate number of course sections can be varied, and the amount of equivalent nonteaching activities can be varied.

The central university administration determines certain upper limits of capacity through its budget allocations for faculty and student enrollments, and leaves of absence and extramural releases reduce the capacity available in the planning period.

The criteria and output listed on the right of Figure 6-13 are in fact the unique conception of the school model. As outputs, they represent the aggregate schedule for courses of different levels to be offered, plus faculty time spent in other

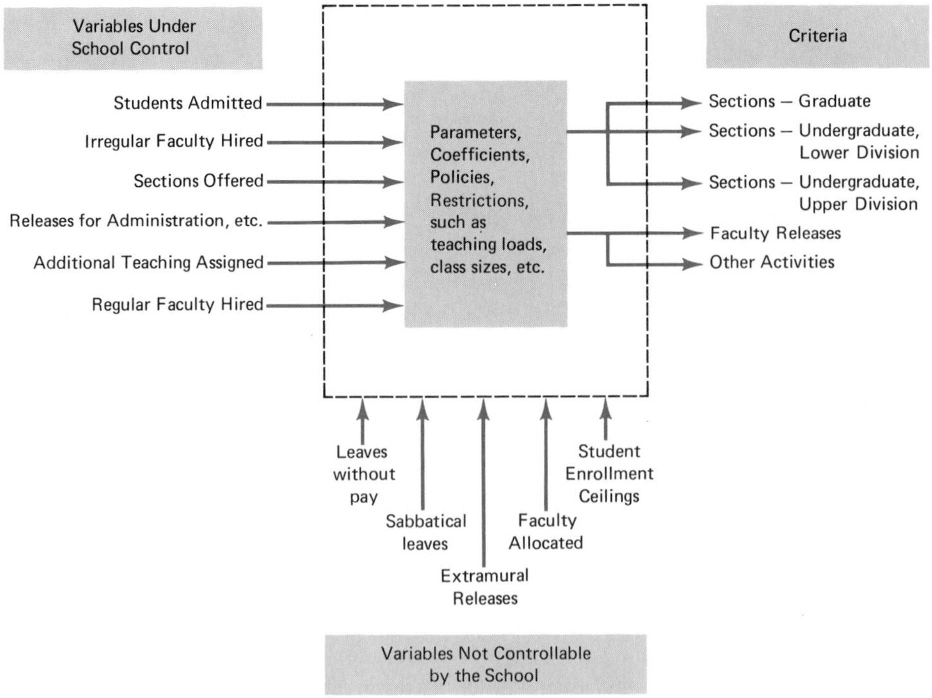

FIGURE 6-13
Aggregate planning model for a school of management.
SOURCE: *Adapted from A. Feinberg, "An Experimental Investigation of an Interactive Approach for Multi-Criterion Optimization with an Application to Academic Resource Allocation." Unpublished PhD dissertation, UCLA, 1972.*

ways. However, they are also the dimensions of a multiple criterion problem, on which the decision maker's preference for a given mix of activities is defined.

The decision model is actually interactive between the decision maker and a computational model in which the decision maker estimates weights for his or her preference for each of the outputs. Given these weights, a mathematical programming algorithm allocates available faculty time to courses at the several levels, releases, research, and additional activities. The computer output of the results is then examined, and the weights are altered if necessary. The weights, in fact, are expressive of the decision maker's preference for the relative emphasis between criteria. They are estimates of marginal substitution rates between each pair of criteria, for example, how many sections of lower-division courses the decision maker is willing to give up for an additional graduate section. This approach to the problem is an attempt to incorporate within the quantitative model dimensions of the multiple criteria for decisions. Given the broad allocations of faculty effort in the aggregate schedule, detailed schedules for actual faculty course assignments must be made, as well as correlated classroom and other facility schedules.

How do you evaluate the school of management aggregate planning system? Would it be really useful to administrators, or would it fall into disuse? What is the criterion in the model?

25. Krajewski and Thompson [1975] developed an aggregate employment planning model for public utilities and applied it in the specific setting of a telephone company. The model has important implications both for management and for the public. The implications for management are for improving resource management and taking advantage of the infrastructure of costs to maximize returns. The implications for the public area in the allowable prices (rates) they must pay.

In the specific setting of a telephone company, the model covered the plant personnel required to install, remove, relocate, repair, and maintain all equipment used for telephones. This equipment ranges from the telephone itself to cable and wires necessary for cross-country transmission. Four classes of employees are involved: PBX repairmen, installer-repairmen, cable splicers, and linemen. In general, the four work groups are regarded as crafts, with the only interchangeability being that linemen can also make telephone connections.

The demand for service is subject to some seasonality for installation and removal of phones (summer peak), and weather conditions can produce seasonal demands for service and cable installation and splicing. On the other hand, telephone repair service is closely related to the number of phones in service. The result of the seasonal nature of the aggregate demand is that the size and timing of hirings, layoffs, and use of overtime work and part-time employees are important decisions. The adroit use of the flexibility available in the employment decisions is a buffer available to management to deal with seasonality, since the output of the system cannot be stored. Two additional variables are under managerial control to deal with demand variations, but both of them interact with the employment decisions: they can vary the level of service and for construction work they can subcontract.

Krajewski and Thompson developed a linear programming model designed to minimize aggregate employment related costs over a planning horizon. The costs were full- and part-time wage costs for regular and overtime work, hiring costs, layoff costs, and subcontracting costs. Within the model, hiring costs were segregated for replacement, a rehire of one laid off, and a hire for expansion. The latter requires additional capital investment, which in turns has rate base implications. The model also contains constraints on employment, production, overtime, and backlog or service delay.

The managerial use of the employment planning model is to prepare annual budgets for the plant department and to determine the monthly operating guidelines for hiring, layoff, overtime by labor class, and for subcontracting.

Some of the results of the study are in the service delay versus cost relationships and the rate of return versus cost to public relationships. For an illustrative simplified example, they showed that the service delay-cost curve was quite shallow; that is, the annual cost difference between zero and a five-day delay was only $11,200. The reason is found in the model structure, since the service delay-cost trade-off is only through overtime and hiring costs, which are a small fraction of total employment costs.

Expansion of the work force to accommodate demand increases requires supporting capital investment, and this investment goes into the *rate base* on which the company is allowed to earn a return. Therefore, Krajewski and Thompson defined an additional cost to expand, involving equity rate of return and depreciation. This cost then enters into the balance of costs for employment planning and leads to an incentive to expand work force in the capital intensive work groups, such as linemen. The result here is that for a demand growth of 10 percent in their example, the model with rate of return yields a 31 percent larger lineman work force for the same service, compared to the results if the capital cost is excluded. The net result is a greater total cost of $5300 to be passed on to the consumer for the same service level.

How do you evaluate the aggregate employment planning system developed for the telephone company? How would you as a manager use the results of the system in monthly planning? Is there an incentive for the telephone company to expand the lineman work force and use less overtime and subcontracting in order to optimize its rate base?

REFERENCES

Bergstrom, G. L., and B. E. Smith. "Multi-item Production Planning—an Extension of the HMMS Rules," *Management Science, 16*(10), June 1970, pp. 614–629.

Bowman, E. H. "Consistency and Optimality in Managerial Decision Making," *Management Science, 9*(2), January 1963, pp. 310–321.

Bowman, E. H. "Production Scheduling by the Transportation Method of Linear Programming," *Operations Research, 4*(1), February 1956, pp. 100–103.

Buffa, E. S. "Aggregate Planning for Production," *Business Horizons,* Fall 1967.

Buffa, E. S., and J. G. Miller. *Production-Inventory Systems: Planning and Control* (3rd ed.). Richard D. Irwin, Homewood, Ill. 1979.

Buffa, E. S., and W. H. Taubert. "Evaluation of Direct Computer Search Methods for the Aggregate Planning Problem," *Industrial Management Review,* Fall 1967.

Damon, W. W., and R. Schramm. "A Simultaneous Decision Model for Production, Marketing, and Finance," *Management Science, 19*(2), October 1972, pp. 161–172.

Eisemann, K., and W. M. Young. "Study of a Textile Mill with the Aid of Linear Programming," *Management Technology, 1,* January 1960, pp. 52–63.

Fabian, T. "Blast Furnace Production—a Linear Programming Example," *Management Science, 14*(2), October 1967.

Feinberg, A. "An Experimental Investigation of an Interactive Approach for Multicriterion Optimization with an Application to Academic Resource Allocation." Unpublished PhD dissertation, UCLA, 1972.

Flowers, A. D., D. L. Dukes, and W. D. Curtiss. "A Dynamic Programming Approach to Workforce Scheduling." Private communication, November 1977.

Flowers, A. D., and S. E. Preston. "Workforce Scheduling with the Search Decision Rule," *Omega, 5*(4), 1977.

Galbraith, J. R. "Solving Production Smoothing Problems," *Management Science, 15*(12), August 1969, pp. 665–674.

Geoffrion, A. M., J. S. Dyer, and A. Feinberg. "An Interactive Approach for Multicriterion Optimization, with an Application to the Operation of an Academic Department," *Management Science, 19*(4), December 1972, pp. 357–368.

Goodman, D. A. "A New Approach to Scheduling Aggregate Production and Work Force," *AIIE Transactions, 5* 2), June 1973, pp. 135–141.

Gordon, J. R. M. "A Multi-Model Analysis of an Aggregate Scheduling Decision." Unpublished PhD dissertation, Sloan School of Management, MIT, 1966.

Greene, J. H., K. Chatto, C. R. Hicks, and C. B. Cox. "Linear Programming in the Packing Industry," *Journal of Industrial Engineering, 10*(5), September-October 1959, pp. 364–372.

Hanssmann, F., and S. W. Hess. "A Linear Programming Approach to Production and Employment Scheduling," *Management Technology, 1,* January 1960, pp. 46–52.

Hausman, W. H., and R. Peterson. "Multiproduct Production Scheduling for Style Goods with Limited Capacity," *Management Science, 18* (7), March 1972, pp. 370–383.

Holloway, C. A. "A Mathematical Programming Approach to Identification and Optimization of Complex Operational Systems with the Aggregate Planning Problem as an Example." Unpublished PhD dissertation, UCLA, 1969.

Holt, C. C., F. Modigliani, and J. F. Muth. "Derivation of a Linear Decision Rule for Production and Employment," *Management Science, 2*(2), January 1956, pp. 159–177.

Holt, C. C., F. Modigliani, J. F. Muth, and H. A. Simon. *Planning, Production, Inventories and Work Force.* Prentice-Hall, Englewood Cliffs, N.J., 1960.

Holt, C. C., F. Modigliani, and H. A. Simon. "A Linear Decision Rule for Production and Employment Scheduling," *Management Science, 2*(2), October 1955, pp. 1–30.

Hooke, R., and T. A. Jeeves. "A 'Direct Search' Solution of Numerical Statistical Problems," *Journal of the Association for Computing Machinery,* April 1961.

Jones, C. H. "Parametric Production Planning," *Management Science, 13*(11), July 1967, pp. 843–866.

Krajewski, L. J. "Multiple Criteria Optimization in Production Planning: An Application for a Large Industrial Goods Manufacturer," *Ohio State University,* WPS 78–88, November 1978.

Krajewski, L. J., and H. E. Thompson. "Efficient Employment Planning in Public Utilities," *Bell Journal of Economics and Management Science,* Spring 1975.

Lee, S. M. *Goal Programming for Decision Analysis.* Auerback, Philadelphia, 1972, Chapter 8.

Lee, S. M., and L. J. Moore. "A Practical Approach to Production Scheduling," *Production and Inventory Management,* 1st Quarter, 1974, pp. 79–92.

Lee, W. B., and B. M. Khumawala. "Simulation Testing of Aggregate Production Planning Models in an Implementation Methodology," *Management Science, 20* (6), February 1974, pp. 903–911.

Milsum, J. H., E. Turban, and I. Vertinsky. "Hospital Admissions Systems: Their Evaluation and Management," *Management Science, 19*(6), February 1973, pp. 646–666.

Moskowitz, H. "The Value of Information in Aggregate Production Planning—a Behavioral Experiment," *AIIE Transactions, 4*(4), December 1972, pp. 290–297.

Redwine, C. N. "A Mathematical Programming Approach to Production Scheduling in a Steel Mill." Unpublished PhD dissertation, UCLA, 1971.

Schwarz, L. B., and R. E. Johnson. "An Appraisal of the Empirical Performance of the Linear Decision Rule for Aggregate Planning," *Management Science, 24*(8), April 1978, pp. 844–849.

Silver, E. A. "A Tutorial on Production Smoothing and Work Force Balancing," *Operations Research, 15*(6), November-December 1967.

Taubert, W. H. "Search Decision Rule for the Aggregate Scheduling Problem," 14(6), February 1968, pp. 343–359.

Taubert, W. H. "The Search Decision Rule Approach to Operations Planning." Unpublished PhD dissertation, UCLA, 1968a.

Tuite, M. F. "Merging Market Strategy Selection and Production Scheduling," *Journal of Industrial Engineering, 19*(2), February 1968, pp. 76–84.

Vergin, R. C. "Production Scheduling Under Seasonal Demand," *Journal of Industrial Engineering, 17*(5), May 1966.

Zimmerman, H. J., and M. G. Sovereign. *Quantitative Models for Production Management.* Prentice-Hall, Englewood Cliffs, N.Y., 1974.

CHAPTER 7 Inventory Replenishment Policies

IN THE PRECEDING CHAPTER ON AGGREGATE PLANNING, INVENTORIES were viewed as a source of short-term capacity. In this role, managers can use inventories as a trade-off against other sources of short-term capacity. Now, however, we wish to focus on inventories themselves and consider policies for their control, item by item.

STOCK POINTS IN A PRODUCTION-DISTRIBUTION SYSTEM

Figure 7-1 identifies the main stock points that occur in a production-distribution system from raw materials and ordering of supplies through the productive process, culiminating in availability for use. At the head of the system, we must have raw materials and supplies in order to carry out the productive process. If we are to be able to produce at minimum cost and by the required schedule, these materials and supplies need to be available. Therefore, we need to develop policies for when to replenish these inventories and how much to order at one time. These issues are compounded by price discounts and by the need to ensure that delays in supply time and temporary increases in requirements will not disrupt operations.

As a part of the conversion process within the productive system we have in-process inventories, which are converted to finished goods inventories. The finished goods inventory levels depend on the policies used for deciding on the production lot sizes and their timing and on the usage rates determined by distributors' orders. High-volume items would justify different policies for production and inventory replenishment than medium- or low-volume items. The production lot size decisions and their timing are very important in relation to the economical use of personnel and equipment and may justify continuous production of a high-volume item. On the other hand, low-volume items will be produced only periodically in economic lots. Again, We will need policy guidelines to determine the size of buffer inventories to absorb the effects of production delays and random variations in demand by distributors.

The functions of distributors and retailers is one of inventorying products to make them available. Distributors and retailers often carry a wide range of items, and they

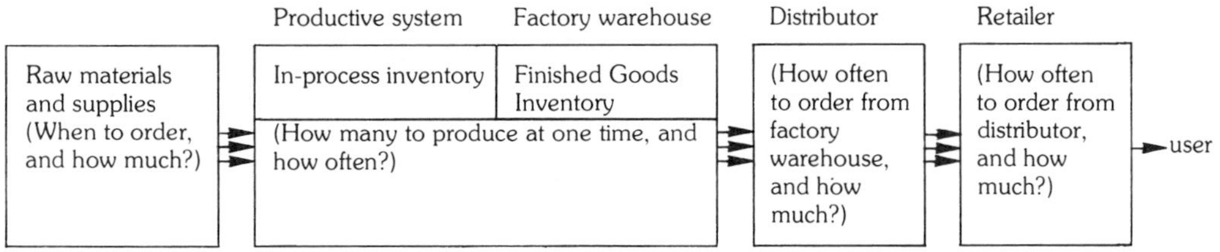

FIGURE 7-1
Main stock points in a production-distribution system.

need replenishment policies that take into account this kind of complexity. They commonly place routine orders periodically, ordering a variety of items from each supplier. Price discounts are often an additional factor to consider.

Although the details of problems may differ at each level in the production-distribution system, note that at each level the basic policy issues are in the inventory replenishment process, focused on the order quantity and when to order. We have a general class of problems for which the concepts of economic order quantities (EOQ) provide important insights. We will first develop the concepts within the framework of the size of purchase orders, and later we will see how the concepts may be adapted for the inventory problems downstream in the system.

SIZE OF PURCHASE ORDERS

Figure 7-2 is a diagram of what happens to inventory levels for a particular item of raw material used in our system. Assume an annual requirement of $R = 2000 \times 52 = 104{,}000$ units, or an average of 2000 units per week. The inventory levels would fluctuate, as in Figure 7-2*a*, if we were to order in lots $Q = 10{,}000$ units. The average inventory level for the idealized situation is one-half the number ordered at one time or $Q/2 = 5000$ units.

If the item is ordered more often in smaller quantities, as shown in Figure 7-2*b*, the inventory level will fall in proportion to the number of units ordered at one time. Inventory level will affect the incremental costs of holding inventory, and therefore these inventory carrying costs will be proportional to the lot size Q, the number ordered at one time.

From Figure 7-2 we see also that the total annual cost of placing orders increases as the number of units ordered at one time decreases. Therefore, we have isolated two types of incremental costs that represent the quantitative criteria for the inventory system. They are the costs associated with inventory level, *holding costs,* and those associated with the number of orders placed, *preparation costs*.

In further defining the system, let us construct a graph that shows the general relation between Q (lot size) and the incremental costs that we have isolated. We noted in Figure 7-2 that if Q were doubled, average inventory level was doubled. It costs, perhaps $c_H = 80$ cents per year to carry a unit of inventory. (Costs include such items as interest, insurance, and taxes.) Since the average inventory level is $Q/2$, and $c_H = 80$ cents, then the annual incremental costs associated with inventory are

$$\frac{Q}{2}(c_H) = \frac{Q}{2}(0.80) = 0.40Q \tag{1}$$

Substituting different values for Q, we can plot the result as in Figure 7-3, curve *a*.

The costs of ordering can be plotted in a similar way. The number of orders placed per year to satisfy requirements is $R/Q = 104{,}000/Q$. If the costs for preparing and following up an order are c_P = \$20, then the total annual incremental costs due to ordering are

$$\frac{R}{Q}(c_P) = \frac{104{,}000}{Q}(20) = 2{,}080{,}000Q \tag{2}$$

Therefore, as Q increases, the annual incremental costs due to ordering decreases. This relationship is plotted in Figure 7-3, curve b, by substituting different values of Q in Equation 2.

Figure 7-3, curve c, shows the resulting total incremental cost curve, determined by simply adding the two previous curves. We have a model that expresses the total incremental cost as a function of the variables that define the system. The equation

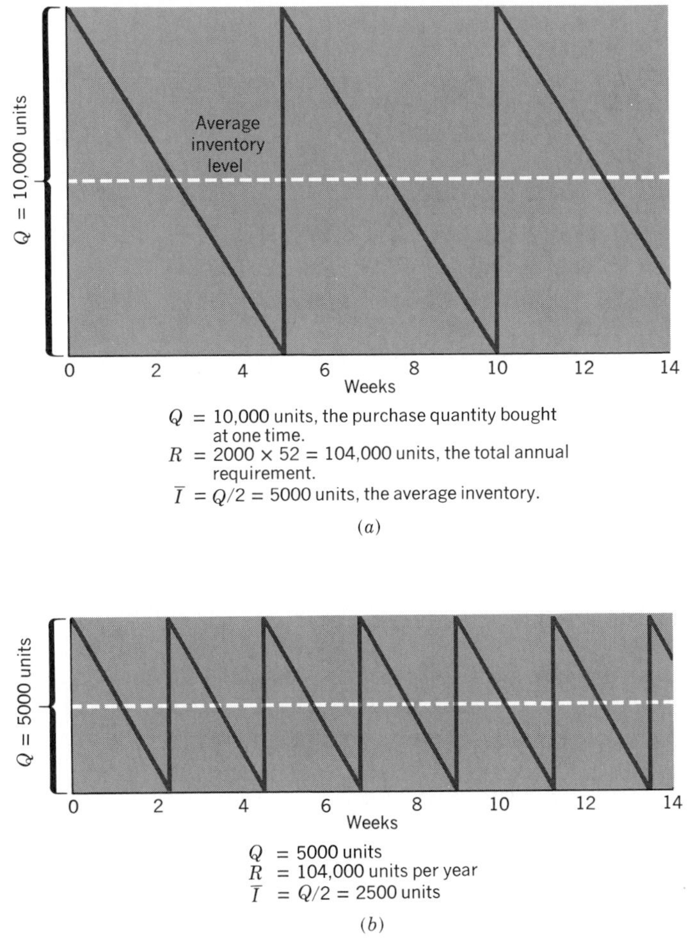

FIGURE 7-2
Simplified model of the effect of lot size on inventory levels.

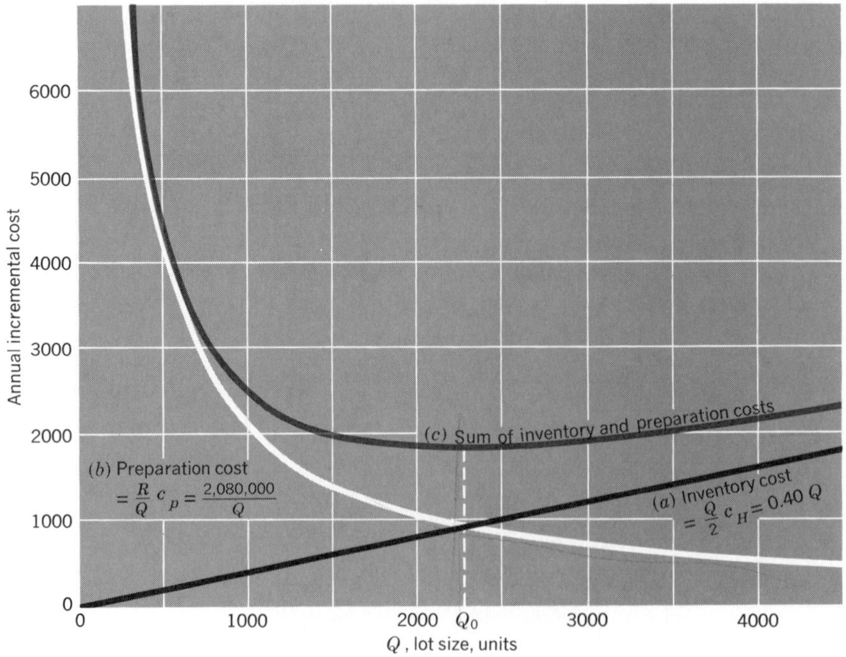

FIGURE 7-3
Graphical model of a simple inventory problem.

for the total cost curve is also determined by adding Equations 1 and 2 for the two separate cost functions.

$$TIC = \frac{Q}{2} c_H + \frac{R}{Q} c_P \tag{3}$$

$$\begin{matrix}\text{Total}\\ \text{incremental}\\ \text{cost}\end{matrix} = \begin{pmatrix}\text{Average}\\ \text{inventory}\end{pmatrix} \begin{pmatrix}\text{Unit inventory}\\ \text{cost per year}\end{pmatrix} + \begin{pmatrix}\text{Number}\\ \text{of orders}\\ \text{per year}\end{pmatrix} \begin{pmatrix}\text{Cost}\\ \text{of an}\\ \text{order}\end{pmatrix}$$

The controllable variable in Equation 3 that can be manipulated by management is Q (lost size). Uncontrollable variables are requirements related to such factors as consumer demand, taxes, and insurance rates. For our example, uncontrollable variables are c_H (inventory costs per unit), R (demand or requirements), and c_P (order preparation costs).

A General Solution

For our simplified model we can select the optimum policy as the minimum point on the total incremental cost curve of Figure 7-3, or $Q_0 = 2280$ (the symbol Q_0 denotes the optimal value of Q). This is a solution to the specific problem with the given values for c_H, R, and c_P. From Equation 3 for the total incremental cost, we may derive a formula for the minimum point on the curve by use of differential calculus. The formula that represents the general solution for the model is

$$Q_0 = \sqrt{\frac{2Rc_P}{c_H}} \tag{4}$$

This formula gives directly the value Q_0, that yields the minimum total incremental cost for the model. Substituting for the values in our example, we have

$$Q_0 = \sqrt{\frac{2 \times 104{,}000 \times 20}{0.80}} = \sqrt{5{,}200{,}000} = 2280.35 \text{ units}$$

In using Equation 4, if it is desired to express economic quantity in dollars, the requirements must also be expressed in dollars. Similarly, if the requirements are expressed in monthly rates, the inventory cost must be expressed as a monthly rate. These and other changes in the units lead to modifications of the formula used in practice. Also, in practice, charts, graphs, and tables based on the formula are often used to minimize computations, or more currently, computing systems automatically issue purchase orders for the quantities computed by the formula.

The incremental cost of the optimal solution, TIC_0, is also simple:

$$TIC_0 = \sqrt{2c_Pc_HR} \tag{5}$$

Substituting the values from our example gives

$$TIC_0 = \sqrt{2 \times 20 \times 0.8 \times 104{,}000} = \sqrt{3{,}328{,}000} = \$1824.28$$

Important Assumptions

The EOQ model is intuitively attractive because it minimizes the obvious incremental costs associated with an inventory replenishment. In applying the model, however, there are three important assumptions:

1. Average demand is continuous and constant, represented by a distribution that does not change with time. Therefore, if there is significant trend or seasonality in the average annual requirements, R in Equation 4, the resulting costs will not be optimal, and there may be other control problems as well.

2. Supply lead time is constant. Although this assumption may be reasonable in many situations, supply lead times are often quite variable. The result of variable lead time is that the timing of the receipt of the order quantity produces excess inventories when lead times are shorter than expected and produces shortages when lead times are longer than expected. The resulting costs would not be optimal.

3. Independence between inventory items. The EOQ model assumes that the replenishment of an inventory item has no effect on the replenishment of any other inventory item. This assumption is valid in many instances, but breaks down when a set of supply items are coupled together by a common production plan.

There are various ways of dealing with the effects of these three assumptions. Indeed, much of the research on inventory models that followed the development of Equation 4 has centered on concepts and techniques for dealing with situations where one or more of the assumptions are not valid in practice.

The Effect of Quantity Discounts

The basic economic order quantity formula assumes a fixed purchase price. When quantity discounts enter the picture, the total incremental cost equation is no longer a continuous function of order quantity but becomes a step function with components of annual inventory cost, ordering cost, and material cost involving the price discount schedule. The total incremental cost equation becomes

$$TIC = \frac{c_H Q}{2} + \frac{c_P R}{Q} + p_i R \tag{6}$$

where p_i is the price per unit for the ith price break, and $c_H = p_i F_H$, where F_H is the fraction of inventory value. The procedure is then one of calculating to see if there is a net advantage in annual ordering plus material costs to counterbalance the increased annual inventory costs.

As an illustration, assume that a manufacturer's requirement for an item is 2000 per year. The purchase price is quoted as \$2 per unit in quantities below 1000, and \$1.90 per unit in quantities above 1000. Ordering costs are \$20 per order, and inventory costs are 16 percent per year per unit of average inventory value, or \$0.32 per unit per year at the \$2 unit price. Equation 4 indicates that the economic order quantity is

$$Q_0 = \sqrt{\frac{2 \times 20 \times 2{,}000}{0.32}} = \sqrt{250{,}000} = 500 \text{ units}$$

Using the preceding data and Equation 6, we can compute *TIC* for each of the three price ranges, as shown in Figure 7-4 (the solid line curves indicate the relationships for valid price ranges):

1. Note that for the \$2 price that applies for $Q < 1000$, $EOQ = 500$ units produces the lowest cost of \$4,160 (Curve 1).

2. However, between order quantities of 1000 and 1999, the price of \$1.90 per unit applies, and when $Q = 1000$, $TIC = \$3992$—a cost saving of \$168 per year, compared with ordering in lots of 500 units (Curve 2).

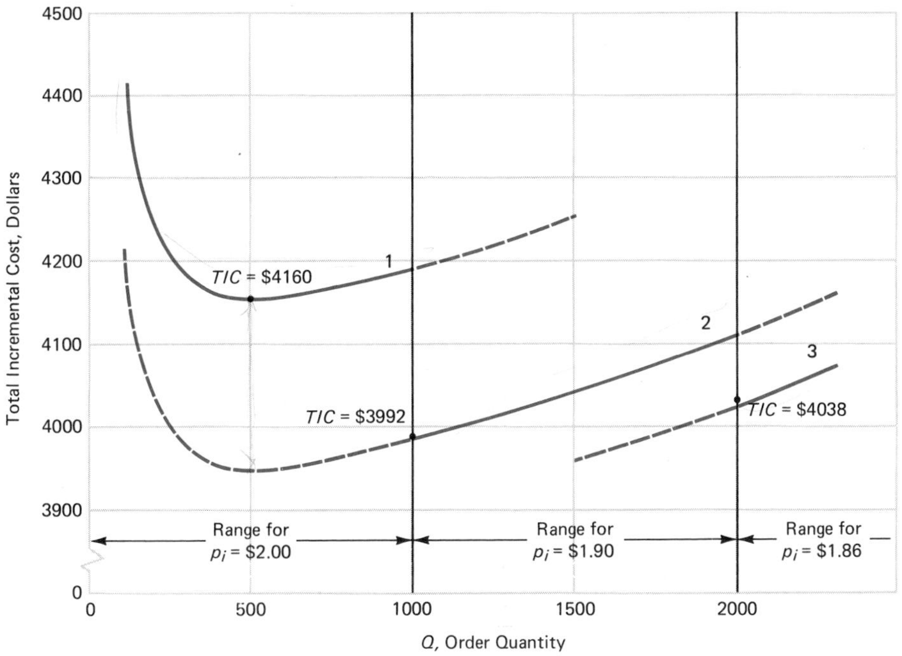

FIGURE 7-4
Total incremental cost curves for inventory model with three price breaks. R = 2000 units per year, c_p = \$20, F_H = 0.16.

3. Finally, at $Q \geq 2000$, the price of \$1.86 per unit applies, and TIC = \$4038 at Q = 2000 units (Curve 3).

The lowest cost ordering policy is to take advantage of the first price break but not the second and to order in lots of Q = 1000 units. Summary calculations are shown in Table 7-1.

With other numerical values, it is possible for Curve 2 to have its optimum occur within the middle price range or for Curve 3 to have its optimum occur in the upper price range. Therefore, a general procedure is needed as follows:

1. Calculate the EOQ for each price.

2. Eliminate EOQs that fall outside of valid price ranges.

3. Calculate TICs for valid EOQs and at price breaks.

4. Select lot size associated with the lowest TIC.

TABLE 7-1 **Incremental Cost Analysis to Determine Net Advantage or Disadvantage When Price Discounts Are Offered**

	Lots of 500 Units Price = \$2.00 per Unit	Lots of 1000 Units Price = \$1.90 per Unit	Lots of 2000 Units Price = \$1.86 per Unit
Purchase of a year's supply ($p_i \times 2000$)	\$4000	\$3800	\$3720
Ordering cost ($20 \times 2000/Q$)	80	40	20
Inventory cost (average inventory × unit price × 0.16)	80	152	298
Total	\$4160	\$3992	\$4038

DEALING WITH VARIABILITY OF DEMAND

In the simple inventory models we assumed that both demand and supply lead times were constant. Yet variability of demand and supply lead time is an element of reality that can be very important. It imposes two-sided risks.

We can cushion the effects of demand and supply lead time variation by absorbing risks in carrying larger inventories, called *buffer stocks*. The larger we make these buffer stocks, the greater our risk, in terms of the funds tied up in inventories, the possibility of obsolescence, and so on. However, we tend to minimize the risk of running out of stock.

On the other hand, although the inventory risk can be minimized by reducing buffer inventories, the risks associated with poor inventory service increase, including the costs of back ordering, lost sales, disruptions of production, and so on. Our objective, then, is to find a rational model for balancing these risks.

Service Levels and Buffer Stocks (Constant Lead Time)

Figure 7-5 shows the general structure of inventory balance when a fixed quantity Q is ordered at one time. When inventory falls to a preset reorder point P, an order for the quantity Q is placed. The reorder point P is set to take account of the supply lead time L, so that if we experience normal usage rates during L, inventory is reduced to minimum levels when the order for Q units is received. But demand may not occur at a constant rate. Inventories may decline to the reorder point P earlier or later than expected. If demand during lead time is greater than expected values, inventory levels may decline below the planned buffer stock level. In the limiting situation, if we experience maximum demand during lead time (as shown in Figure 7-5), inventory levels will decline to zero by the time the order for Q units is received. The size of the needed buffer stock, then, is the difference between the expected or average demand $\overline{D}$ and the maximum demand D_{max} during the supply lead time, or buffer stock = $B = D_{max} - \overline{D}$. The issue concerns how we define D_{max}.

Defining D_{max}. Maximum demand is not a fixed number that we can simply abstract from a distribution of demand; instead, it depends on an analysis of the risks. Assume the record for the distribution of demand that exceeds a given level, shown in Figure 7-6. This figure represents only the random variations; if there were other effects, such as trend and seasonals, they have been removed by standard statistical techniques. Note that average monthly demand was $\overline{D} = 460$ units.

Since the average monthly usage rate is 460 units, and if we assume a lead time of $L = 1$ month, we could be 90 percent sure of not running out of stock by having 620 units on hand when the replenishment order is placed (see Figure 7-6 for the demand rate associated with 10 percent). The buffer stock required for this 90 percent service level is $B = 620 - 460 = 160$ units. Similarly, if we wish to be 95 percent sure of not running out of stock, then $B = 670 - 460 = 260$ units. For a 99 percent service level (1 percent risk of stockout), the buffer stock level must be increased to 320 units.

From the shape of the demand curve, it is clear that required buffer stock goes up rapidly as we increase service level, and therefore, the cost of providing this assurance goes up. These effects are shown by the calculations in Table 7-2, in which we have assumed the demand curve of Figure 7-6, assigning a value of \$100 to the item and inventory holding costs of 25 percent of inventory value. The average inventory required to cover expected maximum usage rates during the one-month lead time is calculated for the three service levels shown. To offer service at the 95 percent level instead of the 90 percent level requires an incremental \$1,250 per year, but to move to the 99 percent level of service from the 95 percent level requires an additional \$2,750 in inventory cost.

Management could define D_{max} at any of the three levels of demand by setting a

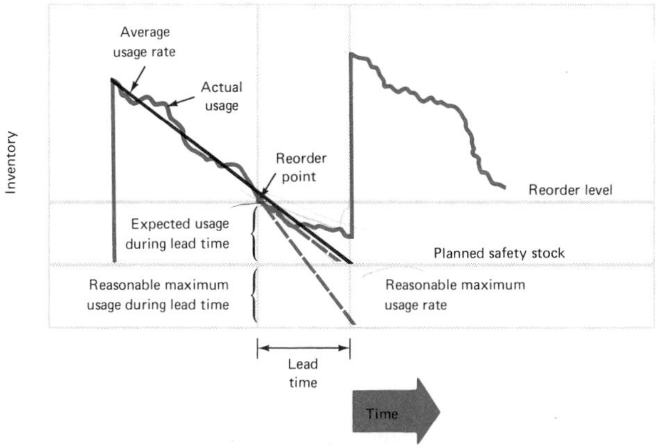

FIGURE 7-5
Structure of inventory balance for a fixed reorder quantity system.

service level policy. Given the service level policy, the buffer stock required to implement that policy is, simply, $B = D_{max} - \overline{D}$. The value for D_{max} sets the order point P which allows for the needed buffer stock B.

Practical Methods for Determining Buffer Stocks

The general methodology that we have discussed for setting buffer stocks is too cumbersome for practical use in systems that may involve large numbers of items. Computations are simplified considerably if we can justify the assumption that the demand distribution follows some particular mathematical function, such as the normal, Poisson or negative exponential distributions.

First, let us recall the general statement for buffer stocks:

$$B = D_{max} - \overline{D} \tag{7}$$

Note, however, that $D_{max} = \overline{D} + n\sigma_D$; that is, the defined reasonable maximum demand is the average demand $\overline{D}$, plus some number of standard deviation units n that is associated with the probability of occurrence of that demand (n now is defined as the safety factor). Substituting this statement of D_{max} in our general definition of B, Equation 7, we have

$$\begin{aligned} B &= D_{max} - \overline{D} = (\overline{D} + n\sigma_D) - \overline{D} \\ &= n\sigma_D \end{aligned} \tag{8}$$

This simple statement allows us to determine easily those buffer stocks that meet risk requirements when we know the mathematical form of the demand distribution. The procedure is as follows:

1. Determine whether the normal, Poisson, or negative exponential distribution approximately describes demand during lead time for the case under consideration. This determination is critically important, involving well-known statistical methodology.

2. Set a service level based on (a) managerial policy, (b) an assessment of the balance of incremental inventory and stockout costs, or (c) an assessment of the manager's trade-off between service level and inventory cost when stockout costs are not known.

3. Using the service level, define D_{max} during lead time in terms of the appropriate distribution.

4. Compute the required buffer stock from Equation 8, where n is termed the safety factor and σ_D is the standard deviation for the demand distribution.

We will illustrate the methodology in the context of the normal distribution.

TABLE 7-2

Cost of Providing Three Levels of Service in Figure 7-5 (Item is valued at $100 each and inventory holding costs are 25 percent)

	Service Level		
	90%	95%	98%
Expected maximum usage for one month replenishment time	620	670	780
Buffer stock required, $B = D_{max} - 460$	160	210	320
Value of buffer stock, $100 \times B$	$16,000	$21,000	$32,000
Inventory holding cost at 25 percent	$ 4,000	$ 5,250	$ 8,000

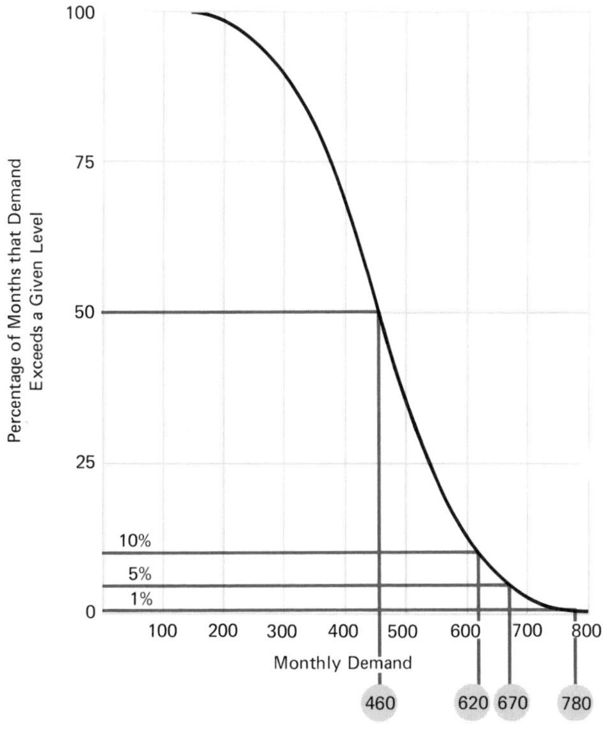

FIGURE 7-6
Distribution showing the percentage of months that demand exceeded a given level.

Buffer Stocks for the Normal Distribution

The normal distribution has been found to describe many demand functions adequately, particularly at the factory level of the supply-production-distribution system [Buchan and Koenigsberg, 1963]. Given the assumption of normality and a service level of, perhaps, 95 percent, we can determine B by referring to the normal distribution tables, a small part of which has been reproduced as Table 7-3. The normal distribution is a two-parameter distribution, which is described completely by its mean value $\bar{D}$ and the standard deviation σ_D. Implementing a service level of 95 percent means that we are willing to accept a 5 percent risk of running out of stock. Table 7-3 shows that demand exceeds $\bar{D} + (n\ \sigma_D)$ with a probability of 0.05, or 5 percent of the time, when $n = 1.645$; therefore, this policy is implemented when $B = 1.645\ \sigma_D$. As an example, if the estimate of σ_D is $s = 300$ units, and $\bar{D} = 1500$ units, assuming a normal distribution, a buffer stock to implement a 95 percent service level would be $B = 1.645 \times 300 = 494$ units. Such a policy would protect against the occurrence of demands up to $D_{max} = 1500 + 494 = 1994$ units during lead time. Obviously, any other service level policy could be implemented in a similar way.

TABLE 7-3

Area Under the Right Tail of the Normal Distribution (showing the probability that demand exceeds $\bar{D} + n\sigma_D$ for selected values of n)

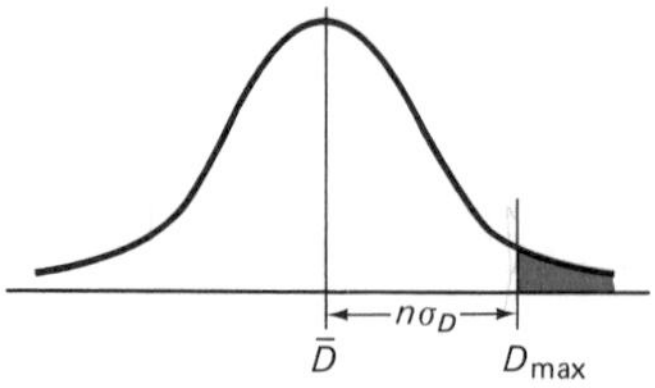

$D_{max} = n\sigma_D$	Probability
$\bar{D} + 3.090\sigma_D$	0.001
$\bar{D} + 2.576\sigma_D$	.005
$\bar{D} + 2.326\sigma_D$	.010
$\bar{D} + 1.960\sigma_D$	.025
$\bar{D} + 1.645\sigma_D$	.050
$\bar{D} + 1.282\sigma_D$	0.100
$\bar{D} + 1.036\sigma_D$	.150
$\bar{D} + 0.842\sigma_D$	.200
$\bar{D} + 0.674\sigma_D$	.250
$\bar{D} + 0.524\sigma_D$	.300
$\bar{D} + 0.385\sigma_D$	0.350
$\bar{D} + 0.253\sigma_D$	.400
$\bar{D} + 0.126\sigma_D$	.450
$\bar{D}$	.500

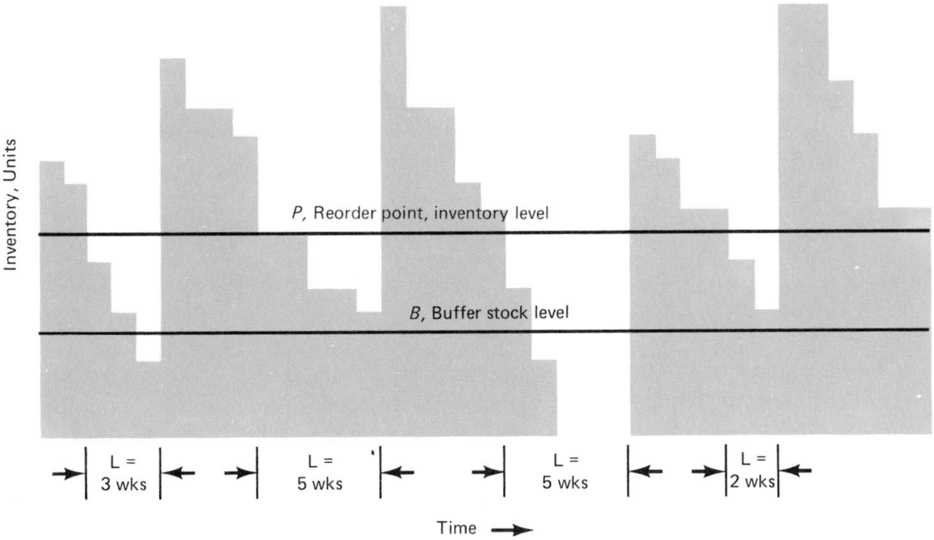

FIGURE 7-7
Inventory balance when both demand and lead time vary. When inventory falls to the reorder point P, the quantity Q is ordered. Inventory falls below the buffer stock level B twice, and a stockout occurs during the third cycle.

Buffer Stocks When Both Demand and Lead Time Are Variable

The problem of determining buffer stocks when both demand and lead time vary is somewhat more complex. When both lead time and demand vary, we have an interaction between the fluctuating demand and the fluctuating lead time, similar to the situation shown in Figure 7-7.

In such situations, a Monte Carlo simulation may determine buffer stocks. To carry out the simulation, we need data describing both the demand and the lead time. Then we can develop buffer stock requirements for the various risk levels of stockout. We can implement whatever risk level we choose by selecting the corresponding buffer stock. The methodology and computed examples are developed in Buffa and Miller [1979] and in McMillan and Gonzalez [1973].

DETERMINING SERVICE LEVELS

The service level states the probability that all orders can be filled directly from inventory during a reorder cycle. As we have stated, the buffer inventory that is designed to provide for the risk of stockout is $B = n\sigma_D$. Assuming a normal distribution and a service level of 95 percent (chance of a stockout is 0.05), then from Table 7-3 the safety factor is $n = 1.645$.

Now let us examine more closely the meaning of this service level statement of policy. It means that there is one chance in twenty that demand during lead time will exceed the buffer stock *when there is exposure to risk.* It does not mean that 5 percent of the demand is unsatisfied but rather that demand during lead time can be expected to exceed buffer stock for 5 percent of the replenishment orders. In other words, the chance that demand will exceed the buffer stock for any given replenishment order is 5 percent.

Effect of Order Size

This interpretation of service level immediately shows us that the expected quantity short over a period of time is proportional to the number of times we order, since we are exposed to shortages only once for each reordering cycle. For our example with buffer stocks designed for a chance of stockout of 0.05, if we ordered Q units 20 times per year, we would expect shortages to occur an average of only once per year ($0.05 \times 20 = 1.0$). If we ordered in quantities of $2Q$ only 10 times per year, we would expect stockouts to occur only once every other year, on the average. Larger orders provide exposure to risk less often and will result in lower annual expected quantities short for the same service level.

Expected Quantities Short

For a given safety factor and distribution of demand during lead time, we can compute the expected quantity short. Assuming a normal distribution of $\bar{D} = 50$ units during lead time and $\sigma_D = 10$ units, let us determine the expected quantity short for service levels of 80, 90, 95, and 99 percent. Based on the safety factors for the stated service levels (Table 7-3), the computed buffer stocks are

$$
\begin{aligned}
B_1 &= 0.842 \times 10 = 8.4, \text{ or } 9 \text{ units} \\
B_2 &= 1.282 \times 10 = 12.8, \text{ or } 13 \text{ units} \\
B_3 &= 1.645 \times 10 = 16.45, \text{ or } 17 \text{ units} \\
B_4 &= 2.326 \times 10 = 23.26, \text{ or } 24 \text{ units}
\end{aligned}
$$

The values of B approximate the stated service levels and give slightly better service because of rounding upwards to integer units.

Brown [1963] has shown that the expected quantity short per order is the product of σ_D and $E(k)$, where $E(k)$ is the partial expectation for a distribution with unit standard deviation. The partial expectation is the expected value of demands beyond some specified level. Brown [1967] developed tables of partial expectations for the normal distribution, a small part of which is reproduced in Table 7-4. Estimates of the expected quantity short per order can be obtained from Table 7-4 for a given safety factor which, in turn, is associated with a given service level. Reading from

Table 7-4, the expected quantities short per order for our example and the four service levels are:

Service Level, Percentage	Expected Quantity Short per Order, Units = $E(k) \times \sigma_D$ (Rounded to two decimals)
80	$0.11156 \times 10 = 1.12$
90	$0.04730 \times 10 = 0.47$
95	$0.02089 \times 10 = 0.21$
99	$0.00441 \times 10 = 0.04$

For each of the service levels indicated, the expected quantity short for each order is as given, and these expected shortages are rather startlingly small. The effect of a given service policy may be misleading unless it is translated into its equivalent expected quantity short per order. While a 90 percent service policy may seem relatively loose, it holds fairly tight control in terms of the expected shortages on each ordering cycle.

TABLE 7-4 **Expected Quantity Short per Order for Values of the Safety Factor *n***

Safety Factor, n	Service Level, Percent	$E(k)$, Expected Quantity Short/σ_D
3.090	99.9	0.00028
2.576	99.5	0.00158
2.~~236~~ 326	99.0	0.00441
1.960	97.5	0.00945
1.645	95.0	0.02089
1.282	90.0	0.04730
1.036	85.0	0.07776
0.842	80.0	0.11156
0.674	75.0	0.14928
0.524	70.0	0.19050
0.385	65.0	0.23565
0.253	60.0	0.28515
0.126	55.0	0.33911
0.0	50.0	0.39894

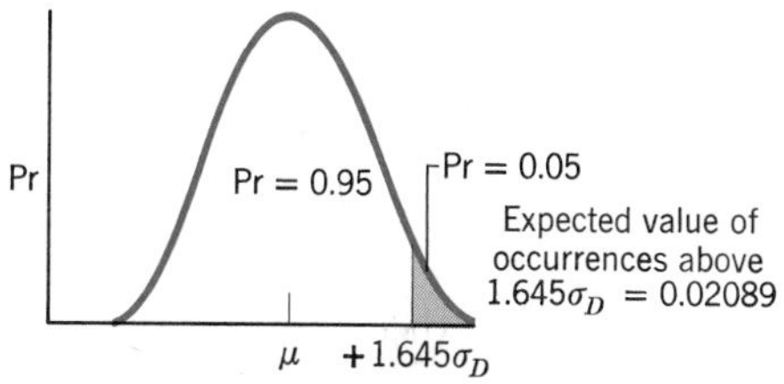

SOURCE. R. G. Brown, *Decision Rules for Inventory Management*. Holt, Rinehart, & Winston, New York, 1967.

Optimal Service Levels, Shortage Costs Known

Now that we have methods for estimating the expected quantity short per order, we can determine the optimal service level if we know the relevant costs. Let us slightly amplify the example that we have been using. Suppose that annual requirements for the example item are $R = 3000$ units per year; inventory holding costs are $c_H =$ \$20 per unit per year; ordering costs are $c_p =$ \$25 per order; and shortage costs are $c_S =$ \$100 per unit short. If the order quantity were $Q = 500$ units, six orders per year would be required.

Let us examine the annual buffer inventory and shortage costs for the four different service levels. We already computed the buffer inventory and expected quantities short per order. Since there are six orders per year, the *annual* expected quantity short is $6\sigma_D\ E(k)$. These values and the relevant costs are summarized in Table 7-5. The service policy that minimizes relevant costs for these data is the 95 percent policy that involves maintaining a buffer of $B = 17$ units and that results in an annual expected quantity short of $6 \times 0.21 = 1.26$ units and a minimum total relevant cost of \$466 per year. What would be the optimal service policy if the cost of shortages was only $c_S =$ \$40?

TABLE 7-5

Annual Buffer Inventory and Shortage Costs for Four Service Levels ($\overline{D} = 50$ units during lead time, $\sigma_D = 10$ units, $c_H =$ \$20 per unit per year, $_S =$ \$100 per unit short, $R = 3000$ units per year, and $Q = 500$ units per order)

	Approximate Service Level, Percentage			
	80	90	95	99
Buffer inventory[a]. $B = n\sigma_D = 10 \times n$	9	13	17	24
Expected quantity short per order[b] $\sigma_D E(k) = 10 \times E(k)$	1.12	0.47	0.21	0.04
Buffer inventory cost. $c_H B = 20 \times B$	\$180	\$260	\$340	\$480
Shortage cost, $c_S(R/Q) \times$ (expected quantity short per order) $= 100 \times 6 \times$ (expected quantity short per order)	\$672	\$282	\$126	\$24
Total incremental costs	\$852	\$542	\$466	\$504

[a] Values of n from Table 7-3 for given service level. Values of B rounded to next highest integer.
[b] $E(K)$ from Table 7-4.

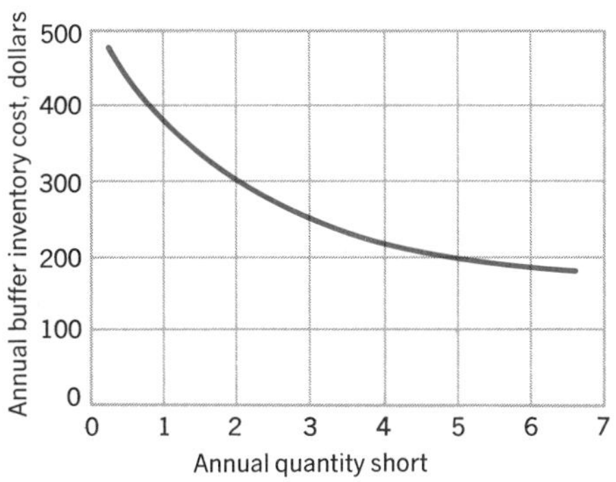

FIGURE 7-8
Annual buffer inventory costs versus annual quantity short for a system where: $D = 50$ units during lead time, $\sigma_D = 10$ units, $c_H = \$20$ per unit per year, $R = 3000$ units per year, and $Q = 500$ units per order.

Optimal Service Levels, Shortage Costs Unknown

It is frequently true that we do not know the value of c_S with any degree of confidence. Many factors in a given situation may affect the true cost of shortages. Some of these factors may be reasonably objective but difficult to measure. For example, although part shortages in assembly processes create costly disruptions and delays, measuring these costs is quite another matter. Also, when shortages occur, it may be necessary to back order or to expedite them with special handling and extra costs. These incremental costs are real, but they are not segregated in cost records. Thus, making realistic estimates of their value would be costly in itself, and these costs could exceed the value of the information. If a shortage definitely results in a lost sale, we can impute a shortage cost equal to the lost contribution. But do we know whether the sale is lost for certain, or must we merely estimate the probability of a lost sale? Finally, the loss may be intangible, such as the loss of goodwill of a valued customer who receives poor service.

For all the preceding reasons, we may not be able to estimate values of c_S with sufficient precision to justify an analysis similar to that given in Table 7-5 as a basis for selecting an optimum service level policy. Nevertheless, in the absence of known shortage costs, we still have valuable data from Table 7-5. We have objective annual buffer inventory costs for various service level policies, and we have the expected annual quantities short that would result from each service level policy. The graphical relationship between buffer inventory cost and quantities short is shown in Figure 7-8. These data provide the basis for determining the manager's utility function so that trade-offs can be made.

FIXED REORDER QUANTITY SYSTEM

The structure of the fixed reorder quantity system was illustrated by Figure 7-5. A reorder level has been set by the point P, which allows the inventory level to be drawn down to the buffer stock level within the lead time if average usage rates are experienced. Replenishment orders are placed in a fixed, predetermined amount (in practice, not necessarily the EOQ) that is timed to be received at the end of the supply lead time. The maximum inventory level becomes the order quantity Q plus the buffer stock B. The average inventory, then, is $B + Q/2$. Usage rates are reviewed periodically in an attempt to react to seasonal or long-term trends in requirements. At the time of the periodic reviews, the order quantities and buffer stock levels may be changed to reflect the new conditions. Buffer stock levels are set based on determinations of the appropriate service level policy. This policy reflects the balancing of buffer inventory costs and shortage costs, or the manager's trade-off between buffer inventory cost and the expected quantity short. But buffer stocks are actually allowed for by setting the reorder point P, since $P = D_{max}$.

The parameters that define a fixed reorder quantity system are Q, the fixed number ordered at one time, and the reorder point P.

Fixed reorder quantity systems are common where a perpetual inventory record is kept or where the inventory level is under sufficiently continuous surveillance that notice can be given when the reorder point has been reached. One of the simplest methods for maintaining this close watch on inventory level is the use of the "two-bin" system. In this system, the inventory is physically (or conceptually) separated into two bins, one of which contains an amount equal to the reorder inventory level, P. The balance of the stock is placed in the other bin, and day-to-day needs are drawn from it until it is empty. At this point, it is obvious that the reorder level has been reached, and a stock requisition is issued. Then stock is drawn from the second bin, which contains an amount equal to the average use during the lead time plus a buffer stock. The stock is replenished when the order is received, and the physical segregation into two bins is made again; the cycle then is repeated.

Fixed reorder quantity systems are common with low-valued items, such as nuts and bolts, and where close surveillance over inventory levels is not necessary.

PERIODIC REORDER SYSTEM

A common alternate system of control fixes the review cycle instead of the reorder quantity. In such systems, the inventory status is reviewed on a periodic basis, and an order is placed for an amount that replenishes inventories to a planned maximum based on usage, as shown in Figure 7-9. The reorder quantity is variable in size, covering anticipated normal usage during the supply lead time, plus usage since the last review. A maximum planned inventory level, I_{max}, is indicated in Figure 7-9.

An economic cycle can be approximated by computing EOQ; the economic cycle

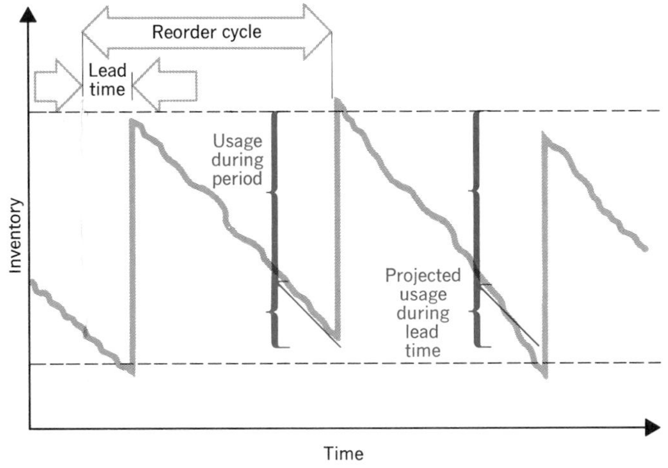

FIGURE 7-9
Periodic reorder system of control. An order is placed at regular intervals of variable size Q. In each instance, Q = usage during the past period + projected usage during lead time.

would then be Q_0/R, where R is the annual requirement. For example, if $Q_0 = 7500$ units and annual requirements were $R = 100{,}000$ units, then the economic cycle would be $7500/100{,}000 = 0.075$ years, or 3.9 weeks. This value would probably be rounded to four weeks.

The periodic reorder system has some advantages in production cycling, where high valued items require close control, in the control of items that may deteriorate with time, and where a number of items may be ordered from the same supplier. In the latter situation, it may be possible to gain shipping cost advantages by grouping orders normally sent to a common supplier. In addition, the periodic system makes possible operating efficiencies by reviewing the status of all items at the same time. Because of these advantages, the review cycle is commonly set by considerations other than the individual item economic cycle.

Perhaps the single most important advantage of the periodic reorder system is that the periodic review of inventory and usage levels provides the basis for adjustments to take account or demand changes. This is particularly advantageous with seasonal items. If demand increases, order sizes increase; if demand decreases, order sizes decrease. Therefore, as actually used, the periodic system does not assume constant demand, as in the EOQ system.

OPTIONAL REPLENISHMENT SYSTEM

Control systems that combine regular review cycles and order points are also used (called s,S systems in the literature). In such systems, stock levels are reviewed on a periodic basis, but orders are placed only when inventories have fallen to a predetermined reorder level, P. Then, an order is placed to replenish inventories to a maximum level denoted by I_{max}, which is sufficient for buffer stocks plus expected needs for one cycle, as in the periodic system.

Such systems have the advantage of the close control associated with the periodic reorder system, resulting in minimum buffer stocks. On the other hand, since replenishment orders are placed only when the reorder point has been reached, fewer orders are placed on the average. Therefore, ordering costs are comparable to those associated with fixed reorder quantity systems.

INVENTORY PLANNING SYSTEMS

The three inventory control systems that we have just discussed are all *reactive* in nature. Action is triggered by inventory status falling to a reorder point or by the inventory status that exists on review. They are backward looking, depending on events that happen.

If we have forecasts, however, why not look forward, making inventory replenishment plans that anticipate events. Forward-looking inventory plans would be particularly valuable when we expect increases or decreases in demand, as is common where demand is seasonal.

Suppose we had an item that was used at the constant rate of $r = 100$ units per week, as shown by the forecast in Table 7-6. The inventory on hand is also shown week by week, beginning initially with 400 units. From the projected schedule of units on hand, we can see that we would run out of stock in the fourth week. Now suppose that lead time is two weeks, that we wish to maintain a buffer inventory of $B = 100$ units, and that we have computed EOQ as $Q_0 = 500$ units. It is easy to see from Table 7-6 that we must plan to place an order for $Q_0 = 500$ units in Week 2 in order to anticipate the projected inventory status of $I = 0$ units in Week 4.

The inventory record is revised in Table 7-7 to reflect the planned order. Following the same rationale, a second order must be placed in Period 7 to be received in Period 9. The second order anticipates the fact that inventory would fall to the buffer level in Week 9. Therefore, observing the two-week lead time, an order must be placed in Week 7.

This example illustrates the basic format and functioning of inventory planning systems. One might observe, however, that an EOQ system would have produced the same results. We have conveniently observed the constant demand assumption in the example. Suppose that the item were seasonal, with demand increasing by 10 units per week during the 10-week horizon. We can still function with the EOQ system, but orders will need to be placed more often and ordering costs will increase. In this situation, periodic review, coupled with an inventory planning system, has advantages.

Table 7-8 shows a revised schedule of forecasts, with demand increasing by 10

TABLE 7-6 **Forecast Versus Projected Inventory on Hand for an Item**

Period (weeks)		1	2	3	4	5	6	7	8	9	10
Forecast		100	100	100	100	100	100	100	100	100	100
Scheduled receipts											
On hand	400	300	200	100	0	(100)	(200)	(300)	(400)	(500)	(600)
Planned orders											

TABLE 7-7 **Orders Placed in Periods 2 and 7 to Counteract Projected Fall in Inventory Level Below the Desired Buffer of B = 100 Units (Q_O = 500 units; supply lead time is 2 weeks)**

Period (weeks)		1	2	3	4	5	6	7	8	9	10
Forecast		100	100	100	100	100	100	100	100	100	100
Scheduled receipts					500					500	
On hand	400	300	200	100	500	400	300	200	100	500	400
Planned orders			500					500			

TABLE 7-8 **Projected Forecasts Versus Inventory on Hand, Forecasts Increasing by 10 Units per Week**

Period (weeks)		1	2	3	4	5	6	7	8	9	10
Forecast		50	60	70	80	90	100	110	120	130	140
Scheduled receipts											
On hand	400	350	290	220	140	50	(50)	(160)	(280)	(410)	(550)
Planned orders											

TABLE 7-9 **Inventory Planning Systems with Periodic Ordering by the Rule, $Q = D(5 \text{ weeks}) - (I - B)$ (Replenishment Scheduled for Weeks 3 and 8, with Supply Lead time of 2 Weeks)**

Period (weeks)		1	2	3	4	5	6	7	8	9	10
Forecast		50	60	70	80	90	100	110	120	130	140
Scheduled receipts				330					750		
On hand	400	350	290	550	470	380	280	170	800	670	530
Planned orders		330					750				

units per week. Beginning inventory is 400 units as before, and we note that inventory would fall to 50 units in Period 5 and would go negative after that. Now suppose that each five weeks we plan to order projected demand for the next five weeks. In placing the order, however, we wish to take account of inventory status and desired buffer inventory of $B = 100$ units. The ordering rule would then be, order an amount to cover five weeks projected demand minus $(I - B)$ or

$$D(5 \text{ weeks}) - (I - B)$$

As an example, suppose that the next ordering period for the increasing demand schedule is Week 3. The projected number on hand in Week 3 is 220 units. Projected demand for the next 5 weeks is $D = 70 + 80 + 90 + 100 + 110 = 450$ units. This five-week quantity must then be adjusted for the projected inventory position of 220 units and the desired buffer of 100 units, $220 - 100 = 120$ units of excess inventory. The order would then be $Q = 450 - (220 - 100) = 330$ units, set back by the two-week lead time. The necessary inventory adjustments are shown in Table 7-9.

Given the planned receipt of the order for 330 units in Period 3, we reexamine the projected inventory level 5 weeks later, in Period 8. Again, an order would be placed to cover forecast demand for the Periods 8 through 12, less the inventory-buffer adjustment. Assuming that forecast demand continues to increase at the same rate, the five-week forecast demand is $D = 120 + 130 + 140 + 150 + 160 = 700$ units. Projected eighth-week inventory would be 50 units, and therefore, the order to be placed in the sixth week is $Q = 700 - (50 - 100) = 750$ units. Note that in this instance, the adjustment for current inventory position and desired buffer stock shows a shortage of 50 units compared to the desired buffer stock, so that the adjustment results in an order that rebuilds the buffer to the desired level of 100 units. Table 7-9 also shows the adjustments to the inventory record to reflect the second order.

The periodic ordering rule coupled with an inventory planning system automatically adjusts for the changing forecast. If demand had been decreasing during the review period, the ordering rule would automatically reduce the size of orders to take account of decreased forecasts. When demand is dependent on the production schedule of a product, it may be quite easy to forecast, but extremely variable. This kind of dependent demand is often termed "lumpy," and inventory planning systems are commonly used. We deal with such systems in Chapter 9 "Material Requirements Planning."

ABC CLASSIFICATION OF INVENTORY ITEMS

Equal control effort for all items is not ordinarily justified. First, the differing value of inventory items suggests that we concentrate our attention on higher-valued items and be less concerned about low-valued items. For example, Figure 7-10 shows a fairly typical relationship between the percentage of inventory items and the percentage of total dollar inventory value. Twenty percent of the items account for 60 percent of the total dollar inventory value in Figure 7-10. The second 20 percent of items account for 20 percent of the value, and finally, the greatest percentage of items (60 percent) accounts for only 20 percent of total inventory value. Since inventory costs are associated directly with inventory value, the potential cost saving from closer

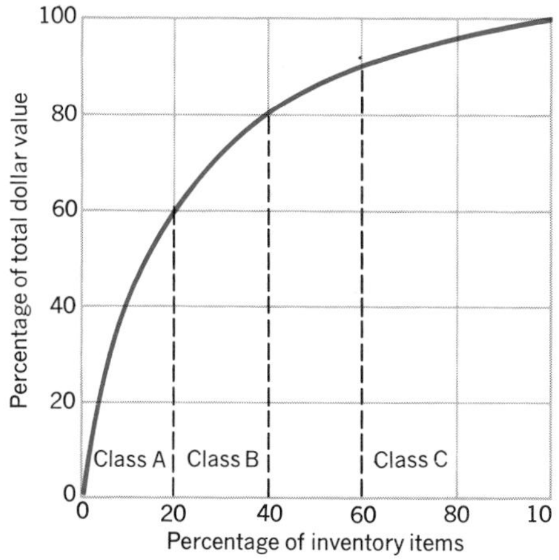

FIGURE 7-10
ABC classification of inventory items versus dollar volume.

control is greater among the first group, which accounts for most of the inventory value.

Second, even though an item may in itself be of low value, it is possible that the cost of a stockout could be considerable. For example, the shortage of a seemingly minor raw material item could cause idle labor costs and loss of production on an entire assembly line. Therefore, for both of the preceding reasons, a classification of inventory items by the degree of control needed allows managers to place their efforts where the returns can be most important.

Controls of Class A Items

Close control is required for inventory items that have high stockout costs, plus those items in Figure 7-10 that account for a large fraction of the total inventory value. The closest control might be reserved for raw materials used continuously in extremely high volume. Purchasing agents may arrange contracts with vendors for the continuous supply of these materials at rates that match usage rates. In such instances, the purchase of raw materials is not guided by either economical quantities or cycles. Changes in the rate of flow are made periodically as demand and inventory position changes. Minimum supplies are maintained to guard against demand fluctuations and possible interruptions of supply.

For the balance of Class A items, periodic ordering, perhaps on a weekly basis, provides the needed close surveillance over inventory levels. Variations in usage rates are absorbed quickly by the size of each weekly order, according to the periodic, optional, or inventory planning systems discussed previously. Also, because of the close surveillance, the risk of a prolonged stockout is small. Nevertheless, buffer stocks that provide excellent service levels will be justified for items having large stockout costs.

Controls for Class B Items

Periodic ordering once or perhaps twice per month could be sufficient for Class B items, again applying the periodic, optional, or inventory planning replenishment systems. Stockout costs for Class B items should be moderate to low, and buffer stocks should provide adequate control of stockout, even though the ordering occurs less often.

Controls for Class C Items

Class C items account for the great bulk of inventory items, and carefully designed but routine controls should be adequate. A reorder point system will ordinarily suffice. For each item, action is triggered when inventories fall to the reorder point. If usage changes, orders will be triggered earlier or later than average, providing the needed compensation. Semiannual or annual reviews of the system parameters should be performed to update usage rates, estimates of supply lead times, and costs, resulting in possible changes in EOQ.

The development of the concepts of EOQ, buffer stocks, and common managerial control systems has been in the context of raw materials and supplies. Recall from Figure 7-1, however, that there are stock points downstream in the process where similar decisions of inventory replenishment are required. The concepts and control systems are transferable with modification to the production phase and to the other stock points.

PRODUCTION ORDER QUANTITIES AND PRODUCTION CYCLING

In the intermittent model of production systems, where production is carried out in batches, management must decide how large these batches should be. The concept of EOQ represented by Equation 4 can be applied directly, recognizing that preparation costs are the incremental costs of writing production orders, controlling the flow of orders through the system, and setting up machines for the batch; inventory costs are associated with holding in-process inventory.

However, the assumption that the order is received and placed into inventory all at one time is often not true in manufacturing. Equation 4 assumes the general inventory pattern shown in Figure 7-11a, where the order quantity Q is received into

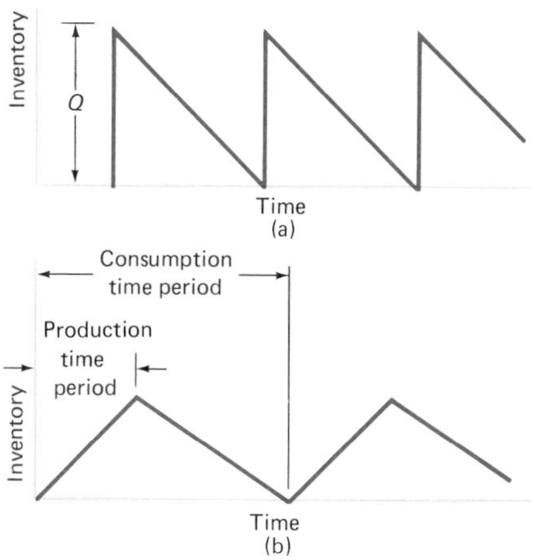

FIGURE 7-11
Comparison of inventory balance: *(a)* when the order quantity, Q, is received all at one time, and *(b)* when Q is received over a period of time.

inventory. The inventory is then drawn down at the usage rate, subsequent orders being placed with sufficient lead time so that their receipt coincides with the minimum inventory level.

For many manufacturing situations the production of the total order quantity takes place over a period of time, and the parts go into inventory in smaller quantities as production continues. This results in an inventory pattern similar to Figure 7-11*b*. The maximum and average inventory levels are reduced, and the optimum cost quantity formula becomes

$$Q_p = \sqrt{\frac{2Rc_p}{c_H\left(1 - \frac{r}{p}\right)}} \tag{9}$$

where r = requirements or usage rate (short term, perhaps daily or weekly)
p = production rate (on same time base as for r)
c_P = ordering and setup cost
c_H = unit inventory cost
Q_p = minimum cost production order quantity
R = annual requirements

The Q_p that results is larger than that which would result from Equation 4. Average inventory is smaller for a given Q_p, and balance between setup costs and inventory costs takes place at a higher value of the order quantity Q. The number of production cycles per year is R/Q_p.

The total incremental cost of the optimal production order quantity is

$$TIC_0 = \sqrt{2c_Pc_HR\left(1 - \frac{r}{p}\right)} \tag{10}$$

There are additional problems related to production cycling that we will cover in Chapter 8. Also, the demand for parts and components is dependent on the demand for the primary product and more specifically on its production schedule. Because of the "lumpy" nature of demand for dependent items, the concepts of requirements planning are important to production lot size policies. Requirements planning will be covered in Chapter 9.

ORDERS FOR REPLENISHMENT OF FINISHED GOODS BY WAREHOUSES, DISTRIBUTORS, AND RETAILERS

Let us dispose of the factory warehouse first. It is often true that no orders for replenishment would be placed by the factory warehouse, since the factory and its warehouse may be considered a single unit in terms of stock points for finished goods. In such situations, products go directly to the factory warehouse for finished goods storage. When the sales organization controls inventories, sometimes orders to the factory for replenishment of types and sizes may occur. Such orders could be on the basis of fixed order quantities, those ordered when inventories fall below the order point, or periodically based on the quantities needed to replenish inventory to some preset level, as in the periodic reorder system. The normal distribution is commonly used as a basis for the design of buffer stocks at this level in the system.

For distributors and retailers, however, orders for replenishment are more likely to be placed periodically, ordering all items needed from a given supplier. By combining orders for all items from a given supplier, it may be possible to effect freight cost advantages. While we can derive an individual "economic order cycle time," Q_0/R, for each item, it may not be worthwhile, as noted previously. Setting different optimal cycle times for different products loses the advantage of reviewing needs for a group of items at one time and possibly loses freight advantages as well. The result is that economic order cycle times are seldom used, the cycle being set by these other considerations.

The negative exponential distribution has been found to describe demand at the distributor level, and the Poisson and negative exponential distributions have been useful descriptions of demand at the retail level. These distributions then provide a basis for the design of buffer stocks.

IMPLICATIONS FOR THE MANAGER

It is not uncommon for inventories to represent 20 to 30 percent of the total assets of a manufacturing organization. Even in nonmanufacturing organizations, inventories of materials and supplies can represent a significant investment. Therefore, inventory replenishment policies are an important aspect of day-to-day managerial control.

The simple concepts of EOQ provide the basis for balancing the costs affected by inventory replenishment decisions. Note that the typical total incremental cost curve shown in Figure 7-3 was shallow near the optimum. Thus, managers need to be more sensitive to operating in the optimum range than to slavishly following an EOQ based policy. In following an EOQ policy, it is important to recognize that the assumption of constant demand is crucial. This assumption is often not true in practice. Seasonal variations and dependent demand in production lot size decisions may favor periodic ordering and other policies coupled with inventory planning.

The concepts underlying the design of buffer stocks and service levels should be of concern to managers. Defining a realistic maximum demand, D_{max}, requires a managerial trade-off of buffer inventory cost versus service, unless good estimates of shortage costs are available. Recognizing the relationship between service level and the expected quantity short is important if managers are to be able to make good judgments about appropriate service levels. Also, recognizing the effect of order size on the expected quantity short, for the same service level, helps managers make appropriate service level decisions.

Most inventory replenishment systems for the Class C low-valued items can be automated through computing systems. On the other hand, Class A and sometimes Class B items may need the attention of a responsible executive, since the decisions can be of crucial significance to operating success.

Inventory replenishment for finished goods, particularly at the distributor and retail levels, can be accomplished by either an order point or periodic reorder system. However, there are often important advantages for the periodic system. The advantages stem from the possible grouping of orders to a given supplier in order to minimize frieght costs and from the close surveillance over inventory and demand levels that is possible in such systems.

REVIEW QUESTIONS AND PROBLEMS

1. What are the relevant costs that management should try to balance in deciding on the size of purchase orders? How do they vary with order size?

2. What is the total incremental cost equation involving
   a. Ordering and inventory costs?
   b. Price discounts?
   c. Shortage costs?

3. Explain the rationale for the derivation of Equation 4.

4. We have the following data for an item that we purchase regularly: annual requirements, $R = 10{,}000$ units; order preparation cost, $c_P = \$25$ per order; inventory holding cost, $c_H = \$10$ per unit per year.
   a. Compute the economic order quantity, Q_0.
   b. Compute the number of orders that must be placed each year and the annual cost of placing the orders.

c. Compute the average inventory if Q_0 units are ordered at one time, and compute the annual cost of the inventory.

5. Suppose that the estimate of c_P in question 4 was in error, being only \$20 per order. What is the value of Q_0? What is the percentage change in Q_0 for the 20 percent decrease in c_P?

6. Suppose that the estimate of c_H in question 4 was in error, being actually \$15 per unit per year. What is the value of Q_0? What is the percentage change in Q_0 for the 50 percent increase in c_H?

7. A price discount schedule is offered for an item that we purchase as follows: \$1 per unit in quantities below 800, \$0.98 per unit in quantities of 800 to 1599, and \$0.97 per unit in quantities of 1600 or more. Other data are $R = 1600$ units per year, $c_P = \$5$ per order, and inventory holding costs are 10 percent per year of average inventory value, or \$0.10 per unit per year at the \$1 per unit price. The value of Q_0, using Equation 4, is 400 units. What should be the size of the purchase quantities in order to take advantage of the price discounts?

8. What are the assumptions involved in the simple EOQ formula, Equation 4?

9. Define the following terms:
   a. Order point
   b. Lead time
   c. Reorder level
   d. Maximum demand
   e. Safety factor
   f. Service level

10. If the average demand during lead time is $\overline{D} = 200$ units, and the buffer stock is $B = 125$ units, what is the D_{max}? What managerial judgment is required to determine D_{max}?

11. Under what conditions can service level be determined objectively?

12. If demand during lead time follows a normal distribution, and we wish to maintain a 95 percent service level, what is the safety factor? If the standard deviation is $\sigma = 100$ units, what is the buffer stock required to implement the 95 percent service level?

13. If we were attempting to decide between an order size policy of $Q = 500$, or $Q = 1000$, which would we choose if our only concern were to minimize the risk of stockouts? Why?

14. Explain the concepts of "expected quantity short" and "annual expected quantity short."

15. What happens to the annual expected quantity short if
    a. The buffer stock is increased?
    b. The order size is increased?
    c. Annual requirements double, same order size?
    d. Through a policy change, service level is increased?
    e. Annual requirements double, EOQ policy?
    f. Variability of demand increases?

16. What "triggers" a replenishment order in each of the following managerial control systems?
    a. Fixed reorder quantity system
    b. Periodic reorder system
    c. Optional replenishment system

17. Explain the concept of the "two-bin" system.

18. Under what conditions would one use the fixed reorder quantity system in preference to the periodic reorder system, and vice versa?

19. As time passes, any inventory control system may become dated as demand, costs, and competitive pressures change. Thus, periodic review of system parameters is important. What are the parameters that should be reviewed for the fixed reorder quantity and periodic reorder systems?

20. What are likely to be the most appropriate replenishment policies for distributors and retailers?

21. Define an economic reorder cycle time. Why is it that organizations that use a periodic reorder policy may not actually use the economic cycle time?

22. Weekly demand for a product, exclusive of seasonal and trend variations, is represented by the empirical distribution shown in Figure 7-12. What safety or buffer stock would be required for the item to ensure that one would not run out of stock more than 15 percent of the time? 5 percent of the time? 1 percent of the time? (Normal lead time is one week.)

23. If the product for which data are given in question 22 has a unit value of \$50, shortage costs of c_S = \$10 each, and an annual inventory carrying cost of 25 percent of the average inventory value, which of the three levels of service would be more appropriate?

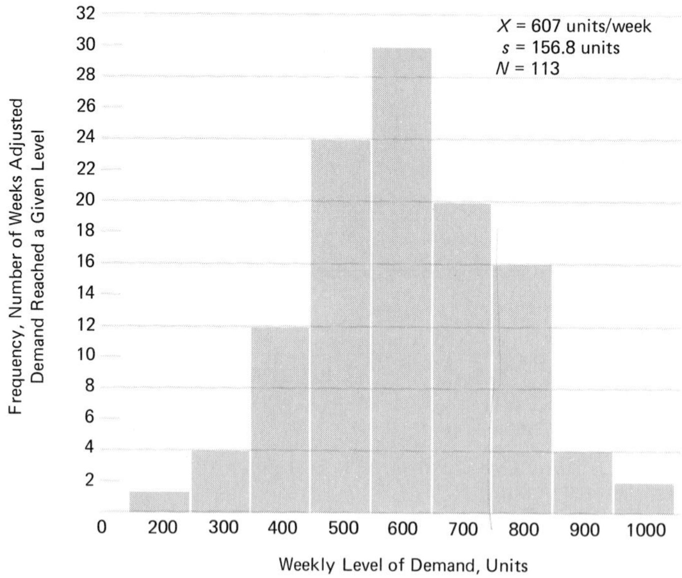

FIGURE 7-12
Distribution representing expected random variation in weekly sales, exclusive of seasonal and trend variations.

24. If the demand distribution is normal with $\overline{D} = 1000$ units per week and the standard deviation is estimated as $s = 200$ units,
    a. What is the buffer stock required for the 90, 95, and 99 percent service levels?
    b. What is the expected quantity short per order for each service level?
    c. If inventory holding costs are $c_H = \$10$ per unit per year, $c_P = \$25$ per order, and $c_S = \$15$ per unit short, which of the three service levels is most economical if 32 orders per year are placed? (EOQ = 1612, or approximately 32 orders per year, i.e., 52,000/1612 = 32.25.)
    d. If order size is reduced to $Q = 1000$ units, how is service policy affected?
    e. If order size is increased to $Q = 2600$ units, how is service policy affected?

25. A manufacturer is attempting to set the production lot size to use for a particular item that is manufactured only periodically. The incremental cost of writing manufacturing orders is approximately \$10. The aggregate cost of setting up machines for the production run is \$40. The inventory holding cost is \$1 per unit per year.

    The annual requirements for the item are 52,000, and the production rate is 5000 units per week.
    a. What is the most economical manufacturing lot size?
    b. How many production runs are needed per year to satisfy requirements?

c. What considerations should influence your choice of the actual lot size selected?

26. In arid and semiarid climates, permanently installed lawn sprinkler systems are used. Automating the control of such systems has been a common procedure for parks and golf courses. More recently, systems have been designed that are low enough in cost so that homeowners can be relieved of the bother of the watering cycle.

One such system was designed with the homeowner in mind. The user can set the watering cycle to come on every one, two, four days and so on, and set the amount of time that the sprinklers run, for example, 3, 11, 22, 45, 60, 90 minutes and so on. In addition, there is a moisture sensor that is implanted in the lawn; if when the cycle is ready to trigger, the moisture sensor indicates that the soil is still moist enough, the cycle will be skipped.

Relate the sprinkler control system to one of the inventory control systems discussed in the text.

27. How do inventory planning systems make it possible to adjust to variations in demand? If forecasts are inaccurate, how valid are inventory planning systems?

28. Assuming an inventory planning system with seasonal demand, compare the performance of ordering in preset fixed quantities with ordering variable quantities periodically.

SITUATIONS

29. The Mixing and Bagging Company produces a line of commercial animal feeds in 10 mixes. The production process itself is rather simple, as the company name implies. There is a variety of basic grain and filler ingredients that are mixed in batches. The mixed product is then fed to an intermediate storage hopper, from which it is conveyed to a bagging operation. The bags of feed are then loaded on pallets and moved to the nearby warehouse for storage.

The operations are supervised by a foreman. Direct labor is provided by a full-time worker who operates the mixing and blending equipment, plus 4 full-time and 10 part-time workers. The foreman is paid $15,000 per year, the mixer-operator $5 per hour, the other full-time workers $4 per hour, and the 10 part-time workers are paid $3 per hour.

The usual routine in making a production run is as follows: the foreman receives job tickets from the office indicating the quantities to be run and the formula. The job tickets are placed in the order in which they are to be processed. At the end of a run, the foreman purges the mixing system and ducts of the previous product. This takes 20 minutes.

Meanwhile, the foreman has directed the mixer-operator and the four full-time employees to obtain the required ingredients for the next product from the storeroom. When the mixing equipment has been purged, the mixer-operator loads the first batch of materials according to the formula in the mixer and gets it started. This takes about 10 minutes. The total time spent by the mixer-operator in obtaining materials and loading the mixer is 30 minutes. The four full-time employees devote the 30 minutes to obtaining materials.

During the previous activities, the foreman has turned his attention to the bagger line, which requires minor changeover for the bag size and the product-identifying label sewed on the top edge of the bag as it is sewed closed.

While the foreman was purging the system, the 10 part-time employees have transferred what is left of the last run to the warehouse, which requires about 15 minutes. They are then idle until the bagging operation is ready to start again.

The inventory in the finished goods warehouse is valued according to the sale price of the item, which is about \$5 per 100 pounds. The cost of placing the material in the warehouse has been calculated as approximately \$0.25 per 100 pounds, based on the time required for one of the part-time workers to truck it to the warehouse and place it in the proper location. The front office has calculated that the storage space in the owned warehouse was worth about \$10 per square foot per year, but since the bags were palletized and stacked 12 feet high, this cost reduced to only \$0.20 per 100 pounds per year. The product mixes were stable, and there was very little risk of obsolescence. There was some loss because uninvited guests (rats, etc.) came to dine. The total storage and obsolescence costs were estimated as 5 percent of inventory value.

The Mixing and Bagging Company has a factory overhead rate that it applied to materials and direct labor. This overhead rate is currently 100 percent and is applied to the average material cost of \$1.87 per 100 pounds, plus the direct labor costs of \$0.13 per 100 pounds. The company earns 8 percent after tax and can borrow at the local bank for an interest rate of 9 percent.

The factory manager is currently reviewing the basis for deciding on the length of production runs for products. He figures that operations are currently at about 85 percent of capacity. He has heard of EOQ as a basis for setting the length of production runs. What values should he assign to c_P and c_H for his operations?

30. Mr. Winston DeBay, president of the Ronald S. Woods Publishing Company, has called in a consultant to help him reconsider his reordering policies and practices. Woods publishes textbooks in business and economics, and once a book is published, a recurring problem is keeping it in stock in order to meet demand. For each book, there is a set of film masters for each page. A reprint order would be placed with a printer who performs all the manufacturing operations and delivers printed books to the publisher's warehouse.

The costs of reordering involve a minimum or fixed charge to cover machine setup and related one-time costs, plus a per unit variable cost to cover paper, ink, book covers, and the like. The typical fixed costs are in the range of \$500 to \$2,000, and the per unit variable costs are in the range of \$1 to \$3, depending on the particular book.

Mr. DeBay's practice had been to spend about three days twice per year reviewing the status of each book in the list to decide if a reprint order should be released and, if so, the order size.

Mr. DeBay described his present policies as follows: "We get two peaks per year, the big one in July, and usually a second one in December. These peaks correspond to the fall ordering by bookstores for schools on both the semester and quarter systems, and the second smaller peak covers needs for the second semester and winter quarters. There are somewhat smaller orderings to cover the spring quarter and summer sessions, but these do not bother us much. Our fiscal year begins March 1, in order to key in with the sales year.

"In May, I order what I think I will need for the fall, and I order again in October for what I think we will need for the balance of the academic year. I keep a minimum stock unless I know a book is being revised, or going out of print. Then I cut it closer, and try to just about use up the minimum stock to coincide with the receipt of the new order."

The consultant asked, "How do you forecast your needs?" DeBay stated, "I watch the progress of sales and adoptions, and take into account the age of the edition. If a book is in the second or third year, used books will cut into sales rather sharply, and I take it into account."

The consultant produced a copy of Figure 7-13, stating, "Here is a forecast of one of your typical books, using a computer program—would forecasts of this nature help you in making decisions regarding the number of copies to reorder?" (Figure 7-13 was produced by the Fourier Series Forecasting Program described in Chapter 5). DeBay responded that he probably could use such a forecast. "It picks out the peaks and valleys pretty well, but it couldn't take the other things into account. It [the program] doesn't know what I know."

The consultant said that he had looked into the costs for book reordering in order to compute an economical order quantity that balances the cost of reordering against the inventory holding costs. Assuming that R = 12,000 units per year, c_P = \$1,500 per reorder, and c_H = \$0.63 per unit per year, then EOQ = 7559 units for the example book. In other words, an order for EOQ lasts for an average of 7559/12,000 = 0.63 years, or 7.6 months. "What do you think of simply ordering EOQ whenever inventory drops to a reorder point that we would set? The reorder point would be that inventory level sufficient to cover sales during the supply lead time, plus a minimum stock."

DeBay responded, "I don't like that idea. To match with our type of business, the reorder point would have to change all the time, depending on the time of year. If we were floating into our July-August peak without a great big inventory, we would be clobbered and lose sales. On the other hand, if we hit the reorder point during slack months, I would not get very excited."

Next, the consultant raised the possibility of setting up a periodic cycle for reordering, perhaps every 3, 6, or 12 months. "Then you would order enough to cover sales since the last reordering, plus projected usage during the lead

* * * LEAST SQUARES FOURIER SERIES FORECAST MODEL * * *

FORECAST MODEL RESULTS

FCST NO.	ACTUAL DEMAND	FORECAST DEMAND	FORECAST ERROR
1	444	561	-117
2	235	402	-167
3	854	711	143
4	391	503	-112
5	3259	3208	51
6	2163	2029	134
7	1055	712	343
8	486	439	47
9	751	969	-218
10	422	1500	-1078
11	622	617	5
12	1098	687	411
13	715	611	104
14	490	300	190
15	553	644	-91
16	556	419	137
17	3187	3222	-35
18	1787	1981	-194
19	466	663	-197
20	263	319	-56
21	906	919	-13
22	2193	1452	741
23	793	496	297
24	876	639	237
25	574	561	13
26	229	252	-23
27	615	667	-52
28	274	299	-25
29	3156	3172	-16
30	2064	2004	60
31	684	830	-146
32	280	271	9
33	1389	1158	231
34	1669	1332	337
35	505	806	-301
36	-201	446	-647
37		440	
38		-12	
39		617	
40		34	
41		3627	
42		1165	
43		1140	
44		-136	
45		1468	
46		924	
47		685	
48		38	

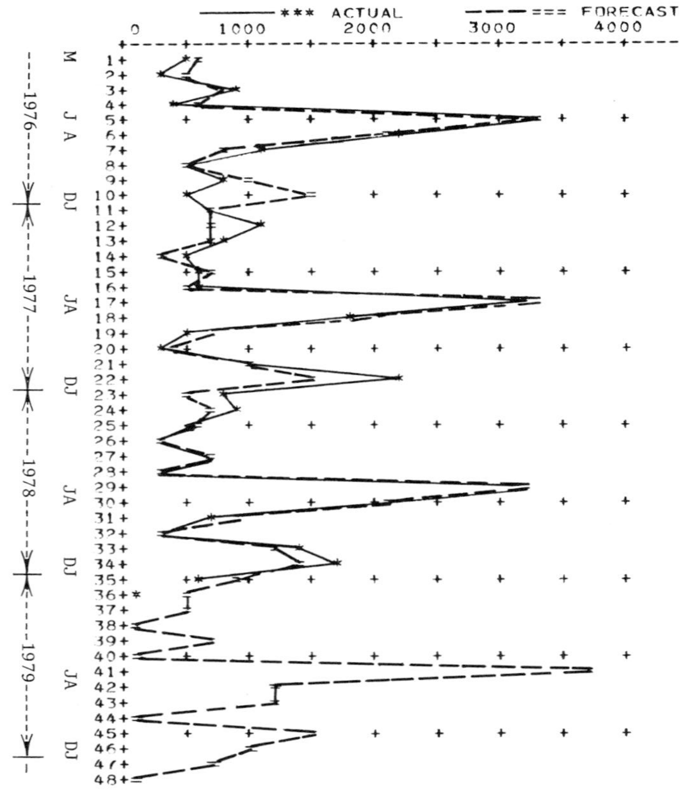

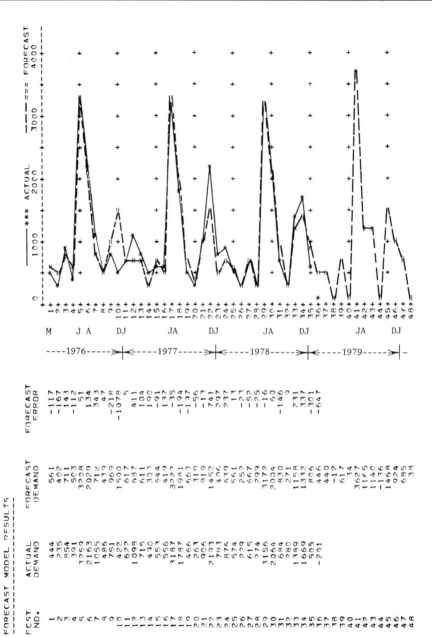

FIGURE 7-13
Least squares Fourier series forecast model for the Ronald S. Woods Publishing Company.

time. We call that a periodic reorder system. Based on your costs, I figure that an optimum cycle would be $Q_0/R = 7559/12{,}000 = 0.63$ years, or 7.6 months. Perhaps we could set an eight-month cycle—would you like that better?"

DeBay responded, "No, I don't like that idea either. A cycle as short as three months would kill us—you said yourself that the reordering costs were significant. I would rather carry some inventory costs. Besides, I won't lose sales if I have stock on hand. Frankly, I don't like any of the regular cycles you suggest. As I said, the business comes in July-August and in December-January. I can't help that, it's the way it is, and I'm not going to experiment with changing the way the business comes to us. Maybe you can help a little with the forecasting, but I don't like your EOQs and your regular cycles. Those things are for somebody else's business that goes by the book. The book business doesn't go by the book!"

Given this encounter, what should the consultant recommend? Would a fixed quantity or a periodic system be appropriate? Why, or why not?

31. The following article is a case study of the installation of "An Integrated Inventory Control System," by Michael J. Lawrence.*

AN INTEGRATED INVENTORY CONTROL SYSTEM

Michael J. Lawrence

Introduction

Inventory control is one of the earliest yet consistently fruitful application areas for the techniques of management science. Rarely are the cash benefits exceptional such as one might experience with management science applications in strategic decision-making areas. However, it is also rare if a well developed and properly used statistically based inventory control system does not enable inventories to be significantly reduced while maintaining or improving the level of customer service. This paper presents a case study of the development of an inventory control system designed to serve the needs of a division of a large U.S. based chemical company.

The usual objective of an inventory control system can be summarized as providing an agreed level of customer service for the cheapest price. There are three fundamental and interrelated aspects in an inventory control system: forecasting future demand, deciding when and how much to re-order and deciding where stocks should be held. The case presented in this paper illustrates the use that can be made of a computer based statistical forecasting system to aid short term sales forecasting. Practical experience has shown that computer aided forecasting which includes routine management review and adjustment provides a better, more reliable and more con-

* Published originally in *Interfaces,* 7(2), February 1977, Copyright © 1977, The Institute of Management Sciences, reprinted with permission.

sistent forecast than either a statistical or a subjective forecast alone. When coupled with a tracking procedure to report original and revised forecasting errors it provides a feedback loop enabling continuous monitoring of the error and measured improvement in forecasting skills. The case also illustrates the inventory reduction possible through the use of a two-stage inventory system comprising a central warehouse feeding smaller branch warehouses.

The Company

The inventory control system was designed for a division whose sales were in excess of $120 million per annum achieved with a product line of about 1500 specialty chemical products sold through eight branches principally to the textile industry. Two U.S. based and two European based affiliated manufacturing facilities supplied the products which were all initially inventoried in a national warehouse in New Jersey. From there the products were, if necessary, blended and mixed, and supplied on requisition to the eight branches, which serviced customers in their area.

Customer service was a very important aspect of the business. Customers generally maintained no inventories to speak of, and expected that orders placed one day would be delivered the following morning. As the product line was largely duplicated by competition, customer service was recognized as the key to maintaining market share.

When products were ranked on the basis of sales volume it was observed that the top 5 percent of products accounted for 35 percent of sales volume, whereas the top 20 percent of products accounted for 85 percent of sales. The inventory level was on an average of 4-1/4 months' supply. With the products partitioned into three groups based on sales volume, the slower moving products and faster moving products had 25 percent above average inventory with the medium selling products below average inventory.

The Earlier Inventory Control System

Before discussing the new inventory control system which was developed, some comments are made on the earlier system and its performance to provide a point of reference and to illustrate the problems being experienced.

Sales forecasting was perceived as the biggest problem with the existing system. Firm manufacturing commitments for overseas supplied goods had to be entered into six months prior to receipt of goods. For the locally manufactured goods the lead time commitment was four months. The long manufacturing lead times were necessitated by production considerations and could not reasonably be reduced. Product sales, depending as they did on the tastes of the fashion business, were quite erratic and marketing mangement, who manually updated a six month rolling forecast every month, tended to build in a generous factor of safety. The forecasts were developed on a branch basis and consolidated nationally at the head office. Together with the current inventory and order position, this gave the necessary data for generating new manufacturing orders. The short end of the branch level forecasts were used by the branches in conjunction with maintaining their inventories through re-orders on the national warehouse. The inventory replenishment parameters (re-order point and re-order quantity) had been developed and used for some years for manufacturing re-orders but had not been maintained for branch re-orders.

As mentioned, the total inventory level was about 4-¼ months' supply. This represented an average safety stock level of about 2.6 months' supply, of which 2.35 months was in the branches and 0.25 months in the national warehouse. The term safety stock is used to refer to the amount of stock on hand when a new shipment arrives.

The national/branch inventory balance was not as planned. Management was constantly attempting to reduce the branch inventory level and increase the national warehouse stock. The situation defied change and appeared to be a result of hoarding by branches as a result of poor service by the national warehouse. While it was believed at the time that this was a very disadvantageous situation, we shall see later in this report that its economic significance was not considerable.

The service level experienced was defined in the study as the ratio of quantity supplied out of stock to quantity ordered. Historically the customers had received a service level of 99 percent. This was accomplished despite a branch/customer service level of 89 percent of referring 5 percent of orders to nearby branches for delivery and 5 percent of orders to other than nearby branches or to the manufacturing plants for air freighting. Needless to say, referring orders involved an additional transportation cost premium, as well as taking additional clerical time. However, the national warehouse/branch warehouse service level was 60 percent, reflecting the lack of inventory at the national warehouse. Hence, a branch re-order on average was 60 percent filled out of stock and 40 percent back-ordered.

The order cycle lead time for the branch re-orders depended on two factors:

- Whether the product was sold in the as-manufactured condition or required blending,
- Whether the product was in stock.

The following table gives the lead time in calendar days:

	In Stock	Out of Stock
Sold as manufactured	8–20 days	34–44 days
Sold blended	11–26 days	41–58 days

The distribution of the lead time appeared due to uncontrollable factors and approximated a uniform distribution.

The head office computer based information system contained the inventory status for each product at each location. The national warehouse stock records were satis-

factory as they were updated concurrently with receipts or shipments. However, because the branches received orders and shipped goods days before the central computer received this data and because of inadequate controls on the recording of interbranch transfers, the central computer data of branch stock levels was not accurate. The branches maintained their own accurate stock card files manually.

A project was underway at the time to develop a real time order entry system tied into each branch by visual display units and teleprinters for producing packing slips, shipping documents, etc. The new system would have the capability to maintain accurately all stock records at the head office. Completion was planned for one year after the planned completion of the inventory control project. It was decided to design both systems with all necessary links to facilitate later integration but not to delay the implementation of the inventory control system.

Features of the Inventory Control System Adopted

Forecasting. *As sales forecasting was a recognized problem area and since a dependable forecasting approach is vital to a sound inventory control system, much effort was expended on studying the forecasting requirements and exploring alternate solutions. After this analysis it was determined that the right solution should give consideration to the following points:*

(a) Divisional management wanted the branch level marketing management to be in control of the forecasting, as their customer and product knowledge was indispensable.

(b) Forecasting trials on a sample of two years of actual sales data revealed that the triple exponential smoothing* forecasting technique was 25 percent more accurate than previous manual forecasts on a branch/product basis (accuracy measured by standard deviation of forecast error).

(c) Despite firm marketing conviction that the product line was seasonally affected, no seasonal component could be statistically detected.

(d) About 10 percent of the sales volume was so called referral orders. These are orders referred to another branch because of lack of stock availability. The system would need to record these sales against the "home" branch.

(e) With roughly 9000 forecasts to be prepared each month, computational speed would be an important consideration in selecting a forecasting technique.

(f) Forecasting trials on national sales data revealed greater accuracy than that obtained by summing the forecasted branch sales.

* Triple exponential smoothing assumes a quadratic trend. For methods, see R. G. Brown, *Smoothing, Forecasting and Prediction*. Prentice-Hall, Englewood Cliffs, N.J., 1963, pp. 136–144.

The forecasting solution adopted was to develop a branch/product level forecast for the next six months using the triple exponential smoothing technique. The appropriate set of forecasted products was sent to each branch for approval or amendment. If the statistical forecast was amended, the salesman would indicate by means of a code why the change was made. Codes covered, for instance, such conditions as new key customer, loss of key customer to competition, fashion change, expected marketing plan impact, etc. The branch adjusted forecasts were consolidated on the computer at the head office by key punching the amendments. The computer program then compared the consolidated branch level forecasts with a separately prepared national level forecast and any deviations greater than a predetermined percentage were printed out for management reconciliation. This would be done bearing in mind the greater accuracy of the national level forecast. The purpose for coding the reason for amending the statistical forecast was to develop a tabulated feedback for marketing management indicating which types of amendments on average improved forecasting accuracy and which types led to a less accurate forecast. The feedback loop created a continuous learning environment for improving forecasting skills with an established reference basis of the exponential smoothing technique to measure progress.

Re-Order Points. *The minimum safety stock sufficient to achieve a given level of branch service to customers needs to be jointly calculated for both the national and the branch warehouses. The factors entering the decision include forecast accuracy, customer service level desired, distribution of lead time for manufacturing re-orders and distribution of lead time for branch re-orders. All these factors are product dependent.*

An approximate analytical solution was obtained to indicate the level of safety stock required at branch and national warehouses in order to achieve various levels of branch customer service. The solutions were then checked by running a series of computer simulations. The results are presented graphically [see Figure 7-14], *showing, for a customer service level, the safety stock levels expressed in months supply plotted against the national warehouse service level.*

The safety stock level for both the branch and the national warehouse are plotted together with the total safety stock in the system. The general shape of the curves is as expected. If the national warehouse service to the branches is increased by raising the stock level in the national warehouse, the branch inventory level can be reduced while still maintaining the same customer service.

For a 90 percent service level the balance of inventories between national and branch warehouses is not very critical. The national warehouse service level can be anywhere from 60 to 90 percent with no real impact on the total level of safety stock. The minimum is achieved by a national warehouse service level of 75 percent and a total safety stock equal to two months' supply. The service level of the national warehouse had historically been about 60 percent with the branch service level at 89 percent, which from this graph is satisfactory.

The goal was to increase the branch service level to 95 percent and so reduce the number of orders referred out of the region. The graph pertaining to this branch service level reveals that the national warehouse service level is now more critical.

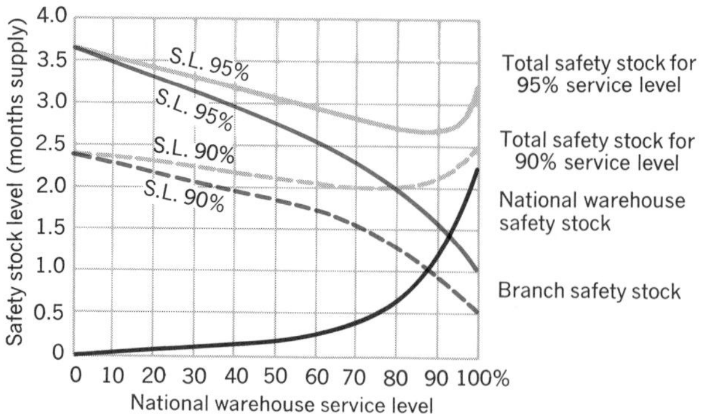

FIGURE 7-14
Safety stocks required at branch and national (central) warehouse level to achieve given customer service.

Operating it at a 60 percent service level would involve a total safety stock of 2.95 months while at the optimum of 85 percent service level the total safety stock is 2.7 months; that is, a saving of one week's inventory. The graphs are not included here, but for higher branch service levels the national warehouse service level is still more critical.

The graphs reveal the value in total safety stock terms of having a two-stage distribution system of a national warehouse and branches. Obviously there are other reasons for having a national warehouse than reduction of safety stock. For instance, in this case study all blending and mixing of the products from the various plants needed to be carried out centrally. On the other hand, a national warehouse does involve more transportation cost in a multiplant environment. As the broad structure of the distribution system was not in the terms of reference of the study this question was not explored.

For a branch level of 90 percent the use of a national warehouse saves in this case about 1-1/2 weeks of safety stock, while for a 95 percent branch service level about three weeks of safety stock are saved. The value of the national warehouse would be greater were the national/branch order lead times less. The graphs are based on the national/branch lead times presented earlier which range from 8-26 days if the product is in stock, to 34–53 days in the case of the product not being in stock. The impact on safety stock of reducing this lead time and its variability was not explored in the project, though it was believed to be high. In addition, consideration was given to not holding national inventories of some imported products which were always sold in the as-manufactured condition. Freight savings could be realized by shipping direct but were not of sufficient magnitude to offset the increased inventory costs.

Consolidation on a regional basis of slower moving stock was investigated and was found to have some attractive benefits. Safety stock reduction was in the vicinity of 40 percent due mainly to lowering of forecasting error. Order processing savings *coupled with average inventory level reduction (excluding safety stock) promised at least an additional $60,000 per year savings.*

To summarize, the safety stock allocation adopted was based on providing a branch service level of 95 percent with a national warehouse service level of 85 percent. This resulted, on average, in one month's safety stock in the national warehouse and 1-3/4 month's safety stock in the branch warehouse. For any given product, its safety stock depended on its forecast accuracy, where manufactured, and whether sold in the as-manufactured or blended condition. Additionally, the desired customer service was product dependent, determined by a service level code associated with each product. These factors were used in the computer based model to arrive at the appropriate national and branch warehouse re-order points for each product in terms of months supply.

The re-order quantity was calculated on an approximate basis using the simple EOQ model. While it was recognized that this approach would lead theoretically to a too low re-order quantity when compared to an "exact" solution, the lack of accuracy of the re-order cost estimate made a more refined computational procedure seem a waste of time.

Implementation of System

The initial implementation of the inventory control system [see Figure 7-15] allowed for:

(i) Statistical forecasts on a branch level of the high and medium volume products. This comprised 40 percent of the 1500 products and included more than 90 percent of the sales volume. On a national basis all products were forecasted. Each branch amended, where necessary, its statistical forecast based on its marketing, product and customer experience and knowledge.

(ii) Branch control of its inventory and re-ordering from the national warehouse, by application of the product specific re-order level and re-order quantity parameters. These values were initially supplied to all branches and then updated quarterly on an exception basis.

(iii) Head office control of reconciling the consolidated branch level forecasts and the national forecast.

(iv) Computer prepared manufacturing re-orders, on an automatic basis once each month.

(v) Gradual shift in the level and balance of inventories to approach the calculated optimum levels while avoiding disruptions.

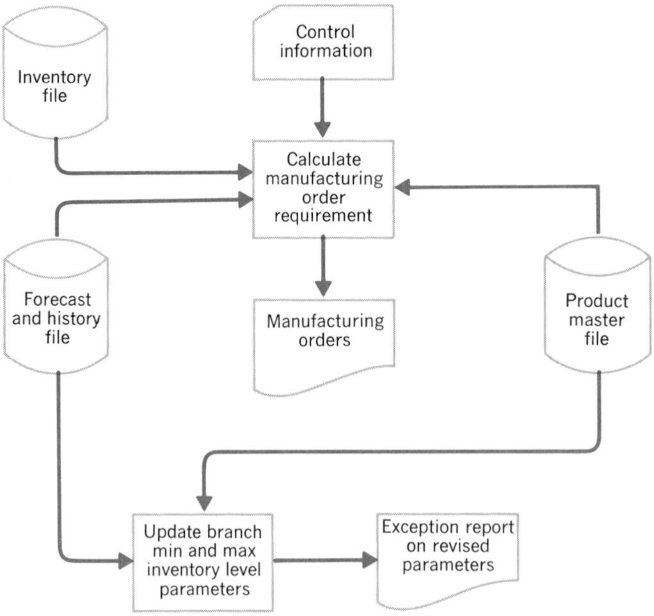

FIGURE 7-15
Inventory control system.

At the time of implementation the national warehouse inventory levels were reliably contained in the EDP inventory file. While branch inventory levels were on the EDP file they were out of date by a week or more. The only reliable up-to-date inventory file sufficient for branch inventory maintenance was a manual system in the branch. This gave a valuable opportunity to allow the branch logistics personnel to control their end of the system and utilize the computer prepared re-order parameters and so gain experience and confidence in the system. It was planned that when the new real time computer order entry system was installed, up-to-date reliable inventory status data would be available and branch re-ordering could then be handled automatically, along the same lines as had been followed manually.

Benefits

Benefits from the implementation of the project, which incidentally took place about 12 months after project inception, arose from three main areas:

(i) Stock reduction from improvement of forecasting accuracy;

(ii) Stock reduction from improvement of inventory re-order point and re-order quantity calculations; and

(iii) By increasing branch service level to 95 percent, a reduction of additional costs incurred in referring orders to other branches or back to manufacturing after taking into account the increased inventory at the branch warehouses.

In addition, considerable benefits flowed from the computer forecasting in the reduction of clerical work load on the sales force and also in the head office where previously the manually prepared forecasts had been consolidated.

Direct measurement of project benefits was not considered possible because of the number of changes introduced. For instance, raising the level of customer service provided by the branch had the combined effect of increasing branch inventory, reducing referred orders together with their attendant additional costs, and providing the customer with better service.

The bulk of the cost benefits were provided by the improved forecasting approach which provided a two week stock usage reduction. The benefits gained by the improved re-order point, re-order quantity parameters were estimated at about one week's usage. In total, this inventory reduction amounted to a little over 10 percent of average operating inventory, to save the company in the order of $300,000 per annum at the 20 percent inventory cost used. Additionally, raising the branch service level was expected to save $50,000 per annum after allowing for the increased inventory. The consolidation of slow moving items into regional warehouses was planned for later implementation and promised a saving of at least $100,000 per annum. In total, savings were believed to be in excess of $450,000 or 15 percent of the inventory carrying cost.

QUESTIONS

a. What are the advantages and disadvantages of the new forecasting system?

b. How do you reconcile the fact that the simulation shows that branch safety stock actually declines rapidly as National Warehouse service level increases? Why would not the reverse policy of keeping Branch safety stock high and National safety stock low be even better?

c. What further improvements do you feel could be made in the recommended system?

d. Do you agree that the basis for inventory replenishment should be EOQ? Throughout the system? For all items?

e. What is your evaluation of the installed system as a whole? As a manager, would you install it or ask for restudy and possible changes? If the latter, what changes?

REFERENCES

Austin, L. M. "Project EOQ: A Success Story in Implementing Academic Research," *Interfaces,* 7(4), August 1977, pp. 1–12.

Brown, R. G. *Decision Rules for Inventory Management.* Holt, Rinehart, & Winston, New York, 1967.

Buchan, J., and E. Koenigsberg. *Scientific Inventory Control.* Prentice-Hall, Englewood Cliffs, N.J., 1963.

Buffa, E. S.,and J. G. Miller. *Production-Inventory Systems: Planning and Control* (3rd ed.). Richard D. Irwin, Homewood, Ill., 1979.

Buffa, F. P. "A Model for Allocating Limited Resources When Making Safety-Stock Decisions," *Decision Sciences, 8*(2), April 1977, pp. 415–426.

Lewis, C. D. *Demand Analysis and Inventory Control.* Lexington Books, London, 1975.

Magee, J. F., and D. M. Boodman. *Production Planning and Inventory Control* (2nd ed.). McGraw-Hill, New York, 1967.

McMillan, C., and R. F. Gonzalez. *Systems Analysis: A Computer Approach to Decision Models,* (3rd ed.). Richard D. Irwin, Homewood, Ill., 1973.

Peterson, R., and E. A. Silver. *Decision Systems for Inventory Management and Production Planning.* John Wiley, New York, 1979.

Starr, M. K., and D. W. Miller. *Inventory Control: Theory and Practice.* Prentice-Hall, Englewood Cliffs, N.J., 1962.

CHAPTER

8 Industrial Scheduling Systems

THERE IS A WIDE VARIETY OF SCHEDULING SYSTEMS IN INDUSTRIAL practice, tailored to fit specific situations. However, the general nature of systems, the impact of downstream inventories, and the system dynamics are common problems.

Scheduling for high volume continuous systems is focused more in the aggregate plan, and translating and implementing the plan, since the entire system operates more as a giant machine. Scheduling intermittent systems requires more detailed planning, since individual orders must be handled in batches.

NATURE OF THE PRODUCTION-DISTRIBUTION SYSTEM

In order to understand scheduling and control problems, we need to understand the overall flow and appreciate the importance of system inventories and the system dynamics. Figure 8-1 represents the production-distribution system for an inventoriable item such as a small appliance. The flow and major functions performed are shown, together with information flow required for the replenishment of inventories. There are 1000 independent retailers, 50 independent distributors, and a single factory in the system. Each of the distribution steps involves a stock point for the finished product. Therefore, one way of looking at the system downstream from the factory is to envision it as a *multistage* inventory system.

We assume that each retailer uses a periodic inventory control system and has a replenishment cycle that involves the review of demand and inventory status, transmission of orders to the supplying distributor, and the filling and shipping of orders by the distributor. Similarly, each distributor has an equivalent replenishment cycle, based on the assessment of demand from retailers, the transmission of orders to the factory warehouse, followed by the filling and shipping of the orders. Similar cycles are required for the factory warehouse in ordering from the factory and for the factory itself in ordering raw materials from vendors.

We have recognized the existence of inventories in previous chapters, but largely as a source of short-term capacity in aggregate planning systems and as individual items for replenishment. Now we consider other vital functions of inventories in a productive system.

In a sense, inventories make possible a rational productive system. Without them we could not achieve smooth flow, obtain reasonable utilization of machines, or expect to give reasonable service in terms of off-the-shelf availability. At each stage of both manufacturing and distribution, inventories *decouple* the various operations in the sequence. Between each pair of activities, inventories make the required operations sufficiently independent of each other so that low-cost operations can be carried out.

Thus, when raw materials are ordered, a supply is ordered that is large enough to justify the out-of-pocket cost of putting through the order and transporting it to the plant. When production orders to manufacture parts and products are released, they

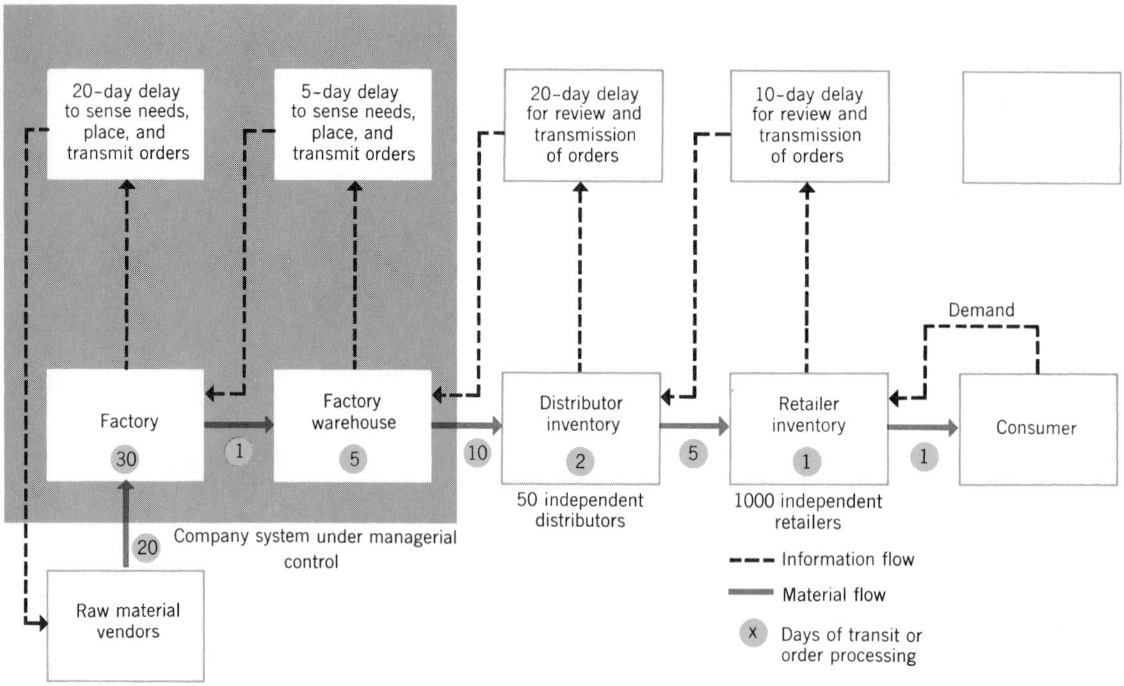

FIGURE 8-1
Production-distribution system for high-volume continuous system showing broad flow of materials and information, transit times, and order processing delays. (System volume averages 4000 units per week.)

are made large enough to justify the cost of writing the orders and setting up machines. Running parts through the system continuously (or in batches) tends to reduce handling costs. Similarly, in distributing finished products to warehouses and other stock points, freight and handling costs per unit are lower if we ship in quantity. Therefore, while inventories are costly, they are vital to low-cost manufacture and distribution.

SYSTEM INVENTORIES

Using Figure 8-1, let us examine the kinds of inventories that occur in the system. Our concern will be mainly with the physical flow and the information system related to inventory replenishment. Some inventory will be required just to fill the flow pipelines of the system, another component to take account of the periodic nature of the ordering cycles, and a third component to absorb fluctuations in demand. Depending on the nature of the policies for factory scheduling, additional seasonal inventories might also be required.

Pipeline Inventories

If the average system volume is 4000 units per week and it takes one day to transport the products from the factory to the warehouse, then there are an average of 4000/7 = 571 units in motion at all times. If the order processing delay at the factory warehouse is five days (from Figure 8-1), then there are an average of 4000 (5/7) = 2857 units tied up at all times because of the delay.

Table 8-1 summarizes the pipeline inventory requirements for finished goods in the system, indicating a minimum of 14,284 units required just to fill the pipelines. These inventories cannot be reduced, unless transit times, delays, and handling times can be reduced. Pipeline inventories are proportional to the system volume and the physical flow time. If system volume increases, pipeline inventories must increase to keep the system functioning.

Cycle Inventories

We will assume that the periodic system of inventory replenishment discussed in Chapter 7 is being used at the factory warehouse, distributor, and retail levels. The average order size is then set by the ordering frequency and requirements at each stage.

The average retailer orders once every two weeks following a review of sales. Note that once an order is placed the information and physical flow cycle follows that shown in Figure 8-1. The ordering cycle, however, is on a regular two-week basis and is not altered by the delays shown. Therefore, when a retailer orders, it must be for a two-week supply, just to meet average demand.

The average retailer sells 4000/1000 = 4 units per week, or eight units during the two-week ordering cycle. No less than eight units must be on hand to service sales during the replenishment period, and the average inventory for this purpose is half

TABLE 8-1 **Summary of Pipeline Inventory Requirements for Finished Goods (average system volume is 4000 units per week)**

	Average Transit Delay Time, Days	Average Pipeline Inventory, Units (Days/7 × 4000)
Factory to factory warehouse	1	571
Delays at factory warehouse	5	2,857
Warehouse to distributors	10	5,714
Delays at distributors	2	1,143
Distributors to retailers	5	2,857
Delays at retailers	1	571
Retailers to customers	1	571
Totals	25	14,284

this amount, or four units. The cycle inventory for the system of 1000 retailers is then $4 \times 1000 = 4000$ units. Table 8-2 summarizes the cycle inventory requirements of the system, showing that an average of 24,000 units are required in inventory because of the periodic nature of ordering.

Buffer Inventories

Buffer inventories are designed to absorb random fluctuations in demand, thereby decoupling supply operations from the effects of demand variability. The size of these inventories depends on the nature of the demand distribution and the established service levels. The buffer stock needed to cushion the effects of greater than expected demand is the difference between the estimated maximum demand and the average demand during the supply lead time, as discussed in Chapter 7.

Suppose, for example, that from the demand distribution for the average retailer, the probability of a demand of 18 units or less is 95 percent during the 17-day supply lead time. This is consistent with management's willingness to run out of stock 5 percent of the time. The required buffer stock would then be $18 - 9.7 = 8.3$ units. The average system buffer stock for 1000 retailers is then $8.3 \times 1000 = 8300$ units. Table 8-3 shows the computation of buffer stocks for finished goods for the system as a whole, 20,229 units.

System Finished Goods Inventories

Summarizing, the finished goods inventories for the system as a whole needed to accommodate all three functions are as follows:

1. Pipeline inventories	14,284
2. Cycle inventories	24,000
3. Buffer inventories	20,229
Total	58,513

TABLE 8-2

Summary of Cycle Inventory Requirements for Finished Goods (average system volume is 4000 units per week)

	Reorder Cycle Time, Weeks	Average Cycle Inventory, Weeks
1000 Retailers	2	4,000
50 Distributors	4	8,000
Factory warehouse	6	12,000
Total		24,000

TABLE 8-3 **Summary of Buffer Inventory Requirements for Finished Goods (average system volume is 4000 units)**

	Average Demand per Week	Lead Time, Days	Average Demand During Lead Time	Maximum Demand During Lead Time	Average System Buffer Stock
500 Retailers	4	17	9.7	18	8.300
50 Distributors	80	35	400.0	550	7,500
Factory warehouse	4000	36	20,571.4	25,000	4,429
Total					20,229

These inventories are required by the structure of the system and the ordering rules and service levels used. They represent the minimum possible amounts necessary to operate the system. Inventories might be larger than this minimum if controls were not effective or if seasonal inventories were also accumulated in the system.

The impact of inventories on the problem of scheduling the factory is emphasized when one notes that ordinarily management has control over only approximately one-third of the total system inventories.

SYSTEM DYNAMICS

What is the impact of activities downstream in the production-distribution system on factory scheduling? To answer this question, we focus our attention again on the multistage system diagrammed in Figure 8-1.

Suppose that consumer demand falls by 10 percent from its previous rate. During the next inventory review the retailer reflects this 10 percent decrease in orders for replenishment sent to the distributor, but 10 days have elapsed. Similarly, the distributor reflects the decrease in the next orders to the factory warehouse for replenishment, but an additional 20 days have elapsed before the factory warehouse will feel the impact of the fall in sales. Adding up all the time delays in the information system, the factory will not learn of the 10 percent fall in demand until 35 days have passed. Meanwhile, the factory has been producing $1.00/0.90 = 1.11$ times the new consumer requirement (111 percent). An excess of 11 percent would have accumulated each day in inventory at the various stock points. The system inventory will have increased to $11 \times 35 = 385$ percent of the usual normal day's supply.

In order to react to the change, retailers, distributors, and the factory warehouse decrease the quantities ordered. To take account of the excess inventory, the factory will now have to cut back by substantially more than 10 percent. The effect of the time lags in the system is to amplify the original 10 percent change at the consumer level. The change in production levels is much greater than would have seemed justified by the simple 10 percent decrease in consumer demand, and inventories have increased instead of decreased. A more direct communication of changes in demand can reduce the magnitude of this amplification.

Figure 8-2 is identical to Figure 8-1 with the exception that a more direct information feedback loop has been added in the form of a system for assessing actual demand and forecasting demand for the upcoming period. The 10-day delay in assembling the actual demand and forecasting information reduces the total delay by 25 days. A 10 percent decrease in sales under this system would mean an excess inventory of only $11 \times 10 = 110$ percent of the normal levels would accumulate before the factory was aware of the demand change.

SCHEDULING AND AGGREGATE PLANNING

The broad outline of the scheduling process for the multistage production-distribution system is shown in Figure 8-3. Forecasts are produced for the planning horizon as

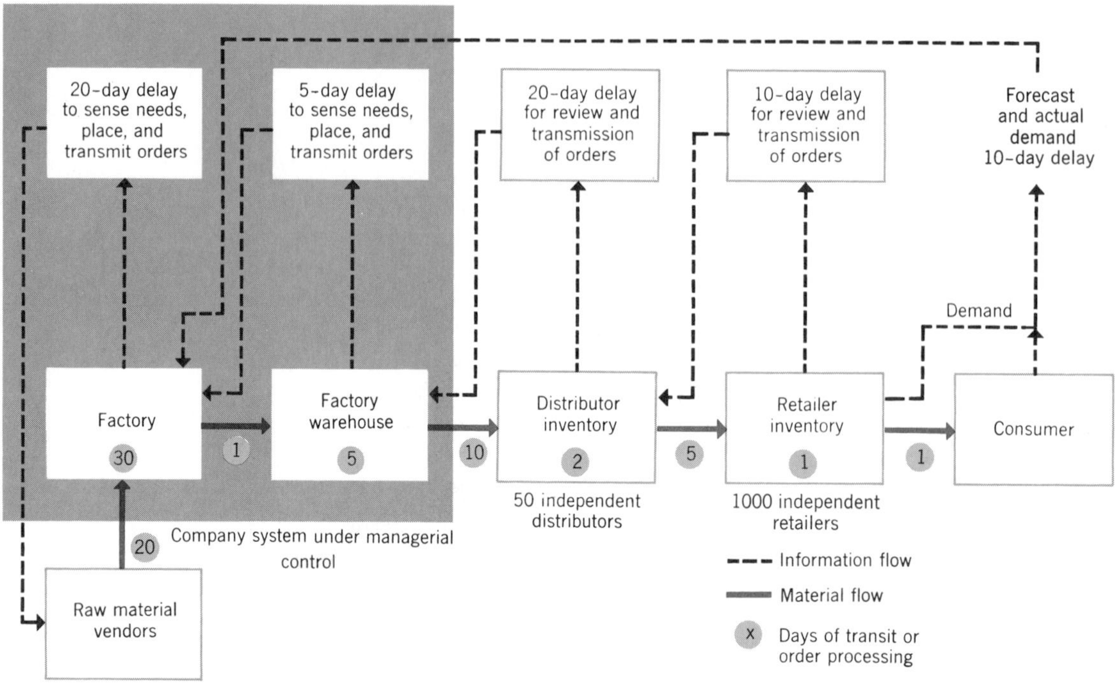

FIGURE 8-2
Production-distribution system with information feedback loop provided by forecasts and up-to-date information concerning actual consumer demand.

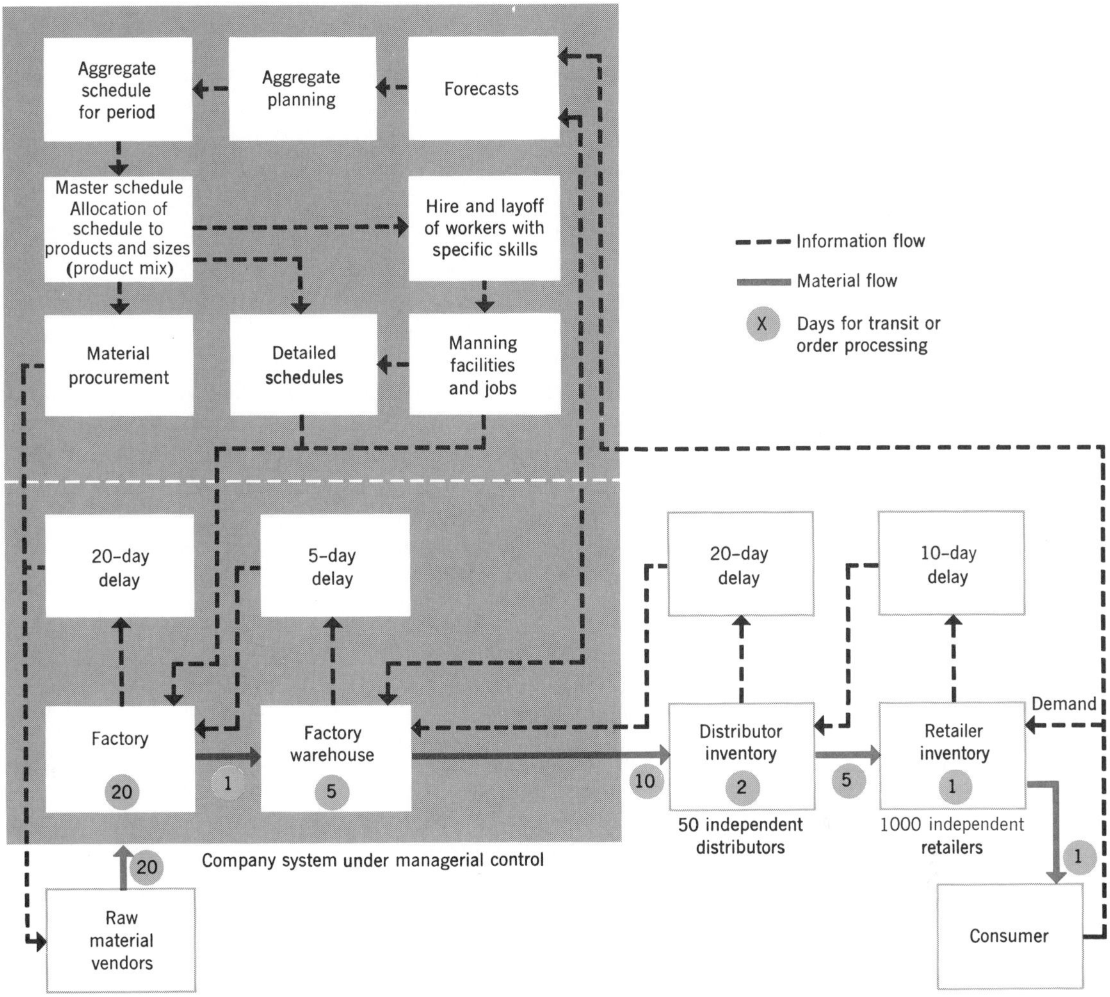

FIGURE 8-3
Relationship of forecasts, aggregate scheduling, and detailed scheduling to production-distribution system.

an input to the aggregate planning process. Basic decisions on production rate and work force levels for the upcoming period are then produced by the aggregate plan. We can either compute the overtime, subcontracting required, and the projected end-of-period inventory, or these data will be produced as a part of the aggregate planning procedure, depending on the model used. Translating these aggregate plans into master schedules and detailed work schedules is the next task.

Master Schedules and Product Mix

Aggregate scheduling produces a set of constraints within which we must operate. These broad-level decisions are important in setting the basic production strategy for the best use of resources. Master schedules are stated more specifically in terms of the quantities of each individual product to be produced and the time periods for production. For example, the aggregate schedule might call for 1000 units in planning period 1, where the planning period might be a month or perhaps a four-week period. If there were three products, the more detailed master schedule would indicate the quantities of each of the three products to be produced in each week of the planning period, consistent with the aggregate schedule.

As an example, assume the aggregate schedule calls for a level production rate during the three-period planning horizon. The periods of planning horizons are often months or four-week periods. The latter is particularly convenient, since it breaks the year into 52/4 = 13 periods of equal size. The weekly master schedule then conveniently fits most work schedules in our society. In our example, we will assume three four-week periods in the planning horizon.

The aggregate plan may have produced the following:

Aggregate Plan

Period (4 weeks)	Initial	1	2	3
Aggregate forecast	—	1200	1100	800
Production	—	1000	1000	1000
Aggregate inventories	800	600	500	700

Now we must disaggregate the forecasts on a weekly basis for the three products. Suppose that the composition of the forecasts is as follows:

Period	1				2				3				
Week	1	2	3	4	5	6	7	8	9	10	11	12	Total
Product 1	100	100	100	100	100	80	75	75	75	75	65	50	995
Product 2	150	125	125	100	75	75	75	65	65	65	60	50	1030
Product 3	75	75	75	75	100	120	125	135	135	60	50	50	1075
Total	325	300	300	275	275	275	275	275	275	200	175	150	3100
Period Total	1200				1100				800				3100

The aggregate schedule by planning periods was to be level at 1000 units per four-week period. The aggregate forecasts are declining, but the forecasts by weeks for the three products follow different patterns. Products 1 and 2 decline, but product 3 increases during the second four-week period. although declining rapidly in the third four-week period.

The master scheduling problem requires the construction of schedules for each of the products, consistent with the aggregate plan, the forecasts, and planned inventories. Table 8-4 shows a master schedule that fits these requirements. The production plan in Table 8-4a indicates an initial cycling between the three products, where productive capacity is allocated each week to one of the products. In the fifth week, the strategy changes to cycle product 3 every other week, to take account of the high forecast requirements of that product during the fifth through ninth weeks. In the tenth week, the plan returns to a three-week cycling of the three products.

The individual product inventory profiles that would result from the master schedule are shown in Table 8-4b. Note that the individual inventory balances, I_t, are end-of-week figures that result from the simple accounting equation, $I_t = I_{t-1} - F_t + P_t$, where F_t is the forecast and P_t the production during the week. For example, in Table 8-4b, the initial inventory for product 3 is $I_0 = 150$ units, requirements were forecast as $F_1 = 75$ units and production is scheduled as $P_1 = 250$ units. Therefore, the projected inventory at the end of week 1 is $I_1 = 150 - 75 + 250 = 325$ units as indicated in Table 8-4 for product 3. The total end-of-week inventory is simply the sum of the inventories for the three periods. The inventory called for by the aggregate plan was an end-of-period (four-week) figure, so there are deviations from plan in the first three weeks of each four-week period, but the deviation is zero in the last week of each four-week planning period.

TABLE 8-4a
Master Schedule for Three Products Consistent with Forecasts and Aggregate Production Plan

Period		1				2				3			
Week	0	1	2	3	4	5	6	7	8	9	10	11	12
Product 1, production		—	—	250	—	—	125	—	125	—	250	—	—
Product 2, production		—	250	—	250	—	125	—	125	—	—	250	—
Product 3, production		250	—	—	—	250	—	250	—	250	—	—	250
Total production		250	250	250	250	250	250	250	250	250	250	250	250
Capacity, aggregate plan		250	250	250	250	250	250	250	250	250	250	250	250
Deviation		0	0	0	0	0	0	0	0	0	0	0	0

TABLE 8-4b
Product End of Week Inventory Profiles Associated with Master Schedule

Period		1				2				3			
Week	0	1	2	3	4	5	6	7	8	9	10	11	12
Product 1, inventory	400	300	200	350	250	150	195	120	170	95	270	205	155
Product 2, inventory	250	100	225	100	250	175	225	150	210	145	80	270	220
Product 3, inventory	150	325	250	175	100	250	130	255	120	235	175	125	325
Total inventory	800	725	675	625	600	575	550	525	500	475	525	600	700
Inventory, aggregate plan	800	600	600	600	600	500	500	500	500	700	700	700	700
Deviation	0	125	75	25	0	75	50	25	0	(225)	(175)	(100)	0

The master scheduling methods indicated by this discussion follow from the simple logic of the disaggregation of aggregate plans. In more complex situations, problems might have occurred because of conflicts in forecast requirements, production capacities, and inventory position, resulting in stockout. Alternate plans could be tried to see if the problem could be solved. It may be necessary to cycle back to a reevaluation of the aggregate plan, with possible changes.

In the master schedule of Table 8-4, no attempt was made to consider the balance of setup and inventory costs. These questions will be considered later in the chapter. Optimizing methods for master scheduling are developed by Hax and Meal [1975] in a hierarchical product planning system.

The master schedule provides a basis for detailed plans of materials flow, equipment, and personnel schedules. These plans are likely to be different for continuous versus intermittent productive systems. Obviously, all materials, components, and subassemblies need to be coordinated with the master schedule, and this is the subject of Chapter 9, "Material Requirements Planning."

DETAILED SCHEDULING—CONTINUOUS SYSTEMS

A result of the allocation to types and sizes gives basic information for detailed scheduling, detailed hiring and layoff instructions, and material procurement schedules, as shown in Figure 8-3. Conflicts in detailed schedules may occur at this point because of capacity limitations for either labor or facilities.

In setting the individual type and size production rates and personnel schedules, the scheduler may be faced with either great rigidity or reasonable flexibility, depending on the nature of the processes and the productive system design. For example, the output rates of production lines are at a fairly fixed hourly rate.

What flexibility does the scheduler have to obtain a certain target weekly or monthly rate of output? There are basically two alternatives. The work force on the line can be scheduled to work shorter or longer hours (including overtime or undertime), or the entire line can be rebalanced to achieve a somewhat higher or lower hourly rate of output. The latter would be used to achieve more drastic changes in output rate, since it involves hiring or laying off personnel. Here, however, the scheduler will be following the basic instructions given by the aggregate plan and master schedule, which presumably have taken into account the relative costliness of changing production rates through changing hours, using overtime, and rebalancing (hiring and laying off).

The aggregate plan establishes the constraints under which the scheduler must perform. Thus we see that assembly line balance is a subject for concern not only for the original design of productive systems but also for continued operation. For example, the use of rebalancing to achieve different hourly rates of output is common in automotive assembly lines.

If the line is completely rigid in design, being mechanically paced, then the scheduler can change total output for the period only by changing the number of hours worked or by changing the line speed. The rigid system is, of course, often used.

The main problems of detailed scheduling are to devise ways of following out the aggregate plan and master schedule as far as possible, since the broadly based optimization has already taken place. Following the allocation of aggregate production to product types and sizes, an iterative process begins that develops a tentative plan and checks back to see if the details of the plan fit the constraints.

The first question raised in the schematic diagram of Figure 8-4 involves possible changes in employment levels called for by the aggregate plan and master schedule. These changes will require rebalancing of personnel to facilities. In many instances, alternate staffing plans may exist for various output rates, based on previous careful studies. The result of rebalancing will be to hire or lay off personnel in specific skill categories, checking to see if the result is within the aggregate plan. If the aggregate planning model has been carefully constructed to reflect the relationship between productivity and personnel, it should be possible to adjust personnel assignments within the constraints of the plan. Balance solutions must deal with assignment of whole worker units, so that cost and capacity increase or decrease in step fashion rather than in a continuous relationship to work force size.

When the hiring-layoff question has been answered with resultant rebalancing, this information is an input to the initial attempt to generate a detailed schedule of products to facilities. The scheduler then generates a detailed assignment of product items to facilities by production days, as well as assignments to subcontractors if that is appropriate. The result is then checked in an iterative fashion to see if assigned levels of overtime, subcontracting, and inventories meet the requirements of the aggregate plan and master schedule. Other constraints must also be observed, such as maintenance schedules and agreed company-union work rules.

In a parallel way the existing schedules are reviewed for possible modification based on new information dealing with the progress of actual sales versus forecast sales. The schedules of production commitments are then updated by adding the newest information and dropping off the most recent period.

Based on the resultant information the scheduler may then develop full schedules for two or three time spans, such as those shown at the bottom of Figure 8-4. For some specific company situations most of the process could possibly be computerized.

HIERARCHICAL PLANNING FOR HIGH-VOLUME CONTINUOUS SYSTEMS

Heirarchical planning is a methodology by which decisions at aggregate levels provide constraints that more detailed levels must meet, and the execution of plans at the detailed levels provide useful feedback that may generate change in higher-level, more aggregated decisions. The result is that hierarchical planning systems are responsive to the organizational structure of the firm and serve to define a framework for the partitioning and linking of the planning activities. Hax and Meal [1975] report an application of such a plan in a process manufacturing firm stated as being analogous to a chemical plant or steel mill. They point out that it is difficult to construct general purpose hierarchical planning systems that could be applied to any kind of

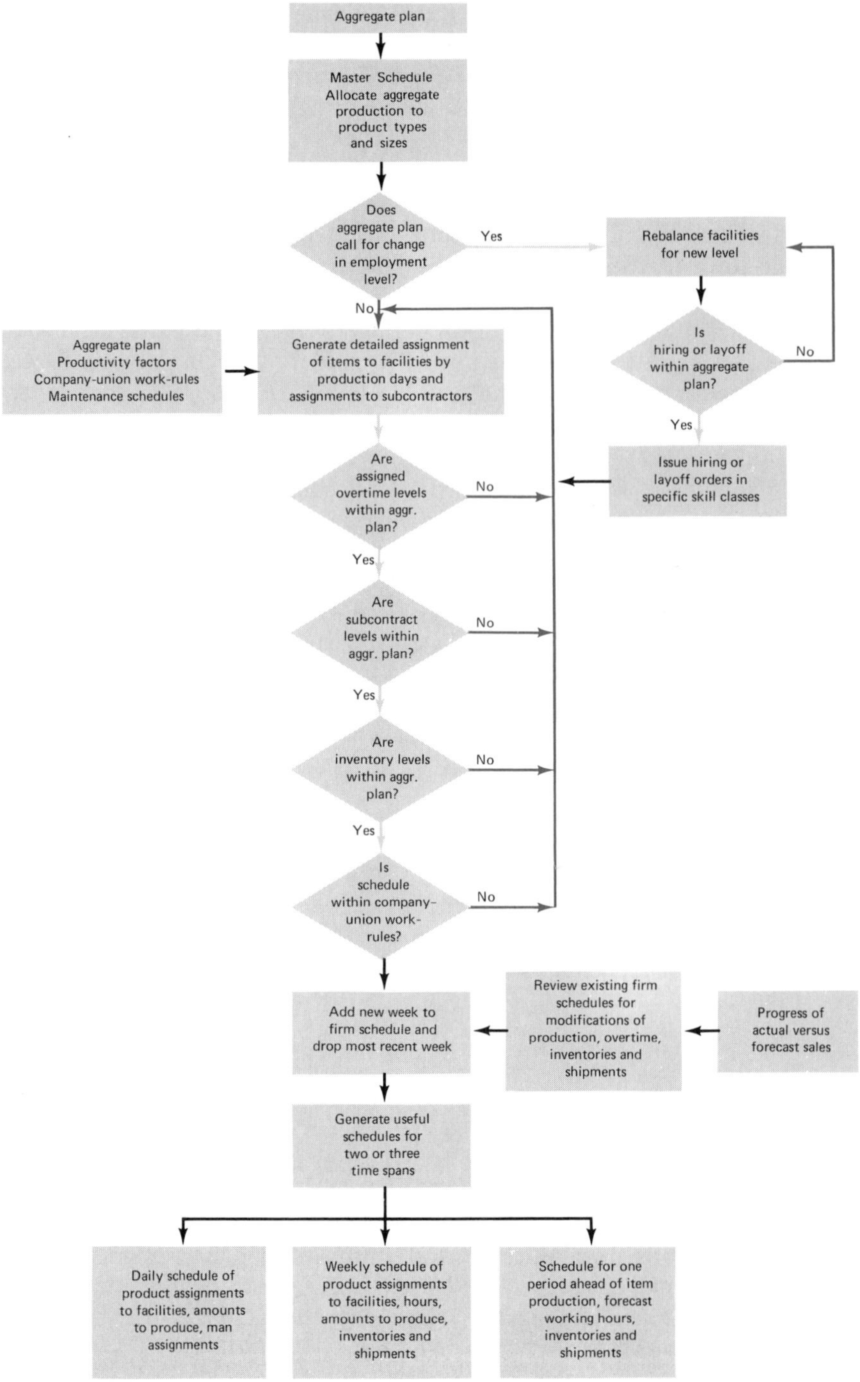

FIGURE 8-4 Flow diagram of operations scheduling for continuous systems.

industry. Instead, the planning effort must be tailored to a specific organization and its idiosyncrasies.

The example firm is a multiplant, multiproduct operation with three distinct seasonal demand patterns. There is a strong incentive to maintain a nearly level manufacturing rate for the following reasons.

1. The capital cost of equipment is very high compared to the cost of shift premium for labor, and the plants normally operate three shifts five days a week with occasional weekend work.

2. The labor union is very strong and exerts pressure to maintain constant production levels throughout the year for employment stabilization.

At the beginning of the study the symptoms of problems were poor customer service, excessive inventory, and high production costs. The high inventory in the face of poor customer service was caused by excessive seasonal stock accumulation for some items and shortages for others. High production costs were primarily due to runs that were uneconomically short, with consequent high setup cost and low productivity. These, in turn, resulted from the high stockout rate and the consequent need to produce a small amount of each of many items in order to satisfy back orders. Hax and Meal state the problem to be solved as:

> *. . . planning aggregate production levels, particularly in allocating available production capacity among several product types with differing seasonal demand patterns, and in the subsequent detailed scheduling of each item belonging to a product type.*

System Structure

The first step in the development of the planning and control system to help solve the stated problem was to define levels of aggregation for the various items produced. This was done by examining the extent to which sets of decisions regarding production were interdependent. If two sets of decisions were found to be independent, they were totally separated in the hierarchy of decisions.

Beginning at the most detailed level, items sharing a major setup cost were grouped into "families." Thus, scheduling decisions for items in a family were very dependent, while the opposite was true for items in different families. It was also found that decisions for a family in one time period were strongly tied to decisions for the same family in other time periods. This time dependence resulted from the need to accumulate seasonal inventories in both product families.

Product families were aggregated into "types," if they shared a common seasonal pattern and production rate. This facilitated seasonal planning, since only the aggregate for all families in the type needed to be considered in developing the plan.

The next step in the process was the development of a hierarchy of decisions based on the relationships developed in the aggregation process. The following steps were developed.

1. Assignment of families to plants
2. Seasonal planning
3. Scheduling of families
4. Scheduling of items

In addition to the preceding steps, basic inventory methods were used to establish minimum run lengths and overstock limits. The complete decision sequence is shown in Figure 8-5. A brief description of the several submodels and the nature of their interaction follows.

Plant/Product Assignment Subsystem (PAS). The PAS determines the plant locations at which each family should be manufactured. The model balances the cost of interterritory transportation against incremental capital investment cost required to manufacture the product family in question. The model is run annually to take account of new products and changes in variable manufacturing cost and demand patterns.

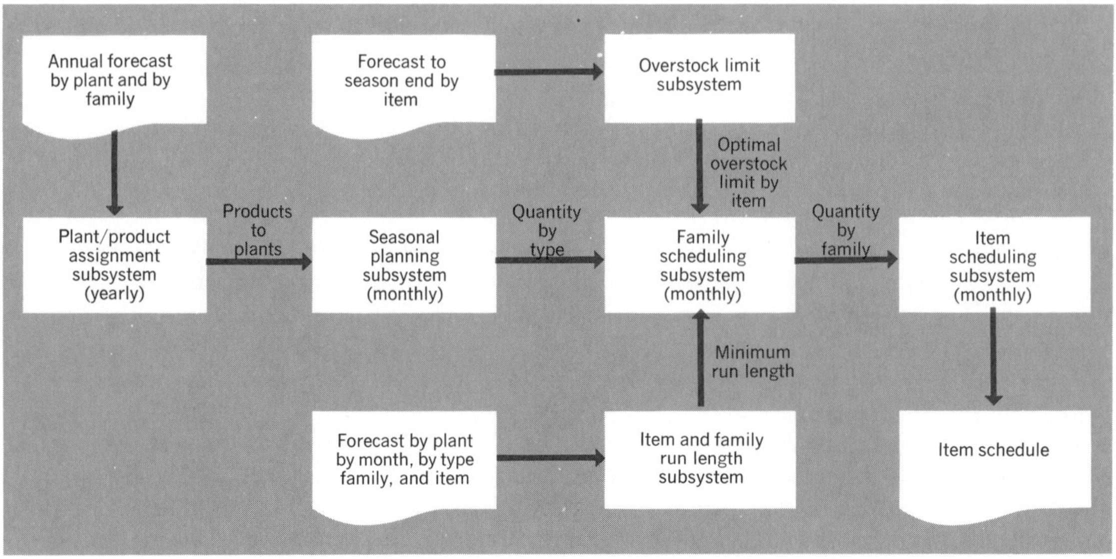

FIGURE 8-5
Decision sequence in planning and scheduling.
SOURCE: *A. C. Hax and H. C. Meal, "Hierarchical Integration of Production Planning and Scheduling," in* Studies in Management Sciences, Vol. I, Logistics, *edited by M. A. Geisler. North Holland-American Elsevier, Amsterdam, 1975.*

Seasonal Planning Subsystem (SPS). The SPS is the aggregate planning subsystem, and it determines the production requirements and seasonal stock accumulations by product type for each plant. The objective is to minimize total regular and overtime production costs plus inventory holding costs, subject to constraints on available regular and overtime labor, meeting demand, and safety stock requirements. The labor force size was not included as a decision variable, since employment stability was assumed. Linear programming was used as the solution technique.

Family Scheduling Subsystem (FSS). The FSS is used to schedule enough production for the families in a product type to use the time allocated to the type by the seasonal planning subsystem. This includes the accumulation of the necessary seasonal stock.

Item Scheduling Subsystem (ISS). The ISS determines the production quantities for each item within the constraints of the family schedules determined by the family scheduling subsystem. As in FSS, overstock limits are observed, and an attempt is made to maximize customer service. In order to carry out this task, Hax and Meal developed heuristics that equalize the expected runout times for the items in the family.

Results

Hax and Meal report a total development cost in the range of $150,000 to $200,000. Although exact benefits were not reported, they stated that cost reductions from smoother production, fewer emergency interruptions, and reduced inventory carrying costs were expected to be more than $200,000 per year in each plant.

BATCH PRODUCTION FOR INTERMITTENT SYSTEMS

As discussed in the Part II Introduction, most often it is not possible to produce continuously because the quantities needed do not justify the full use of facilities. In order to obtain fuller utilization of machines and personnel, a variety of parts and sizes may be produced on the same equipment. Machines are grouped by generic type as shown in Figures II-1*a* and *b*, and individual orders or jobs are processed in batches. In such systems, the basis for planning and control is the individual job order. The use of facilities is intermittent, and orders may follow different paths through the system, depending on the individual processing requirements of the product or part.

Capacity Versus Lot Size

Many batch processing systems have very limited choice of the order size or lot size. These kinds of systems are normally those that produce a noninventoriable or a custom output. The lot size is dictated by the customer order in such instances, and since the item is custom in design, there is little or no opportunity to group orders for more efficient processing.

An extremely common type of system is one in which the output is inventoriable and produced in substantial volume, even though the volume may not justify continuous production. In these situations, the manager must determine the lot size to be produced at one time. For example, average annual requirements may be R = 12,000 units, but production time for this quantity might require only 160 hours. Therefore, there are many options available.

Suppose that once machines are set up for our example, the variable or "run time" is only 0.80 minutes per unit, but five hours are required to set up machines for the lot. Then, for different lot sizes, output in units per hour including setup, and total hours for 12,000 units is as follows:

Q, Units per Lot	Units per Hour, Including Setup	Total Hours for 12,000 Units
1,000	54.5	220.0
2,000	63.2	190.0
5,000	69.8	172.0
12,000	72.7	165.0

The total machine and worker time requirements for the lot of 1000 units are 33.3 percent greater than the lot of 12,000 units.

On the other hand, when setup time is somewhat smaller in relation to run time, perhaps only 60 minutes per setup, comparable calculations indicate the following:

Q, Units per Lot	Units per Hour, Including Setup	Total Hours for 12,000 Units
1,000	69.8	172.0
2,000	72.3	166.0
5,000	73.9	162.4
12,000	74.5	161.0

In this instance, total machine and worker time requirements for the lot of 1000 units are only 6.8 percent greater than for the lot of 12,000 units. When setup times are large in relation to run times, system capacity is very sensitive to the lot size chosen. Conversely, for the same lot size, if setup times can be reduced, capacity increases. However, setup times are not usually the decision variable. Instead, lot size is the decision variable under managerial control. Of course, our example deals only with one item in showing the capacity-lot size relation. It is lot size policy that we must be concerned about, where the cumulative effects of large or small lots affect capacity and the cost of providing capacity.

Independent EOQ Scheduling

Why not determine lot sizes economically, according to the EOQ Equations 4 or 9 from Chapter 7? EOQs would be computed independently for each item and processed through the system as a lot. Sometimes this decision may be a good one, but more often than not, Equations 4 or 9 are oversimplifications of the true situations, and improved decision policies can be used. Some of the complexities that commonly intrude on the simplicity of Equations 4 and 9 are as follows:

1. Because of differing requirements, setup costs, and inventory carrying costs for each job, inventories that result from EOQ lots may not last through a complete cycle. Because of stockouts, special orders of smaller size may then be needed, resulting in capacity dilution.

2. When operating near capacity limits, competition for machine and/or worker time may cause scheduling interference. In order to maintain scheduled commitments, lots may be split. Again, a side effect is to reduce capacity.

3. Rush jobs may require the running of special orders of nonoptimal size.

4. Sometimes there is a "bottleneck" machine or process through which all or most jobs must be sequenced. The limited capacity may exert pressure toward smaller capacity, diluting lot sizes in order to meet scheduled commitments on at least a part of job orders.

5. Where parts or products are produced in regular cycles, the individual lot sizes are constructed to fit in with the cycling, rather than from the balance of setup and inventory holding costs for each individual item.

6. The assumption of constant demand is not met, either as a result of seasonal usage or sales or because demand is *dependent* on the production schedules of other parts, subassemblies, or products. This point will be dealt with in Chapter 9.

Most of the previous reasons for deviating from the concepts of Equations 4 and 9 lead to smaller lot sizes and to reductions of effective capacity. Under these conditions, relatively larger fractions of available machine and worker time are devoted to machine setup. Note that items 1 through 5 in the preceding list all indicate some kind of dependence of the individual lot size on the other orders in the system.

An Example. Table 8-5 gives data on requirements, costs, and production for 10 products that are processed on the same equipment. The capacity of the equipment is limited to 250 days usage per year. When the daily production rates for each product listed in column 4 are converted to required production days in column 5, we see that the total annual production requirement of 241 days is within the 250-day maximum (setup times are included). Our particular interest is in columns 8

TABLE 8-5

Requirements, Costs, and Production Data for Ten Products Run on the Same Equipment

(1) Product Number	(2) Annual Requirements, R_i	(3) Sales per Production Day (250 days per year), Col. 2/ 250, r_i	(4) Daily Production Rate, p_i	(5) Production Days Required, Col. 2/ Col. 4	(6) Inventory Holding Cost per Unit per Year, c_{H_i}	(7) Machine Setup Cost per Run, c_{P_i}	(8) EOQ, Equation 9 from Chapter 7	(9) Number OF Runs per Year, Col. 2/ Col. 8	(10) Production Days per Lot, Col. 8/ Col. 4	(11) TIC_0, Equation 10 from Chapter 7
1	9,000	36	225	40	$0.10	$40	2928	3.1	13.0	$245.93
2	20,000	80	500	40	0.20	25	2440	8.2	4.9	409.88
3	6,000	24	200	30	0.15	50	2132	2.8	10.7	281.42
4	12,000	48	600	20	0.10	40	3230	3.7	5.4	297.19
5	16,000	64	500	32	0.02	50	9578	1.7	19.2	167.04
6	15,000	60	500	30	0.50	40	1651	9.1	3.3	726.64
7	8,000	32	1000	8	0.35	30	1190	6.7	1.2	403.27
8	9,000	36	900	10	0.05	60	4743	1.9	5.3	227.68
9	2,000	8	125	16	0.55	25	441	4.5	3.5	226.89
10	3,000	12	200	15	0.20	20	799	3.8	4.0	150.20
				241		$380			70.5	$3136.14

through 11 of Table 8-5. The lot sizes are computed using Equation 9 from Chapter 7, and the number of runs per year, production-days per lot, and costs are computed for independent EOQ scheduling of the 10 products.

Since only one product can be produced at a time, there are problems that result from an attempt to schedule the 10 products independently (see Table 8-6). Each EOQ lot provides inventories to be used at the daily usage rate of r_i of that product and will be depleted in EOQ/r_i days, as shown in column 4. Therefore, the cycle

TABLE 8-6

Calculation of Production Days Required, Peak Inventory, and Number of Days of Sales Requirements Met by an EOQ for Ten Products

(1) Product Number	(2) Production Days Required, EOQ/p_i	(3) Peak Inventory, Production Days $\times (p_i - r_i)$	(4) Days to Deplete EOQ, EOQ/r_i
1	13.0	2457	81.3
2	4.9	2058	30.5[a]
3	10.7	1883	88.8
4	5.4	2980	67.3[a]
5	19.2	8371	149.7
6	3.3	1452	27.5[a]
7	1.2	1162	37.2[a]
8	5.3	4579	131.8
9	3.5	410	55.1[a]
10	4.0	752	66.6[a]
	70.5		

[a] Items that stock out, inventory lasts less than 70.5 days.

inventory must last long enough for that product to be recycled. The total number of production days for all 10 products is shown in column 2 as 70.5 days. Scanning column 4, we see that the inventory for products 2, 4, 6, 7, 9, and 10 will be depleted before the cycle can be repeated. The situation is shown graphically for product 2 in Figure 8-6 and is called *scheduling interference*.

Clearly, independent scheduling will not provide a feasible solution. This infeasibility is not surprising, because the schedule of each product is not independent of the other products since they are all processed on the same equipment which has limited capacity. We must treat the 10 products as a system. Before considering alternatives, note that our system is operating near capacity, with 241 production days scheduled, or 241 × 100/250 = 96.4 percent of the days available. Scheduling interference would probably not occur if we were operating at relatively low loads. The slack capacity under conditions of low load would be available to make the system work.

Common Cycle Scheduling

Perhaps the simplest way to ensure a feasible solution to scheduling interference problems is to adopt some common cycle for all products and produce lot quantities for each that cover usage rates for the cycle. Scanning column 2 of Table 8-6, the number of runs per year ranges from 1.7 to 9.1 (the average is 4.55 runs). A cycle of four or five runs per year might be reasonable. Table 8-7 gives key results for a

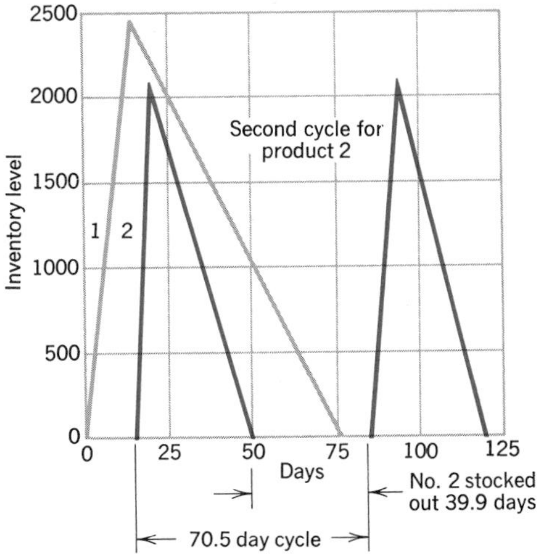

FIGURE 8-6
Inventory levels versus days for products 1 and 2. Product 2 is produced in the quantity Q_p = 2058 units, which lasts only 30.6 days at the usage rate of r_2 = 80 units per day.

TABLE 8-7
Calculation of Inventory and Setup Costs When Ten Products Are Produced Four Times per Year in a Common Cycle of Length, 250/4 = 62.5 days. Each Product Is Produced in a Lot Size Q_i Sufficient to Cover 62.5 Days' Supply

(1) Product Number	(2) Daily usage Rate, r_i	(3) Production Lot Quantity, $62.5 \times r_i =$ $62.5 \times$ Col. 2	(4) Production Days Required, $Q_i/p_i =$ Col. $3/p_i$	(5) Peak Inventory, Col. 4 $\times$ $(p_i - r_i)$	(6) Inventory Holding Cost per unit per year, c_{H_i}	(7) Annual Inventory Holding Cost $\frac{(\text{Col. } 5 \times \text{Col. } 6)}{2}$	(8) Annual Setup Costs, $4 \times c_{P_i}$
1	36	2250	10.00	1890	$0.10	$94.50	$160.00
2	80	5000	10.00	4200	0.20	420.00	100.00
3	24	1500	7.50	1320	0.15	99.00	200.00
4	48	3000	5.00	2760	0.10	138.00	160.00
5	64	4000	8.00	3488	0.02	34.88	200.00
6	60	3750	7.50	3300	0.50	825.00	160.00
7	32	2000	2.00	1936	0.35	338.80	120.00
8	36	2250	2.50	2160	0.05	54.00	240.00
9	8	500	4.00	468	0.55	128.70	100.00
10	12	750	3.75	705	0.20	70.50	80.00
			60.25			$2203.38	$1520.00

TIC = inventory + setup costs = 2203.38 + 1520.00 = $3723.38

cycle of four runs per year. Each product is scheduled for production of a quantity to provide 250/4 = 62.5 days supply.

At the stated usage rates, the inventory of each product will not be depleted before it is recycled. These production lot quantities are computed in column 3 of Table 8-7 as simply $62.5 \times r_i$. The graphs of inventory build up and depletion follow the same general form of Figure 8-5. The inventory costs are computed in column 7, and column 8 gives the costs for four setups per year for each product. The total incremental costs for the common cycle scheduling system are shown at the bottom of the table as $3723.38, $587.24 greater than for independent scheduling. But independent scheduling did not provide a feasible solution. The total number of production days required is 60.25 days in column 4 of Table 8-7, leaving slack capacity of 62.50 − 60.25 = 2.25 days, or 2.25 × 100/62.5 = 3.6 percent.

Economic Common Cycle Scheduling

Since a common cycle can provide a feasible solution, as long as the total load is within system capacity, the next question is, which cycle should be used, N = 4, 5, 6, etc.? Calculations similar to those in Table 8-7 can be made for various cycles. Intuitively, we know that as we increase the number of cycles per year, decreasing lot sizes, annual inventory costs will decrease, but annual setup costs will increase in proportion to the number of cycles per year. Table 8-8 summarizes costs for common cycles of N = 4, 5, 8, and 10 runs per year for the same 10 products. For the alternatives computed, the lowest total incremental cost is associated with $N = 5$ runs to cover $250/5 = 50$ days supply. If one were to select any of the larger number of runs per year as a plan, involving more rapid cycling, it would be important to consider the extent to which effective capacity would be reduced. If setup times were relatively large, the system capacity limit would be reached at these high loads. Remember that setup *times* are not isolated in our example.

Formal models for economic common cycles for a number of products can be derived by methods similar to those used to develop EOQ formulas. They are slightly more complex, but they are quite parallel in concept. The total incremental cost equation is developed for the entire set of products and similar mathematical operations produce the number of production runs that jointly minimize annual inventory plus setup costs for all products,

$$N_0 = \sqrt{\frac{\sum_{i=1}^{m} c_{H_i} R_i (1 = r_i/p_i)}{2 \sum_{i=1}^{m} c_{P_i}}} \tag{1}$$

Equation 1 requires the multiplication of c_{H_i}, R_i, and $(1 - r_i/p_i)$ for each individual product in the numerator, which are then summed. The denominator is simply two times the sum of the setup costs for all products. The total incremental cost of an optimal solution is

$$TIC_0 = \sqrt{2 \sum_{i=1}^{m} c_{P_i} \sum_{i=1}^{m} c_{H_i} R_i \left(1 - \frac{r_i}{p_i}\right)} \tag{2}$$

Table 8-9 shows the calculations applying Equations 1 and 2 to the 10-product example. The optimal number of production runs is $N_0 = 4.82$ at a cost of $TIC_0 =$ \$3660.13. As a practical matter, one would probably select a number of runs close to the optimal number, based on other considerations, since the total incremental cost differences are small near the optimum, as indicated in Table 8-8. In the absence of other overriding considerations, one would probably simply round the number of

TABLE 8-8 **Inventory and Setup Costs for Common Cycles of N = 4, 5, 8, and 10 Runs per Year for Ten Products**

N, Cycles per Year	Number of Days Supply Produced	Average Annual Inventory Cost	Annual Setup Costs	Total Incremental Costs
4	62.50	\$2203.38	\$1520.00	\$3723.39
5	50.00	1762.70	1900.00	3662.70
8	31.25	1101.69	3040.00	4141.69
10	25.00	881.35	3800.00	4681.35

TABLE 8-9

Calculation of the Economic Number of Production Runs per Year, using Equation 1

(1) Product Number	(2) $(1 - r_i/p_i)$ from Table 8-5, (1 − Col. 3/Col. 4)	(3) $c_{H_i}R_i(1 - r_i/p_i)$, Col. 6(Table 8-5)x Col. 2(Table 8-5)x Col. 2(this table)	(4) c_{P_i}, from Col. 7 of Table 8-5
1	0.840	756.00	$40
2	0.840	3,360.00	25
3	0.880	792.00	50
4	0.920	1,104.00	40
5	0.872	279.04	50
6	0.880	6,600.00	40
7	0.968	2,710.40	30
8	0.960	432.00	60
9	0.936	1,029.60	25
10	0.940	564.00	20
		17,627.00	$380

$$N_0 = \sqrt{\frac{17{,}627}{2 \times 380}} = 4.82 \text{ runs per year}$$

$$TIC_0 = \sqrt{2 \times 380 \times 17{,}627} = \$3660.13$$

cycles to $N = 5$ runs per year in our example, since the annual cost difference between $N = 5$ and $N_0 = 4.82$ is only $2.57.

The lot size scheduling problem has been the subject of a great deal of research, and improvements over the costs that result from applying Equation 1 have been developed. Lot size scheduling assumes that the inventory position of all items is such that the formulas can be used without encountering back ordering. These assumptions are most troublesome when sales are either increasing or decreasing, creating inventory imbalance. Matthews [1978] develops models that take these dynamic features into account.

IMPLICATIONS FOR THE MANAGER

Some of the most important managerial problems come into focus when the production-distribution system is seen as a chain of replenishment cycles. Even though much of the inventory may not be under the control of the manager of the system that produces the product, these inventories can have extremely important effects on the producer.

The size of system inventories required is surprisingly large. It is important for managers to understand how these inventories vary as system volume changes. First, the pipeline inventories are directly proportional to system volume. However, cycle and buffer inventories are more likely to vary in proportion to the square root of volume changes. Thus, if volume doubles, these inventory components would need to increase by only a factor of $\sqrt{2} = 1.4$. There is a definite economy of scale as volume increases.

Aggregate planning systems can help managers to cope with the system dynamics by making plans that stabilize rather than simply react to demand changes. Time lags in the information system need to be as short as possible, but these time lags cannot be completely eliminated.

The relationship between capacity and lot size is important to recognize. Capacity can be diluted by responding to the pressures to meet commitments on rush jobs through lot splitting. Scheduling the use of time-shared equipment often results in pressures to reduce lot sizes in order to meet commitments on a variety of orders. The intensity of these kinds of problems is greatest when operating near capacity limits. But the dilution of capacity in such situations has the effect of intensifying the original problem.

Finally, while EOQ concepts can help managers think in terms of balancing the relevant costs, they are seldom applicable in their simple forms. The assumptions of constant demand and lead time are very often not valid, and the assumption of independence between orders fails in most situations unless the system is operating at low load where the interdependent effects are minimal. Managers who understand these effects seek scheduling systems that treat orders as a part of an interdependent system.

REVIEW QUESTIONS AND PROBLEMS

1. Suppose that we have a simple two-station assembly line. The output of each station is equalized so that we have perfect balance, theoretically. If the first operation breaks down or runs out of raw material, the second operation must also stop. If the second operation breaks down, the first can continue, but a large inventory of partially completed items will be built up between the two operations. How can we decouple the two operations?

2. What is the function of cycle inventories at each of the following stages in a production-distribution system?
   a. Raw material storage
   b. Within a factory
   c. Finished goods at a factory warehouse
   d. Distributor warehouse
   e. Retailers

3. A production-distribution system for dry dog food is made up of a factory, a factory warehouse, 10 regional distribution warehouses, and 10,000 retailers. The average handling, delay, processing, and transit times are as follows:

2 days—production process

1 day—transit to factory warehouse

2 days—handling delays at factory warehouse

7 days—transit to distributor

4 days—handling and delays at distributor

4 days—transit to retailers

2 days—handling and delays at retailers

The average retailer's volume is 200 pounds per day, and all other stages are geared to this aggregated flow.

Compute the system finished goods pipeline inventory.

4. The finished goods inventory replenishment process for the system described in problem 3 follows. Compute the average system cycle inventory.
   a. The production process is continuous with output flowing to the factory warehouse.
   b. The 10 regional distributors order every two weeks to cover needs.
   c. The 10,000 retailers order weekly to cover their needs.

5. Buffer stocks are held at each of the major stock points in the system described by problems 3 and 4. The factory warehouse has set the maximum demand during the five-day replenishment time from the factory at $D_{max} = 6000$ tons.
   The average supply time from the factory warehouse to distributors is 10 days, and distributors feel that the reasonable maximum demand during this exposure time is $D_{max} = 1200$ tons.
   The average retailer supply time from distributors is six days, and retailers feel that the maximum demand they experience during the six-day supply time is $D_{max} = 1500$ pounds.
   Compute the system buffer stock.

6. Compute the total system finished goods inventories for the dog food company described in problems 3, 4, and 5.

7. Suppose that system volume for the dog food company described in problems 3 through 6 were to increase from 1000 tons per day to 1500 tons per day. How much inventory would be required for the entire system at the new demand rate?
   Assume that the replenishment and buffer policies were optimal previously

and that under the new rate, management holds to optimal policies. For example, the ordering frequencies might be adjusted to reflect the new rates. (Note that you need not recompute as in problems 3 through 6. An estimate of the quantitative answer is required. Adjust the inventories already computed.)

8. Suppose that for the production-distribution system of Figure 8-1, retail demand suddenly increases 20 percent. Explain what happens to inventory levels for distributors and at the factory warehouse.

9. Using Figure 8-3 as a background, discuss the aggregate plan as a constraint to operations scheduling.

10. Under what circumstances is rebalancing of production lines necessary when adjusting to a new aggregate plan?

11. Given the aggregate plan as a constraint to operations scheduling in continuous systems, what flexibility is left to planners in developing detailed schedules?

12. What flexibility is left to management in continuous systems for adjusting deviations in actual output from planned levels within a given period?

13. Note in the model of the production-distribution system of Figure 8-2 that information concerning demand at the retail level is fed back directly to the factory. As a result, there is a considerable reduction in the magnitude of oscillations in orders and production activity. What are the implications of this result for organizing and establishing lines of authority and responsibility for a multistage production-distribution system?

14. Data for a particular job order are that setup time is 10 hours, and run time is 1 minute per unit once the setup is made; c_P = \$150 per setup, c_H = \$0.12 per unit per year, R = 10,000 units, and EOQ = 5000 units.

    What percent dilution of capacity would take place if production were actually in lots of Q = 1000 units instead of EOQ?

15. Suppose that the setup time for problem 14 were only one hour, and setup costs only c_P = \$15 per order. EOQ is then Q_0 = 1581 units.

    What percent dilution in capacity takes place if production were actually in lots of Q = 300 units instead of EOQ?

16. What are the advantages and disadvantages of the practice of lot splitting?

17. Table 8-10 gives data on five products that are produced on the same equipment, assuming 250 productive days available per year. The EOQs and associated costs were computed using Equations 4 and 5 from Chapter 7.

    If the products were produced in EOQ lots, what problems would result?

TABLE 8-10 **Data for Five Products to Be Produced on the Same Equipment**

(1) Product Number	(2) Annual Requirements, R_i	(3) Daily Requirements, $R_i/250 = r_i$	(4) Daily Production rate, P_i	(5) Annual Inventory Holding Costs, c_{H_i}	(6) Setup Costs per Run, c_{P_i}	(7) EOQ, Equation 4, Chapter 7	(8) TIC_0, Equation 5, Chapter 7
1	5,000	20	400	$1.00	$40	648.89	$616.44
2	12,500	50	300	0.90	25	912.87	684.65
3	7,000	28	200	0.30	30	1,275.89	329.18
4	16,250	65	300	0.75	27	1,222.14	718.01
5	4,000	16	160	1.05	80	822.95	777.69
					$202		$3,125.97

18. Using the data in Table 8-10, compute lot sizes that would result from using a common cycle for all products of $N = 6$ and 8. How many days supply must be produced for each product? Which common cycle do you prefer? Why?

SITUATIONS

19. The Aircraft Maintenance Company (AMCO) performs jet engine maintenance on contract for airlines in five branch locations. The company maintains branch inventories at each of the maintenance locations. At its national warehouse, AMCO obtains the required parts from manufacturers and maintains an inventory of these parts, using a fixed reorder quantity control system. Also at the national warehouse, parts are assembled into kits for each type of engine maintained, and an inventory of kits is retained. The national warehouse kit inventory is also controlled using a fixed reorder quantity system.

Each branch maintains an inventory of kits and reorders weekly from the national warehouse, based on usage. Thus, there are three levels of inventories: parts and kits at the national warehouse and kits at the branches. The general physical and information flow is shown in Figure 8-7.

The nature of engine maintenance operations requires fast turnaround time, and contracts include a penalty for each hour beyond standard that an engine spends in the maintenance operation. Thus, kit availability is very important to profitable operations. On the other hand, aircraft engine parts represent high value, and controlling inventory levels is also important.

The company has been studying its inventory system, and has simulated the basic elements of the three-stage structure. Figure 8-8 shows the simulated response of the system to a 10 percent increase in usage at branches. Management was surprised by the volatile reaction of inventories to the 10 percent

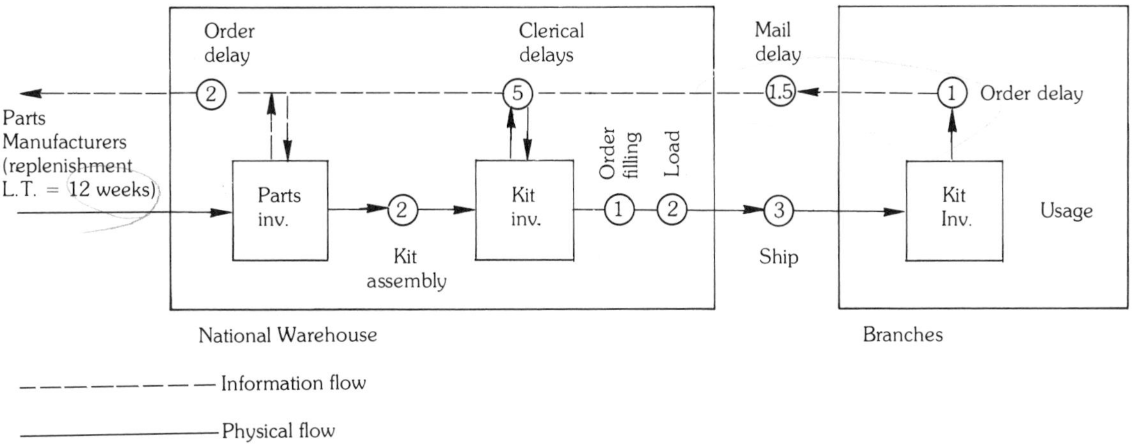

FIGURE 8-7
Aircraft Maintenance Company, information and physical flow.

increase in demand, particularly those at the central warehouse. Management felt that they should be able to smooth out the response somewhat.

What changes in policies or structure do you recommend?

20. The Mixing and Bagging Company discussed as situation 29 in Chapter 7 has carried through its plan to install EOQ scheduling of its products. The idea expressed by the factory manager is to use an order point trigger for each product. He set the order point, P_i, for each product to cover average demand

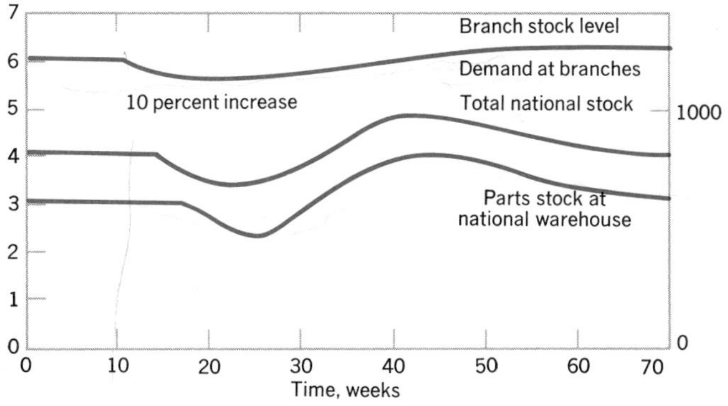

FIGURE 8-8
Aircraft Maintenance Company, simulated response to a 10 percent increase in demand.

for two weeks. When warehouse inventory for an item declines to the order point, an order for the approximate EOQ would be written and released to the foreman.

The delay between actual recognition that the inventory level had triggered an order and the order release to the foreman was three days, and the production lead time of two days was felt to be normal. The factory manager throught that it should be possible to get an inventory item replenished in mixed and bagged form in the warehouse within the lead time, because of the simple production process. Thus, five working days (one week) of the two-week supply of inventory was planned to be used in replenishment. This left a one-week supply as a buffer against the possibility of especially high demand. The factory manager felt that the extra one-week supply should cover most situations.

The factory manager had classified the 10 product types into *practical* production lot sizes, based on his EOQ calculations. There were three high-demand items that were run in lots of 4000 bags (100 pounds per bag), four intermediate-demand items that were run in lots of 2000 bags, and three low-demand items that were run in lots of 1000 bags. These lot sizes were reasonably close to the EOQs and corresponded to units of raw materials and supplies that were easy to deal with. Also, having only three sizes of runs made it a simple system for the foreman and production workers.

Using the factory manager's classification, the product demand, lot sizes, and runs per year are summarized in the following table:

Product	Average Demand, 100 sacks/year	Lot Size, 100-pound sacks/run	Average No. of Runs/ Year/Product
3 high demand	160,000	4,000	40
4 medium demand	40,000	2,000	20
3 low demand	10,000	1,000	10

A run of 4000 sacks required about 590 minutes including setup, a run of 2000 sacks about 310 minutes, and a run of 1000 required about 170 minutes. Thus the factory manager figured that the average number of runs with the average mix of order sizes required only about 34 hours per week of production time on the equipment, including setup. In other words, the plant was operating at about 85 percent of capacity.

After a short time operating under the new EOQ system, the factory manager was puzzled by the results. He was stocking out of some of the high-demand items before the completion of production runs. He examined demand figures and found that the demand for these items had been greater than the average because of some seasonality in demand, but he could not understand why the one-week supply should not have taken care of the problem.

The foreman said that he had the place "humming," but complained that the factory manager was always and forever telling him to produce a given order first, because of short supply. "Every order can't be run first," he said.

The number of orders that he had in his list to produce seemed to be growing, and he processed them strictly in the order in which they were received, unless told to expedite a particular order. When the foreman was asked what he thought the problem was, he said, "The problem is that the runs are too short for the high-demand items. I spend my time constantly changing over to a new product. I suggest that we make all of the lot sizes larger, but particularly the high-demand items. Perhaps the lot sizes should be 10,000, 5,000 and 1,000. The small runs of 1000 are OK for the low demand items, since I can sandwich them in easily."

What is the problem? What are your recommendations to the factory manager?

REFERENCES

Bomberger, E. "A Dynamic Programming Approach to a Lot Size Scheduling Problem," *Management Science, 12*(11), July 1966.

Buffa, E. S., and J. G. Miller. *Production-Inventory Systems: Planning and Control* (3rd ed.). Richard D. Irwin, Homewood, Ill., 1979.

Doll, C. L., and D. C. Whybark. "An Iterative Procedure for the Single-Machine Multi-Product Lot Scheduling Problem," *Management Science, 20*(1), September 1973.

Elmaghraby, S. E. "The Economic Lot Scheduling Problem (ELSP): Review and Extensions," *Management Science, 24*(6), February 1978, pp. 587–598.

Forrester, J. *Industrial Dynamics.* MIT Press, Cambridge, 1961.

Forrester, J. "Industrial Dynamics: A Major Breakthrough for Decision Makers," *Harvard Business Review*, July-August 1958, pp. 37–66.

Goyal, S. K. "Scheduling a Multi-Product Single-Machine System," *Operational Research Quarterly, 24*(2), June 1973.

Hax, A. C., and H. C. Meal. "Hierarchical Integration of Production Planning and Scheduling," in *Studies in the Management Sciences, Vol. I, Logistics*, edited by M. A. Geisler., North Holland-American Elsevier, Amsterdam, 1975.

Madigan, J. C. "Scheduling a Multi-Product Single-Machine System for an Infinite Planning Period," *Management Science, 14*(11), July 1968.

Magee, J. F., and D. M. Boodman. *Production Planning and Inventory Control* (2nd ed.). McGraw-Hill, New York, 1967.

Matthews, J. P. "Optimal Lot Sizing for the Process Cycling Problem," *International Journal of Production Research, 16*(3), May 1978, pp. 201–214.

Peterson, R., and E. A. Silver. *Decision Systems for Inventory Management and Production Planning.* John Wiley, New York, 1979.

Saipe, A. L. "Production Runs for Multiple Products: The Two-Product Heuristic," *Management Science, 23*(12), August 1977, pp. 1321–1327.

Stankard, M. F., and S. K. Gupta. "A Note on Bomberger's Approach to Lot Size Scheduling: Heuristic Proposed," *Management Science, 15*(7), March 1969.

CHAPTER 9 Material Requirements Planning

THE DIFFERENCES IN PLANNING, SCHEDULING, AND CONTROL BEtween high-volume and intermittent systems are very substantial. Although both *may* be producing finished products for inventory, the periodic nature of production of the latter produces an immensely more complex detailed scheduling problem. The nature of forecasting and planning production lot sizes is unique, because of the dependent nature of the demand for parts and components. Their demand is dependent on the production schedules of primary products, which in turn are dependent on market demand. In the case of complex assembled products that have important subassemblies that are produced for inventory, this dependence may be second or third order.

REQUIREMENTS PLANNING CONCEPTS

Figure 9-1*a* is an illustration of a simple dinette table showing the parts or components required. It consists of a plywood top covered with Formica, four metal legs, and some miscellaneous hardware.

Figure 9-1*b* is a simplified operation process chart showing the sequence of major operations required to fabricate and assemble the table. The glue, screws, plastic feet, and paint are noted as being purchased outside; unlike the other raw materials of plywood, metal tubing, and so on, they are not processed before use.

Now, if the table is to be a "one-of-a-kind" custom table, the operation process chart of Figure 9-1*b* specifies what must be done and the sequences required. However, if a substantial number of tables are to be made, we have alternatives that make the problem more complex but offer planning and scheduling opportunities for efficient manufacture.

Bills of Materials

We can abstract from Figure 9-1*a* a "bill of materials." This document is not simply a materials list but is constructed in a way that reflects the manufacturing process. Of course, the operation process chart of Figure 9-1 does this also, but we want information useful to a material requirements system in a form that can be maintained in a computerized file.

The information that will be particularly significant, in addition to the simple list of parts, will be the dependency structure. For example, if we examine operation 11 in Figure 9-1, there is a subassembly; the plywood top, skirts, Formica top, and Formica skirts came together in operation 11 to form a unit. Also, the metal tubing, brackets, paint, and feet form a second subassembly. These two subassemblies can be produced separately in batches and then be assembled in operation 12 into finished tables in batches as needed. Indeed, an important reason for dealing with these subassemblies as separate units is that the same leg is used in another product, as we shall see. Therefore, the lot size and timing of the production of legs takes on broader significance.

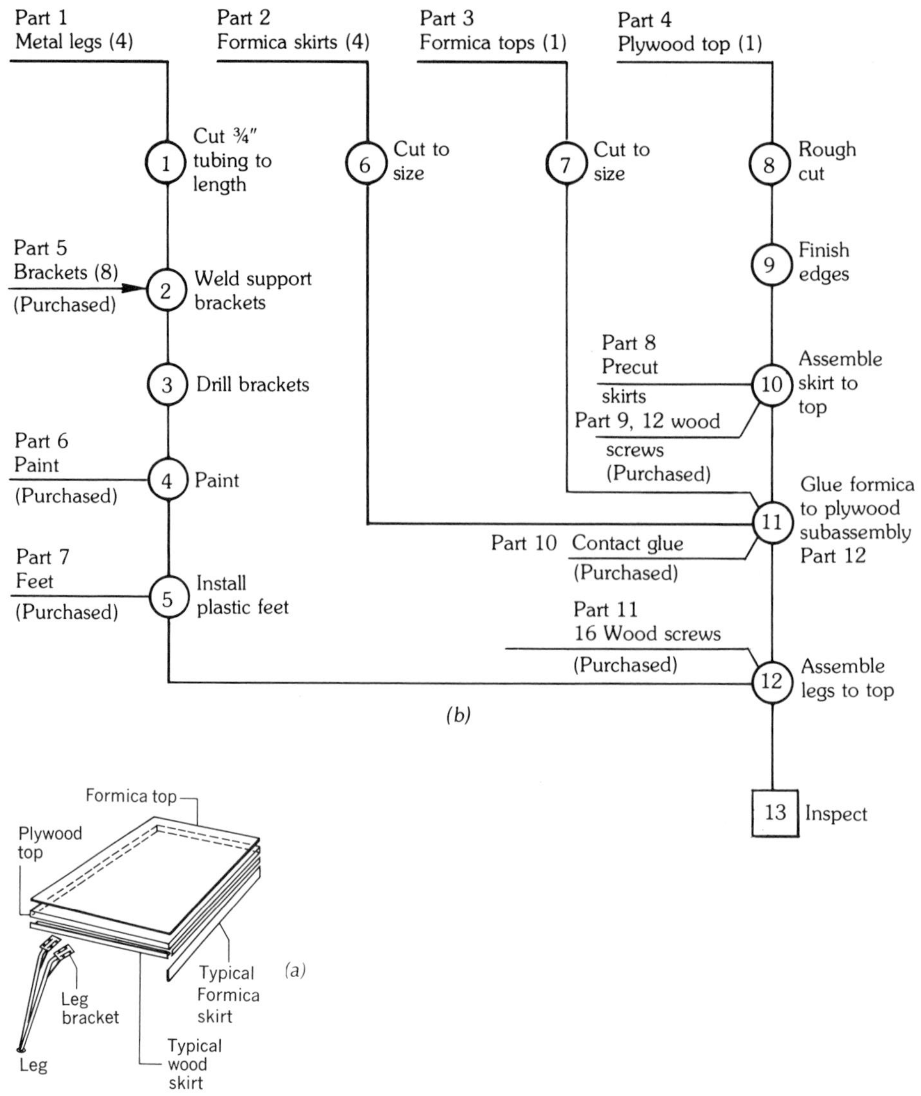

FIGURE 9-1
(*a*) Part components for dinette table; and (*b*) operation process chart.

As a result, one important input to a material requirements planning (MRP) system is a bill of materials constructed in a way that recognizes the dependence of certain components on subassemblies, which in turn are dependent on the final product. In more complex products, there could be several levels of dependency, since subassemblies could contain sub-assemblies and so on.

Figure 9-2 is a bill of materials for the square dinette table. It is called an indented bill because the dependence of parts and components is indicated by indenting items in the list. The final product is listed first, preceded by the number "1". All the numbers preceding the materials indicate the quantity of that item required for one unit of the final product. Then, if we wish to have a bill of materials for any larger quantity of the final product, a multiplier may be used and the list printed indicating the quantity of each item for the larger lot.

Demand Dependence

Assume that we have translated the current demand for dinette tables into a master schedule and that they are to be produced in lots of 100 every two weeks. The production rates for each of the operations in Figure 9-1*b* are relatively fast, therefore the tables are not produced continuously, since low utilization of workers and machines would result. Presumably, the enterprise uses the workers and machines for other products, including other table sizes and designs, chairs, and related products. Therefore, the dinette table will be produced periodically in lots to satisfy demand.

There are several alternatives. We can consider the table as a unit and produce enough legs, tops, and the like to assemble 100 tables every two weeks. However, since setup costs and variable production costs for the various operations are different, we may be able to produce more efficiently by considering the manufacture of each component individually. For example, we might produce legs every four weeks in lots of 800 to key in with the master schedule.

Let us consider the schedule of producing tables every two weeks in lots of 100 and legs every four weeks in lots of 800. Because the demand for the legs is entirely dependent on the *production schedule* for tables, the time phasing of the leg lots with respect to table lots has a very important impact on the in-process inventory of legs. Tables are produced in lots of 100 in weeks 3, 5, 7, and so on. In Figure 9-3*a*

1 Square dinette table					#1–80
	1 Tabletop subassembly				#12
		1 Plywood top			#4
			4 Wood skirts		#8
			12 Wood screws		#9
		1 Formica top			#3
		4 Formica skirts			#2
		6 oz Contact cement			#10
	4 Metal leg Subassemblies				#1
		8 Metal brackets			#5
		2 oz Paint			#6
		4 Plastic feet			#7
	16 Wood screws				#11

FIGURE 9-2
Bill of materials for square dinette table #1–80.

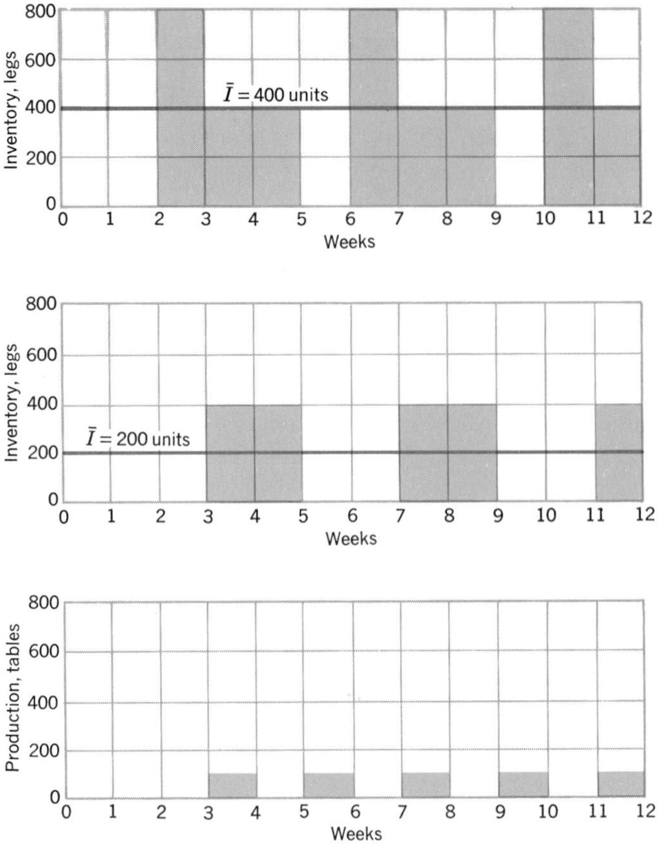

FIGURE 9-3
In-process inventory of table legs when *(a)* produced in weeks 1,5,9, etc., and *(b)* produced in weeks 2, 6, 10, etc. *(c)* Tables are produced in weeks 3, 5, 7, etc.

legs are produced in lots of 800 every four weeks in weeks 1, 5, 9, and so on. Legs go into inventory when the lot is completed and are available for use the following week. They are used in table assembly in weeks 3, 5, 7, and so forth; the average in-process inventory of legs is $\bar{I}$ = 400 units.

If we produce legs in weeks 2, 6, 10, and so on, as in Figure 9-3*b*, the in-process inventory is reduced to $\bar{I}$ = 200 units. Therefore, the problem is not simply to produce legs in lots of 800 every four weeks but to time phase the production of legs with respect to the production of the table, the primary item. The demand for tables is presumably dependent on market factors. However, the production of legs must become a *requirement* as soon as the table production schedule is set. If the proper time phasing is ignored, the price will be paid in higher in-process inventory of components.

Let us follow through the structure of requirements determination in Figure 9-4.

Square Table

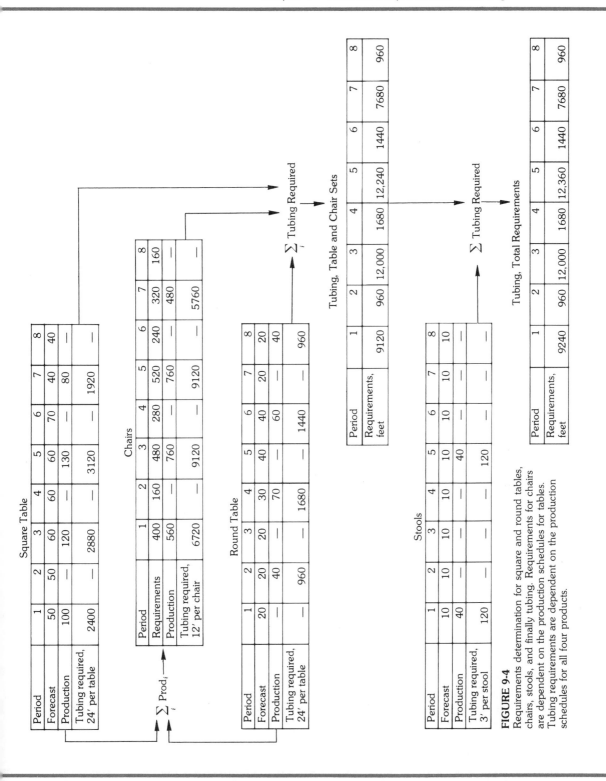

Period	1	2	3	4	5	6	7	8
Forecast	50	50	60	60	60	70	40	40
Production	100	—	120	—	130	—	80	—
Tubing required, 24′ per table	2400	—	2880	—	3120	—	1920	—

$\sum_i \text{Prod}_i$

Chairs

Period	1	2	3	4	5	6	7	8
Requirements	400	160	480	280	520	240	320	160
Production	560	—	760	—	760	—	480	—
Tubing required, 12′ per chair	6720	—	9120	—	9120	—	5760	—

Round Table

Period	1	2	3	4	5	6	7	8
Forecast	20	20	20	30	40	40	20	20
Production	—	40	—	70	—	60	—	40
Tubing required, 24′ per table	—	960	—	1680	—	1440	—	960

$\sum_i$ Tubing Required

Tubing, Table and Chair Sets

Period	1	2	3	4	5	6	7	8
Requirements, feet	9120	960	12,000	1680	12,240	1440	7680	960

Stools

Period	1	2	3	4	5	6	7	8
Forecast	10	10	10	10	10	10	10	10
Production	40	—	—	—	40	—	—	—
Tubing required, 3′ per stool	120	—	—	—	120	—	—	—

$\sum$ Tubing Required

Tubing, Total Requirements

Period	1	2	3	4	5	6	7	8
Requirements, feet	9240	960	12,000	1680	12,360	1440	7680	960

FIGURE 9-4
Requirements determination for square and round tables, chairs, stools, and finally tubing. Requirements for chairs are dependent on the production schedules for tables. Tubing requirements are dependent on the production schedules for all four products.

Figure 9-4 culminates in total requirements for metal tubing needed to produce two different models of dinette tables with accompanying sets of four chairs and a stool. Each product involves the use of metal tubing from which the legs are fabricated. First, the primary demand for the two tables and the stools are indicated by the forecasts in each period. Production schedules to meet demand for each of the three primary products are set to anticipate demand for two periods ahead for the tables and for four periods ahead for stools (we will return to the question of lot size at a later point). The chairs are used for both the square and round tables in dinette sets, four chairs per table. Therefore, requirements for chairs are dependent on the production schedules for the two table styles. The chair requirements are derived through a period-by-period summing from the production schedules for the two tables. For example, the requirement for chairs in period 1 is the sum of the production schedules for the two tables $(100 + 0) \times 4 = 400$.

Tubing requirements are, in turn, the period-by-period summation of the tubing requirements derived from the production schedules of tables and chairs, plus the tubing requirements from the production schedule for stools. Note that tubing requirements, even though representing aggregated usage in four different products, have a very uneven usage pattern; that is, 9240 in period 1, 960 in period 2, 12,000 in period 3, and so on. Professionals term such a usage pattern "lumpy" demand. Figure 9-4, then, is a requirements plan for the two tables, chairs, stools, and the tubing raw material. A complete plan would provide similar information for all components and raw materials.

Forecasting Versus Requirements

The required use of metal tubing is indicated in Figure 9-4. Suppose that we wish to set up an inventory control system for tubing so it can be reordered as needed to maintain stock. If the requirements for tubing in Figure 9-4 represent demand, can we apply standard forecasting techniques?

Figure 9-5 shows the application of an exponential forecasting system with $\alpha = 0.1$ to the requirements as demand data. Average actual demand for the eight periods is 5790 feet, and this is taken as the initial value for S_{t-1}. The forecast has the usual smoothing effect.

Now suppose that tubing is ordered on the basis of the smoothed forecast. If D_{max} were defined as the maximum demand observed in our small sample, then the buffer stock required with be $B = D_{max} - \bar{D} = 12{,}360 - 5790 = 6570$ feet, and the average inventory would be $\bar{I} = Q/2 + B = 5790/2 + 6570 = 9465$ feet.

But, in fact, we do not want to smooth demand, because the requirements schedule for tubing is the *best* forecast that we can obtain. Using exponential smoothing (or any other forecasting technique) is a misapplication of forecasting here, because the demand for tubing is dependent. Compare the forecast line with actual demand in Figure 9-5. The forecast errors are very large, MAD = 5049. Note that if we provide inventory according to the requirements schedule, the average inventory would be only 5790 feet, or $9465 - 5790 = 3675$ feet less than if we attempted to cover

Period	1	2	3	4	5	6	7	8
Actual requirements	9,240	960	12,000	1,680	12,360	1,440	7,680	960
Forecast, $\alpha = 0.1$	5,790	6,135	5,618	6,256	5,798	6,454	5,953	6,126

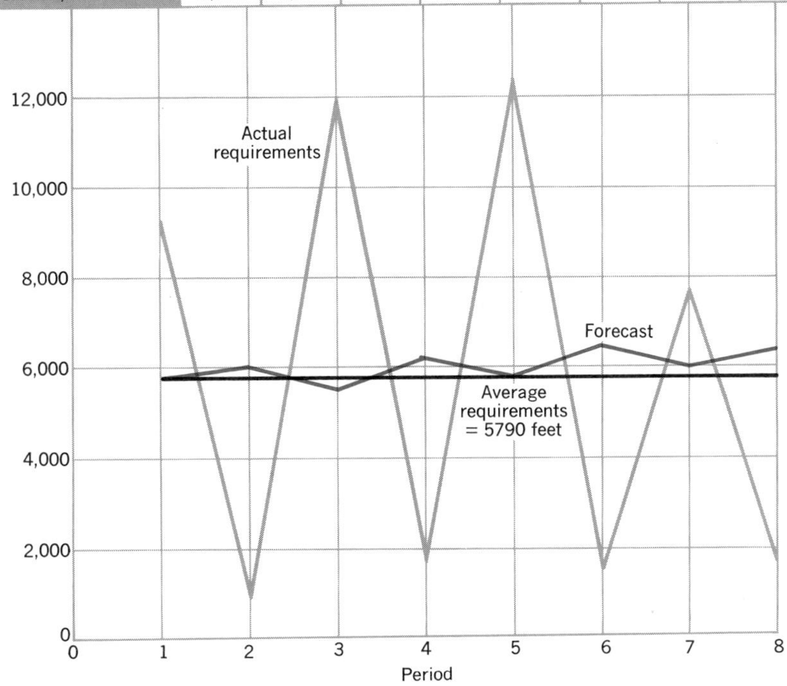

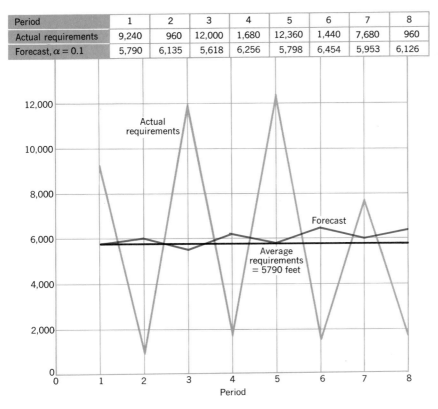

FIGURE 9-5
Effect of using an exponential forecast ($\alpha = 0.1$) for dependent demand of tubing.

requirements based on the exponential forecasting methodology. Using the forecasting methodology, we must provide a very large buffer stock, and while this buffer stock cannot be completely eliminated in requirements systems, the time phasing of inventory receipts can drastically reduce it (we will discuss buffer stocks for requirements systems at a later point).

Where, then, is forecasting applicable in requirements systems? Forecasting is applicable to the primary demand for the tables and stools, not to the dependent components and raw materials. Their requirements are derived directly from the production schedules that have taken into account primary demand and other factors. Therefore, inventory control systems for dependent items will use the requirements schedules directly instead of interposing a forecasting system.

The basic ideas of requirements generation developed in connection with the simple example of the tables, chairs, and stools carry forward into larger scale for more complex systems. Figure 9-6 shows that the form of the plan normally includes the master schedule for the independent item at the head, followed by requirements

MASTER ASSEMBLY SCHEDULE—Table 1-80

Week	1	2	3	4	5	6
Planned requirements	100	—	120	—	130	—
Requirements, Part 1 (legs, 4 per assembly) Production lead time, 2 weeks						
Expected usage	400	—	480	—	520	240
Planned deliveries	400	—	480	—	520	—
Planned stock on hand, end of week	60	60	60	60	60	60
Planned production order release	480	—	520	—	320	—
Requirements, Part 4 (plywood top, 1 per assembly) Production lead time, 1 week						
Expected usage	100	—	120	—	130	—
Planned deliveries	100	—	120	—	130	—
Planned stock on hand, end of week	20	20	20	20	20	20
Planned production order release	—	120	—	130	—	80

FIGURE 9-6
Requirements plan for master schedule and two dependent components.

plans for dependent components that are keyed in with the master schedule. In addition, requirements plans normally indicate the production lead time, a planned stock record, and the timing of production order releases phased to take account of production lead times.

LOT SIZE DECISION POLICIES

In the examples discussed to this point the size of the production run or lot size has been assumed. We now consider some of the alternate decision policies that might be used and how they fit in with requirements planning.

Before considering the alternate policies, however, let us review some key elements concerning requirements systems that may be helpful in deciding the nature of appropriate lot size policies. First, we know that demand for components is dependent and must be thought of as requirements generated to key in with the master schedule of the final product. The nature of the demand distributions that result is not uniform and continuous, as may be true for primary items, such as those discussed in Chapter 8, where demand resulted from the aggregation of independent orders from multiple sources. Demand is lumpy because it is dependent, and demand variations are not the result of random fluctuation. Therefore, some of the assumptions important in traditional inventory control theory are questionable for dependent items. We must keep these assumptions in mind as we discuss alternate lot size policies. The following comparison of three lot size policies is meant to show how the policies react to lumpy demand and is not meant as a valid test of the three policies in MRP systems.

Economic Order Quantity (EOQ)

In the past the EOQ policy or the nonoptimal fixed order quantity alternatives have been used most. In such systems, the order point is set to cover expected usage during lead time, an order for a fixed quantity being placed at that time. If the EOQ is used, it is computed by the equation, $\text{EOQ} = \sqrt{2Rc_P/c_H}$

Table 9-1 shows the performance of an EOQ model for the requirements schedule given. Instantaneous replenishment is assumed. Seven orders were placed to meet requirements. The order quantity was increased to $2 \times$ EOQ in those periods where demand exceeds EOQ plus the inventory carried forward. The ordering cost is then $7 \times 300 = \$2100$. Inventory costs are based on the average for beginning and ending inventory in each period. For example, average inventory for period 1 is $(166 + 156)/2 = 161$ units. The inventory cost for the 12 periods is the sum of the period average inventories multiplied by the unit inventory cost, or $2 \times 12 \times 89.88 = \$2157$. The total incremental cost for the EOQ model for 12 weeks is \$4257. As Berry points out, this example illustrates several problems with the EOQ procedure:

> *When the demand is not equal from period to period, as is often the case in requirements forecasts, one of the assumptions underlying the EOQ formula is violated. Since demand does not occur at a constant rate, as is assumed by the EOQ formula, the restriction of fixed lot sizes results in larger inventory carrying costs. This occurs because of the mismatch between the order quantities and the demand values, causing excess inventory to be carried forward from week to week.* [Berry, 1972]

TABLE 9-1

Performance of an EOQ Model for a Given Requirements Schedule ($\bar{R}$ = 92.1 units per week, c_P = \$300 per order, c_H = \$2 per unit per week, and Q^O = 166 units)

Week number	1	2	3	4	5	6	7	8	9	10	11	12
Requirements	10	10	15	20	70	180	250	270	230	40	0	10
							166		166			
Quantity ordered	166	—	—	—	—	166	166	166	166		—	—
Beginning inventory	166	156	146	131	111	41	27	109	5	107	67	67
Ending inventory	156	146	131	111	41	27	109	5	107	67	67	57

Ordering cost:	\$2100
Inventory carrying cost:	2157
Total incremental cost:	\$4257

Periodic Reorder System

Recall that we can approximate the economic time interval between replenishment orders by dividing the EOQ by the mean demand rate. For our example, EOQ/$\bar{R}$ = 166/92.1 = 1.8, or approximately two weeks. When this procedure is applied to the same requirement schedule used to illustrate the EOQ policy, the reorder and inventory pattern of Table 9-2 results, where the order quantity is the sum of requirements for the next two weeks. The system requires only six orders, but with lot sizes ranging from 20 to 520 units. Therefore, average inventory is 89.4 units, resulting in very slightly lower inventory carrying costs of $2145, and total incremental costs of $3,945. The periodic system results in a $312 reduction in total incremental cost, or 7.3 percent. By being able to vary order size in response to lumpy demand, the periodic system saves one order.

As Berry points out:

Like the EOQ procedure it, too, ignores much of the information contained in the requirements schedule. That is, the replenishment orders are constrained to occur at fixed time intervals, thereby ruling out the possibility of combining orders during periods of light product demand, e.g., during weeks 1 through 4 in the example. If, for example, the orders placed in weeks 1 and 3 were combined and a single order was placed in week 1 for 55 units, the combined costs can be further reduced by $160.

Part-Period Total Cost Balancing

Another procedure is to attempt to balance the total incremental costs in each ordering decision. This procedure is called the "part-period algorithm." It uses all the information provided by the requirements schedule and attempts to equate the total costs

TABLE 9-2
Performance of a Periodic Reorder Model for a Given Requirements Schedule ($\bar{R}$ = 92.1 units per week, c_P = $300 per order, c_H = $2 per unit per week, T_O = 2 weeks)

Week number	1	2	3	4	5	6	7	8	9	10	11	12
Requirements	10	10	15	20	70	180	250	270	230	40	0	10
Quantity ordered	20		35		250		520		270			10
Beginning inventory	20	10	35	20	250	180	520	270	270	40	0	10
Ending Inventory	10	0	20	0	180	0	270	0	40	0	0	0

Ordering cost:	$1800
Inventory carrying cost:	2145
Total incremental cost:	$3945

SOURCE. W. L. Berry, "Lot Sizing Procedures for Requirements Planning Systems: A Framework for Analysis," *Production and Inventory Management*, 2nd Quarter, 1972.

of placing orders and carrying inventories. The procedure considers the alternate lot size choices available at the beginning of week 1, that is, placing an order to cover the requirements for

Week 1 only

Weeks 1 and 2

Weeks 1, 2, and 3

etc.

The inventory carrying cost for these alternatives is as follows, assuming that average inventory is centered within each week.

Week 1: $c_H (\bar{I}_1) = 2 \times 0.5 \times 10 = \$10$

Weeks 1 and 2: $c_H (\bar{I}_1 + \bar{I}_2) = 2(0.5 \times 10 + 1.5 \times 10) = \$40$

Weeks 1, 2, and 3: $c_H (\bar{I}_1 + \bar{I}_2 + \bar{I}_3) = 2(0.5 \times 10 + 1.5 \times 10 + 2.5 \times 15) = \$115$

Weeks 1, 2, 3, and 4: $c_H (\bar{I}_1 + \bar{I}_2 + \bar{I}_3 + \bar{I}_4) = 2(0.5 \times 10 + 1.5 \times 10 + 2.5 \times 15 + 3.5 \times 20) = \$225$

Weeks 1, 2, 3, 4, and 5: $c_H (\bar{I}_1 + \bar{I}_2 + \bar{I}_3 + \bar{I}_4 + \bar{I}_5) = 2(0.5 \times 10 + 1.5 \times 10 + 2.5 \times 15 + 3.5 \times 20 + 4.5 \times 70) = \$885$

By scanning the preceding set of calculations, we see that the fourth alternative, ordering 55 units to cover the demand for the first four weeks, approximates the ordering cost of $300. The result is that an order should be placed at the beginning of the first week to cover the first four weeks' requirements.

Applying this procedure to the same requirements schedule results in Table 9-3. The total incremental cost is reduced by an additional $460 or 12 percent, compared to the results for the periodic reorder system in Table 9-2. The part-period total cost-balancing procedure allows both the lot size and the time between orders to vary, so that in periods of light demand smaller lot sizes and longer time intervals between orders result, compared to periods of high demand.

Obviously, the part-period balancing procedure has performed best of the three reordering policies reported. The policy performs extremely well because of its flexibility in considering replenishments involving both variable reorder quantity and variable reorder frequency. It considers several different possible horizons and selects the one in which order and inventory costs are approximately equated. Thus, in periods of low requirements for several periods, it will group requirements, saving possible high ordering costs. If demand increases, the policy reacts by closing down

TABLE 9-3

Performance of a Total Cost-balancing Model (Part-period Balancing) for a Given Requirements Schedule ($\bar{R}$ = 92.1 units per week, c_P = \$300 per order, c_H = \$2 per unit per week)

Week number	1	2	3	4	5	6	7	8	9	10	11	12
Requirements	10	10	15	20	70	180	250	270	230	40	0	10
Quantity ordered	55				70	180	250	270	270			10
Beginning inventory	55	45	35	20	70	180	250	270	270	40	0	10
Ending inventory	45	35	20	0	0	0	0	0	40	0	0	0

Ordering cost:	\$2100
Inventory carrying cost:	1385
Total incremental cost:	\$3485

SOURCE. W. L. Berry, "Lot Sizing Procedures for Requirements Planning Systems: A Framework for Analysis," *Production and Inventory Management,* 2nd Quarter, 1972.

its horizon, saving incremental inventory costs. In short, the policy uses the information provided in the requirements plan fully to its advantage.

The part-period balancing procedure, however, does not evaluate *all* the possible alternatives for lot sizes, and therefore it does not always produce the optimal solution. The Wagner-Whitin [1958] dynamic programming algorithm produces an optimal solution; however, it is much more complex and requires greater computer time. The result is that it is not used a great deal in practice. Berry included the Wagner-Whitin Algorithm in his study, and it produced a solution that was \$240, or 7 percent, lower in cost than the part-period balancing policy. The part-period procedure did not consider the possibility of combining orders placed in weeks 9 and 12. By carrying an extra 10 units in inventory for three weeks at a cost of \$60, the Wagner-Whitin policy saves placing an order in week 12, and the net cost reduction is \$300 − \$60 = \$240.

Lot Size Policies in Multistage Systems

The discussion of lot size policies to this point has examined each of the three policies in isolated systems. The brief experiments reported are not an adequate test of alternate policies. However, the results of the experiments do give insight into the differing performance characteristics of the policies.

Biggs, Goodman, and Hardy [1977] developed a multistage production-inventory system model that involved a hierarchical system of part and component manufacture, subassembly, and final assembly. The hierarchical system makes it possible to test alternate lot size policies in an operating system where part and component manufacturing schedules are dependent on subassembly and assembly schedules and where subassembly is dependent on final assembly schedules. The final assembly schedule is set in relation to product demand. Thus the structure permits the testing of lot size policies as they would function in the multistage system. Five lot size policies were tested:

Economic order quantity (EOQ)

Periodic reorder system, using the EOQ to determine the reorder time cycle

Part-period total cost balancing

Lot for lot, in which an order is placed for exactly what is needed in the next period

Wagner-Whitin dynamic programming model

System performance of the lot size models was evaluated using the following four criteria:

1. Total number of stockouts for final products

2. Total units of stockouts for final products

3. Total number of setups, total system

4. Average dollar value of inventory, total system

The results indicate that the part-period total cost balancing and EOQ were consistently the best performers in simulation experiments. The dominance of one policy over others depends on the weighting given the four criteria. The reemergence of the EOQ policy as a good performer in a multistage system is explained in part by its excellent performance with respect to final product stockouts. The EOQ policy places larger orders fewer times per year and is therefore exposed to the risk of stockout less often. Based on the simulation experiments, managers may select policies based on their preference for a given weighting of the criteria.

These results must be regarded as preliminary. As with most early simulation experiments, actual shop conditions are only approximated. In this instance, the simulated shop was loaded well under capacity. There was no lot splitting permitted, and when the "desired" production exceeded capacity, a priority index was used to determine which lots would not be processed. It is difficult to tell how the four lot size policies would have performed if these decision rules had not been imposed and if the load had been varied over a range, including heavy loads. Nevertheless, testing alternate lot size rules within a multistage environment is an important condition and represents a step toward resolving the issue of how different lot size policies actually perform.

BUFFER STOCKS IN REQUIREMENTS SYSTEMS

We have already noted that dependent items are not subject to the kinds of random variations in demand that characterize primary product demand. The demand variability is largely planned and is lumpy in nature. We do not need buffer stocks to absorb these kinds of fluctuations. The nature of the requirements plan is designed to counter the variations with production orders, as shown in Figure 9-6. These kinds of variation are under managerial control.

There are sources of variation, however, for which buffer stocks are a logical countermeasure. Buffer stocks in requirements systems are designed to absorb variations in the supply schedule. The time required for processing orders through an intermittent system is variable because of such factors as delays, breakdowns, and plan changes. In addition, the actual quantity delivered from production is variable because of scrap. The result is that we need a cushion to absorb variations in the supply time and the quantity actually delivered.

Buffer Stock Levels

The nature of requirements systems changes the concept of reorder levels in both the EOQ and part-period methods of control. In Chapter 7 we noted that the reorder level was set to cover normal usage during the supply lead time *plus* the buffer stock (see Chapter 7, Figure 7-4). In requirements systems, however, the buffer stock level becomes the trigger or reorder level. This is true because we work with future inventory levels in a coordinated plan. A projected fall in inventory level to or below the buffer level can be anticipated by a production order release phased by the lead time to be delivered in time to eliminate the impending shortage. Table 9-4 shows an example of how a production order release is scheduled in the third period, triggered by a projected decrease in inventories, below the buffer level of 1000 units.

The determination of buffer stock levels can be based on the general experience with supply lead times and an estimate of the maximum usage likely to occur per period. There are other variables, however, that have the equivalent effect of buffer stocks. First, "safety factors" may be involved in the lead time estimates. If the lead time estimate is, in fact, the time that included 90 percent of the cases and most production orders are received prior to that time, then we have a buffer in effect. Inflated requirements schedules also have an equivalent buffering effect. All of these techniques are used and when used in combination, the equivalent buffer can be large and, unfortunately, partially hidden.

TABLE 9-4
Projected Inventory Falls Below the Buffer Stock Level of 1000 Units in Period 5, Triggering a Production Order Release in the Third Period

Period	1	2	3	4	5	6	7	8
Expected usage	—	2400	—	—	2000	1500	—	2000
Planned deliveries	—	—	—	—	2000			
Planned stock on hand, end of week	5000	2600	2600	2600	2600 ~~600~~			
Planned production order release	—	—	2000					

Buffer stock = 1000 units; lead time = 2 periods; EOQ = 2000 units

TABLE 9-5 **Requirements for Table Legs from Production Schedules for Square and Round Tables shown in Figure 9-4 (Production lead time = 3 weeks)**

Period, weeks	1	2	3	4	5	6	7	8
Leg requirements, square	400	—	480	—	520	—	320	—
Leg requirements, round	—	160	—	280	—	240	—	160
Total leg requirements	400	160	480	280	520	240	320	160
Production requirements	1320	—	—	—	1240	—	—	—

CAPACITY REQUIREMENTS PLANNING

The requirements plans we have discussed have shown ways of exploiting the knowledge of demand dependence and product structure in order to develop production order schedules. These schedules take account of the necessary timing of production orders, but they assume that the capacity is available when needed. However, capacity constraints are a reality that must be taken into account. Since the MRP system contains in its files information concerning the processing of each production order, we should be able to use that information to determine whether or not capacity problems will exist.

As an example, consider just the processing of metal legs beginning with the information contained in Figure 9-4. There are two dinette tables that use the same legs, and Table 9-5 summarizes the leg requirements from Figure 9-4. The bottom line of Table 9-5 gives production requirements for legs, if we accumulate four weeks' future requirements as production orders. Therefore, it is necessary to receive a lot of 1320 legs in period 1 and 1240 legs in period 5. Since there is a three-week production lead time, these production orders must be released three weeks ahead of the schedule shown in Table 9-5.

From the operation process chart of Figure 9-1*b*, let us consider only the load requirements on the fabrication operations of (1) cut to length, (2) weld support brackets, and (3) drill brackets. The time requirements for the three operations are shown in Table 9-6 for each of the two lot sizes we must consider.

If the production orders are released three weeks before being needed for assembly,

TABLE 9-6 **Process Time Requirements for Legs in Lots of 1320 and 1240**

Process	Setup Time, Minutes	Run Time, Minutes	Total Time, Hours, in Lots of	
			1320	1240
1. Cut to length	5	0.25	5.58	5.25
2. Weld brackets	10	1.00	22.16	20.83
3. Drill brackets	20	0.75	16.83	15.50
4. Paint	10	0.25	5.66	5.33
5. Install feet	5	0.10	2.28	2.15

assume that the cutoff operation is planned for the first week, the welding operation for the second week, and the drilling and other minor operations for the third week. Then, the machine-hour load on each of the three processes may be projected, as shown in Figure 9-7.

Following the same rationale that produced the load effects on the three processes shown in Figure 9-7, the computer system can pick up load from all orders for all parts and products in the files of the requirements program and print out a projected load for each work center. For example, the projected weekly load on the drill press work center is shown in Figure 9-8. The accumulated load by weeks is shown as "released load," in hours. The capacity for the eight-week horizon is shown as 80 hours, the equivalent of two available machines. The available hours in Figure 9-8 then indicate whether or not capacity problems are projected to occur. In periods 4,

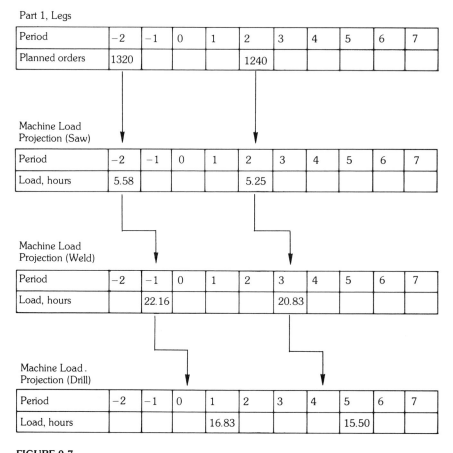

Part 1, Legs

Period	−2	−1	0	1	2	3	4	5	6	7
Planned orders	1320				1240					

Machine Load Projection (Saw)

Period	−2	−1	0	1	2	3	4	5	6	7
Load, hours	5.58				5.25					

Machine Load Projection (Weld)

Period	−2	−1	0	1	2	3	4	5	6	7
Load, hours		22.16				20.83				

Machine Load. Projection (Drill)

Period	−2	−1	0	1	2	3	4	5	6	7
Load, hours				16.83				15.50		

FIGURE 9-7
Load generation for three processes, based on production schedule for legs. Time requirements from Table 9-6.

Projected Weekly Machine Load Report
Work Center 21, Drill Presses Date: 02/01/80

Period	1	2	3	4	5	6	7	8
Released load, hours	65	71	49	90	81	95	48	62
Capacity, hours	80	80	80	80	80	80	80	80
Available hours	15	9	31	−10	−1	−15	32	18

FIGURE 9-8
Sample projected load report for one work center.

5, and 6, there are projected overloads. Given this information, we may wish to anticipate the problem by changing the timing of some orders, meet the overload through the use of overtime, or possibly subcontract some items. In the example shown in Figure 9-8, there is substantial slack projected in periods 1 and 3, and it may be possible to smooth the load by releasing some orders earlier.

COMPUTERIZED MRP

The concepts of MRP clearly require computers for implementation. When there are many assembled products, perhaps with subassemblies, the number of parts can easily be in the thousands. Requirements generation, inventory control, time phasing of orders, and capacity requirements all need to be carefully coordinated. This is a job for computers. Indeed, MRP developed in the computer age for good reasons.

Benefits

It is estimated that somewhat more than a thousand manufacturers are using computerized MRP systems with excellent results. Some of the benefits are obvious. Think of changing schedules as a result of market shifts, changed promises to customers, cancellations, or whatever. A computerized MRP system can reflect immediately the effects of changed order quantities, cancellations, delayed material deliveries, and so on. A manager can change the master schedule and quickly see the effects on capacity, inventory status, or the ability of the system to meet promises to customers.

One of the important advantages is in capacity requirements adjustments. When planners examine work center load reports, such as Figure 9-8, they can see immediately possibilities for work load smoothing. It may be possible to pull some demand forward to fill in slack load. Such actions may reduce idle time and may eliminate or reduce overtime.

MRP Programs

The structure of MRP computer programs is shown in Figure 9-9. The master schedule drives the MRP program. The other inputs are product structures, bills of materials, and inventory status. The outputs of the MRP program are open and planned orders, net requirements of parts and materials, load reports, and updated and projected inventory status. A variety of other reports can be generated to suit individual needs, since the files are maintained so that they can be formulated in a variety of formats.

Subprograms include a means of reflecting the effects of engineering design changes on product structures and bills of materials. Also, inventory status requires the reflection of receipts and withdrawals to maintain an up-to-date file of inventory status.

Figure 9-10 shows a flow diagram for the IBM Requirements Generation System.

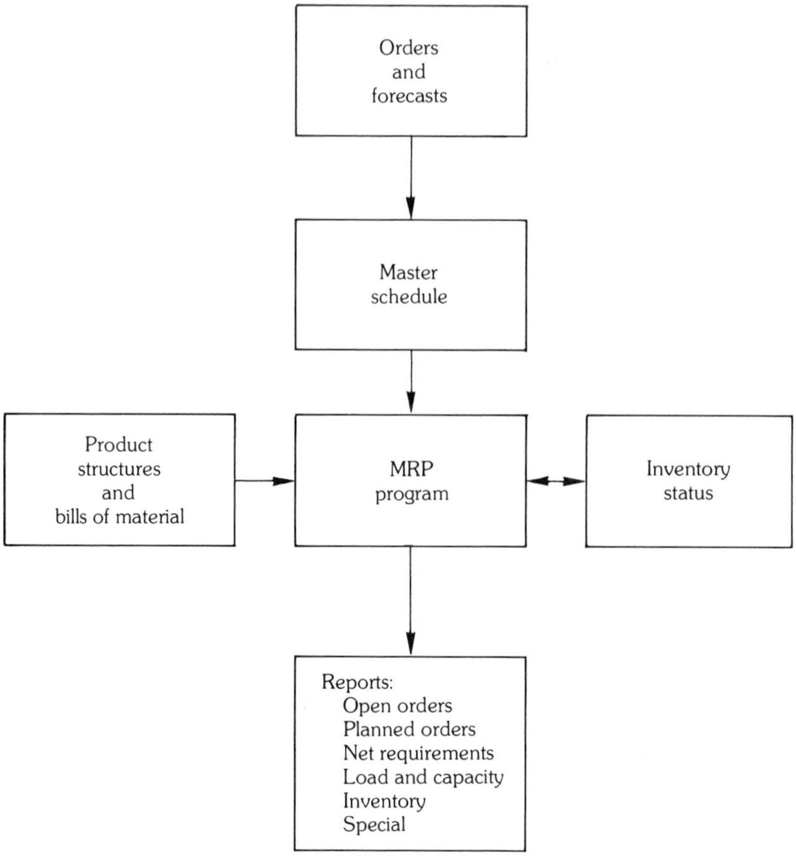

FIGURE 9-9
Structure of MRP computer programs.

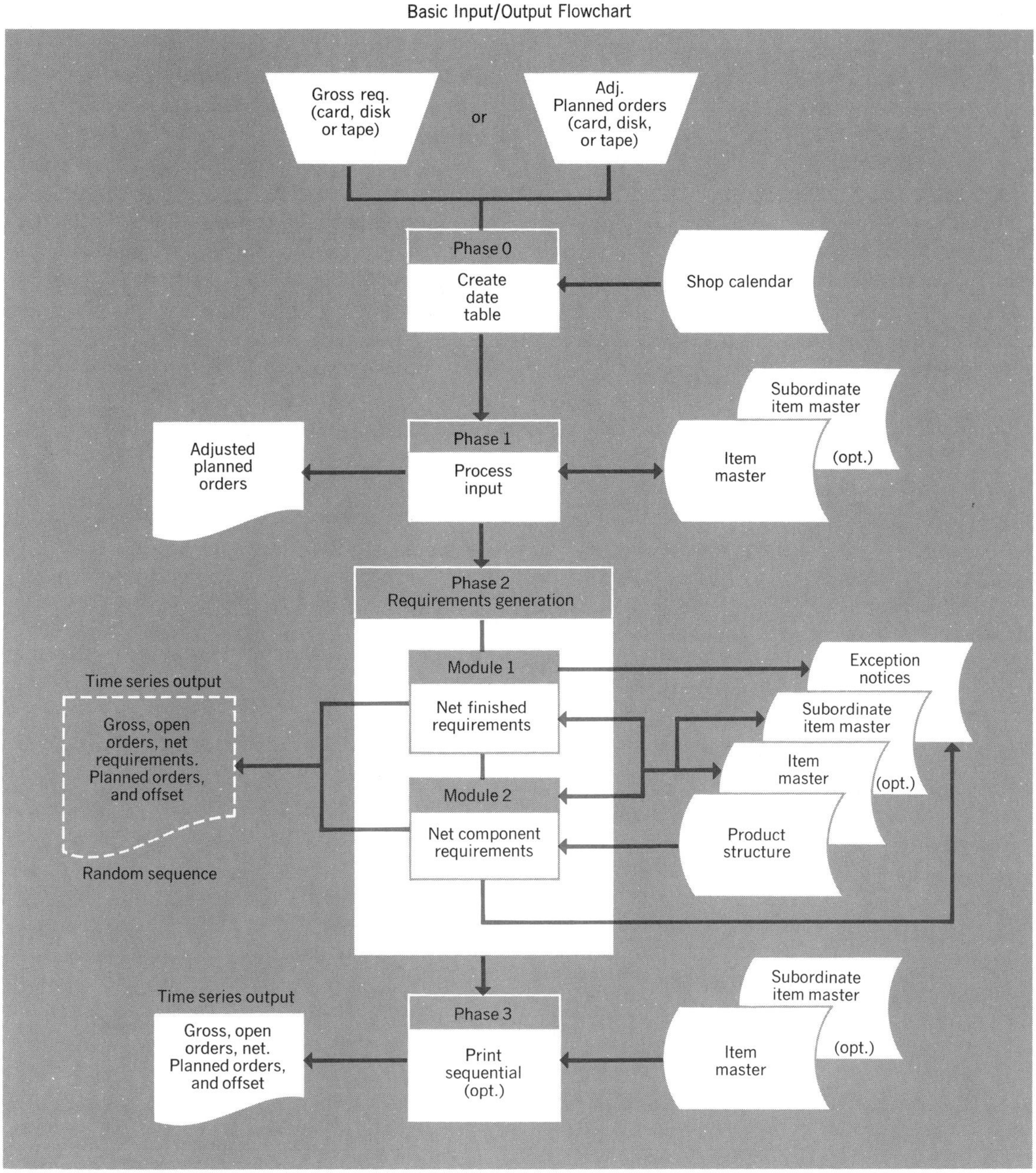

FIGURE 9-10
Computerized requirements generation system. (Courtesy International Business Machines Corporation.)

Options are contained for lot size policies that can be called, including the three lot size policies that we discussed.

SHOP LOADING AND SCHEDULING

The requirements plan is also a general schedule of production orders to be processed through the shop. Each order is for a lot of a specific part or component and may require a sequence of processing through a number of different functional work centers. Orders go to each work center according to the sequence of operations required and enter a queue of orders to be processed by that work center. The orders are processed according to some priority sequence, the simplest of which is first come-first served.

Looking at the system as a whole, we have a set of service centers with orders that require various sequences of processing and variable process or service times. We have, in effect, a network of queues, and much of the research work has viewed the scheduling problem for intermittent systems in this context, using large-scale simulation models to test alternate policies. The queue discipline, that is, the sequence of processing orders, has been the focus of many simulation studies. In shop parlance, the queue discipline is the "priority system." These are the priority rules used to dispatch an order to be processed within a work center or department.

Priority Dispatching Decision Rules

Rule testing occurred largely during the 1960s, using simulation methods. The main results of these studies indicate that the SOT rule (shortest operation time) provides the lowest mean flow time through the shop, but with a large variance. The SOT rule sequences jobs at each work center according to their estimated operation times, taking the shortest jobs first. The SOT rule is good in that it performs well on the average. But some jobs (the ones with long operation times) take a long time to get through the system, causing delivery problems.

The other main result of rule testing was the emergence of rules that emphasized getting jobs out on time. The DS/RO is effective in this regard, as is COVERT. DS/RO means dynamic slack divided by the remaining number of operations. Dynamic slack is defined as the time remaining until due date in excess of expected remaining flow time.

COVERT sequences jobs according to the largest ratio of downstream waiting time c, to operation process time, t (called c over t, or COVERT). In simulation tests, COVERT was extremely effective in meeting due dates.

Heuristic Scheduling Procedure (HSP)

The priority dispatching decision rules just discussed may be classified as "single pass" rules. This means that priorities are established by a single examination of certain characteristics of each job, and these priority indexes are retained. In general, these rules have assumed fixed machine capacity and required all process sequencing requirements to be rigidly satisfied in attempting to minimize lateness of jobs.

Holloway and Nelson [1973] have proposed alternate formulations of the job shop problem that attempt to minimize the extent of process sequence violations subject to meeting constraints on fixed machine capacity and to meeting all job due dates. Within this framework they have designed Heuristic Scheduling Procedure (HSP) that is multipass in nature, allowing backtracking to reexamine previous scheduling decisions to see if improvements can be obtained.

Briefly, HSP establishes earliest possible start times (EPST) and latest possible start times (LPST) for each operation based on the relationships of process sequences and times and on due dates for all operations. These times are hard constraints that cannot be violated, thus ensuring the final technological sequencing restrictions. In addition, a second set of early and late start times are established that are based on the relationships of adjacent operations and possible slack times available. These start times are regarded as soft constraints that can be violated. In the procedure, the violation of one of the soft constraints indicates the need to reschedule an adjacent operation for that job. Complex heuristic rules are then brought into play to reschedule in such a way that violations of the soft constraints are minimized.

Sample Results. Holloway and Nelson report results on problems of various sizes. In addition, they provide comparative results with other dispatching rules, some of which are reported in Table 9-7. The sample problem involved the scheduling of 12 jobs through six machines with a set of due dates that could be satisfied. Note that HSP provides very favorable performance in terms of both the time-in-systems statistics and job lateness.

HSP must still be regarded as being in the research stage, although it appears to have potential value for application. While the heuristic procedures are complex, present-day computing systems can apply complex algorithms if the computing cost is justified by advantageous results.

TABLE 9-7 **Comparative Performance of HSP with SOT and SLACK Priority Dispatching Rules**

	Time in System			Lateness = Scheduled Completion − Due Date		
Scheduling Procedure	Mean	Variance	Maximum	Mean	Variance	Maximum
SOT	45.5	168	61	0.2	152	22
SLACK (time remaining before due date minus all remaining processing time)	52.6	170	69	7.3	50	19
HSP	45.3	44	50	0	0	0

SOURCE: C. A. Holloway and R. T. Nelson, "Alternative Formulation of the Job Shop Problem with Due Dates," *Management Science, 20*(1), September 1973, pp. 65–75.

Tracking the Progress of Orders

Unfortunately, we cannot predict the flaws in our original plans. Machines may break down, work may pile up behind some critical machine, and dozens of other unexpected production troubles may occur that interfere with original schedules. With hundreds or even thousands of current orders in a shop, the only way to be sure that orders ultimately will meet schedules is to provide information feedback and a system for corrective action that can compensate for delays. Returned job tickets, move orders, and inspection reports can provide formal information that can be compared with schedules.

To be of value, this information flow must be rapid, so that action can be taken on up-to-date reports. Usual company mail systems are much too slow to serve the needs of production control systems. Therefore, special communication systems are in common use, such as special mail services, intercommunication systems, teletype writers interconnected to central offices, pneumatic tube systems, and remote data collection centers tied directly in with automatic computing systems. These systems in combination with rapid data processing can grind out current reports that can be used as a basis for expediting and rescheduling orders.

COMPUTERIZED PLANNING, SCHEDULING, AND CONTROL SYSTEMS

Computerized systems are relatively common today. Landmark systems in the past were developed by Bulkin et al. [1966] and LeGrande [1963] at the Hughes Aircraft Company, and a system was developed by Reiter [1966] in a gear manufacturing company. Another system developed by Godin and Jones [1969] at the Western Electric Company is of significant interest because of its interactive nature and application in a small shop. In the Western Electric system the production scheduler or supervisor worked in concert with a scheduling program at a computer terminal. By interacting with the schedule program, one can test the effects of various combinations of possible schedules, simulating effects before selecting a final schedule.

Advanced systems that include the MRP concepts discussed in this chapter have been installed at Markem Corporation and Xerox Corporation. These installations are discussed extensively in Buffa and Miller [1979].

IMPLICATIONS FOR THE MANAGER

Managers of intermittent systems face extreme complexities of planning, scheduling, and control. These complexities are maximum when the productive system must cope with different levels of dependency of items in assembly, subassembly, and component manufacture. Managers who understand the dependency relationships are better able to create effective systems of planning and control.

The conceptual framework of requirements planning recognizes that the nature of demand for primary products is substantially different than for parts and components of the primary products. Demand for parts and components is dependent on the production schedules of primary items. As is often true, parts and components may be used in more than one primary product, so that the requirements schedule for such an item is derived by a complex summing of the needs of primary products. The demand dependence of parts and components has profound effects on the policies used to determine the timing of production orders and the size of production lots.

The demand for parts and components is dependent not only in terms of quantities needed but also on the timing of supply. Since we are dealing with an interlocking structure, the components must be ready for use at a precise time. If they are too late, production of primary items will be disrupted and delayed. If they are too early, in-process inventory costs will increase.

Since demand for components is dependent and usually lumpy, some of the assumptions in traditional inventory models are not valid. For example, demand is not at a constant rate, nor is it the result of the aggregation of independent demands from multiple sources. In short, demand variations for components are not due to random fluctuations. The result is that the economic order quantity policy sometimes belies its name.

The flow of production orders through the shop can be visualized as a network of queues, and a great deal of research has focused on the queue discipline as a variable under managerial control. There are many different criteria by which to judge the effectiveness of priority dispatch decision rules, and when they are weighted equally, the SOT rule gives the best results. When the criteria that place emphasis on getting orders out on time are given heavier weight, then a due-date-oriented rule is favored. Rules that anticipate downstream waiting time cost are particularly effective in meeting due dates. Some of the most recent research suggests that complex heuristic rules perform somewhat better than the simple priority rules. These heuristic procedures allow schedule changes and improvements as scheduling problems arise.

With the availability of present-day computers, integrated systems of planning and control are possible, such as the Markem or Xerox systems. Such systems load work centers based on some priority rule and produce updated daily status reports on department schedules and weekly reports on shop load. Interactive systems are also now available, such as the one at the Western Electric Company.

REVIEW QUESTIONS AND PROBLEMS

1. Figure 9-11 is a cross-classification chart showing the subassemblies, parts, and raw materials that are used in each of nine primary products. For example, reading horizontally, product 1 requires subassembly 11 and part 28; subas-

		Subassembly										Part										Purchased part or raw material						
	Item	10	11	12	13	14	15	16	17	18	19	20	21	22	23	24	25	26	27	28	29	30	31	32	33	34	35	36
Finished product	1		1																	1								
	2			1																		1		1				
	3				2						1																	
	4					1						1																
	5						2					2		1														
	6	1						1													1							
	7								1	1			2															
	8		1															1										
	9								1					1														
Subassembly	10															1	2											
	11											1								1								
	12												1								2							
	13													2		1		2										
	14														3		3		4					1			2	
	15													3		2				2					1			
	16																2				1							1
	17																					1						1
	18												1					2					1		1			
	19													1						1				1		1		
Part	20																										1	
	21																											1
	22																						1			1		
	23																							1				
	24																						1				1	
	25																							1				1
	26																								1			
	27																									1		
	28																						1		2		1	
	29																						1			2		1
	30																							1		2		1

FIGURE 9-11
Cross-classification chart showing the complete explosion of a product line.

sembly 11 requires parts 20 and 28; and parts 20 and 28 require purchased parts or raw materials 20(35) and 28(31, 33, 35).

a. Prepare an indented bill of materials for one unit of product 7.
b. If one of each of the nine products were produced, how many of part 20 would be required?

2. Using the dinette table shown in Figure 9-1 as an example for requirements planning, why not produce enough components every two weeks to key in with a schedule of 100 completed tables per two weeks? What are the disadvantages?

3. Still using the dinette table as an example, suppose that they are produced in lots of 400 in weeks 3, 6, 9, and so on. Legs are produced in lots of 3200 every six weeks. How should the leg production be phased with respect to the table assemblies? Why?

4. What are the definitions of "dependent" and "independent" demand items? What kinds of forecasting methods are appropriate for each as a basis for production and inventory control?

5. The requirements for a motor drive unit to be assembled into a dictating machine follow the assembly schedule for the completed unit. The assembly schedule requires motor drive units with the timing shown in Table 9-8. Other data for the motor drive unit are: average requirements are $\overline{R} = 116.7$ units per week, $c_P = \$400$ per lot, and $c_H = \$4$ per unit per week. What is the inventory record and total incremental cost under each of the following lot size policies? 115.8
   a. Economic lot size
   b. Economic periodic reorder model
   c. Part-period total cost balancing

6. Account for the differences in performance of the three lot size policies in problem 5.

7. The requirements for the motor drive unit described in problem 5 have been stabilized considerably by compensatory promotion of the dictating machine and by designing a line of portable tape recorders that have general use and a counterseasonal cycle. Table 9-9 shows the new requirements schedule. What is the inventory record and total incremental cost of the same three lot size policies? Account for the differences in performance of the three policies.

8. What kinds of variation in demand for dependent items are not taken into account by the straightforward computation of requirements? How can the effects of these kinds of variation be absorbed?

TABLE 9-8
Requirements Schedule for a Motor Drive Unit

Week number	1	2	3	4	5	6	7	8	9	10	11	12
Requirements, units	25	30	75	125	200	325	400	100	0	100	0	10

Total requirements for 12 weeks, 1390 units.

TABLE 9-9

Stabilized Requirements Schedule for a Motor Drive Unit

Week number	1	2	3	4	5	6	7	8	9	10	11	12
Requirements, units	300	300	300	300	350	350	400	400	350	350	350	325

Total requirements for 12 weeks, 4075 units

9. What is a dispatching decision rule? How does it relate to the structure of a waiting line model?

10. If management's objectives place a high value on low in-process inventories, what kind of priority dispatching rule seems best?

11. If management's objectives place a high value on on-time delivery of orders, what kind of priority dispatching decision rule seems best?

12. How do you account for the excellent performance of the COVERT rule from the point of view of on-time delivery?

13. Job orders are received at a work center with the characteristics indicated by the data in Table 9-10. In what sequence should the orders be processed at the work center if the priority dispatch decision rule is:
    a. FCFS (first come-first served)?
    b. SOT (shortest operation time)?
    c. SS (static slack, i.e., due date less time of arrival at work center)?
    d. FISFS (due date system, first in system-first served)?
    e. SS/RO (static slack/remaining number of operations)?

    Compute priorities for each rule and list the sequence in which orders would be processed. Which decision rule do you prefer? Why?

TABLE 9-10 **Order and Processing Data for Six Jobs**

Order Number	Due Date	Date and Time Received at Center	Operation Time, Hours	Remaining Operations
1	May 1	Apr. 18, 9 A.M.	6	3
2	Apr. 20	Apr. 21, 10 A.M.	3	1
3	June 1	Apr. 19, 5 P.M.	7	2
4	June 15	Apr. 21, 3 P.M.	9	4
5	May 15	Apr. 20, 5 P.M.	4	5
6	May 20	Apr. 21, 5 P.M.	8	7

SITUATIONS

14. The Wheel-Pump Company was originally the Pump Company, but many years ago the owner had the opportunity to bid on a contract to produce steel wheels for one of the smaller automobile companies. The Pump Company was successful in this initial venture into a new field, and became a supplier of wheels to the auto industry.

The Wheel Business

The basic process for wheel production involved stamping hub parts from coil strip steel, rolling rim shapes from steel stock, welding the rolled stock into a circular shape, welding hub parts to the rim, and painting. The fabrication processes were connected by conveyor systems, and once set up for a given type and size, the system ran smoothly. The main production control problems seemed to be in maintaining material supply and quality control and in scheduling types and sizes to meet customer requirement schedules on two parallel line setups.

The Pump Business

The original pump business was still flourishing. The company manufactured a line of water pumps used in a variety of applications. There were 10 different models that had some common parts. There was a machine shop that fabricated parts which were stocked as semifinished items. Production orders were released for these parts based on a reorder from two to six weeks, depending on the item. The intention was to maintain a buffer stock of six weeks' supply. A forecast of demand for each item was updated monthly, using an exponential smoothing system with $\alpha = 0.3$. The relatively large value of α was used to provide fast reaction to demand changes for inclusion in EOQ computations. Purchase parts were handled on an EOQ-reorder point basis, again with a buffer stock of six weeks' supply.

The basic schedule for the assembly of pumps was set monthly in a master schedule. The master schedule was projected for three months, with the first month representing a firm schedule. The most important input to the master scheduling process was a forecast for each model based on an exponential smoothing system merged with knowledge of contracts, orders, and sales estimates from the salespeople. The basic production rule was to produce one month's estimated requirements for the four models with heavy demand and two months' estimated requirements for the six models with smaller demand. These estimates were then adjusted for backorders or excess inventory. The intention was to maintain a buffer inventory of finished goods of two weeks' supply.

Part shortages at assembly were common, and it was the job of expeditors to get rush orders through the shop to keep the assembly of pumps on schedule. Nevertheless, it was often true that pumps could not be assembled completely because of part shortages and had to be set back in the assembly schedule, resulting in stockouts of finished products.

Material Requirements Planning

The two product lines have always been operated as two separate businesses, both in sales and manufacturing, although the manufacturing facilities were located on the same site. The systems common to both product lines were in the accounting and finance functions. On retirement of the original owner, the firm was sold to a conglomerate, and a new professional manager was installed as president. The new president was satisfied that the manufacturing facilities for the two product lines had to be separate but felt that the functions of production planning and control could and should be merged. He attended seminars on MRP at a local university and was impressed with the integrating nature of the computer-based systems available. He felt that the requirements generation system available as software for the company computer could generate material requirements for both the wheel and pump product lines. He believed that production order releases keyed to sales requirements would be beneficial to both product lines and that the use of the capacity requirements concepts would enable forward planning on the use of overtime and the prevention of finished goods stockouts.

The president called in the vice-president–production and suggested the use of the MRP concept. The vice-president–production had served under the previous owner for 10 years and objected. "They are two different businesses, it won't work," he said.

What do you recommend?

REFERENCES

Berry, W. L. "Lot Sizing Procedures for Requirements Planning Systems: A Framework for Analysis," *Production and Inventory Management*, 2nd Quarter, 1972, pp. 13–34.

J. R. Biggs, S. H. Goodman, and S. T. Hardy. "Lot Sizing Rules in a Hierarchical Multi-Stage Inventory System," *Production and Inventory Control Management,* 1st Quarter, 1977.

Buffa, E. S., and J. G. Miller. *Production-Inventory Systems: Planning and Control* (3rd ed.). Richard D. Irwin, Homewood, Ill., 1979.

Bulkin, M. H., J. L. Colley, and H. W. Steinhoff. "Load Forecasting, Priority Sequencing, and Simulation in a Job-Shop Control System," *Management Science, 13*(2), October 1966, pp. 29–51.

Carroll, D. C. "Heuristic Sequencing of Single and Multiple Component Jobs." Unpublished PhD dissertation, MIT, 1965.

GH20-0751-1, OS/360 Requirements Planning Application Description, *IBM* (2nd ed.), June 1972.

Godin, V., and C. H. Jones. "The Interactive Shop Supervisor," *Industrial Engineering,* November 1969, pp. 16–22.

Holloway, C. A., and R. T. Nelson. "Job Shop Scheduling with Due Dates and Operation Overlap," *AIIE Transactions, 7*(1), March 1975, pp. 16–20.

Holloway, C. A., and R. T. Nelson. "Alternative Formulation of the Job Shop Problem with Due Dates," *Management Science, 20*(1), September 1973, pp. 65–75.

Holstein, W. K. "Production Planning and Control Integrated," *Harvard Business Review,* May–June 1968.

LeGrande, E. "The Development of a Factory Simulation System Using Actual Operating Data," *Management Technology, 3*(1), May 1963.

Magee, J. F., and D. M. Boodman. *Production Planning and Inventory Control* (2nd ed.). McGraw-Hill, New York, 1967.

"Material Requirements Planning by Computer," *American Production and Inventory Control Society,* 1971, 86 pp.

Moodie, C. L., and D. J. Novotny. "Computer Scheduling and Control Systems for Discrete Part Production," *Journal of Industrial Engineering, 19*(7), July 1967, pp. 648–671.

Moore, F. G., and R. Jablonsky. *Production Control* (3rd ed.). McGraw-Hill, New York, 1969.

Nelson, R. T., C. A. Holloway, and R. M. Wong. "Centralized Scheduling and Priority Implementation Heuristics for a Dynamic Job Shop Model," *AIIE Transactions, 9*(1), March 1977, pp. 95–102.

New, C. *Requirements Planning.* Gower Press Ltd., Epping, Essex, Great Britain; Halsted Press, New York, 1973.

Orlicky, J. *Material Requirements Planning.* McGraw-Hill, New York, 1975.

Peterson, R., and E. A. Silver. *Decision Systems for Inventory Management and Production Planning,* John Wiley, New York, 1979.

Plossl, G., and O. Wight. *Production and Inventory Control: Principles and Techniques.* Prentice-Hall, Englewood Cliffs, N.J., 1967.

Reiter, S. "A System for Managing Job Shop Production," *Journal of Business, 39*(3), July 1966, pp. 371–393.

Rowe, A. J. "Sequential Decision Rules in Production Scheduling." Unpublished PhD dissertation, UCLA, August 1958.

Thurston, P. H. "Requirements Planning for Inventory Control," *Harvard Business Review,* May-June 1972.

Wagner, H. M., and T. M. Whitin. "Dynamic Version of the Economic Lot Size Model," *Management Science, 5*(1), October 1958.

CHAPTER 10 Large-Scale Projects

OUR SOCIETY PRODUCES BUILDINGS, ROADS, DAMS, MISSILES, ships, and other products of large-scale projects. The problems of managing such projects stem from their great complexity and the nonrepetitive nature of the required activities; when the lot size is only one, there is little chance to take advantage of learning effects. Projects are produced by intermittent systems.

PROJECTS AND INTERMITTENT SYSTEMS

In the previous two chapters we have dealt with problems of industrial planning and scheduling. Recall from the Part II Introduction, Table II-1, that intermittent systems were divided into those with both inventoriable and noninventoriable outputs.

Systems with inventoriable outputs commonly produce a line of products in various types and sizes. The emphasis in planning and scheduling is on cycling the products through common facilities, lot size determination, the coordination of material requirements with production schedules, capacity requirements planning and control, and so forth. MRP represents a system for dealing with the planning and control problems. The focus in these kinds of enterprises is on the system and process for handling a large number of material and production orders. The nature of the complexity requires that we rely on a system, rather than focusing attention on each individual production order. A well-designed system is likely to produce good results.

When the output is not inventoriable, many of the problems shift to the custom order. Each order or job is likely to be unique or at least have some unique characteristics. In the case of a jobbing machine shop, we have a general-purpose productive system that processes a large volume of custom orders for metal parts and components. Like the intermittent system for inventoriable products, the focus must still be on the system and process for handling a large number of orders. Here, either centralized or local shop floor control relies on priority decision rule systems. Good performance is measured in part in terms of order tardiness, cost, and the like. A good system will provide good performance, given other enabling factors such as requisite capacity and skills. But an additional factor may be present in some of these custom systems—an explicit penalty for missing deadlines. Most systems, in fact, involve costs that result from poor scheduling and control (idle labor, assembly out of sequence, loss of customers and goodwill). However, when explicit penalties for poor scheduling and control are present, emphasis shifts to more detailed planning, scheduling, and control of each activity to ensure timely completion.

The large-scale project also employs an intermittent system. By its nature, the product is either custom designed or has many features that are custom designed. The activities or operations required flow from the unique design. Because of the large scale, complexity results in terms of the number of activities, their sequence and timing. The risks that result from failure to meet project completion deadlines are high, and large penalties for missing completion dates are common. These penalties may be explicit or may be in the form of higher costs. Thus the focus for managerial effort in project systems is on detailed planning, scheduling, and control of each major activity in relation to the project as a whole. As we will see, the interdependent nature of the sequence and timing of activities can be exploited to provide project managers with the crucial information they need.

ORGANIZATIONAL STRUCTURES

Given a single project, the manager organizes around the needs of that project, with all functional organizational units focused on achieving the project objectives. The organizational structure is then comparable to the functional organization used commonly in industry. Organizational problems begin when we add a second project. Multiple projects suggest resource sharing, with obvious advantages of better utilization. But how will resources be shared? By what schedule? Who will decide these issues, if there are two project managers? The problems can be solved by simply duplicating resources and making the two projects independent, but the potential advantages of economies of the larger scale of operations are lost.

The common organizational form for multiple projects is the matrix organization. Figure 10-1 shows the concept of matrix organization, with the functional managers holding the resources and each project manager coordinating the use of designated resources for each project. The advantages of matrix organization are in the efficient use of resources, the coordination across functional departments, and the visibility of project objectives. The disadvantages are in the need for coordination between functional and project managers and the often unbearable situation of the person in the middle (the project representative) who is working for two bosses. Vertically, this person reports to the functional manager, but horizontally the project manager is the boss. If conflicts occur, the person in the middle may be caught there.

THE PROJECT MANAGER

The nature of organizational structures for project management and the one-time nature of projects create a difficult situation for project managers. The project manager has less than full authority over the project, being in the position of bargaining for resources and their timing from functional managers. At the same time, the uniqueness of each project creates problems that often cannot be foreseen. These problems commonly result in schedule slippage, replanning, and possibly reallocation of resources in order to maintain the project completion schedule.

If the project manager focuses attention in one critical area to the neglect of other important areas, the secondary areas may become critical problems. In short, the project manager must cope with short deadlines, changing situations, rapid-fire decision making, incomplete information, and skimpy authority. The position has all the potential for being one of simply "putting out fires." The issues are then to devise means for remaining in control.

Thanhain and Wilemon [1975] report on interviews with 100 project managers concerning the primary sources of conflict during a project. The results indicate the following sources of conflict in three broad phases of a project:

1. *Planning Stage:* Priorities, the required activities, and their sequencing.

2. *Buildup Stage:* Priorities, with the scheduling of activities becoming very important.

3. *Execution:* Scheduling of activities the primary source of conflict, with issues concerning the trade-off of the use of resources versus time performance also being very important.

The foregoing typical issues focus on problems of planning, scheduling, and control of activities. For project managers to remain in control, they need mechanisms that provide a clear logic for planning, a detailed schedule for all activities that can be easily updated, and mechanisms for resource trade-off. Network planning and scheduling methods have been developed to meet these special needs.

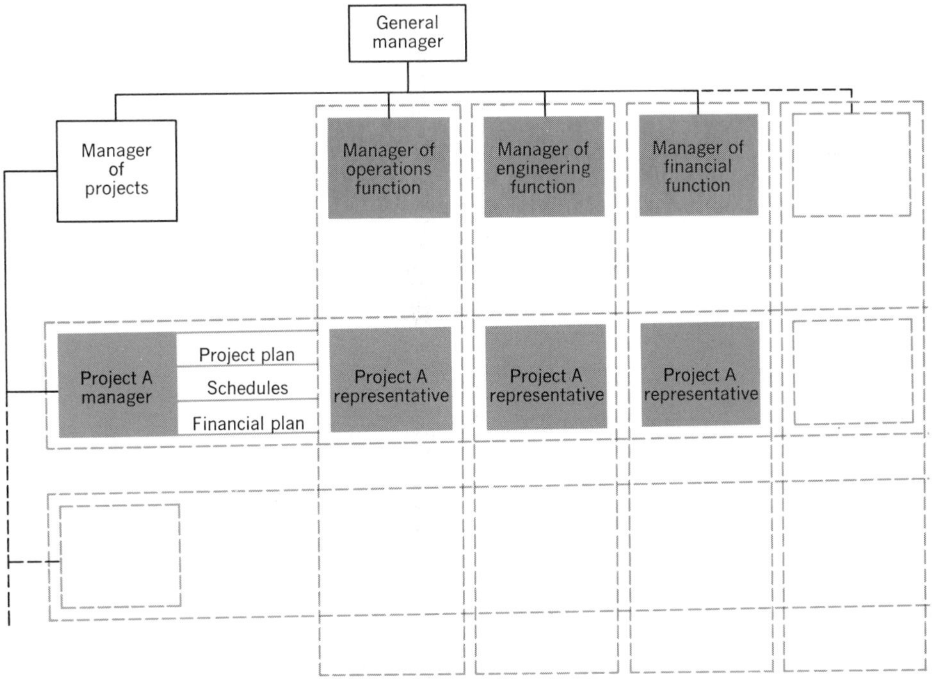

FIGURE 10-1
Structure of matrix organization for project management.

ORIGIN OF NETWORK PLANNING

Network planning methods were developed independently by two different groups. As an internal project of the DuPont Company, Critical Path Methods (CPM) were developed to plan and control the maintenance of chemical plants. They were subsequently widely used by DuPont for many engineering functions.

Parallel efforts were undertaken by the U.S. Navy at about the same time to develop methods for planning and controlling the Polaris Missile Project. The project involved 3000 separate contracting organizations and was regarded as the most complex of projects experienced to that date. The result was the development of the Performance Evaluation and Review Technique (PERT) methodology.

The immediate success of both the CPM and PERT methodologies may be gauged by the following facts. DuPont's application of their technique to a maintenance project in their Louisville works resulted in reducing downtime for maintenance from 125 to 78 hours. The PERT technique was widely credited with helping to shorten by two years the time originally estimated for the completion of the engineering and development program for the Polaris missile.

PERT and CPM are based on substantially the same concepts. As originally developed, PERT was based on probabilistic estimates of activity times that resulted in a probabilistic path through a network of activities and a probabilistic project completion time. CPM, however, assumed constant or deterministic activity times. Actually either the probabilistic or the deterministic model is equally applicable to and usable by either technique.

PERT/CPM PLANNING METHODS

We will use a relatively simple example, the introduction of a new product, to develop the methods used in generating a network representation of a project. The development of a project network may be divided into (1) activity analysis, (2) arrow diagramming, and (3) node numbering.

Activity Analysis

The smallest unit of productive effort to be planned, scheduled and controlled is called an "activity." For large projects, it is possible to overlook the need for some activities because of the great complexity. Therefore, while professional planning personnel are commonly used, the generation of the activity list is often partially done in meetings and round-table discussions that include managerial and operating personnel. Table 10-1 is an activity list for the production of a new product.

TABLE 10-1

Precedence Chart Showing Activities, Their Required Sequence, and Time Requirements for the New Product Introduction Project

Activity Code	Description	Immediate Predecessor Activity	Time, Weeks
A	Organize sales office	—	6
B	Hire salespeople	A	4
C	Train salespeople	B	7
D	Select advertising agency	A	2
E	Plan advertising campaign	D	4
F	Conduct advertising campaign	E	10
G	Design package	—	2
H	Set up packaging facilities	G	10
I	Package initial stocks	H, J	6
J	Order stock from manufacturer	—	13
K	Select distributors	A	9
L	Sell to distributors	C, K	3
M	Ship stock to distributors	I, L	5

Network Diagramming*

A network is developed that takes account of the precedence relationships among activities and must be based on a complete, verified, and approved activity list. The important information required for these network diagrams is generated by the following three questions:

1. Which activities must be completed *before* each given activity can be started?

2. Which activities can be carried out in *parallel?*

3. Which activities immediately *succeed* other activities?

The common practice is simply to work backwards through the activity list, generating the immediate predecessors for each activity listed, as shown in Table 10-1. The estimated normal time for each activity is also shown in the table, although it is not necessary at this point. The network diagram may then be constructed to represent the logical precedence requirements shown in Table 10-1. It is not particulary

* Activities will be diagrammed as occurring on the arcs, or arrows. An alternate network diagramming procedure, where activities occur at the nodes, will be discussed later in the chapter. We can refer to the first as an "arcs" network, and the second as a "nodes" network.

important whether or not arcs cross. However, if the network diagram is to be presented to others there is a value to developing an uncluttered final diagram

Dummy Activities

Care must be taken in correctly representing the actual precedence requirements in the network diagram. For example, in house construction, consider the immediate predecessor activities for activity *s*, sand and varnish flooring, and activity *u*, finish electrical work. Activity *s* has immediate predecessors *o* and *t*, finish carpentry and painting, respectively, while *u* has a predecessor of only activity *t*, paint. The relationship shown in Figure 10-2*a* does not correctly represent this situation because it specifies that the beginning of *u* is dependent on both *o* and *t*, and this is not true.

To represent the situation correctly, we must resort to the use of a dummy activity that requires zero performance time. Figure 10-2*b* represents the stated requirement. The finish electrical work, *u*, now depends only on the completion of painting, *t*. Through the dummy activity, however, both finish carpentry and painting must be completed before activity *s*, sand and varnish flooring, can be started. The dummy activity provides the logical sequencing relationship. But since the dummy activity is assigned zero performance time, it does not alter any scheduling relationships that may be developed.

Another use of the dummy activity is to provide an unambiguous beginning and ending event or node for each activity. For example, a functionally correct relationship may be represented by Figure 10-3*a*, with two activities having the same beginning and ending nodes. If Figure 10-3*a* were used, however, it would not be possible to identify each activity by its predecessor and successor events because both activities *m* and *n* would begin and end with the same node numbers. This is particularly important in larger networks employing computer programs for network diagram generation. The computer is programmed to identify each activity by a pair of event numbers. The problem is solved through the insertion of a dummy activity, as shown in Figure 10-3*b*. The functional relationship is identical, since the dummy activity

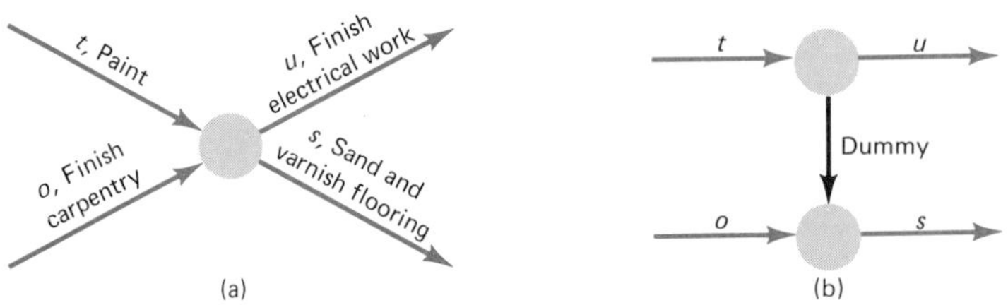

FIGURE 10-2
(a) Diagram does not properly reflect precedence requirements, since *u* seems to be dependent on the completion of both *o* and *t* but actually depends only on *t*. *(b)* Creating two nodes with dummy activity between provides the proper predecessors for both activities *s* and *u*.

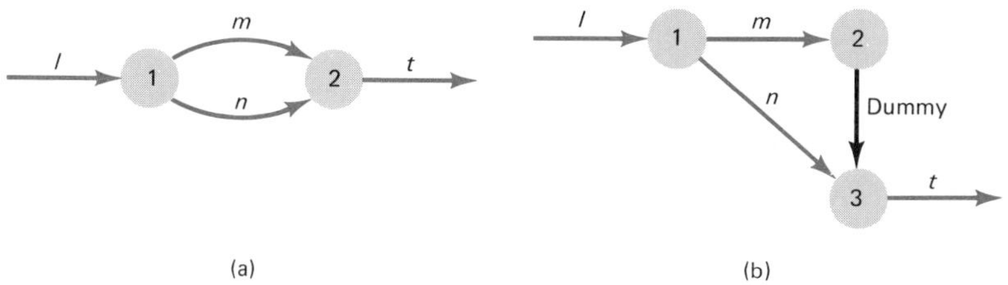

FIGURE 10-3
(a) Activities m and n may be carried out in parallel but result in identical beginning and end events. *(b)* Use of dummy activity makes possible separate ending event numbers.

requires zero time, but now both m and n are identified by different combinations of node numbers.

Figure 10-4 shows the completed network diagram for the new product introduction project. Activities are identified with their required times in weeks, and all the nodes are numbered. The activity times were not used to this point and were not necessary for the construction of the diagram. However, the activity times will have great significance in the generation of schedule data, critical path determination, and the generation of alternatives for the deployment of resources.

Node Numbering

The node numbering shown in Figure 10-4 has been done in a particular way. Each arc, or arrow, represents an activity. If we identify each activity by its tail (i) and head (j) numbers, the nodes have been numbered so that for each activity, i is always less than j, $i < j$. The numbers for every arrow are progressive, and no backtracking through the network is allowed. This convention in node numbering is effective in

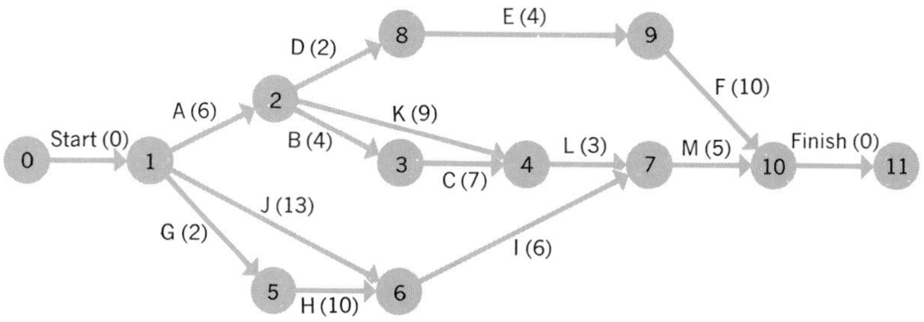

FIGURE 10-4
Arcs network diagram for the new product introduction project.

computing programs to develop the logical network relationships and to prevent the occurrence of cycling or closed loops.

A closed loop would occur if an activity were represented as going back in time. This is shown in Figure 10-5, which is simply the structure of Figure 10-3*b* with the activity *n* reversed in direction. Cycling in a network can result through a simple error, or, when developing the activity plans, if one tries to show the repetition of an activity before beginning the next activity. A repetition of an activity must be represented with additional separate activities defined by their own unique node numbers. A closed loop would produce an endless cycle in computer programs, without a built-in routine for detection and identification of the cycle. Thus, one property of a correctly constructed network diagram is that it is *noncyclical.*

CRITICAL PATH SCHEDULING

With a properly constructed network diagram, it is a simple matter to develop the important schedule data for each activity and for the project as a whole. The data of interest are the earliest and latest start and finish times, the available slack for all activities, and the critical path through the network.

Earliest Start and Finish Times

If we take zero as the starting time for the project, then for each activity there is an earliest starting time (*ES*) relative to the project starting time. This is the earliest possible time that the activity can begin, assuming that all the predecessors also are started at their *ES*. Then, for that activity, its earliest finish (*EF*) is simply *ES* + activity time.

Latest Start and Finish Times

Assume that our target time for completing the project is "as soon as possible." This is called the "latest finish time" (*LF*) of the project and of the finish activity. The latest

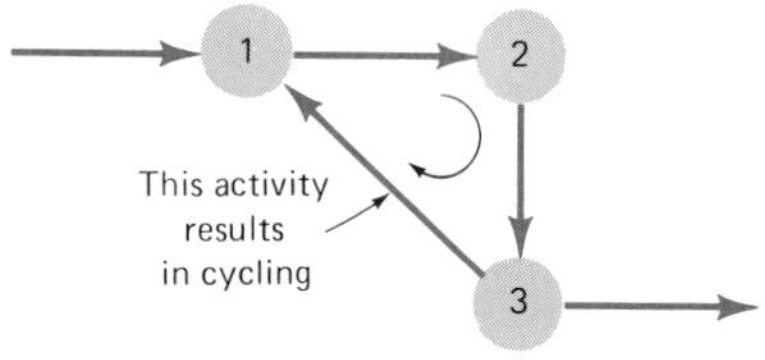

FIGURE 10-5
Example of a closed loop or cycling in a network diagram.

THE CRITICAL PATH IS

START → A → B → C → L → M → FINISH

THE LENGTH OF THE CRITICAL PATH IS 25

NODE	DURATION	EARLY START	EARLY FINISH	LATE START	LATE FINISH	TOTAL SLACK	FREE SLACK
START	0.00	0.00	0.00	0.00	0.00	0.00	0.00
A	6.00	0.00	6.00	0.00	6.00	0.00	0.00
B	4.00	6.00	10.00	6.00	10.00	0.00	0.00
C	7.00	10.00	17.00	10.00	17.00	0.00	0.00
D	2.00	6.00	8.00	9.00	11.00	3.00	0.00
E	4.00	8.00	12.00	11.00	15.00	3.00	0.00
F	10.00	12.00	22.00	15.00	25.00	3.00	3.00
G	2.00	0.00	2.00	2.00	4.00	2.00	0.00
H	10.00	2.00	12.00	4.00	14.00	2.00	1.00
I	13.00	0.00	13.00	1.00	14.00	1.00	0.00
J	6.00	13.00	19.00	14.00	20.00	1.00	1.00
K	9.00	6.00	15.00	8.00	17.00	2.00	2.00
L	3.00	17.00	20.00	17.00	20.00	0.00	0.00
M	5.00	20.00	25.00	20.00	25.00	0.00	0.00
FINISH	0.00	25.00	25.00	25.00	25.00	0.00	0.00

FIGURE 10-6
Sample computer output of schedule statistics and critical path for the new product introduction project.

start time (LS) is the latest time at which an activity can start, if the target or schedule is to be maintained. Thus, LS for the finish activity is LF – activity time. Since the finish activity requires zero time units, $LS = LF$.

Existing computer programs may be used to compute these schedule data automatically, requiring as inputs the activities, their performance time requirements, and the precedence relationships established. The computer output might be similar to Figure 10-6, which shows the schedule statistics for all activities when no slack has been allowed in the overall project completion time. Note, then, that all critical activities have zero total slack in their schedules. All other activities have greater slack.

The total schedule slack is simply the difference between computed late and early start times ($LS - ES$) or between late and early finish times ($LF - EF$). The significance of total slack (TS) is that it specifies the maximum time that an activity can be delayed without delaying the project completion time. If TS for an activity is used up, then the critical path changes.

Free slack (FS) shown in Figure 10-6 indicates the time that an activity can be delayed without delaying the ES of any other activity. FS is computed as the difference between EF for an activity and the earliest of the ES times of all immediate successors. For example activity F has FS = 3 weeks. Its LF = 25 weeks, but its EF = 22 weeks. If its finish time is delayed up to three weeks no other activity is affected, nor is the project completion time affected. Note also that activity K can be delayed 2 weeks without affecting activity L, its successor.

On the other hand, total slack is shared with other activities. For example, activities D, E, and F all have $TS = 3$. If activity D uses the slack, then E and F no longer have slack available. These relationships are most easily seen by examining the network diagram, where the precedence relationships are shown graphically.

Actually, there are five different paths from start to finish through the network. The shortest path requires 22 weeks by the sequence START-A-D-E-F-FINISH and the longest or limiting path requires 25 weeks by the critical sequence START-A-B-C-L-M-FINISH. In a small problem such as this one, we could enumerate all the alternate paths to find the longest path, but there is no advantage in doing so, since the critical path is easily determined from the schedule statistics, which are themselves useful.

Manual Computation of Schedule Statistics

Manual computation is appropriate for smaller networks and helps to convey the significance of the schedule statistics. To compute *ES* and *EF* manually from the network, we proceed as follows, referring to Figure 10-7:

1. Place the value of the project start time in both the *ES* and *EF* positions near the start activity arrow. See the legend for Figure 10-7. We will assume relative values, as we did in the computer output of Figure 10-6, so the number 0 is placed in the *ES* and *EF* positions for the start activity. (Note that it is not necessary in PERT to include the start activity with zero activity duration. It has been included to make this example parallel in its activity list with the comparable

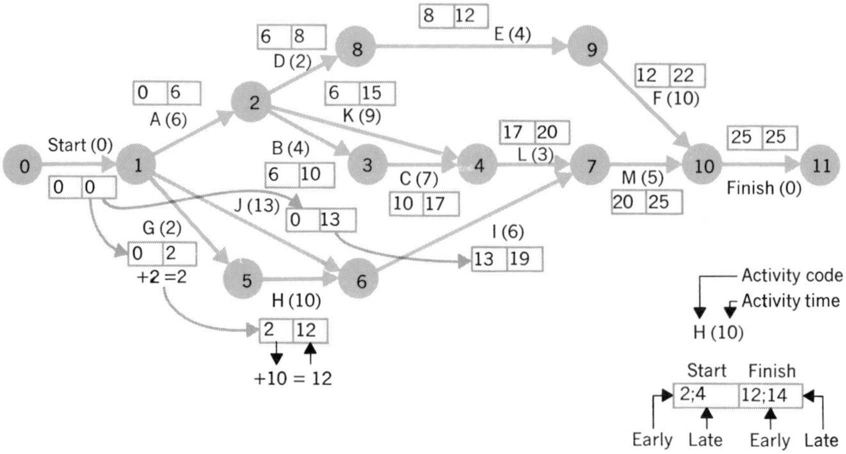

FIGURE 10-7
Flow of calculations for early start *(ES)* and early finish *(EF)* times.

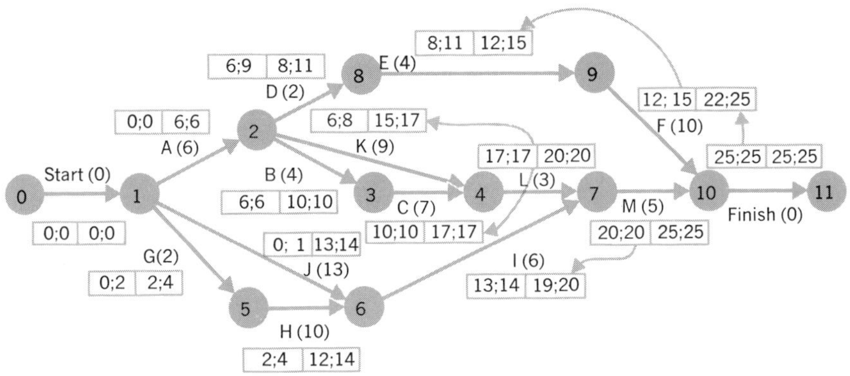

FIGURE 10-8
Flow of calculations for late start *(LS)* and late finish *(LF)* times.

"Activities on nodes" example of Figure 10-9. The start and finish activities are often necessary in nodes network.)

2. Consider any new unmarked activity, *all of whose predecessors have been marked* in their *ES* and *EF* positions, and mark in the *ES* position of the new activity the *largest* number marked in the *EF* position of any of its immediate predecessors. This number is the *ES* time of the new activity. For activity A in Figure 10-7, the *ES* time is 0, since that is the *EF* time of the preceding activity.

3. Add to this number the activity time, and mark the resulting *EF* time in its proper position. For activity A, $ES + 6 = 6$.

4. Continue through the entire network until the "finish" activity has been reached. As we showed in Figure 10-6, the critical path time is 25 weeks, so $ES = EF = 25$ for the finish activity.

To compute the *LS* and *LF*, we work backwards through the network, beginning with the finish activity. We have already stated that the target time for completing the project is as soon as possible, or 25 weeks. Therefore, $LF = 25$ for the finish activity without delaying the total project beyond its target date. Similarly, the *LS* time for the finish activity is *LF* minus activity time. Since the finish activity requires 0 time units, $LS = LF$. To compute *LS* and *LF* for each activity, we proceed as follows, referring to Figure 10-8.

1. Mark the value of *LS* and *LF* in their respective positions near the finish activity.

2. Consider any new unmarked activity, all of whose successors have been marked, and mark in the *LF* position for the new activity the smallest *LS* time marked for

any of its immediate successors. In other words, *LF* for an activity equals the earliest *LS* of the immediate successors for that activity.

3. Subtract from this number the activity time. This becomes the *LS* for the activity.

4. Continue backwards through the chart until all *LS* and *LF* times have been entered in their proper positions on the network diagram. Figure 10-8 shows the flow of calculations, beginning with the finish activity backwards through several activities.

As discussed previously, the schedule slack for an activity represents the maximum amount of time that it can be delayed beyond its *ES* without delaying the project completion time. Since critical activities are those in the sequence of the longest time path, it follows that these activities will have the minimum possible slack. If the project target date coincides with the *LF* for the finish activity, all critical activities will have zero slack. If, however, the project date is later than the *EF* of the finish activity, all critical activities will have slack equal to this time-phasing difference. The manual computation of slack is simply *LS* − *ES* or, alternately, *LF* − *EF*. As noted previously, free slack is computed as the difference between *EF* for an activity and the earliest of the *ES* times of all immediate successors.

ACTIVITIES ON NODES—NETWORK DIAGRAM DIFFERENCES

Thus far, we have been using the "activities on arcs" network diagramming procedures. The "activities on nodes" procedure results in a slightly simpler network system by representing activities as occurring at the nodes, with the arrows showing the sequences of activities required. The advantage in this methodology is that it is not necessary to use dummy activities in order to represent the proper sequencing. Figure 10-9 shows the network for the new product introduction project, which may be compared with the comparable "activities on arcs" network shown in Figure 10-4.

The analysis developing the early and late start and finish times and slack times is identical with the procedure previously outlined. The net results of both systems are the schedule statistics that are computed. Since these are the data of interest and since the entire procedure is normally computerized for both methodologies, the choice between the two may fall to other criteria, such as the availability and adaptability of existing computer routines, or the choice may be simply a matter of personal preference.

PROBABILISTIC NETWORK METHODS

The network methods that we have discussed so far may be termed "deterministic," since estimated activity times are assumed to be the expected values. No recognition

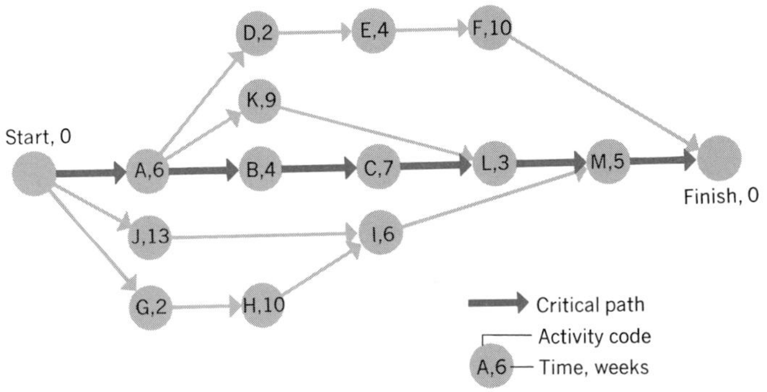

FIGURE 10-9
Project graph of activities on nodes for the new product introduction project.

is given to the fact that the mean or expected activity time is the mean of a distribution of possible values that could occur. Deterministic methods assume that the expected time is actually the time taken.

Probabilistic network methods assume the more realistic situation in which activity times are represented by a probability distribution. With such a basic model of the network of activities, it is possible to develop additional data important to managerial decisions. Such data help in assessing planning decisions that might revolve around such questions as: What is the probability that the completion of activity A will be later than January 10? What is the probability that the activity will become critical and affect the project completion date? What is the probability of meeting a given target completion date for the project? What is the risk of incurring cost penalties for not meeting the contract date? The nature of the planning decisions based on such questions might involve the allocation or reallocation of personnel or other resources to the various activities in order to derive a more satisfactory plan. Thus, a "crash" schedule with extra resources might be justified to ensure the on-time completion of certain activities. The extra resources needed are drawn from noncritical activities or activities where the probability of criticality is small.

The discussion that follows is equally applicable to either method of network diagramming. The probability distribution of activity times is based on three time estimates made for each activity.

Optimistic Time

Optimistic time, a, is the shortest possible time to complete the activity if all goes well. It is based on the assumption that there is no more than one chance in a hundred of completing the activity in less than the optimistic time.

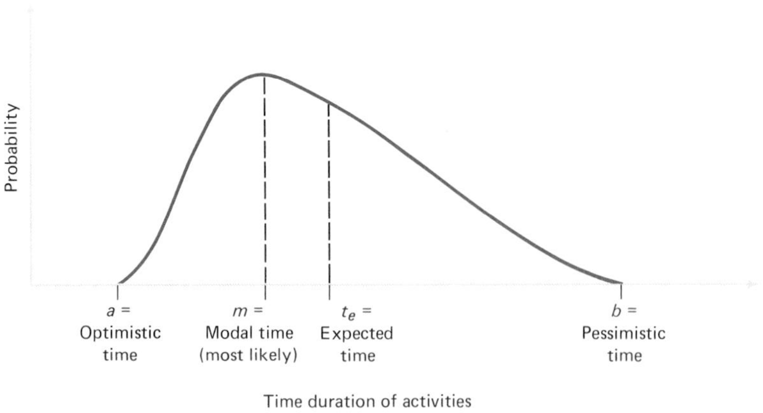

FIGURE 10-10
Time values in relation to a distribution of activity time.

Pessimistic Time

Pessimistic time, b, is the longest time for an activity under adverse conditions but barring acts of nature. It is based on the assumption that there would be no more than one chance in a hundred of completing the activity in a time greater than b.

Most Likely Time

Most likely time, m, is the modal value of the activity time distribution.

The three time estimates are shown in relation to an activity completion time distribution in Figure 10-10. The computational algorithm reduces these three time estimates to a single average or expected value, t_e, which is actually used in the computing procedure.* The expected value is also the one used in computing schedule statistics for the deterministic model. The example distribution in Figure 10-10 represents only one possibility. Actually, the time distributions could be symmetrical or skewed to either the right or the left.

With a probabilistic model we can see that there is a probability that seemingly noncritical activities could become critical. This could happen by the occurrence either

* The usual model assumes that t_e is the mean of a beta distribution. The estimates of the mean and variance of the distribution may be computed as follows:

$$\bar{x} = 1/6[A + 4M + B]$$
$$s^2 = [1/6(B - A)]^2$$

where A, B, and M are estimates of the values of a, b, and m, respectively, and $\bar{x}$ *and* s^2 *are estimates of the mean and variance,* t_e *and* σ_t^2.

of a long performance time for an activity or of short performance times for activities already on the critical path. This is a signal that the schedule plans developed are likely to change. As actual data on the progress of operations come in, it may be necessary to make changes in the allocation of resources in order to cope with new critical activities.

Probability theory provides the basis for applying probabilistic network concepts. First, the sum of random variables is itself a random variable, but is normally distributed, even if the individual random variables are not. Second, the variance of the new normally distributed random variable is the sum of the variances of the original random variables.

Translated for network scheduling concepts, we have the following useful statements: (1) the mean project completion time is the mean of a normal distribution that is the simple sum of the t_e values along the critical path, and (2) the variance of the mean project completion time is the simple sum of the variances of the individual activities on the critical path.

Therefore, we may use the normal tables to determine probabilities for the occurrence of given project completion time estimates. For example, the probability that a project would be completed in the mean time is only 0.50. The probability that a project would be completed in less time than the mean plus one standard deviation of the mean is about 0.84; the mean plus two standard deviations, 0.98 and so on.

DEPLOYMENT OF RESOURCES

Given the activity network, the critical path, and the computed schedule statistics, we have a plan for the project. But is it a good plan? We can abstract from the plan some additional data on the demand for resources for the early start schedule. By using the schedule flexibility available through slack in certain activities, we can generate alternate schedules, comparing the use of important resources with the objective of *load leveling*.

Another way to look at the initial or raw plan is in terms of activity costs. The initial activity duration estimates are based on an assumed level of resource allocation. Is it possible to alter activity times by pouring in more or less resources? Activity times for some activities can be directly affected in this way. For example, adding carpenters will usually shorten the time to frame a house. Would it be worthwhile to add more personnel on the critical framing and allocate less to the noncritical brickwork? Would the alternate plan be more or less expensive? Would shortening the critical path be advantageous? *Least-costing* considerations are worth examining.

Finally, in some situations we may be faced with a demand for some critical resource that is limited in supply. The raw plan may, in fact, not be feasible if it schedules the use of the only available power shovel in two places at the same time. The raw plan must be examined with the objective of the feasible scheduling of *limited resources*, again using available slack time where possible or even lengthening the project in order to generate a feasible plan.

Load Leveling

What are the costs of not attempting to level loads in an already feasible schedule? Some factors that enter the problem occur in the following example of a major oil refinery repair and overhaul project. After a raw plan was developed, a series of computer runs was made to examine personnel requirements for the refinery project. In the first run, it was found that the schedule required 50 boilermakers for the first 4 hours, 20 for the next 6 hours, and 35 for the immediately following period. Similar fluctuations in requirements were found for other crafts. In terms of costs associated with this fluctuation, there is the possible cost of idle labor.

For example, in the first 10 hours of the refinery project the peak requirement of 50 boilermakers will probably mean productive work of 50 × 4 hours + 20 × 6 hours = 320 worker-hours. But the likelihood is that it will be difficult to assign the extra 30 workers for the balance of the 8-hour day, so in the first 10 hours of the project the payroll may reflect 50 × 8 + 20 × 2 = 440 worker-hours, 120 of which are idle labor. Figure 10-11 shows the deployment of personnel *after* leveling. Other costs that may be implicit in personnel fluctuation are hiring and separation costs in projects that extend over long periods. Load leveling has the objective of reducing idle labor costs, hiring and separation costs, or the cost of any resource that may be affected by fluctuations in the demand for its use, such as an equipment rental.

For very large and complex projects, a computer-based leveling model may be

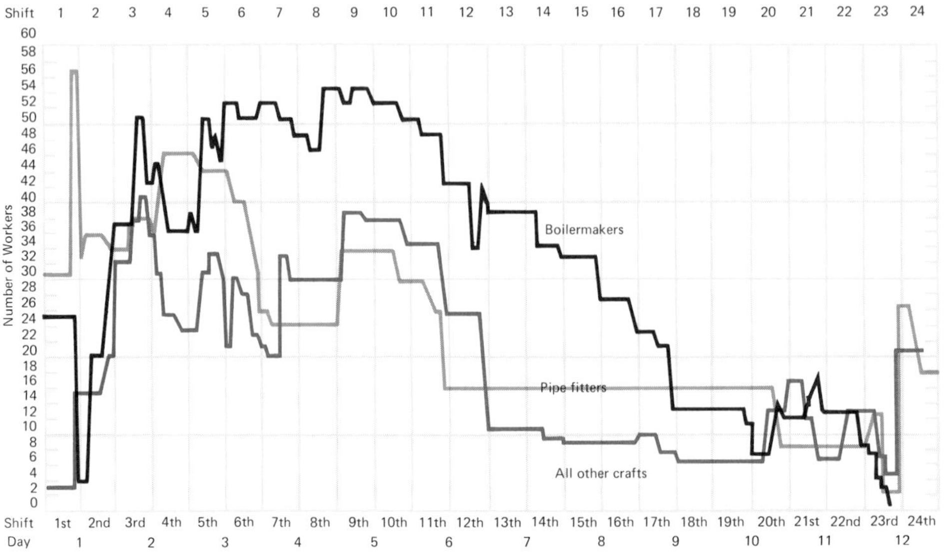

FIGURE 10-11
Personnel usage chart after leveling.
SOURCE: *R. D. Archibald and R. L. Villoria,* Network-Based Management Systems, *John Wiley, New York, 1967, p. 274.*

required. Simulation methodology commonly is used to generate alternate solutions. The starting solution might be the early start schedule, and a first attempt at leveling could then set a maximum of the resource in question just below the highest peak level recorded in the raw plan. The simulation program would then proceed as indicated by the network diagram, beginning all activities leaving node 1, keeping track of the amount of resources used and available. As the calendar is advanced and as activities are completed, resources are returned to the "available" pool. As new activities are started, resources are drawn from the pool. Simulation then proceeds until an activity requires resources from a temporarily exhausted pool. Depending on the decision criteria used by the simulator, the activity may be delayed, even past its latest starting time, until resources are available. Other decision criteria "bump" noncritical jobs and reassign resources to the delayed job when the latest starting time has been reached. By a progressive lowering of the resource limits in such a simulation program, the leveling effect takes place until a satisfactory deployment of resources is achieved.

Least Costing

Least-costing concepts are based on cost versus activity time curves, such as in Figure 10-12. Different activities respond differently to changes in the application of resources, and some of the activities may not be responsive to changes in resources. Figure 10-12*a* may be typical of an activity, such as house framing, as we discussed previously, where crash, normal, and slow schedules are progressively less costly. A curve similar to Figure 10-12*b*, where the slow schedule is more costly than the normal schedule, could be typical where the meager resources associated with a slow schedule might force the use of inefficient methods. The cost trade-offs are possible partially because of the differential cost-time characteristics of different activities.

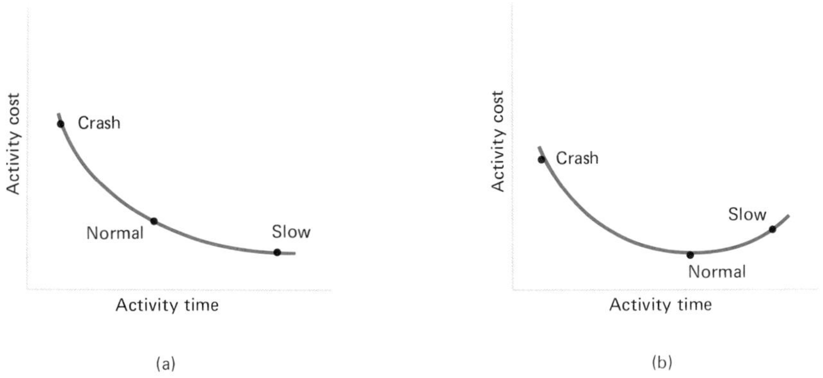

FIGURE 10-12
Typical activity time-cost curves.

Limited Resources

A limited resource model called "SPAR" (Scheduling Program for Allocation of Resources) has been developed by Wiest [1967]. SPAR is a heuristic scheduling model for limited resources designed to handle a project with 1200 single resource activities, 500 nodes, and 12 shops over a span of 300 days. The model focuses on available resources that it allocates, period-by-period, to activities listed in order of their early start times. The most critical jobs have the highest probability of being scheduled first, and as many jobs are initially scheduled as available resources permit. If an available activity fails to be scheduled in one period, an attempt is made to schedule it in the next period. Finally, all jobs that have been postponed become critical and move to the top of the priority list of available activities.

Wiest applied the SPAR program to a space vehicle project that required large block engineering activities with up to five different types of engineers and involving 300 activities. Figure 10-13 shows an overall personnel loading chart for the program. The unlimited resources line resulted from a conventional PERT schedule with all activities at their early start times. The limited resources line results from the SPAR schedule where peak personnel requirements were considerably reduced. The total length of the project was shortened by five months, and the number of gross hirings of personnel was reduced by 30 percent as a result of the SPAR schedule.

IMPLICATIONS FOR THE MANAGER

Large-scale projects present managers with unique problems. Although they are intermittent systems, the focus of managerial issues are rather different from the typical

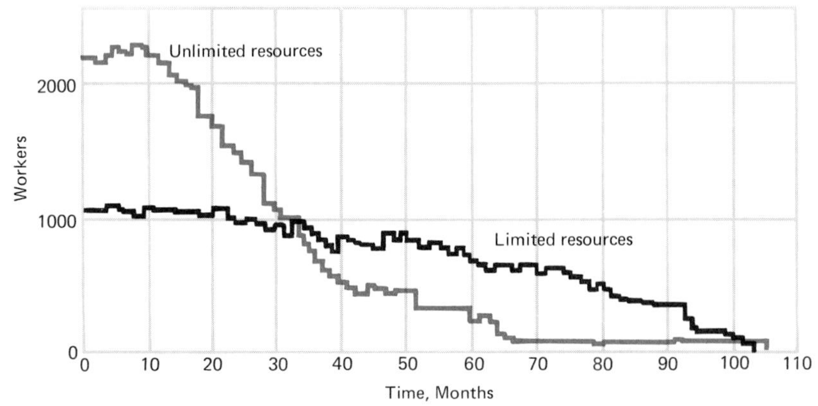

FIGURE 10-13
Personnel loading schedule for the space vehicle project.
SOURCE: *I. D. Weist, "A Heuristic Model for Scheduling Large Projects with Limited Resources,"* Management Science, 13*(6), February 1967, p. B-373.*

manufacturing intermittent system. The project manager's problems center on the detailed schedule of activities and a preoccupation with completion dates of activities and with the project completion date.

The position of project manager requires a particular "breed of cat." A project manager must be someone who can deal with changing deadlines, immediate decisions, and skimpy information all in a general environment of divided authority. The managerial situation requires special concepts and techniques, and network planning and control systems are the best available today.

Davis [1974*a*] surveyed the top 400 construction firms and found over 80 percent of the 235 respondents were using network planning methods. Thus, it seems clear that many managers have recognized the value of network methods; they are not simply theoretical concepts.

Managers need to be involved in formulating the initial network. Perhaps a majority of the benefits of network methods comes at the planning stage. Given the network and time estimates, managers can have a number of computer runs made to assess alternate plans that involve cost-time trade-offs and personnel smoothing. This interaction between the manager and the network model can have enormous benefits.

The control phases of network methods are the most costly. Smaller projects may not justify the cost of obtaining updated information and the periodic rescheduling needed for control. Davis also reported that the primary use of network methods in the construction industry was for project planning rather than for control.

REVIEW QUESTIONS AND PROBLEMS

1. What are the characteristics of large-scale projects that focus managerial effort on the detailed scheduling of activities and on project completion dates?

2. Contrast the large project system with the manufacturing intermittent systems discussed in previous chapters. Why does the manufacturing system focus managerial effort on process and system instead of on detailed activity scheduling?

3. Why is the matrix type of organization structure used in project management?

4. In the context of the "activities on arcs" planning methods, define the following terms: activity, event, node, and critical path.

5. For "arcs" planning methods, discuss and interrelate the three phases: (a) activity analysis, (b) network diagramming, and (c) node numbering.

6. What are the functions of dummy activities in an "arcs" network diagram?

7. What is the convention for numbering nodes in an "arcs" network? Why is this convention used?

8. Why must activity networks be noncyclical?

9. Define the following terms: Early start (*ES*), early finish (*EF*), latest start (*LS*), latest finish (*LF*), and slack.

10. Outline the procedure for manual computation of schedule statistics.

11. What are the differences in the construction of the network diagram between the "arcs" and "nodes" methodologies? How can the probabilistic network model provide additional data helpful for managerial decisions?

12. Table 10-2 provides data for the project of installing a gas forced air furnace.
    a. Develop an arcs network diagram for the project.
    b. Identify any dummy activities necessary. Why are they necessary?
    c. Number the nodes of the network so that no cycling would result.

13. For the data of Table 10-2, develop a nodes network diagram. Do you feel that the nodes network is simpler to interpret than the arcs network? If computer programs were available to compute schedule statistics for both models, how would you choose between them?

TABLE 10-2

Activities, Sequence, and Time Requirements for the Installation of a Gas-Forced Air Furnace

Activity Code	Activity Description	Immediate Predecessor Activity	Time, Days
A	Start	—	0
B	Obtain delivery of furnace unit	A	10
C	Delivery of piping	A	5
D	Delivery of dampers and grilles	F	14
E	Delivery of duct work	F	10
F	Design duct layout	A	2
G	Install ducts and dampers	D, E	12
H	Install grilles	G	1
I	Install furnace unit	B	1
J	Install gas piping	C	5
K	Connect gas pipes to furnace	I, J	0.5
L	Install electric wiring	B	2
M	Install controls and connect to electrical system	I, L	1
N	Test installation	H, K, M	0.5
O	Clean up	N	0.5

TABLE 10-3 **Activities, Sequence Requirements, and Times for the Renewal of a Pipeline**

Activity	Activity Code	Code of Immediate Predecessor	Activity Time Requirement (days)	Crew Requirements per Day
Assemble crew for job	A	—	10	—
Use old line to build inventory	B	—	28	—
Measure and sketch old line	C	A	2	—
Develop materials list	D	C	1	—
Erect scaffold	E	D	2	10
Procure pipe	F	D	30	—
Procure valves	G	D	45	—
Deactivate old line	H	B,D	1	6
Remove old line	I	E,H	6	3
Prefabricate new pipe	J	F	5	20
Place valves	K	E,G,H	1	6
Place new pipe	L	I,J	6	25
Weld pipe	M	L	2	1
Connect valves	N	K,M	1	6
Insulate	O	K,M	4	5
Pressure test	P	N	1	3
Remove scaffold	Q	N,O	1	6
Clean up and turn over to operating crew	R	P,Q	1	6

14. Using the manual computation algorithms, compute the following schedule statistics for the furnace installation project.
    a. *ES, EF, LS, LF* for each activity
    b. Total slack
    c. Free slack
    d. Critical path

15. Define the following terms: optimistic time, pessimistic time, most likely time, and expected time in probabilistic PERT networks.

16. What is meant by load leveling? How may it be accomplished?

17. Discuss the concepts of least costing in relation to crash, normal, and slow schedules.

18. What is the nature of the SPAR limited resource model?

19. Listed in Table 10-3 is a set of activities, sequence requirements, and estimated activity times for a pipeline renewal project. Figure 10-14 provides computer

THE CRITICAL PATH IS

START → A → C → D → G → K → O → Q → R → FINISH

THE LENGTH OF THE CRITICAL PATH IS 65

NODE	*DURATION*	*EARLY START*	*EARLY FINISH*	*LATE START*	*LATE FINISH*	*TOTAL SLACK*	*FREE SLACK*
START	0.00	0.00	0.00	0.00	0.00	0.00	0.00
A	10.00	0.00	10.00	0.00	10.00	0.00	0.00
B	28.00	0.00	28.00	16.00	44.00	16.00	0.00
C	2.00	10.00	12.00	10.00	12.00	0.00	0.00
D	1.00	12.00	13.00	12.00	13.00	0.00	0.00
E	2.00	13.00	15.00	43.00	45.00	30.00	14.00
F	30.00	13.00	43.00	16.00	46.00	3.00	0.00
G	45.00	13.00	58.00	13.00	58.00	0.00	0.00
H	1.00	28.00	29.00	44.00	45.00	16.00	0.00
I	6.00	29.00	35.00	45.00	51.00	16.00	13.00
J	5.00	43.00	48.00	46.00	51.00	3.00	0.00
K	1.00	58.00	59.00	58.00	59.00	0.00	0.00
L	6.00	48.00	54.00	51.00	57.00	3.00	0.00
M	2.00	54.00	56.00	57.00	59.00	3.00	3.00
N	1.00	59.00	60.00	62.00	63.00	3.00	0.00
O	4.00	59.00	63.00	59.00	63.00	0.00	0.00
P	1.00	60.00	61.00	63.00	64.00	3.00	3.00
Q	1.00	63.00	64.00	63.00	64.00	0.00	0.00
R	1.00	64.00	65.00	64.00	65.00	0.00	0.00
FINISH	0.00	65.00	65.00	65.00	65.00	0.00	0.00

FIGURE 10-14
Computer output showing critical path and schedule statistics for the pipeline renewal project.

output for the project. Which activities can be delayed beyond *ES* times without delaying the project completion time of 65 days? Which activities can be delayed without delaying the *ES* of any other activity?

20. For the data of the pipeline renewal project, suppose that activity H is delayed 12 days. What is the total slack remaining for activity I? What is the remaining free slack for activity I?

21. Suppose that activity H is not delayed, but I is delayed 14 days. Which activity or activities will be affected? How, and by how many days?

22. Activity K in the pipeline renewal project is delayed by 2 days. Which activities are affected? How, and by how many days?

23. In Table 10-4 there is additional information in the form of optimistic, most likely, and pessimistic time estimates for the pipeline renewal project. Compute variances for the activities. Which activities have the greatest uncertainty in their completion schedules?

SITUATIONS

24. The manager of the Pipeline Removal Company has always operated on the basis of having detailed knowledge of the required activities. The manager had learned the business from the ground up, seemingly having faced virtually all possible crisis types. The PERT/CPM network diagram made good sense, but the manager was really intrigued with the schedule statistics that could be derived from the network and associated activity times.

The manager was bidding on the project for which Table 10-3 represented the basic data. The schedule statistics shown in Figure 10-14 were developed. The estimated times were felt to be realistic and achievable, based on past experience. The bidding was competitive, and time performance was an important factor, since the pipeline operator would lose some revenue, in spite of the fact that an inventory was developed in activity B, since storage was limited. As a result of these facts, it was common to negotiate penalties for late performance in pipeline renewal contracts.

Because of the uncertainties and risks, the manager estimated values for *a*, *m*, *b*, and computed values for t_e in Table 10-4. The manager decided that an attractive completion date would be likely to win the contract. Therefore, the bid promised completion in 70 days, with penalties of $100 per day if actual completion time was greater than 65 days and $200 per day if greater than 70 days. The contract was awarded. Has the manager got a good deal? How likely are the penalties?

TABLE 10-4 **Time Estimates for the Pipeline Renewal Project**

Activity Code	Optimistic Time Estimate of, *a*	Most Likely Time Estimate of, *m*	Pessimistic Time Estimate of, *b*	Expected Time Estimate of, t_e
A	8	10	12	10
B	26	26.5	36	28
C	1	2	3	2
D	0.5	1	1.5	1
E	1.5	1.63	4	2
F	28	28	40	30
G	40	42.5	60	45
H	1	1	1	1
I	4	6	8	6
J	4	4.5	8	5
K	0.5	0.9	2	1
L	5	5.25	10	6
M	1	2	3	2
N	0.5	1	1.5	1
O	3	3.75	6	4
P	1	1	1	1
Q	1	1	1	1
R	1	1	1	1

After the contract was awarded, the manager became depressed because five days were immediately lost in activity A as the result of a strike. The manager generated the following possible actions to recover schedule time:

a. Shorten t_e of activity B by four days at a cost of $100.
b. Shorten b of activity G five days at a cost of $50.
c. Shorten t_e of activity 0 by two days at a cost of $150.
d. Shorten t_e of activity 0 by two days by drawing resources from activity N, thereby lengthening its t_e by two days.

What should the manager do?

25. The manager of the Pipeline Renewal Company is now concerned with resource utilization. Along with other data, Table 10-3 gave the normal crew requirements per day for each of the activities. When related to the *ES* schedule given by Figure 10-14, the manager developed the labor deployment chart shown in Figure 10-15.

Based on past experience, the manager had always been aware of the "lumpy" nature of the demand for worker-days on renewal projects. Figure 10-15 verified and dramatized the situation.

The nature of most of the tasks and skill levels was such that a generalized crew was used that could do all the tasks, with few exceptions. For example, activity L has a crew requirement of 25 workers for 6 days, or 150 worker-days. These 150 worker-days may be allocated over many chosen activity times, such as 10 workers for 15 days, or vice versa. The generalized crew provided great flexibility, allowing reallocation of labor between projects being carried on simultaneously. Also, adjustments for absence or vacations were relatively simple.

The manager was now thinking of alternates to Figure 10-15 as ways of allocating labor to projects. There were a number of reasons for being dissatisfied with the current mode of operation. First, the lumpy nature of the typical situation shown in Figure 10-15 often resulted in idle crew time because of the mismatch of crew sizes and activity times. The company absorbed this idle time.

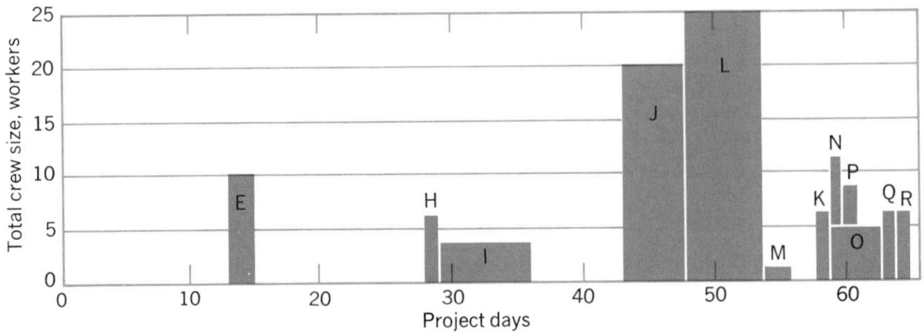

FIGURE 10-15
Worker deployment for the *ES* schedule using normal crew requirements for the pipeline renewal project.

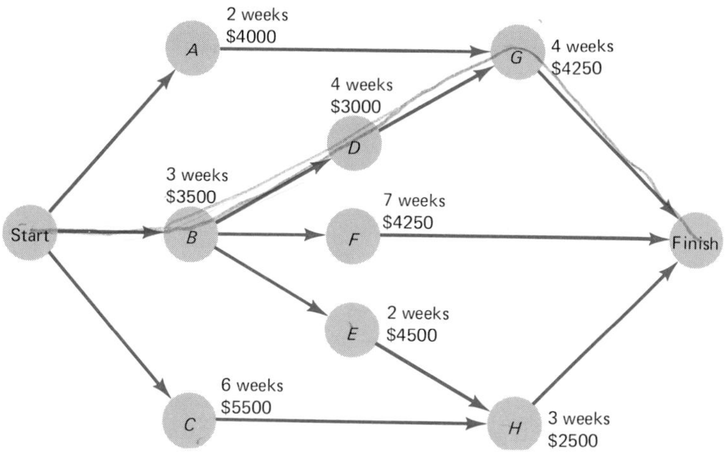

THE CRITICAL PATH IS

START → B → D → G → FINISH

THE LENGTH OF THE CRITICAL PATH IS 11

NODE	*DURATION*	*EARLY START*	*EARLY FINISH*	*LATE START*	*LATE FINISH*	*TOTAL SLACK*	*FREE SLACK*
START	0.00	0.00	0.00	0.00	0.00	0.00	0.00
A	2.00	0.00	2.00	5.00	7.00	5.00	5.00
B	3.00	0.00	3.00	0.00	3.00	0.00	0.00
C	6.00	0.00	6.00	2.00	8.00	2.00	0.00
D	4.00	3.00	7.00	3.00	7.00	0.00	0.00
E	2.00	3.00	5.00	6.00	8.00	3.00	1.00
F	7.00	3.00	10.00	4.00	11.00	1.00	1.00
G	4.00	7.00	11.00	7.00	11.00	0.00	0.00
H	3.00	6.00	9.00	8.00	11.00	2.00	2.00
FINISH	0.00	11.00	11.00	11.00	11.00	0.00	0.00

(b)

FIGURE 10-16
Construction contract project: (*a*) CPM network diagram showing time in weeks for the completion of each activity, and normal schedule activity costs, and (*b*) the computer output indicating the critical path, the overall project-time of 11 weeks, and the schedule statistics.

Second, because of similar mismatches and schedule slippage, it was often necessary to work crews overtime to meet deadlines. Third, the lumpy nature of demand often resulted in layoffs or in short-term hiring, with the usual costs of workforce size fluctuation. In addition, the layoffs were a source of discontent among the crews.

In working with alternatives, the manager became aware that alternatives resulted in changes in activity times that sometimes affected the critical path. The manager was impressed, however, that schedule slack could be used to good effect in generating alternatives to Figure 10-15.

What alternate labor deployment do you recommend? Is the critical path affected by your recommendations?

26. Figure 10-16*a* is a nodes network diagram for a construction contract that shows the time in weeks to complete each activity and the normal schedule estimated cost for each activity. Figure 10-16*b* shows the computer output and indicates an overall project time of 11 weeks, the critical path, and the schedule statistics. The total contractor's cost is $31,500, but the contract price is $45,000.

The contractor's problem is that the costs are based on a normal completion of 11 weeks, but the customer insists on a 10-week time and a penalty for late performance of $2000 per week. The price is attractive, since it provides the contractor with a $13,500 profit. Therefore, the contractor is interested in alternatives that might achieve the 10-week delivery schedule.

The contractor develops cost-time data for several of the activities; the following list indicates the reduction in weeks and the incremental cost:

A One week, $1500

B One week, $6000

D One week, $2000

G One week, $1500; second week, $2000

The contractor is aware that changes in the activity times are sometimes "tricky," changing the critical path. What action should the contractor take?

REFERENCES

Archibald, R. D., and R. L. Villoria. *Network-Based Management Systems.* John Wiley, New York, 1967.

Buffa, E. S., and J. G. Miller. *Production-Inventory Systems: Planning and Control* (2nd ed.). Richard D. Irwin, Homewood, Ill., 1979.

Burgess, A. R., and J. B. Killebrew. "Variation in Activity Level on a Cyclical Arrow Diagram," *Journal of Industrial Engineering, 13*(2), March–April 1962, pp. 76–83.

Davis, E. W. "CPM Use in Top 400 Construction Firms," *Journal of the Construction Division,* ASCE, *100*(01), Proc. Paper 10295, March 1974, pp. 39–49.(a)

Davis, E. W. "Networks: Resource Allocation," *Industrial Engineering,* April 1974, pp. 22–32.(b)

Davis, E. W. "Project Scheduling Under Resource Constraints—Historical Review and Categorization of Procedures," *AIIE Transactions, 5*(4), December 1973, pp. 297–313.

Dewitte, L. "Manpower Leveling in PERT Networks," *Data Processing Science/Engineering,* March–April 1964.

Fulkerson, D. R. "A Network Flow Computation for Project Cost Curves," *Management Science, 7*, 1961.

Kelly, J. E. "Critical Path Planning and Scheduling: Mathematical Basis," *Operations Research, 9*(3), 1961.

Levin, R. I., and C. A. Kirkpatrick. *Planning and Control with PERT/CPM.* McGraw-Hill, New York, 1966.

Levy, F. K., G. L. Thompson, and J. D. Wiest. "Multi-Shop Work Load Smoothing Program," *Naval Research Logistics Quarterly,* March 1963.(a)

Levy, F. K., G. L. Thompson, and J. D. Wiest. "The ABCs of the Critical Path Method," *Harvard Business Review,* September–October 1963, pp. 98–108.(b)

MacCrimmon, K. R., and C. A. Ryavec. "An Analytical Study of the PERT Assumptions," *Operations Research,* January–February 1964.

Malcolm, D. G., J. H. Roseboom, C. E. Clark, and W. Fazar. "Application of a Technique for Research and Development Program Evaluation," *Operations Research,* 7(5), September–October 1959.

Meyers, H. B. "The Great Nuclear Fizzle at Old B. & W.," *Fortune,* November 1969, p. 123.

Moder, J. J., and C. R. Phillips. *Project Management with CPM and PERT* (2nd ed.). Reinhold, New York, 1970.

Shaffer, L. R., J. B. Ritter, and W. L. Meyer. *The Critical Path Method.* McGraw-Hill, New York, 1965.

Smith, L. A., and P. Mahler. "Comparing Commercially Available CPM/PERT Computer Programs," *Industrial Engineering, 10*(4), April 1978, pp. 37–39.

Thamhain, H., and D. Wileman. "Conflict Management in Project Life Cycles," *Sloan Management Review, 16*(3), 1975, pp. 31–50.

Wiest, J. D. "A Heuristic Model for Scheduling Large Projects with Limited Resources," *Management Science, 13*(6), February 1967, pp. 359–377.

Wiest, J. D. "Some Properties of Schedules for Large Projects with Limited Resources," *Operations Research, 12*(3), May–June 1964.

Wiest, J. D., and F. K. Levy. *A Management Guide to PERT/CPM* (2nd ed.). Prentice-Hall, Englewood Cliffs, N.J., 1977.

Woodworth, B. M., and C. J. Willie. "A Heuristic Algorithm for Resource Leveling in Multi-Project, Multi-Resource Scheduling," *Decision Sciences, 6*(3), 1975, pp. 525–540.

CHAPTER 11 Service Systems and Scheduling Personnel

SERVICE AND NONMANUFACTURING OPERATIONS HAVE AT LAST been recognized as interesting and challenging arenas in which to work. The great diversity of service and nonmanufacturing operations immediately raises the question of whether or not they have enough in common that we can generalize. Do they have enough in common with manufacturing systems so that some of what we have learned in the manufacturing arena can be transferred? Although service and nonmanufacturing operations have some special characteristics, they are sensitive to similar costs and pressures found in manufacturing. The problems of aggregate, personnel, and detailed planning are as important as they are in manufacturing.

SPECIAL CHARACTERISTICS OF SERVICE SYSTEMS

The special characteristics of service and nonmanufacturing operations that need to be taken into account in problems of planning and scheduling are as follows:

1. The output cannot be inventoried.
2. Extremely variable demand on a short-term basis, but often weekly and seasonal variations as well. Backlogging and/or smoothing of demand is often difficult or impossible.
3. Operations are usually labor intensive.
4. The location of service operations is dictated by the location of users.

Problems Resulting from a Noninventoriable Output

In manufacturing systems, the decoupling function of inventories allows us to carry on each set of activities relatively independently. We use buffer inventories to decouple demand variations from disrupting operations and to provide good delivery service. We use seasonal inventories to buffer the effects of seasonal variations in demand on employment levels and to minimize peak physical capacity requirements.

Unlike physical products, services are consumed in the process of their production. Therefore, in service systems, one of the most important strategies for manipulating short-term capacity in aggregate planning and scheduling is absent. The degrees of freedom that remain in aggregate planning are: hiring and layoff of personnel, use of part-time employees, overtime, and in some instances actually scheduling demand to conform to resources. With no inventory buffering, it is likely that the full impact of demand variations is transmitted to the system.

Problems Resulting from Demand Variability

In the telephone industry, it is not uncommon for calls during the busiest hour of each week during a year to vary by a factor of 1.38 to 1 (peak to valley); the Saturday and Sunday call load to constitute only about 55 percent of that of the typical week-day; and the peak-to-valley variation during a typical day to vary by a factor of 128 to 1. The demand for fire protection in New York City on an average day reveals that the peak is 7.7 times the minimum fire alarm rate. The demand for emergency medical service in Los Angeles has been shown to vary from a low of 0.5 calls per hour at 6 A.M. to a peak of 3.5 calls per hour at 5:30 P.M., a ratio of 7 to 1. As a result of day-end mailing practices by business, 40 to 60 percent of letters brought to the post office plus that collected from local mail boxes is received between 4:00 and 8:00 P.M.

The previous examples represent some of the most extreme cases. However, unlike most manufacturing systems, inventories cannot buffer between the demand for service and the service-producing system. Therefore, one of the really significant problems for management is to devise buffering systems. In many medical service systems, this may be accomplished through scheduling appointments and by priority systems. In hospitals, scheduled admissions can be manipulated to smooth occupancy level. In most instances, however, the dominant available strategies used to cope with demand variability are queuing, when possible, and personnel scheduling.

Problems Resulting from Labor-Intensiveness

Traditionally, manufacturing has enjoyed a continuous increase in productivity, stemming from the substitution of other forms of energy for human energy and from mechanization and automation. By contrast, education, medical service, the post office, and most other service-oriented activities have recorded very little productivity increases. For example, during the period of 1956 and 1967, productivity in U.S. industry increased 34 percent, while productivity in the U.S. Postal Service increased only 4 percent. Wage increases in service operations translate almost directly into increased costs and prices.

Although the overall demand for services has increased tremendously, systems could expand only by hiring more labor. The result has often been massive problems of scheduling an army of individual workers. There have been few economies of scale, nor is there a likelihood that there will be future economies of scale. Mechanization and automation are unlikely mechanisms for reducing the strains imposed by greater and greater demands for services. The logical alternatives are more efficient use of personnel through scheduling and other management techniques.

Problems Resulting from System Location

The very nature of service and nonmanufacturing operations is such that they must be decentralized and available to users. This means that each unit is relatively small in scale, and so suffers all the economic consequences of small-scale operations. One of these consequences has been that the aggregate and detailed scheduling techniques used have often been those appropriate for the handicraft industries.

EXAMPLE OF A SERVICE SYSTEM

We will use as an example a report of an outpatient clinic at the University of Massachusetts by Rising, Baron, and Averill [1973]. During the period of the study (1970–1971) the Outpatient Clinic treated an average 400 to 500 patients per day with a staff of 12 full-time physicians. Because of a variety of other duties, only 260 physician-hours per week were available during regular clinic hours, or nearly 22 hours per physician per week. Only about half of the patients were seen by a physician on an appointment or walk-in basis. The others were treated by nurses under a physician's supervision or in specialized subclinics, involving tests or immunizations.

In aggregate terms, for the fall 1969 period, approximately 178 patients per day needed access to an average of 52 available physician-hours. Thus the average time with a physician was about 17.5 minutes.

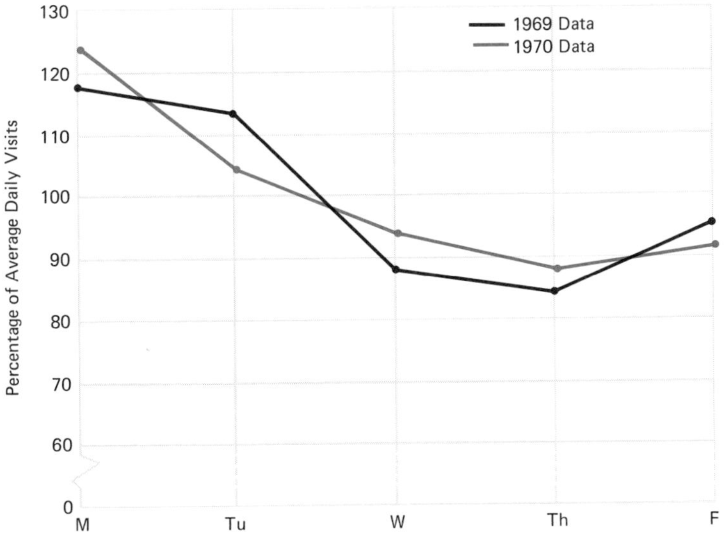

FIGURE 11-1
Percentage of patients arriving at a university health service to see either a physician or a nurse.
SOURCE: *E. J. Rising, R. Baron, and B. Averill, "A Systems Analysis of a University-Health-Service Outpatient Clinic,"* Operations Research, *21(5), September-October 1973, p. 1030-1047.*

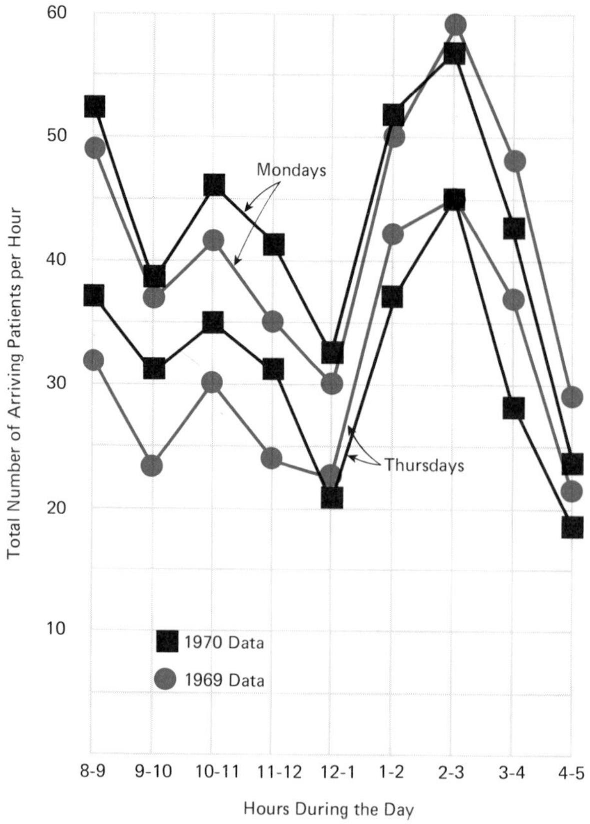

FIGURE 11-2
Hourly arrivals at the Student Health Service (Monday and Thursday averages for the fall semesters in 1969 and 1970).
SOURCE: *E. J. Rising, R. Baron, and B. Averill, "A Systems Analysis of a University-Health-Service Outpatient Clinic,"* Operations Research, *21(5), September-October 1973, pp. 1030-1047.*

Demand for Service

Part of the difficulty in rendering service is seen in Figure 11-1. Demand is not uniform through the week, being approximately 20 percent above the average on Mondays, 84 to 88 percent of the average on Thursdays, and increasing slightly on Friday.

Furthermore, the daily variation is significant. Figure 11-2 shows arrival data for Monday and Thursday (the days with the heaviest and lightest loads, respectively) highlighting great demand variation during the day, with peaks at 8 A.M., 10 A.M., and 2 P.M. When the arrival data are placed on an interarrival time basis (time between arrivals), they exhibit a negative exponential distribution, as shown in Figure 11-3.

Time for Service

The amount of time physicians spent with patients was measured in three separate categories: walk-in, appointment, and second-service times. Figure 11-4 shows histograms of service times recorded for the three categories. The second-service category represents a return of the patient to the physician following diagnostic tests or other intervening procedure. Although the three distributions are different, they share the common general property of being skewed to the right and having relatively large variances. Thus the average appointment service time is only 12.74 minutes, but the variance is nearly 10 minutes, and the maximum recorded time is 40 minutes. These typical service time distributions reflect the variety of tasks involved in a consultation, depending on the nature of the complaint.

SYSTEM DESIGN PROBLEMS

With variable arrival patterns on both a day-of-the-week and an hour-of-the-day basis, and with highly variable service times depending on the type of patient, important problems result. First, what can be done to schedule appointments in order

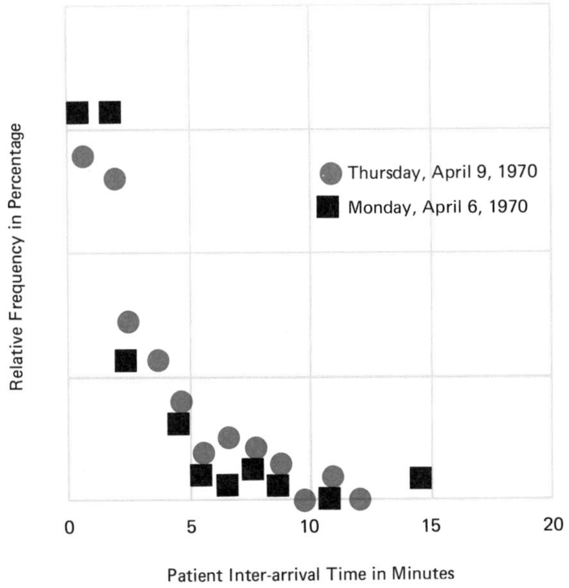

FIGURE 11-3
The frequency distribution of patient interarrival times. Monday, April 6, 1970: $\overline{x} = 2.167$, $s = 2.402$, $n = 237$. Thursday, April 9, 1970: $\overline{x} = 2.626$, $s = 2.838$, $n = 202$.
SOURCE: *E. J. Rising, R. Baron, and B. Averill, "A Systems Analysis of a University-Health-Service Outpatient Clinic,"* Operations Research, *21(5), September-October 1973, pp. 1030–1047.*

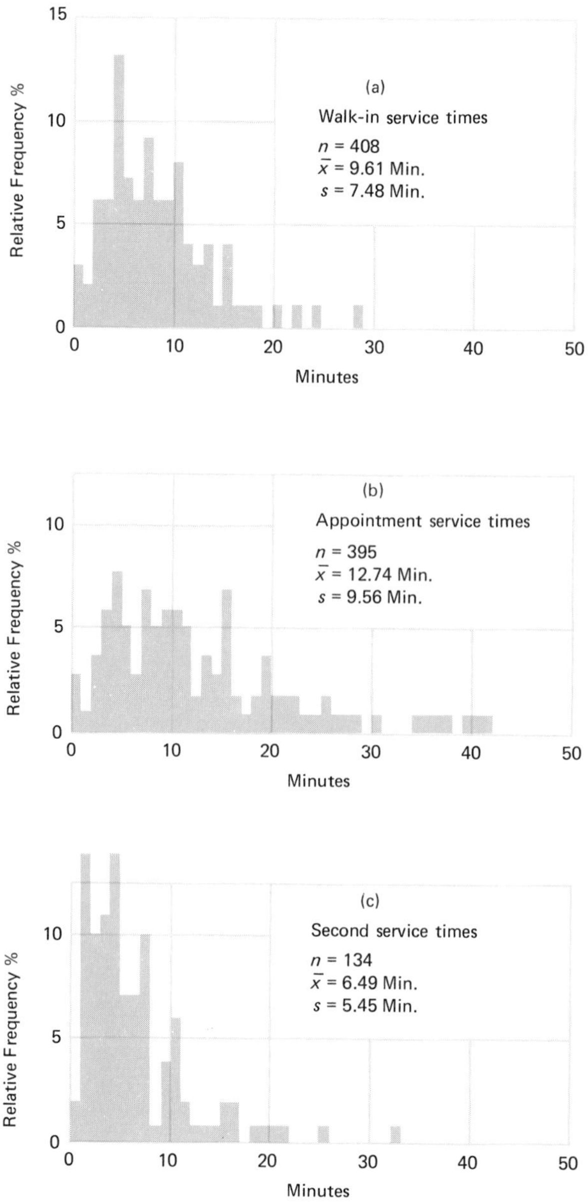

FIGURE 11-4
Histograms of service time for *(a)* walk-in, *(b)* appointment, and *(c)* second-service patients.
SOURCE: *E. J. Rising, R. Baron, and B. Averill, "A Systems Analysis of a University-Health-Service Outpatient Clinic,"* Operations Research, *21(5), September-October 1973, pp. 1030–1047.*

to smooth the patient load on physicians? How should appointment schedules be arranged through the week and the day in the light of demand variation? What overall capacity for service is really needed? How long can patients reasonably be expected to wait for service? Is physician idle time justified? Would a system of priorities help level loading?

SCHEDULES FOR SERVICE SYSTEMS

Service-oriented organizations face unique scheduling problems. In all of these kinds of systems, demands for service and the time to perform the service may be highly variable. It often appears that no sensible schedule can be constructed if arrivals for service and service time are random. The result would be to maintain the capability for service at capacity levels sufficient to keep the waiting line to certain acceptable average levels—the service facility being idle for some fraction of time in order to provide service when needed. In a sense, scheduling of the personnel and physical facilities is simple in such situations, being controlled by policies for the hours during which the service is to be available and for the service level to be offered. Schedules are then simple statements of "capacity for service," and personnel and other resources are keyed to these levels. The design for the size of maintenance crews have often been on this basis, for example.

Usually, however, we can improve on the system response of simply "keeping hours." Sometimes overall performance can be improved by a priority system, taking arrivals on other than a first-come first-served basis. Also, improvements often result from examining the demand to see if there is a weekly and/or daily pattern. When a pattern exists, it may be possible to schedule more effectively to improve service facility utilization, shorten average waiting time, or both. Thus, we have three broad groups of situations: the one described by random arrivals at a service center that performs a service requiring variable time, one where priority systems are the basis for improved scheduling, and one in which arrivals follow some dominant pattern.

The random arrival, variable service time case is the classic waiting line or queuing problem. When the distribution of arrivals and service times follows certain known mathematical functions, fairly simple equations describe flow through the system. These computations can be performed for simple situations such as a single chair barbershop, a multiple channel system as in supermarket check-out stands, or a serial set of operations such as an assembly line. Waiting line models are discussed in Appendix D, and Monte Carlo simulation in Appendix E.

One of the variables that may be controllable in waiting line systems is the order of processing of arrivals. For example, in medical facilities, patients will tolerate a priority system that allows emergency cases to be taken first. In machine shops, priority systems are often used to determine the sequence of processing jobs through a service center, as discussed in Chapter 9. Recall that computer simulation of alternate priority rules have shown that certain rules are more effective in getting work through the system on schedule.

When arrivals follow a dominant pattern, we can use that information to schedule personnel and facilities. For example, if arrivals of patients at a clinic followed the weekly pattern shown in Figure 11-1, we could use an appointment system to counterbalance the pattern and smooth the load over the week. Similarly, if the typical daily pattern for physicians' services in a clinic followed Figure 11-2, it could be counterbalanced both by an appointment system and by having a larger number of physicians on duty during the afternoon hours. If physicians could work different periods or shifts, we may be able to find quite good capacity-service solutions. Thus the entire subject of work shift scheduling becomes an important issue.

SCHEDULING PERSONNEL AND WORK SHIFTS

The objective in scheduling personnel and work shifts is to minimize labor costs, given service standards, or to establish some happy compromise between labor cost and service performance. Although our emphasis will be on personnel and shift scheduling itself, it is important to recognize that shift scheduling is a part of a larger process. The demand for the service must be forecast and converted to equivalent labor requirements by the hour of the day, day of the week, and so forth. Scheduling is then done in relation to these requirements, and finally individual workers must be assigned to the work days and shifts.

Weekly Schedules for Seven-Day Operations

Many service and some manufacturing operations must operate on a seven-day per week basis, employing labor that normally works approximately 40 hours. Legal requirements and company-union work rules result in constraints concerning the permissible work periods. One of the simplest constraints is that each worker must be provided with two consecutive days off each week. There are often additional constraints concerning the number of weekend days off, lunch periods, rest breaks, and so on.

Figure 11-5*a* shows a typical situation in which the number of workers required involves a peak requirement of three workers on Wednesday, Thursday, and Friday, with only two workers being required the other four days. The weekly labor requirement is the sum of the daily requirements, or 17 worker-days. If each worker is guaranteed five days of work per week, then the minimum number of workers to staff the operation is four, or 20 worker-days, resulting in three worker-days of slack capacity.

Figure 11-5*b* shows a configuration of the schedules for four workers that meets requirements, with each worker having two consecutive days off. Figure 11-5*c* shows the comparison of workers scheduled versus requirements, indicating the location of the slack associated with the peak days of Wednesday, Thursday, and Friday. Obviously, if part-time workers were available and fit in with other possible work rule constraints, a better solution would be to employ three full-time workers plus one

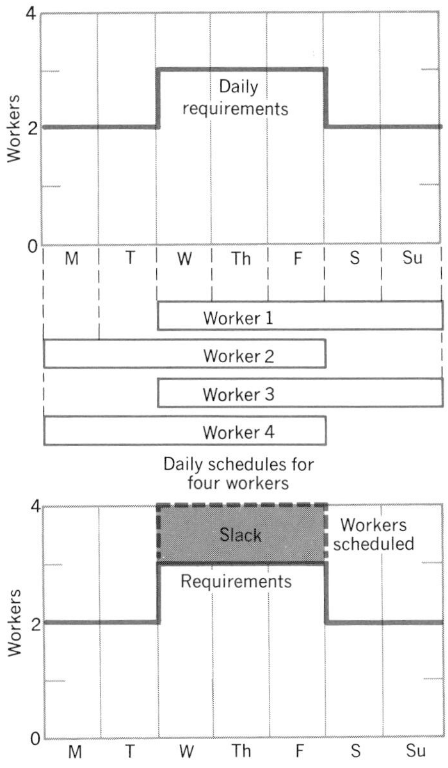

FIGURE 11-5
(a) Requirements for a seven day per week operation, *(b)* daily schedule for four workers to meet requirements, and *(c)* comparison of requirements and workers scheduled with slack shown.

part-time worker for two days per week. Such a solution would minimize labor costs but provide no slack in the system during the peak load days.

Figure 11-5 provides a framework for the general situation that we wish to analyze. Solutions to that simple situation are fairly obvious, but we need a formal model to analyze such situations efficiently when the problems are larger and more complex.

Cyclic Personnel Schedules

Suppose that we must staff an operation 7 days per week with personnel requirements each day by the following schedule:

	M	T	W	Th	F	S	Su
Required (R) workers	2	2	2	2	2	2	3

If each worker is assigned a five-day week with two consecutive days off, we have a seemingly difficult problem. However, if we sum the requirements, they total 15 worker-days. Theoretically, three workers can be assigned to meet requirements, and in this case the objective can be met by the schedule:

	M	T	W	Th	F	S	Su
Worker 1	O	O	W	W	W	W	W
Worker 2	W	W	O	O	W	W	W
Worker 3	W	W	W	W	O	O	W

Days off are indicated by O's, and W's are days worked. This schedule of three workers meets requirements and is optimal. The schedule is unique, there being no other way of scheduling the three workers to meet requirements, other than a simple reversal of the days off for workers 1 and 3.

The optimum schedule was generated by the use of simple rules that can be applied to larger problems with somewhat more variable requirements schedules. The computerized algorithm was developed by Tibrewala, Philippe, and Browne [1972, 1975]. The algorithm starts with the requirements schedule and assigns days off for worker 1, subtracts the requirements satisfied by worker 1 from the original schedule, and repeats the process until all workers have been assigned days off. Days worked for each worker are simply the five days that remain after days off are assigned.

We shall use another example to explain the process. The requirements schedule at the time we assign days off to worker 1 is:

	M	T	W	Th	F	S	Su
R_1	3	3	4	3	3	1	2

Note that the total requirements for the week are 19 worker-days. The best possible solution will then require only four workers.

Rule 1

Examine the schedule in descending order of requirements (high requirements to low) to see if there is a unique pair of consecutive days off that could be assigned at the low end of the descending schedule, not involving the higher level requirements.

The unique pair for our example is S, Su. That pair is circled as the assigned days off for worker 1. The unique pair could have been in the middle of the week and could have had requirements greater than 0; for example, if the requirements for the Monday through Sunday schedule were 5-5-2-1-4-4-5, the unique days off would be W, Th.

The requirements for the work days are reduced by 1 to produce a requirements schedule used to assign days off for worker 2. Since none of the S, Su requirements have been satisfied, these requirements are carried forward to R_2. For our example, the original and reduced schedules are:

	M	T	W	Th	F	S	Su
R_1	3	3	4	3	3	1	2
R_2	2	2	3	2	2	1	2

In R_2, the maximum requirement is three workers, followed by tied requirements of two on Su-M-T, and Th-F, leaving no unique pair. Therefore we must have a second rule to resolve the assignment.

Rule 2

If there is no unique pair and ties exist, choose the days-off pair with the lowest requirements on a day adjacent to one of the tied requirement days, and following the tied pair.

F, S meets this requirement. The assignment of F, S as the days off for worker 2 is made by circling these days in R_2, and one day is subtracted from worker 2's work days to produce R_3:

	M	T	W	Th	F	S	Su
R_1	3	3	4	3	3	1	2
R_2	2	2	3	2	2	1	2
R_3	1	1	2	1	2	1	1

In R_3, the maximum requirement is two workers on W and F, followed by ties of one worker required on all other days. Neither rule 1 nor rule 2 resolves the choice of days off for R_3, so a third rule is needed.

Rule 3

If there are ties not resolved by rule 2, choose the first available tied pair following higher-level requirements as a days-off assignment.

The first tied pair in R_3 following in a cycle after a higher-level requirement is S, Su which is assigned to worker 3. One day is subtracted for each work day in R_3 to produce R_4. Rule 3 applied to R_4 results in the unique consecutive pair M, T. Reduction of R_4 results in 0 requirements for all days, completing the schedule.

	M	T	W	Th	F	S	Su	Rule
R_1	3	3	4	3	3	1	2	1
R_2	2	2	3	2	2	1	2	2
R_3	1	1	2	1	2	1	1	3
R_4	0	0	1	0	1	1	1	3

The work days and days off are summarized as follows:

Worker	M	T	W	Th	F	S	Su
1	W	W	W	W	W	O	O
2	W	W	W	W	O	O	W
3	W	W	W	W	W	O	O
4	O	O	W	W	W	W	W
Workers, W_i	3	3	4	4	3	1	2
Slack, $S = W_i - R_i$	0	0	0	1	0	0	0

The solution is optimal, since requirements have been met with four workers. The total slack is one day on Thursday. The slack can be used to generate alternate solutions. For example, worker 2 could take Th, F off instead of F, S. The slack would then shift to Saturday, with both workers 2 and 4 available compared to a requirement of only one. There are also other alternate optimum solutions for this simple example.

Weekend Peak Example. Now suppose that the personnel requirements emphasize a weekend load. The following schedule has a total weekly requirement of 20 worker-days and can theoretically be satisfied by four workers, each working five days, with two consecutive days off:

	M	T	W	Th	F	S	Su
R_1	3	2	2	2	3	4	3

Applying the three choice rules, we obtain the following schedule:

	M	T	W	Th	F	S	Su
R_1	3	(2)	(2)	2	3	4	4
R_2	2	2	(2)	(1)	2	3	3
R_3	(1)	(1)	2	1	1	2	2
R_4	1	1	1	(0)	(0)	1	1
W_i	3	2	2	2	3	4	4
S_i	0	0	0	0	0	0	0

The solution is optimal with no slack. There are variations on the schedule because of the overlapping days off, but these are really variations of the same basic schedule, with pairs of workers exchanging days off. The following days-off patterns are also optimal.

	Days Off Schedule			
Worker	1	2	3	4
1	TW	TW	MT	WTh
2	MT	MT	TW	MT
3	WTh	ThF	WTh	TW
4	ThF	WTh	ThF	ThF

Rotating Schedules. The schedules illustrated in our examples are assumed fixed. Each employee works a specific cyclic pattern with specified days off. Because of the desire for free weekends, there is the possibility of workers rotating through the several schedules. This creates a different problem, since the number of work days between individual schedules will vary. For example, the schedule that we used to explain the three choice rules resulted in:

	Work Days	Days Off
Worker 1	M-F	S, Su
Worker 2	Su-Th	F, S
Worker 3	M-F	S, Su
Worker 4	W-Su	M, T

If one rotates through these schedules, the number of work days between days off is variable. Shifting from the first schedule to the second, there are only four work days between days off; from the second to the third, five work days; from the third to the fourth, zero work days, there being two consecutive sets of days off. If the sequence of the schedules is changed, patterns of work days between days off will also change, but the new pattern will probably also have problems. These variations in numbers of work days between rotating schedules are often unacceptable, even though they average out over a period of time.

Additional Work Rule Constraints. Baker and Magazine [1977] provide algorithms for days-off constraints in addition to the two consecutive days-off situation that we have discussed. These more constraining situations include the following:

1. Employees are entitled to every other weekend off and to four days off every two weeks.

2. Employees are entitled to every other weekend off and to two pairs of consecutive days off every two weeks.

When part-time workers can be used, the problem of scheduling personnel is eased somewhat. The scheduling of the part-time workers themselves then becomes an interesting problem.

Using Part-Time Workers

When demand for service varies significantly but follows a fairly stable weekly pattern, the use of part-time employees can give managers added flexibility. Mabert and Raedels [1976] reported such an application involving eight branch offices of the Purdue National Bank. Typical demand for service in the branches is shown in Figure 11-6; Mondays and Fridays usually exhibit peak demand. The problem also involved the anticipation of increased demand for service on paydays.

Traditionally, the bank employed only full-time tellers to meet requirements. This

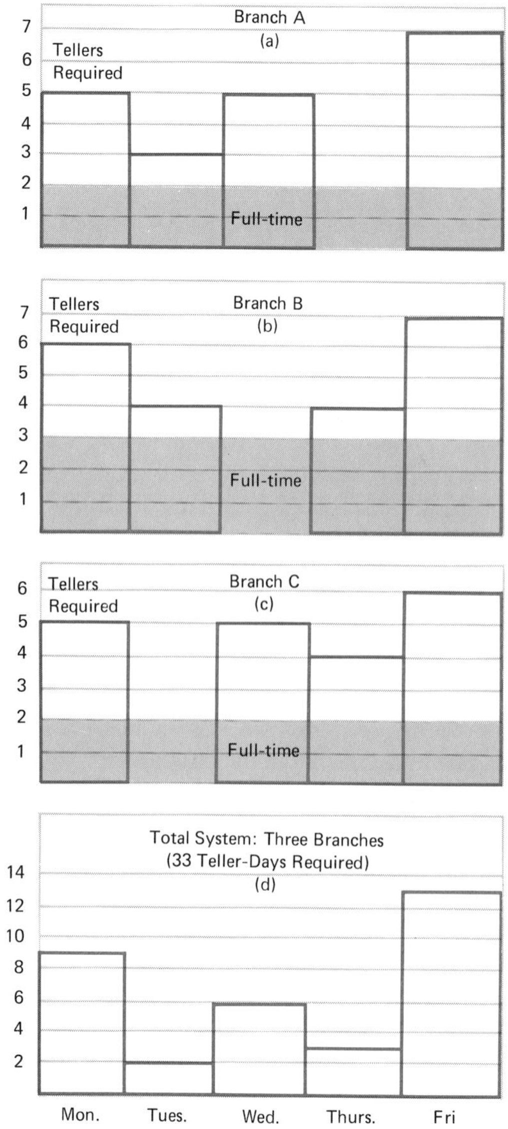

FIGURE 11-6
Teller requirements for three branches and aggregate system part-time teller requirements.
SOURCE: *V. A. Mabert and A. R. Raedels, "The Detail Scheduling of a Part-Time Work Force: A Case Study of Teller Staffing,"* Decision Sciences, 8, *January 1977, pp. 109–120.*

policy required staffing to meet peak demand and resulted in relatively poor average staff utilization during the week. Bank management later changed the staffing policy by employing full-time tellers equal to minimum expected demand and by using part-time tellers to staff the peak needs.

Problem Formulation. The basic problem involved determining the minimum cost teller assignments for the eight-branch system and, secondarily, minimizing (1) the number of part-time tellers, and (2) the teller transfers between branches. Although Mabert and Raedels used an integer programming formulation of the problem, the size of the problem limits the use of this technique. The scale of such a mathematical programming problem increases rapidly with the number of branches because of the increasing number of possible alternate work assignments. Therefore, other techniques were needed for a practical solution.

Heuristic Assignment Procedure. The Heuristic Assignment Procedure (HAP) starts with a requirements schedule (such as in Figure 11-6*d*). Since the Friday schedule requires 13 tellers, this is the feasible minimum number of tellers. A decreasing demand histogram, shown in Figure 11-7, is then developed from Figure 11-6*d*. For example:

Friday—13 tellers required

Monday—9 tellers required

Wednesday—6 tellers required

Thursday—3 tellers required

Tuesday—2 tellers required

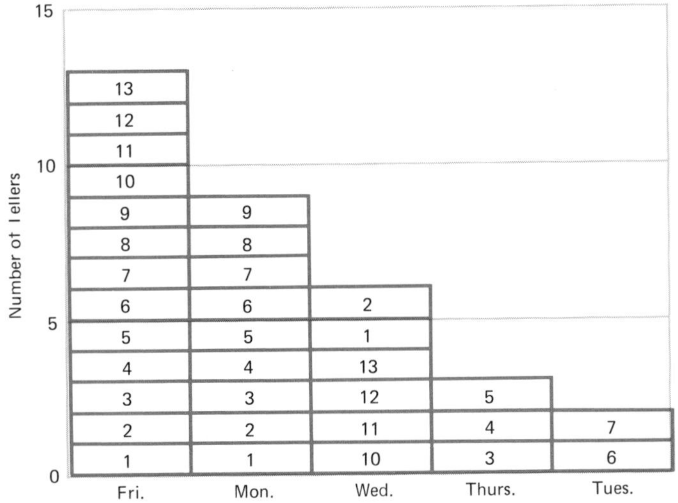

FIGURE 11-7
Assignment of part-time tellers to decreasing-demand histogram. Beginning with the highest-demand day, tellers are assigned in sequence from 1 to 13 in cycles until all 33 teller blocks have been assigned.
SOURCE: *B. A. Mabert and A. R. Raedels, "The Detail Scheduling of a Part-Time Work Force: A Case Study of Teller Staffing,"* Decision Sciences, 8, *January 1977, pp. 109–120.*

The 13 tellers are then assigned in sequence, beginning with Friday and progressing through the daily requirements, according to the preceding sequence of days. The teller numbers simply are repeated until all 33 requirement blocks have been assigned a teller number. A feasible set of two- and three-day teller schedules can be abstracted from Figure 11-7, as follows:

Teller	Daily Work Schedule
1, 2	M, W, F
3, 4, 5	M, Th, F
6, 7	M, Tu, F
8, 9	M, F
10, 11, 12, 13	W, F

The aggregate schedule is disaggregated by assigning schedules to branches by means of a heuristic procedure. First assignments are made within branches until branch requirements are met, as far as possible, without transfers. Then, unassigned tellers are assigned between branches to cover the remaining requirements. For the sample three-branch problem, Figure 11-8 shows that tellers 12 and 13 need to be transferred between branches. Teller 12 transfers between branches B and C, and teller 13 between A and C.

According to Mabert and Raedels, the Purdue National Bank has been using the HAP technique on a manual basis so successfully that the scheduling of part-time tellers saves the eight-branch system about $30,000 per year.

WORK SHIFT SCHEDULING

Thus far, we have considered methods of scheduling requirements that vary on a weekly basis. There are many situations where operations are required on a 24-hour basis and where the variations in labor requirements are severe during the work period. Much of the development of methods for dealing with the problem have occurred in the telephone industry, but the applications occur in a wide variety of other service and manufacturing operations also. The following description of a system described the problem and a practical way of dealing with it.

GENERAL TELEPHONE—AN INTEGRATED WORK SHIFT SCHEDULING SYSTEM

The general concepts of work shift scheduling have been applied at the General Telephone Company of California in an integrated, computerized system. The company has used the system since 1973 to schedule approximately 2600 telephone operators in 43 locations in California. The size of installations ranges approximately from 20 to 220 operators.

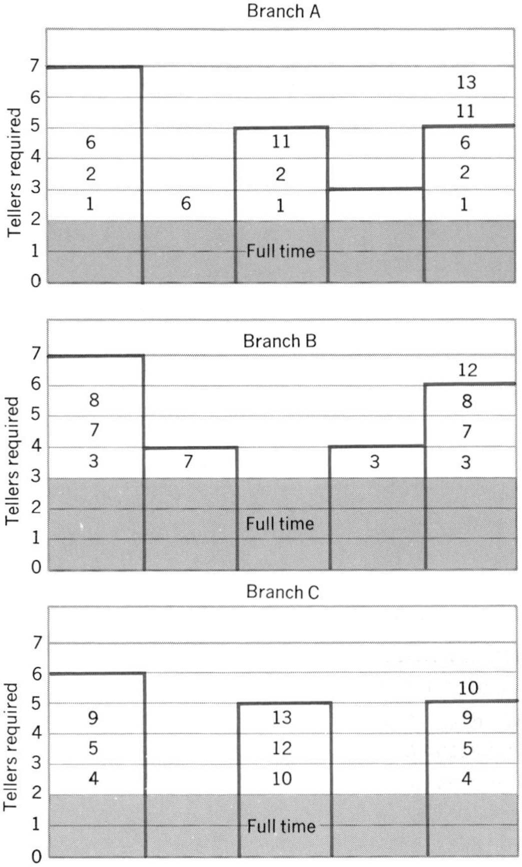

FIGURE 11-8
Assignments of part-time tellers to branches using the HAP procedure.
SOURCE: *V. A. Mabert and A. R. Raedels, "The Detail Scheduling of a Part-Time Work Force: A Case Study of Teller Staffing,"* Decision Sciences, 8, *January 1977, pp. 109–120.*

The system combines a computerized forecasting system and conversion to operator requirements, the scheduling of tours or shifts, and the assignment of operators to shifts.*

Demand for Service

The service offered is the telephone exchange: operators are assigned to provide directory assistance, coin telephone customer dialing, and toll call assistance. The

* The materials in this section are from E. S. Buffa, M. J. Cosgrove, and B. J. Luce, "An Integrated Work Shift Scheduling System," *Decision Sciences, 7* October 1976.

standard for service is supplied by the Public Utilities Commission in unusually specific terms: service must be provided at a resource level such that an incoming call can be answered within 10 seconds 89 percent of the time. The difficulty in implementing the response standard lies in the severe demand variability of incoming calls.

Figures 11-9, 11-10, 11-11, and 11-12 show typical call variations during the year, the week, the day, and within a peak hour. Figure 11-9 shows the annual variation, highlighting the two sharp peaks that were made during the busiest hour in each of the 52 weeks. The minimum occurred in the twenty-eighth week (3200) calls, and the maximum occurred during Christmas (4400 calls). The peak-to-valley ratio is 1.38 to 1. Translating the seasonal scheduling problem, the company must provide about 38 percent more capacity at Christmas time than in the twenty-eighth week, and in general the summer months involve a somewhat lighter load.

Figure 11-10 shows the daily call load for January 1972 at one location. The weekly pattern is very pronounced, and the Saturday and Sunday call load constitutes only about 55 percent of the typical load through the week. Although the telephone company offers somewhat lower weekend toll call rates to help smooth the load, the resultant weekly variation still is very large.

Figure 11-11 shows the half-hourly variation for a typical 24-hour period. Peak call volume is in the 10:30–11:00 A.M. period (2560 calls), and the minimum occurs at 4:30 A.M. (about 20 calls). The peak-to-valley variation for the typical half-hourly load

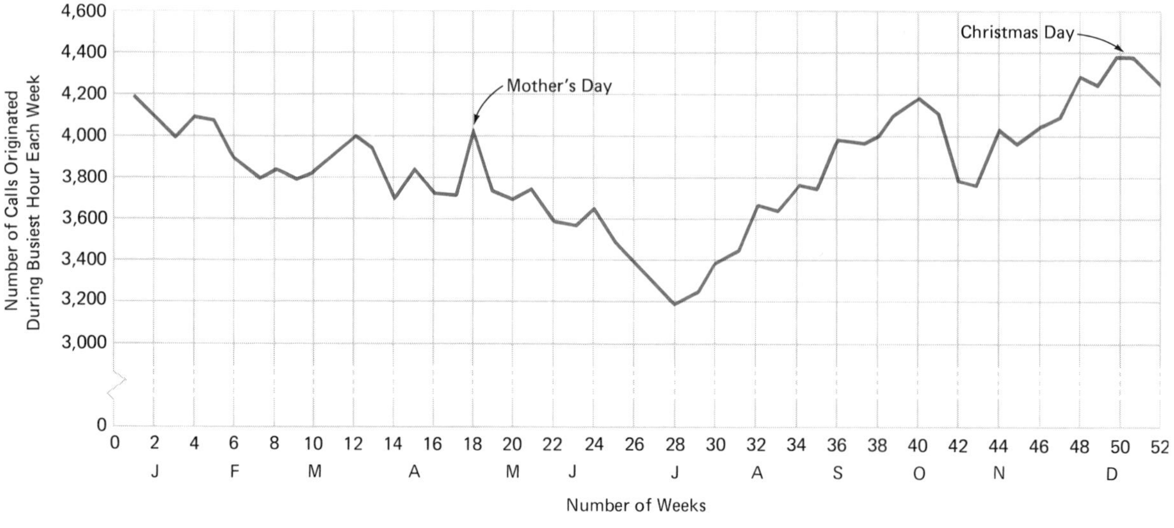

FIGURE 11-9
Typical distribution of calls during the busiest hour for each week during a year.
SOURCE: *Figures 11-9 to 11-17 are from E. S. Buffa, M. J. Cosgrove, and B. J. Luce, "An Integrated Work Shift Scheduling System,"* Decision Sciences, *7, October 1976. (Courtesy, General Telephone Company of California).*

is 128 to 1. Again, the telephone company offers somewhat lower nighttime toll call rates. Figure 11-11 suggests the daily problem of scheduling operator shifts to meet the load.

Finally, Figure 11-12 shows the typical intrahour variation in call load, indicating the number of simultaneous calls by one-minute intervals. This variation is random. In fact, continuous tracking of the mean and standard deviation of calls per minute indicates that the standard deviation is equal to the square root of the mean (a reasonable practical test for randomness) or that the arrival rates are described by the Poisson distribution. For the sample of Figure 11-12, $\overline{x} = 15.75$ calls per minute, $s = 4.85$ calls per minute and $\sqrt{15.75} = 3.99$. Therefore, the variation within the hour is taken as random. We cannot cope with this variation by planning and sched-

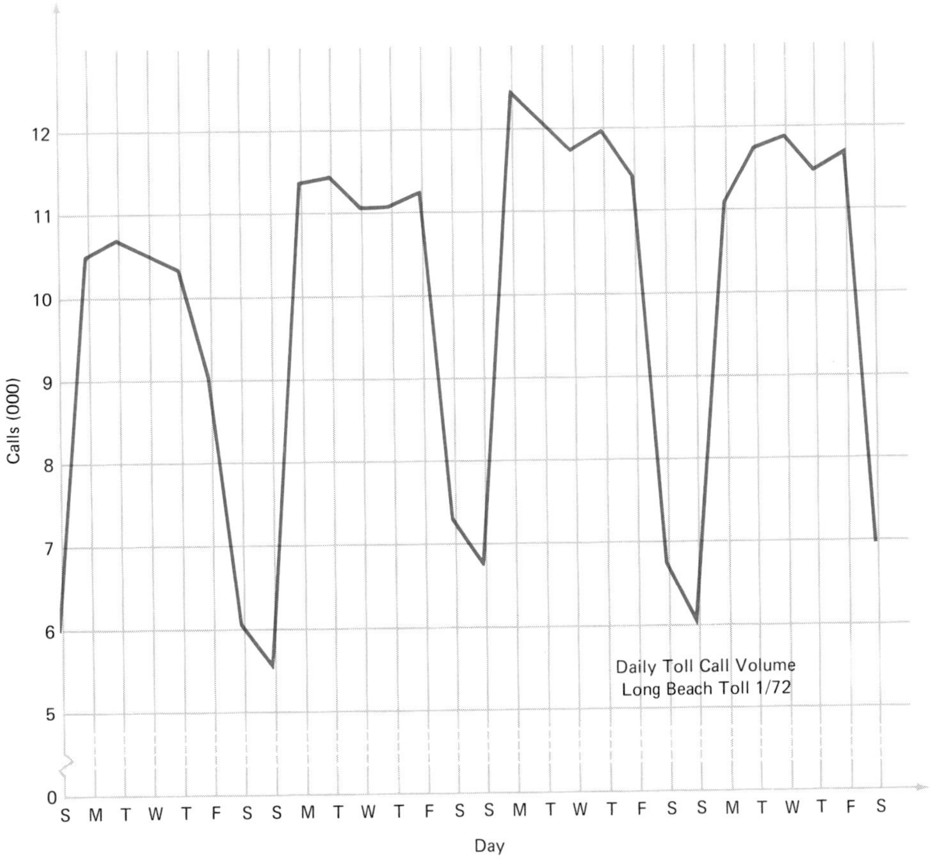

FIGURE 11-10
Daily call load for January.

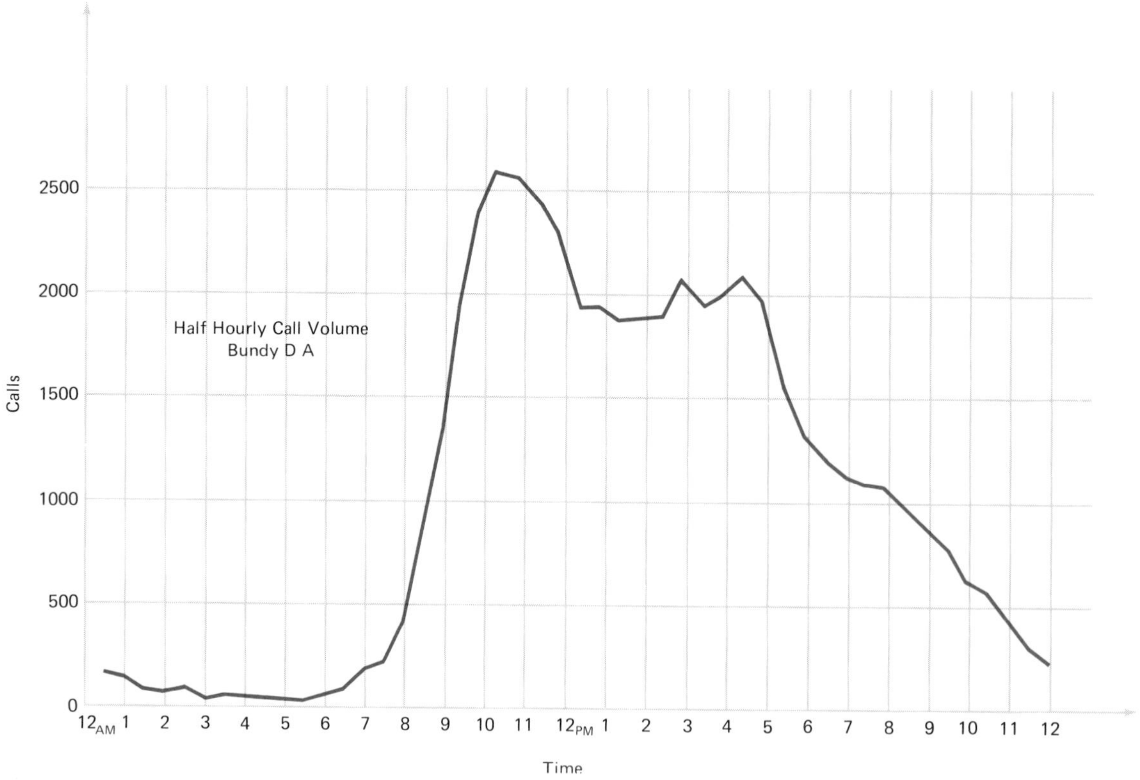

FIGURE 11-11
Typical half-hourly call distribution.

uling. We must simply accept it and provide enough capacity to absorb the random variations.

The overall situation that results from the typical distributions of Figures 11-9, 11-10, 11-11, and 11-12 is that a forecastable pattern exists for seasonal, weekly, and daily variation. In addition, the call rate at any selected minute is described adequately by a Poisson distribution.

The Integrated System

Given the description of the demand for service, Figure 11-13 indicates the system developed at General Telephone. There are basically three cycles of planning and scheduling, which involve information feedback concerning actual experience. The forecast of daily calls is the heart of the system. As we will see, the forecast considers seasonal and weekly variation as well as trends. The forecast is converted to a dis-

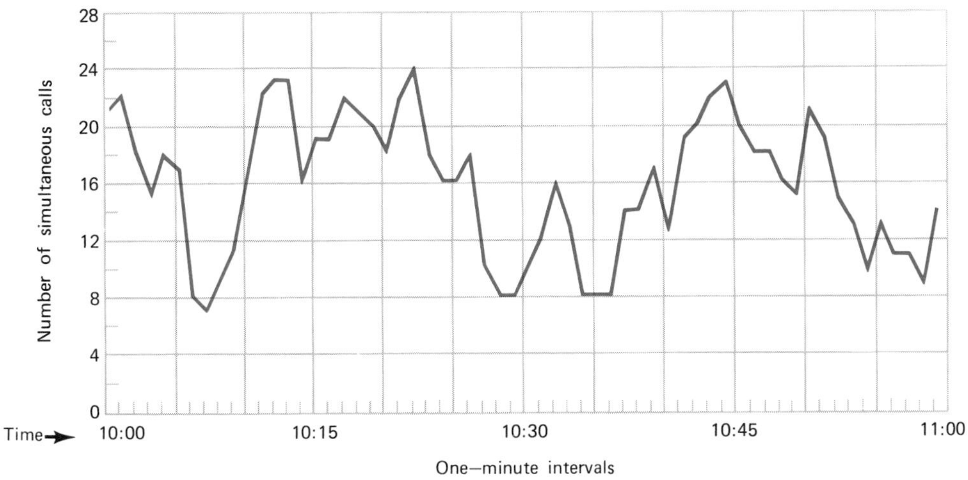

FIGURE 11-12
Typical intrahour distribution of calls, 10:00–11:00 A.M.

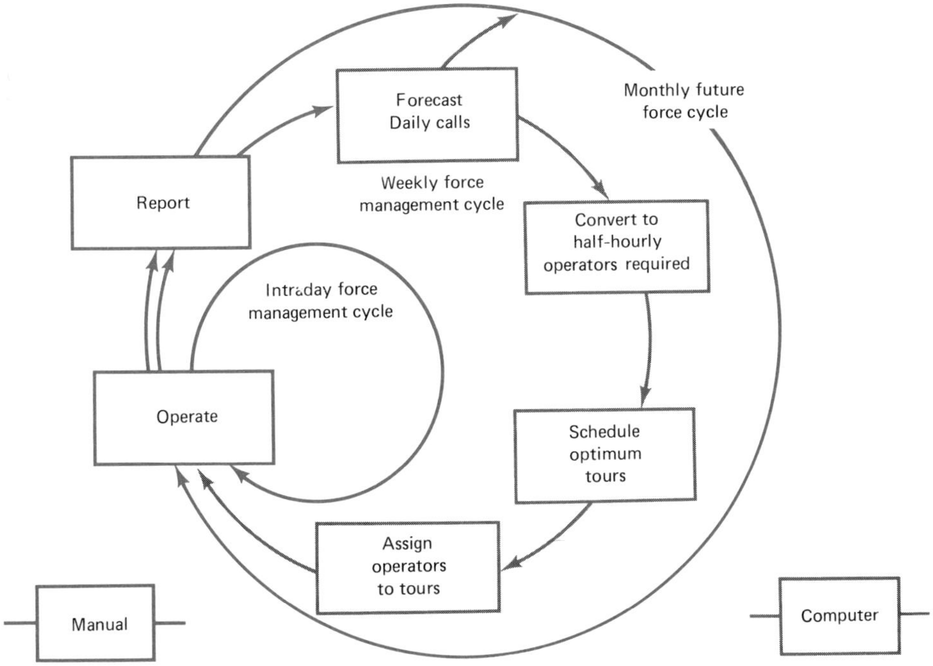

FIGURE 11-13
"Force Management System."

tribution of operator requirements by half-hour increments. Based on the distribution of operator requirements, a schedule of tours or shifts is developed, and finally, specific operators are assigned to tours. This sequence of modules is entirely computerized, as indicated in Figure 11-13.

Given the operator schedule, there are two additional cycles that operate on a manual basis. First, a schedule for "today" may be affected by unintended events, such as operator illness or an emergency increase in call load. Supervisors in local installations cope with such events, and this is the "Intraday Management Cycle" shown in Figure 11-13. In addition, there is the "Monthly Future Force Cycle," in which management can make higher level adjustments based on reports of actual operations and on forecasts involving particular trend and seasonal factors. The hiring and training of operators is planned in the future cycle, up to 12 months in advance.

Forecasting Demand

The demand forecasting system is based on a Box-Jenkins [1970] model. The system involves the following major terms if we are attempting to forecast the number of calls at a specific location for next Monday:

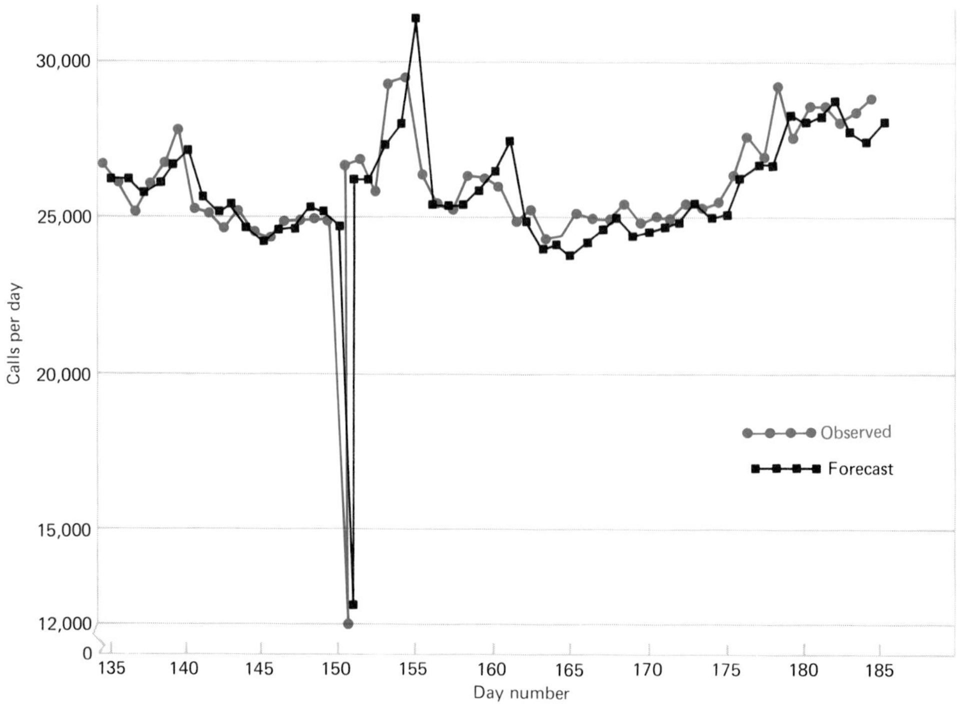

FIGURE 11-14
Sample of forecast versus observed numbers of calls at Santa Monica.

$$\begin{aligned}\text{Calls next Monday} = &\text{ calls last Monday}\\ &+ \text{weekly growth at this time last year}\\ &\quad (\text{Monday}_{-52} + \text{Monday}_{-53})\\ &- \text{error last week} \times \theta\\ &- \text{error 52 weeks ago} \times \phi\\ &+ \text{error 53 weeks ago} \times \phi \times \theta\end{aligned}$$

where θ is a nonseasonal moving average parameter, and ϕ is a seasonal moving average parameter.

In terms of actual operation, the computer inputs are: last week's calls by day and type of service (toll, assistance, directory service); coefficients (work units per call) for the forecasted week by day and type of service; and board load (productivity) by day for the forecasted week. The computer outputs are forecasts of daily calls for up to five weeks in advance and a translation of the forecast into required board hours by day (also for up to five weeks in advance).

Forecast Errors. Figure 11-14 shows a typical record of comparison between forecasted and observed numbers of calls for Santa Monica. The uncanny forecast for day 151 is for Thanksgiving, and people predictably are more interested in dinner and family affairs than in communication. The average error for the forecasting system as a whole is 3.5 percent.

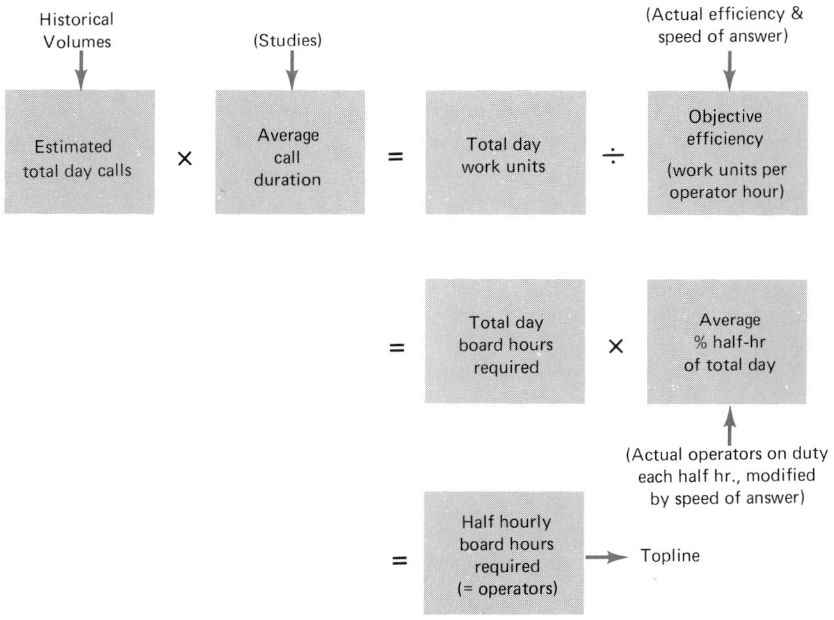

FIGURE 11-15
Model for conversion of calls to half-hourly operator requirements (topline).

Conversion to Half-Hourly Operator Requirements

The objective at this point is to produce a daily schedule of operator requirements. The profile formed by the requirements curve is called the "topline," and the program required to generate it is called the "topline program." The program itself produces a printout of half-hourly operator requirements for each day in a week, and Figure 11-15 shows the formula for the conversion. The parameter that defines the model is average call duration, based on studies of actual times, efficiency, and the response time standard. The response time standard (speed of answer) is a constraint. The result for each day is the information for the topline profile.

Actual half-hourly staffing is based on a percentage of total daily requirements. Exponential smoothing of each half-hourly percent is used to develop the topline program. A table based on a queuing model is used to adjust the actual half-hourly staffing to account for the speed of answering.

Scheduling of Shifts

The graphical representation of a topline profile is shown in Figure 11-16. The problem in assigning tours or shifts involves fitting in shifts so that they aggregate to the topline profile (also shown in Figure 11-16).

The Shift Set. In order to build up shifts so that they aggregate to the topline profile, we need flexibility in shift types, and we get it in the shift lengths and in the positioning of lunch hours and rest periods. A large shift set provides flexibility. On the other hand, the set of shifts is constrained by state and federal laws, union agreements, company policy, and practical considerations. Shifts in the set are actually selected based on California state restrictions, company policy, and local management input concerning the desirability of working hours by their employees.

Each shift consists of two working sessions separated by a rest period, which may be the lunch period. Each working session requires a 15-minute rest period near the middle of the session. The following rules enumerate the admissible shift set:

1. Shifts are 6.5, 7, or 8 hours.
2. Work sessions are in the range of 3 to 5 hours.
3. Lunch periods either are a half hour or an hour.
4. Split work periods are in the range of 3.5 to 5 hours (split work periods are separated by more substantial nonwork periods).
5. Eight-hour shifts end before 9 P.M.
6. Seven-hour shifts end from 9:30 to 10:30 P.M.
7. Six-and-a-half-hour shifts end at 11:00 P.M. or later.
8. Earliest lunch period is at 10:00 A.M.

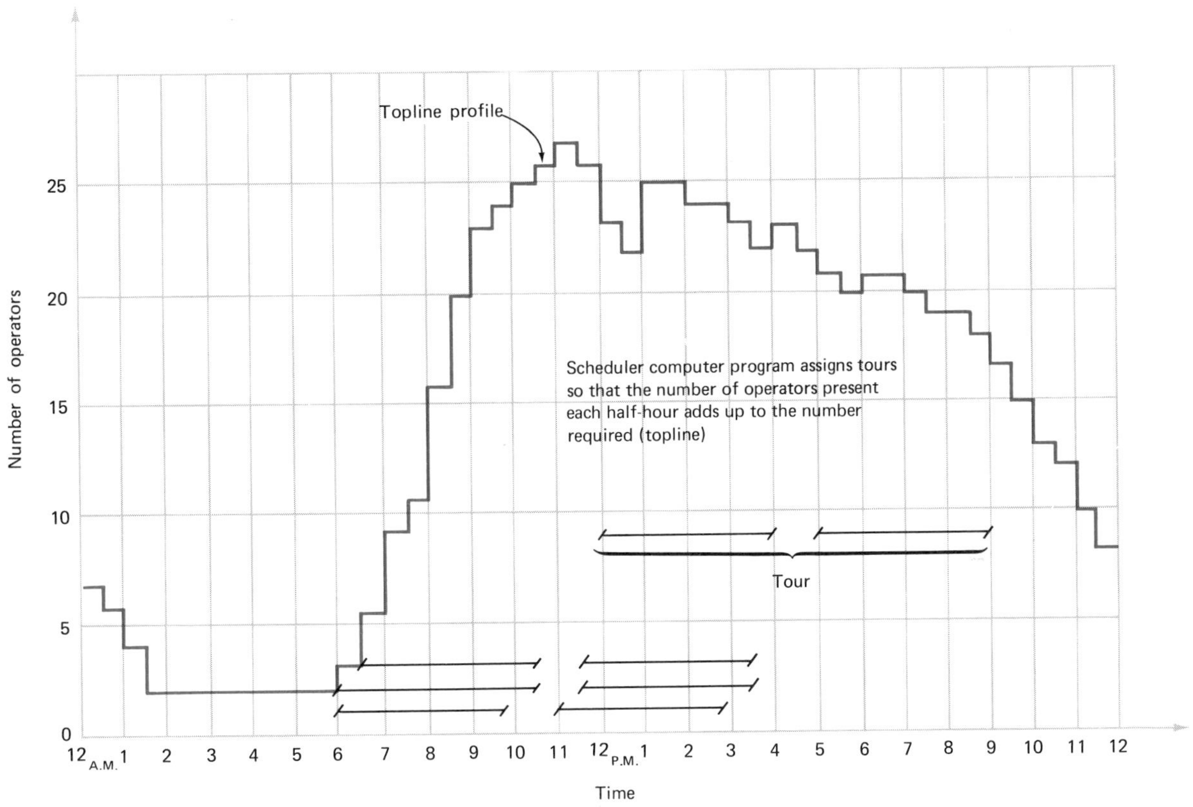

FIGURE 11-16
Topline profile and concept for assigning tours to add up to the topline.

The Scheduling Algorithm. Luce [1973] developed a heuristic algorithm for assigning shifts from the approved set so as to minimize the absolute differences between operators demanded by the topline profile in period i, D_i, and the operators provided, W_i, when summed over all n periods of the day; that is,

$$\text{Minimize} \sum_{i=1}^{n} |D_i - W_i| \tag{1}$$

The strategy is to build up the operator resources in the schedule, one shift at a time, drawing on the universal set of approved shifts. The criterion stated in Equation 1 is used to choose shifts at each step. As the schedule of W_i values is built up, conceptually we attempt to minimize the distance between the schedules of demand and the number of operators supplied, as illustrated by Figure 11-17.

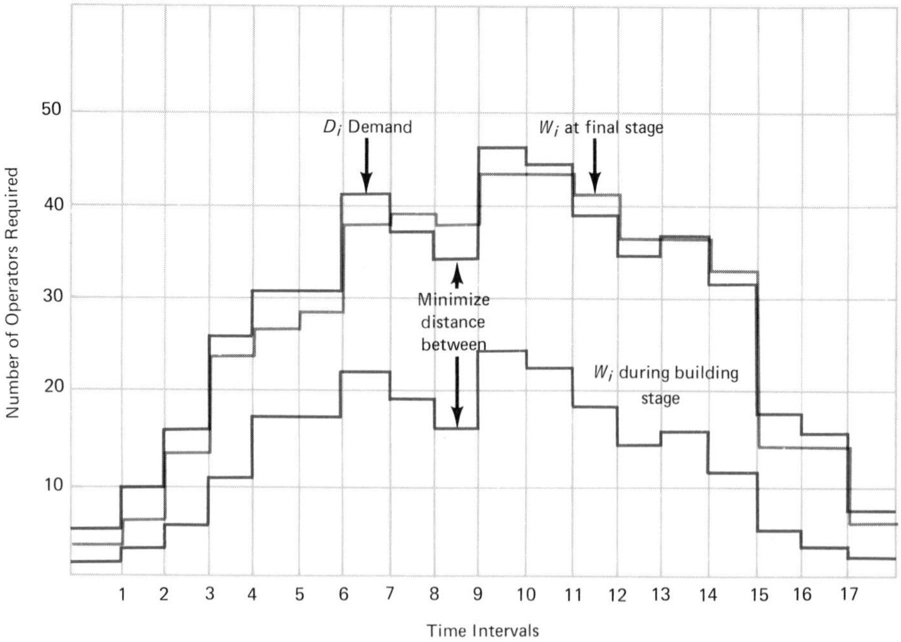

FIGURE 11-17
Concept of the scheduling building process, using the criterion, minimize $\sum_{i=1}^{n} |D_i - W_i|$.

At each stage in building up the schedule, some remaining distance between D_i and W_i exists. The criterion for the choice of the next shift is the following test on each alternate shift: add the contributions of the shift to W_i (1 for all working periods and 0 for idle periods, such as lunch and rest periods), and recalculate Equation 1. Choose the shift that minimizes (1). In order to counteract the shorter length shifts that the preceding rule would favor, weight the different shorter shifts by the ratio of the working times. Thus, if the longest shift is eight hours, then a seven-hour shift would be weighted $8/7 = 1.14$.

As the number of time intervals and shift types increase, the computing cost increases. Luce states that computing costs are moderate when the number of time intervals is less than 100 and the number of shifts is less than 500.

As indicated in Figure 11-17, the final profiles for D_i and W_i do not coincide perfectly in any real case. Operators provided by the algorithm will be slightly greater or less than the demand, and the aggregate figures are a measure of the effectiveness of a given schedule.

Assigning Operators to Shifts

Given a set of shifts that meets the demand profile, the next step is to assign operators to shifts. The 24-hour day, seven-day week operation complicates this process: important questions of equity arise regarding the timing of days off and the assignment of overtime work (which carries extra pay). Employee shift and other preferences, as well as seniority status, also must be considered.

Luce [1974b] developed a computing algorithm that makes "days-off" assignments within the following general rules:

1. Give at least one day off in a week.
2. Days off are one or two.
3. Maximize consecutive days off.
4. If days off cannot be consecutive, maximize the number of work days between days off.
5. Treat weekends separately on a rotational basis in order to preserve equity, because:
   a. Overtime pay is given for weekend work.
   b. Weekends are the most desirable days off.
6. Honor requests for additional days off on a first-come–first-assigned basis.

The days-off procedure must be carried out to assure that a final feasible schedule will result. Trading off for work days is allowed. The actual assignment of operators to shifts considers employee shift preferences. Each operator makes up a list of shifts in rank order. The list can have different preferences for each day of the week. The order of satisfying preferences is determined on a seniority basis, and in the matching process, assignments are made to the highest ranked shift available for each operator.

The final employee schedule for each day is a computer output. The schedule specifies for each operator the beginning and end of two work periods (separated by lunch) and the time for each of two rest periods.

During the first year, the company realized a net annual savings of over $170,000 in clerical and supervisory costs, as well as achieving a 6 percent increase in work force productivity. The company continues to use the system.

IMPLICATIONS FOR THE MANAGER

Level staffing of many activities, such as nursing and fire protection, has been the pattern in the past. Certainly, level staffing is the easiest strategy for managers to employ, the peak requirement determining the staffing level. But level staffing will be expensive in most cases, so managers need to consider alternatives. Scheduling to meet variable requirements is a way that managers can counter the system and remain in control.

Our society seems to demand more and more services to be available throughout the week and often on a 24-hour basis. Supermarkets, health services, food service, and even banks are faced with demand for extended hours. At the same time, we are beginning to consider seriously a variety of alternate work patterns, such as the four-day week, 30 hours per week, 10 hours per day patterns now being discussed. Some of these proposals complicate personnel scheduling, while others may provide flexibility. Formal scheduling models will become even more valuable as the variety of work patterns increases. Fortunately, in most instances, rather simple algorithms can be used to schedule personnel within the constraints of work rules.

Work shift scheduling has been formulated as an integer programming problem, but the current state of the integer programming art does not permit the solution of real-world problems. Heuristic solutions have been used to obtain very good, although not optimal solutions. Integrated systems for scheduling work shifts have been developed in the telephone industry [Buffa et al., 1976], in the postal service [Krajewski et al., 1976], and in nurse scheduling [Smith and Wiggins, 1976]. These integrated systems make it possible to schedule shifts and personnel for fairly large operations, based on forecasts. Managers of such systems can meet service performance requirements at minimum cost on a routine basis.

REVIEW QUESTIONS AND PROBLEMS

1. What are the reasons why the personnel scheduling problem presents unique problems in service-oriented systems?

2. Think about the general logic of the three rules used to assign days off. For example, why assign days off to the lower level requirements rather than the higher level requirements?

3. In applying rule 3, why choose the first available tied pair? Why not choose the last available tied pair?

4. The following schedule of requirements is similar to the example used in the text, with the exception that the requirement for Thursday is four workers. The result is that no slack would be available if four workers could be scheduled with consecutive days off:

$$R_1\text{: } 3\ 3\ 4\ 4\ 3\ 1\ 2$$

   It is still possible to use only four workers to meet requirements? If so, are there alternate solutions?

5. A service operation is offered seven days per week. Demand for the service converts to four workers required throughout the week. Work rules require that

each worker be given two consecutive days off each week. How many workers will be required to staff the operation? Develop the schedule of days worked and days off for each worker.

6. Assume that the schedule of requirements in the previous problem is altered only in that the requirement for Sunday is six workers. Is it still possible to schedule the operation with the same number of workers? Does it make any difference whether or not the requirement of six workers occurs on Sunday or any other day of the week?

7. A service operation requires five workers per day, seven days per week. Total weekly requirements are $5 \times 7 = 35$ worker-days per week. Theoretically, the 35 worker-days can be met with $35/5 = 7$ workers, assuming that each worker must have two consecutive days off. Such a solution would have no slack. Is it possible?

8. State the nature of the formulation of the personnel scheduling problem in the bank teller study.

9. For the Mabert-Raedels bank teller study, define the Heuristic Assignment Procedure (HAP).

10. Summarize the results achieved by shift scheduling using part-time workers in the Purdue National Bank.

11. Define the important characteristics of the work shift scheduling problem.

12. What are the various work shift types that may be used? Do they represent constraints to scheduling solutions, or do they make the problem easier?

13. Describe the formal statement or formulation of the work shift scheduling problem. What solution techniques have been proposed?

SITUATIONS

14. The following situation is based on a paper by Berry, Mabert, and Marcus [1975] dealing with forecasting of teller window demand at the Purdue National Bank and the conversion of demand to teller requirements. An exact count of customer traffic at the teller windows was not normally recorded. However, the system maintained information on the number of transactions processed, such as checks, cash tickets, deposit slips, and the like. It was decided to see if these transaction data could be related to the number of customers served per day.

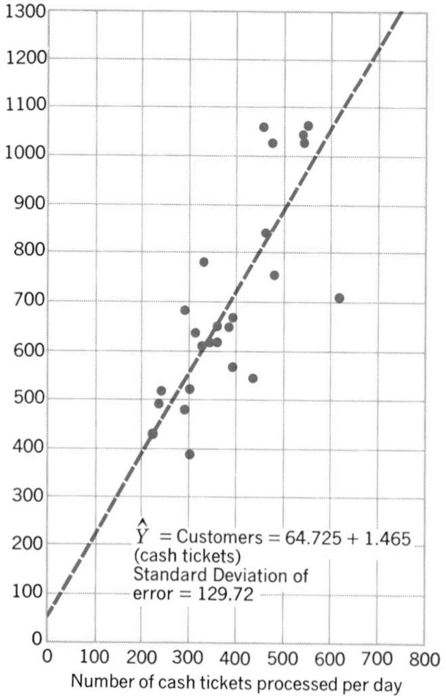

FIGURE 11-18
Customer versus cash ticket regression for the Purdue National Bank, Reserve Square Branch.
SOURCE: *W. L. Berry, V. A. Mabert, and M. Marcus, "Forecasting Teller Window Demand with Exponential Smoothing." Paper No. 536, Institute for Research in the Behavioral, Economic, and Management Sciences, Purdue University, November 1975.*

Then, through a knowledge of the average time spent with customers and the time that the tellers' windows were open for service, the number of tellers needed could be computed. The number of cash tickets processed was selected as the transaction, and a nine-week traffic survey was conducted to determine the actual number of customers requiring teller window service. Figure 11-18 shows the results of the survey with a linear regression line fitted to the points. The resulting regression equation is

$$\text{Number of customers} = 64.725 + 1.465 \text{ (cash tickets)}$$

with a standard deviation of error of 129.72. The correlation coefficient was $r = 0.84$, and $r^2 = 0.71$.

In addition, the time to service the average customer was measured, based on a sample of 3000 customer transactions, indicating that a teller spent an average of $T = 1.5$ minutes per customer. The number of tellers required was then computed as

$$N = \frac{CT}{L}$$

where C = Number of customers forecast by regression equation
T = Average service time per customer in minutes
L = Length of time the teller windows are open for service during the day, in minutes

How do you evaluate this system for converting teller window demand to the number of tellers required? The actual time per transaction varied from as little

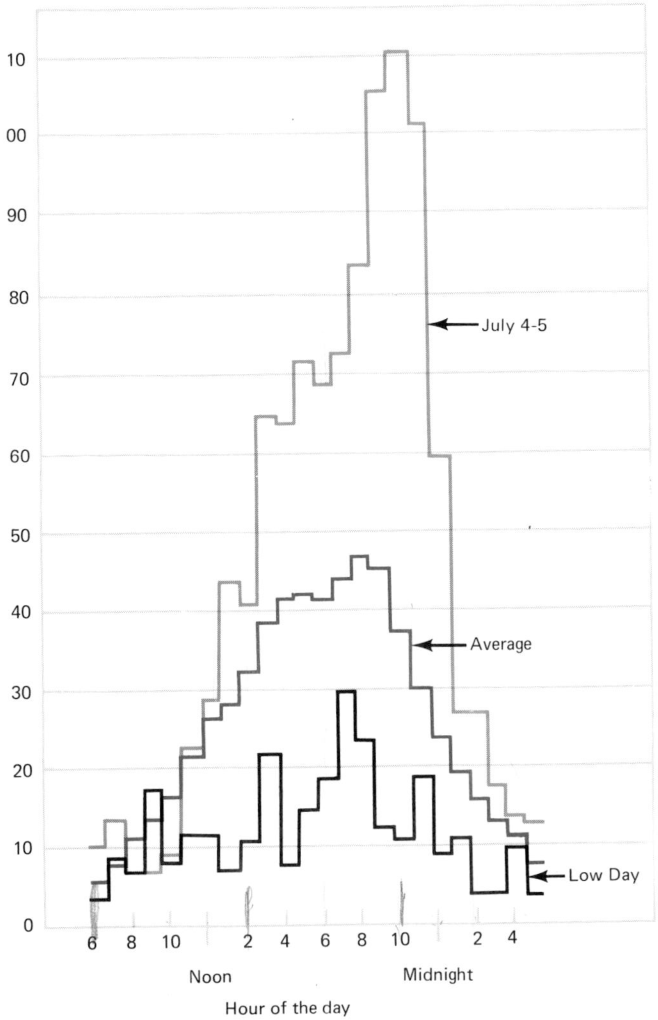

FIGURE 11-19
Total fire alarms received in New York City by hour—1968 data.
SOURCE: E. H. Blum, Deployment Research of the New York City Fire Project, *The New York City Rand Institute, R-968, May 1962.*

as 30 seconds to 5 minutes, depending on the nature of the transaction. How can this variability be taken into account? How good would the service be if the average time were used?

15. Figure 11-19 shows the hourly variation in demand for fire service for three typical days in New York City. On the average day the 8 to 9 P.M. peak is 7.7 times the low point that occurs at 6 A.M. The peak of July 4–5 is 3.7 times the peak in the low day distribution. Traditional deployment policies in fire departments have been to keep the same number of fire fighters and units on duty around the clock. Also, traditional policies have tried to maintain a "standard response" of personnel and equipment to alarms in most areas at all times.

    What staffing strategy should the fire department adopt in attempting to meet the demand for fire protection service? What risks should the department take with respect to the extreme demand days?

16. New York City installed an emergency telephone number (911) that citizens can use. The system has large "trunk" capacity, so that the probability of a busy signal is very small. A call to the system is automatically assigned either to an idle operator or to a first-come–first-served queue, and the first available operator is given the call automatically. The existence of the automatic system provides data concerning the frequency and timing of calls. Figure 11-20 shows the daily variation in the number of calls in July and August. The high for the two months is approximately 22,000 calls per day and the low is 13,000, or a peak-to-valley variation of 1.69 to 1 (41 percent). Such variations have an

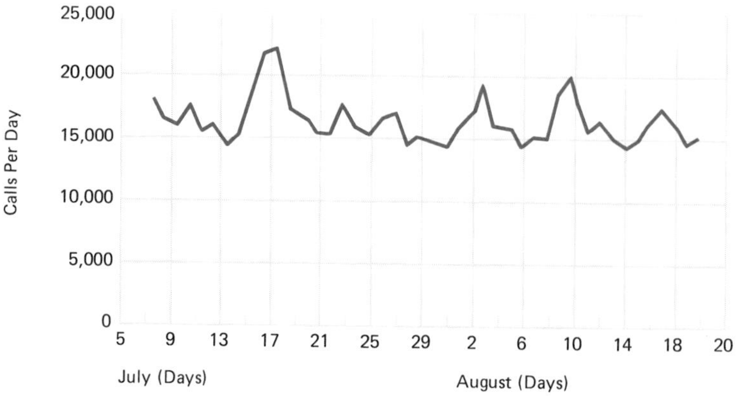

FIGURE 11-20
Number of calls received per day.
SOURCE: *Adapted from R. C. Larson, "Improving the Effectiveness of New York City's 911," in* Analysis of Public Systems, *edited by A. W. Drake, R. L. Keeney, and P. M. Morse. MIT Press, Cambridge, Mass., 1972.*

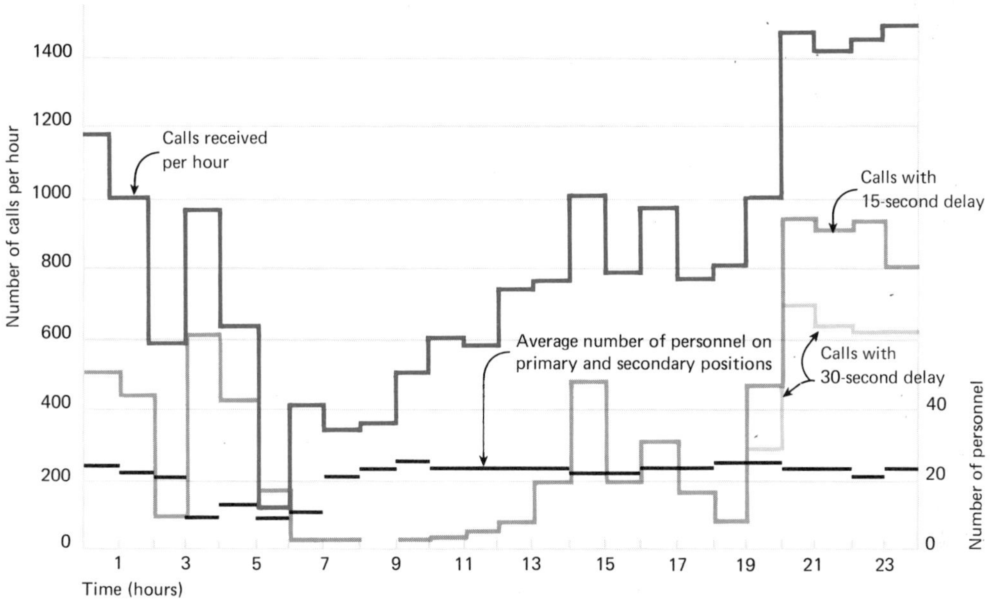

FIGURE 11-21
Distribution of calls, delays, and manning levels (Saturday, August 10).
SOURCE: *R. C. Larson, "Improving the Effectiveness of New York City's 911," in* Analysis of Public Systems, *edited by A. W. Drake, R. L. Keeney, and P. M. Morse. MIT Press, Cambridge, Mass., 1972.*

enormous effect on the personnel planning and scheduling problem and on the resultant costs.

Figure 11-21 shows the hourly variations for Saturday, August 10 (note that most of the peaks in Figure 11-20 occur on Saturdays). Calls are most numerous from approximately 8 P.M. until midnight, occurring at about 1400 per hour, and lowest at 5 A.M., at about 100 calls per hour. This results in a peak-to-valley variation of 14 to 1 (93 percent). Note that the staffing levels for receiving the emergency calls vary only approximately 60 percent. The authors give similar distributions for other days of the week that have lower peaks, and each day the load was met by a somewhat different staffing schedule. Figure 11-21 also contains two additional records of the number of calls per hour that involved a 15-second delay and a 30-second delay.

How do you appraise the staffing strategy indicated by the number of personnel on duty in Figure 11-21? How long a delay in answering is appropriate, or tolerable? What staffing strategy should the police department use for this situation?

17. For this situation, refer to the discussion of the university outpatient clinic near the beginning of the chapter. Figures 11-1 and 11-2 provide data on weekly,

daily and hourly variations of arrivals at the Student Health Service. In addition, Figure 11-4 provides histograms for the service time for three kinds of patients. Reread the materials describing the situation of the outpatient clinic.

What staffing strategy should the outpatient clinic maintain for its physicians? How can one take account of the kinds of weekly, daily, and hourly variations shown? How is this situation different from those described for fire and police protection in situations 15 and 16?

18. After reading the description of the General Telephone Company integrated shift scheduling system in the text, consider the following questions:
    a. If in labor negotiations, union and management agreed on a standard eight-hour shift for all personnel, what would the impact be on the work scheduling system?
    b. Do you think that the methods of assigning operators to shifts and to days off are fairly typically acceptable throughout industry? If not, why are they acceptable in the telephone industry?
    c. How do you appraise the integrated work shift scheduling system installed at the General Telephone Company of California?

REFERENCES

Abernathy, W. J., N. Baloff, and J. C. Hershey. "The Nurse Staffing Problem: Issues and Prospects," *Sloan Management Review, 13*(1), Fall 1971, pp. 87–109.

Abernathy, W. J., N. Baloff, J. C. Hershey, and S. Wandel. "A Three-Stage Manpower Planning and Scheduling Model—a Service-Sector Example," *Operations Research, 21*(3), May–June 1973, pp. 693–711.

Ahuja, H., and R. Sheppard. "Computerized Nurse Scheduling," *Industrial Engineering, 7*(10), October 1975, pp. 24–29.

Baker, K. R. "Scheduling a Full-Time Workforce to Meet Cyclic Staffing Requirements," *Management Science, 20*(12), August 1974, pp. 1561–1568.

Baker, K. R., and M. Magazine. "Workforce Scheduling with Cyclic Demands and Days-Off Constraints," *Management Science, 24*(2), October 1977, pp. 161–167.

Bennett, B. T., and R. B. Potts. "Rotating Roster for a Transit System," *Transportation Science, 2*(1), February 1968, pp. 25–34.

Berry, W. L., V. A. Mabert, and M. Marcus. "Forecasting Teller Window Demand with Exponential Smoothing." *Academy of Management Journal,* March 1979.

Box, G. E. P., and G. M. Jenkins. *Time Series Analysis, Forecasting, and Control.* Holden-Day, San Francisco, 1970.

Browne, J. J., and R. K. Tibrewala. "A Simple Method for Obtaining Employee Schedules," *Proceedings, Conference on Disaggregation,* Columbus, Ohio, 1977.

Browne, J. J., and R. K. Tibrewala. "Manpower Scheduling," *Industrial Engineering,* August 1975, pp. 22, 23.

Buffa, E. S., M. J. Cosgrove, and B. J. Luce."An Integrated Work Shift Scheduling System," *Decision Sciences, 7*(4), October 1976, pp. 620–630.

Burman, D., and M. Segal. "An Operator Scheduling System." ORSA/TIMS National Meeting, March–April 1976.

Butterworth, R. W., and G. T. Howard. "A Method of Determining Highway Patrol Manning Schedules." ORSA 44th National Meeting, November 1973.

Church, J. G. "Sure Staf: A Computerized Staff Scheduling System for Telephone Business Offices," *Management Science, 20*(4) Part 2, December 1973, pp. 709–720.

Harveston, M. F., B. J. Luce, and T. A. Smuczynski. "Telephone Operator Management System—TOMS." ORSA/TIMS/AIIE Joint National Meeting, November 1972.

Healy, W. E. "Shift Scheduling Made Easy," *Factory, 117*(10), October 1969.

Henderson, W. B., and W. L. Berry. "Determining Optimal Shift Schedules for Telephone Traffic Exchange Operators," *Decision Sciences 8*(1), January 1977.

Henderson, W. B., and W. L. Berry. "Heuristic Methods for Telephone Operator Shift Scheduling: An Experimental Analysis," *Management Science, 22*(12), August 1976, pp. 1372–1380.

Hill, A. V., and V. A. Mabert. "A Combined Projection-Causal Approach for Short Range Forecasts." Paper No. 527, Purdue University, Krannert Graduate School of Industrial Administration, Institute for Research in the Behavioral, Economic, and Management Sciences, September 1975.

Jelinek, R. C. "Tell the Computer How Sick the Patients Are and It Will Tell How Many Nurses They Need," *Modern Hospital,* December 1973.

Keith, E. G. "Operator Scheduling," *AIIE Transactions,* 11(1), March 1979, pp. 37–41.

Krajewski, L. J., J. C. Henderson, and M. Showalter. "Selecting Optimal Employee Compliments in a Post Office: A Resolution of Multiple Objectives." Working Paper 76–66, Ohio State University, December 1976.

Krajewski, L. J., L. P. Ritzman, and P. McKenzie. "Schift Scheduling in Banking Operations: A Case Application," *College of Administrative Science,* The Ohio State University, WPS 79–49, May 1979.

Larson, R. C. "Improving the Effectiveness of New York City's 911." In *Analysis of Public Systems,* edited by A. W. Drake, R. L. Keeney, and P. M. Morse. MIT Press, Cambridge, Mass., 1972.

Linder, R. W. "The Development of Manpower and Facilities Planning Methods for Airline Telephone Reservations Offices," *Operational Research Quarterly, 20*(1), 1969, pp. 3–21.

Luce, B. J. "Dynamic Employment Planning Model." ORSA/TIMS Joint National Meeting, April 1974. (a)

Luce, B. J. "Employee Assignment System." ORSA/TIMS Joint National Meeting. April 1974. (b)

Luce, B. J. "A Shift Scheduling Algorithm." ORSA 44th National Meeting, November 1973.

Mabert, V. A., and A. R. Raedels. "The Detail Scheduling of a Part-Time Work Force: A Case Study of Teller Staffing," *Decision Sciences, 7*(4), October 1976.

Maier-Roth, C., and H. B. Wolfe. "Cyclical Scheduling and Allocation of Nursing Staff," *Socio-Economic Planning Sciences, 7*, 1973, pp. 471–487.

Monroe, G. "Scheduling Manpower For Service Operations," *Industrial Engineering*, August 1970.

Murray, D. J. "Computer Makes the Schedules for Nurses," *Modern Hospital*, December 1971.

Paul, R. J., and R. E. Stevens. "Staffing Service Activities with Waiting Line Models," *Decision Sciences, 2*(2), April 1971.

Ritzman, L. P., L. J. Krajewski, and M. J. Showalter. "The Disaggregation of Aggregate Manpower Plans," *Management Science, 22*(11), July 1976, pp. 1204–1214.

Rothstein, M. "Hospital Manpower Shift Scheduling by Mathematical Programming," *Health Services Research*, Spring 1973, pp. 60–66.

Rothstein, M. "Scheduling Manpower By Mathematical Programming," *Industrial Engineering*, April 1972, pp. 29–33.

Sasser, W. E., R. P. Olsen, and D. D. Wyckoff. *Management of Service Operations: Text, Cases, and Readings.* Allyn & Bacon, Boston, 1978.

Segal, M. "The Operator-Scheduling Problem: A Network Flow Approach," *Operations Research, 22*(4), July–August 1974, pp. 808–823.

Smith, L. D., and A. Wiggins. "A Computer-based Nurse Scheduling System," *Computers and Operations Research, 4*, 1977, pp. 195–212.

Tibrewala, R. K., D. Philippe, and J. J. Browne. "Optimal Scheduling of Two Consecutive Idle Periods," *Management Science, 19*(1), September 1972, pp. 71–75.

Trivedi, V. M., and D. M. Warner. "A Branch and Bound Algorithm for Optimum Allocation of Float Nurses," *Management Science, 22*(9), May 1976, pp. 972–981.

Walsh, D. S. "Computerized Labor Scheduling: Supermarkets Jumping on the Bandwagon." Twenty-fifth Annual Conference and Convention, American Institute of Industrial Engineers, May 1974.

Warner, D. M., and J. Prawda. "A Mathematical Programming Model for Scheduling Nurses," *Management Science, 19*(4), December 1972, pp. 411–422.

CHAPTER 12 Maintaining System Reliability

IF THE OUTPUT OF A PRODUCTIVE SYSTEM MAINTAINS STANDARD quality, quantity, and cost, we think of it as being reliable: that is, it continues to do what it was designed to do. In our discussion of operations planning and control systems (see Part III Introduction, Figure III-2), we established basic control loops for performance. Control was accomplished by monitoring the output, comparing it with standards, interpreting differences, and taking action to readjust processes so that they conformed to standards. We also discussed broader level control systems with which to seek a system optimum.

It is not sufficient to think only in terms of average standards of performance; if the system is erratic in its performance, it is unreliable. It is somewhat harder to maintain system reliability with complex systems than it is with simple ones, and an analogy to machines will explain why.

RELIABILITY OF MACHINES

Complex machines may break down often, even though they are designed and manufactured according to the highest standards. We can view machines as being made up of a sequence of components, each of which performs a function. For example, when we strike a key on an electric typewriter, an electric switch is closed. This actuates a solenoid, which causes the mechanical action of the type to strike the ribbon, which in turn transfers carbon to the paper. The type linkage returns to its normal position, and the carriage indexes one space to make ready for the next cycle. The number of individual and mechanical-electrical components required to perform correctly in sequence (series) is surprisingly large. If any single component fails to function correctly—owing to either breakdown or faulty adjustment—the system as a whole fails to function correctly. We are interested in how this affects system reliability.

If we are dealing with a machine that is made up of $n = 50$ components in series, each with an average reliability of 99.5 percent, the reliability of the machine as a whole is only about 77 percent (see Figure 12-1). In other words, if the average probability of breakdown of each of the 50 components were $1.00 - 0.995 = 0.005$, then the probability that the machine might break down because any one of the components broke down is 0.23. As complexity increases, in terms of the number of components in series, the reliability of the system as a whole declines very rapidly.

The reliability of a system in which the components are in parallel differs. With parallel components, there are two or more components that perform the same function. Therefore, if the system is to fail, both parallel components must fail, thus increasing the system's reliability. Increasing reliability through parallel components and systems is expensive, but when the possible losses are great, it may be justified. In space vehicle systems, for example, parallel systems are often used because of the potential costs of system failure. In productive systems, a way to provide parallel paths is by having more than one machine that can perform the same operation.

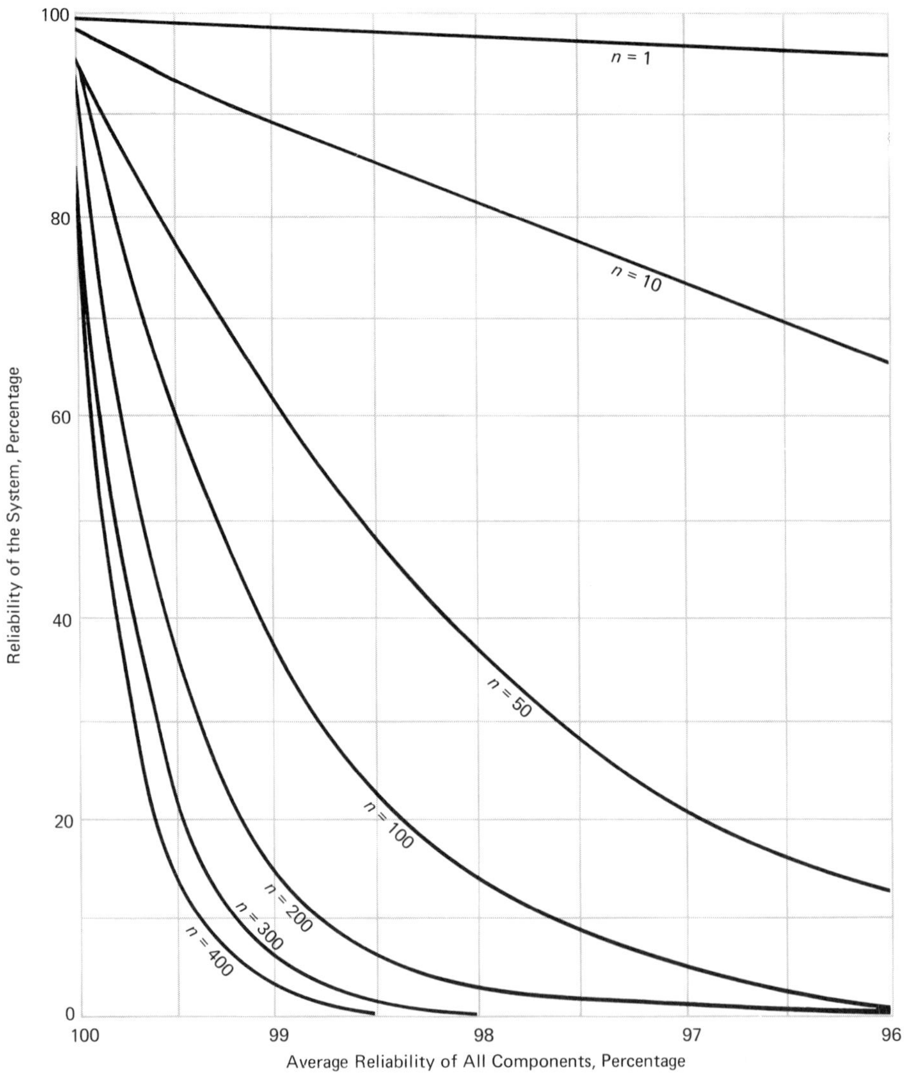

FIGURE 12-1
Overall system reliability as a function of complexity (number of components) and component reliability with components in series.
SOURCE: *R. Lusser, "The Notorious Unreliability of Complex Equipment,"* Astronautics, 2, *February 1958.*

RELIABILITY IN PRODUCTIVE SYSTEMS

There are close parallels between the reliability of machines and of organizations producing services and products. When the *number* of sequential operations required is small and the reliability of each operation is high, the reliability of the system as a whole can be quite good. When the number of sequential operations is large and an error or defect in quality can come from any one operation, then there is a potential for unreliable performance, just as with machine components. We can extend the analogy to include the functioning of entire organizations. Complexity creates the possibility of unreliable performance. This is one of the most important reasons why larger, more complex organizations must use formal procedures in dealing with many kinds of standardized problems. Such organizations become bureaucratic to become more reliable, although the bureaucracy may create problems of its own.

One characteristic of machines is that they can perform repetitive operations fairly consistently, reproducing the same activities. Variations in quality and output quantity tend to be minimized. On the other hand, operations dominated by humans exhibit relatively wide variation in measures of performance. When we combine all the wide variations that are found in manual operations into a complex sequence (as is common in some service operations), the maintenance of system reliability becomes difficult. In such service operations, consistent output quality may depend on individuals following exacting procedures, and variation can have important consequences. For example, if a nurse fails to identify the patient before administering medication and a mixup occurs, the results can be disastrous.

We try to control the reliability of output through general schemes, as diagrammed in Figure 12-2. The output quality and quantity are monitored in some way, and the results are compared with standards. While we generally are interested in quality measures, changes in output quantity also may be symptomatic of reliability problems.

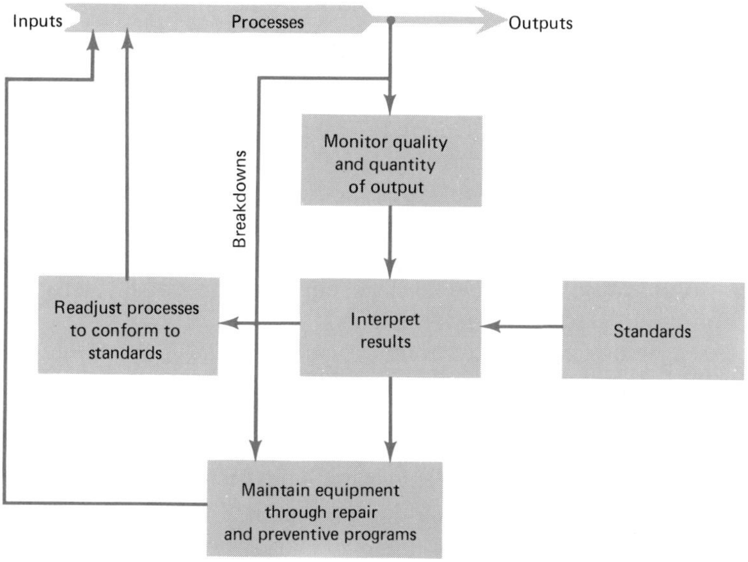

FIGURE 12-2
Control loops for maintaining system reliability by monitoring quality and quantity of output.

Associated costs of quality and quantity control are derivatives. When the results are interpreted, we may conclude that the processes are out of adjustment or that something more fundamental is wrong, requiring machine repair or possibly retraining in manual operations. If equipment actually breaks down, then the maintenance control loop is called directly. Information on output quality and quantity also may be used to form preventative maintenance programs that are designed to anticipate breakdown. Thus, while other important interactions have their effects, the control of system reliability centers on quality control and equipment maintenance.

THE RELIABILITY SYSTEM

Figure 12-2 suggests the nature of control loops for quality and maintenance control. However, it is local in nature and leaves a great deal unsaid. We must ask: Where did the standards come from? What is the nature of the productive system, and is it appropriate?

Figure 12-3 places the reliability system in context. The organization must set policies regarding the desired quality in relation to markets and needs, investment requirements, return on investment, potential competition, and so forth. For profit-making organizations, this involves the judgment of where in the market they have a relative advantage.

For nonprofit organizations, policy setting may involve meeting standards set by legislation and/or by the organization to maximize its position. For example, one reason that universities acquire superior academic staffs is to raise funds, improve the physical plant, obtain outstanding students, and enhance research output. Hospitals may set high standards of care partially to attract the best-known physicians, support fund-raising programs, and attract interns with the best qualifications. The post office may set standards for delivery delay that accommodate current budgetary levels. The policies set by management in box 1 of Figure 12-3 provide the guidelines for the design of the organization's products and services. This design process is an interactive one, in which the productive system design is both considered in and influenced by the design of products and services, as shown in boxes 2 and 3. For manufacturing systems, the design of products in this interactive fashion is termed *production design*. The interaction affects quality considerations, since equipment capability must be good enough to produce at least the intended quality.

Out of the process of the design of products/services and productive system design comes specifications of quality standards, as shown in box 4 of Figure 12-3. Here, we are dealing with a system of quality standards for materials that are consumed in processes, as well as raw materials; for the standards for the output of processes, such as the specification of dimensions, tolerances, weights, and chemical compositions; and for the performance standards for the outputs. The nature of performance standards for products is well known. Manufacturers state the capabilities of their product: the fidelity range of an amplifier, the acceleration of an auto, the waterproof finish of a table surface, and the like.

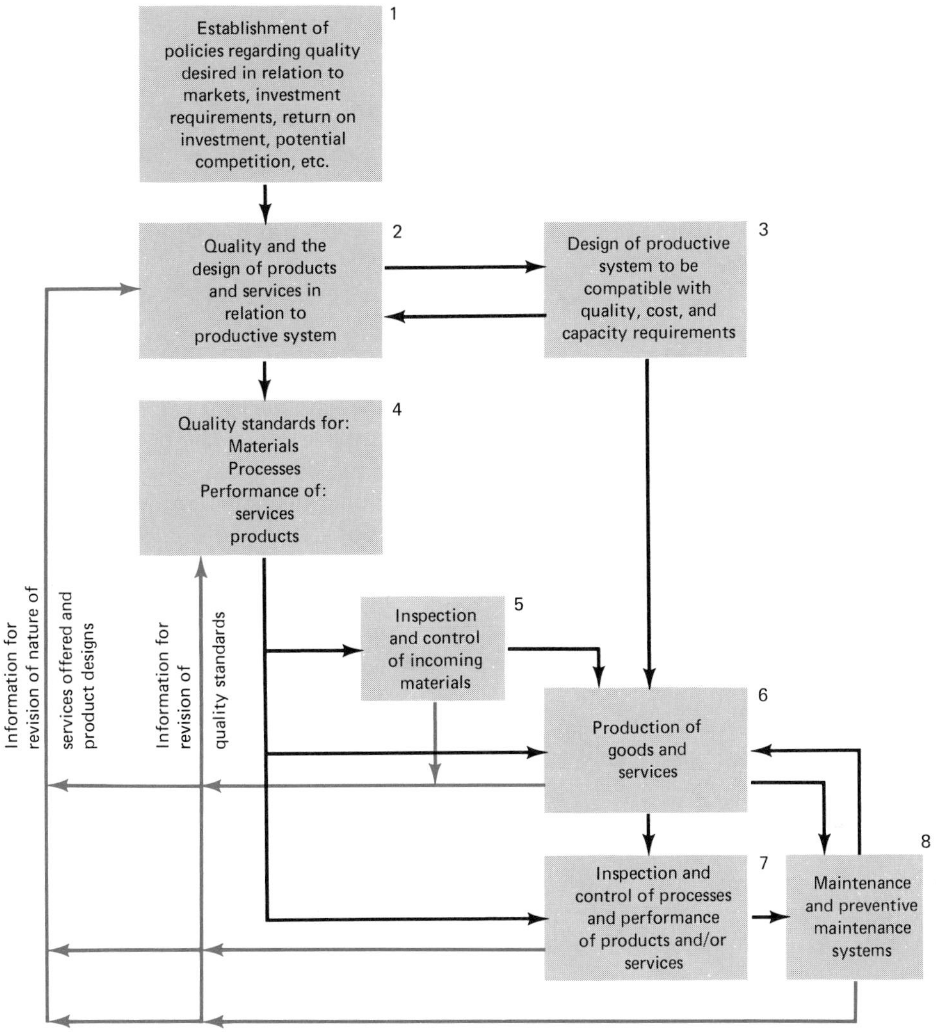

FIGURE 12-3
Schematic representation of the relationships between policies, design of products and services, design of productive system, and the maintenance of the system reliability for quality and quantities.

The performance standards of services seem (so far) to be somewhat less formalized. What performance do we expect from postal systems, educational systems, police and fire protection, and medical care systems? This relative lack of formal standards, and the difficulty of defining them, may be the reason for the paucity of our knowledge about the control of the quality of services.

Given standards, however, we can set up controls for incoming materials in box 5, and for the processes and performance of products and services in box 7. An interrelated control loop concerns the maintenance of the capabilities of the physical system in box 8, through repair and preventive maintenance programs.

Secondary control loops are shown in Figure 12-3 that seek a system optimum. These appear as information flow from boxes 5, 6, 7, and 8 to boxes 2 and 4. Their function is to influence the nature of services and products offered and to help revise quality standards, respectively.

CHOICE OF PROCESSES AND RELIABILITY

Management has basic choices to make in balancing processing costs and the costs to maintain reliability. The process choices may involve more or less expensive equipment. The less expensive equipment may not be capable of holding as close quality standards. Or it may hold adequate quality standards, but only at a higher maintenance cost or by more labor input. So the balance of costs may involve low process cost but higher maintenance and quality control costs, and perhaps lower quality (more rejected product and poorer market acceptance of low quality).

The opposite balance of costs may occur with more expensive processes and equipment. The better processes and equipment may be able to hold better quality standards, resulting in fewer rejected products and perhaps less equipment maintenance and labor input. Some of the best examples of these process choices are in the mechanical industries. A lower-quality lathe may be capable of holding tolerances within ± 0.001 inch. But a precision lathe may be capable of holding tolerances within ± 0.0001 inch or better. However, the choice is certainly not always for the precision lathe. It depends on the product requirements and the balance of costs. These kinds of choices exist generally, although they are not always as clear-cut as the lathe example. Sometimes a more expensive process involves greater labor input with more steps in the productive process.

Figure 12-4 shows the balance between the costs of process choice and the costs of maintaining reliability. The manager's choice among alternatives should be in the middle range shown, near the minimum cost of the total incremental cost curve. The reason for not stating that the choice should be simply to minimize cost is that the manager's choice should be influenced by nonquantifiable factors such as market image and acceptance, flexibility of operations, availability of labor with skills to match equipment requirements, and so on.

CONTROL OF QUALITY

Our general block diagram for control calls for a measurement system to generate information on which to base control actions. In industry, this is the inspection function. Inspectors make measurements that are called for by the quality standards, thereby separating acceptable from nonacceptable units. However, no control or

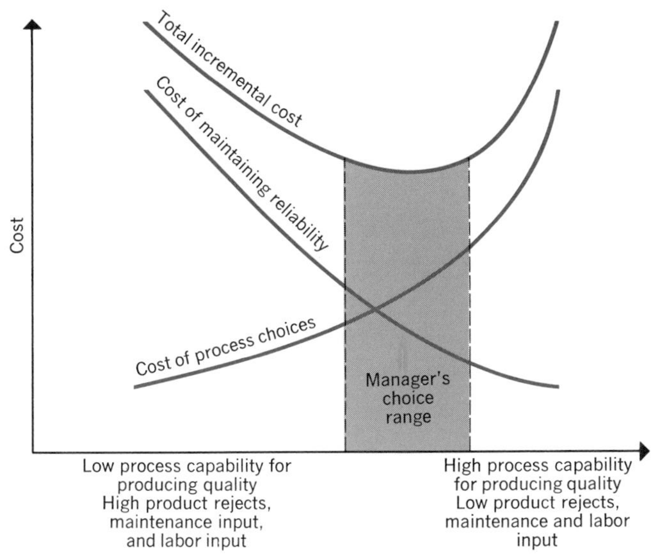

FIGURE 12-4
Cost basis for manager's choice of processes and system reliability.

corrective action is implied. When we link measurement, investigation to determine why an unacceptable product has been produced, and corrective action, we have completed the control loop.

The measures of quality in productive systems are diverse. These measures are perhaps most obvious in manufacturing systems where quality characteristics can be related to objective standards of dimensions, chemical composition, and actual performance tests. Standards for these measures can then be established and regular procedures used to measure the critical characteristics to see if standards are maintained. In service-oriented systems, measures of quality are often not as objective. The personal contact required may emphasize the *way* service is given, even though the service was technically adequate. Waiting time is often a criterion for service quality. Following are typical quality measures of the outputs of productive systems:

Type of System	Measure of Output
Manufacturing	Dimensions
	Tolerances on dimensions
	Chemical composition
	Surface finish
	Performance tests
Medical service	False positive diagnosis
	False negative diagnosis

Postal service	Waiting time at post office Errors in delivery Overall delivery time
Banks	Waiting time at windows Clerical errors

Liability and Quality

Liability for poor product quality has been well established. Although negligence on the part of the manufacturer is central to liability, the concept in legal practice extends to include foreseeable use and misuse of the product. Product warranty includes both that expressed by the manufacturer (written and oral) and the implied warranty that the product design will be safe for the user. The uses are not legally restricted to those specified in warranties but include those that may be foreseen. The latter concept of "foreseeable usage" has often been interpreted to mean that if the product *was* misused, then such use was foreseeable. These legal doctrines place a particularly heavy responsibility on the quality control function, for liability suits can have an important bearing on enterprise survival. See Bennigson and Bennigson [1974] and Eginton [1973] for further discussion of product liability.

Medical malpractice liability has become an important factor in health care costs. In California alone, the number of liability settlements through insurance companies increased from 503 in 1971 to 869 in 1974. The claims in 1971 were more than $15 million, but they more than doubled to almost $33 million in 1974. Insurance premiums have skyrocketed, and physicians fees have reflected the increases. Controversy over the control of health care quality has resulted, with emphasis on establishing standards.

Variation and Control

All processes exhibit variation, and the manager's task is to distinguish between tolerable variation that is representative of the stable system and major changes that result in an unacceptable product. The manager must be aware of the system's inherent capability in order to know when system behavior is abnormal. Thus, since we are dealing with systems that exhibit variation, the manager's control model must be a probabilistic one.

Sampling Information

Because of the ongoing nature of processes and their inherent variability, we must base quality control decisions on samples. First, we cannot usually examine all the data, because the process is continuous and, at best, we have access to a sample at a particular point in time. Second, even if the entire universe of data were available, it might be uneconomical to analyze it. Third, measurement and inspection sometimes require destruction of the unit; and fourth, with some products, any additional handling is likely to induce defects and therefore should be avoided. Thus the sampling of information about the state of incoming raw materials and of control of processes is the common approach on which to base decisions and control actions.

The amount of sampling justified represents another managerial choice. As the amount of sampling increases, approaching 100 percent inspection, the probability of passing defective items decreases, and vice versa. The combined incremental cost curve is again dish-shaped. The manager's range of choices is near the minimum cost but is a range because nonquantifiable factors must influence the decision.

Risks and Errors

Because we normally must use samples of data drawn from a system that naturally exhibits variation, we can make mistakes, even in controlled experiments. Figure 12-5 summarizes the nature of errors and risks taken; here, we classify the actual state of the system and the decision taken. The process either is in control or it is not; or, similarly, we have a batch of parts or materials that have been generated by a system that either was or was not in control.

As Figure 12-5 shows, we can decide either to accept or reject the output. If the process is in control—and if, based on our information, we would reject the output—then we have made an error that is called a *Type I error*. We, the producer, risk making such an erroneous decision on the basis of the probabilities that are associated with the inherent variability of the system and the sample size. Logically, this risk is called the *producer's risk*, since—if the decision is made—it is the producer who absorbs the loss.

True state of system	Decision	
	Reject output as bad	Accept output as good
Process is in control	Type-I error (Producer's risk)	Correct decision
Process is out of control	Correct decision	Type-II error (Consumer's risk)

FIGURE 12-5
Errors and risks in quality control decisions.

Similarly, there is a risk that we may accept output as a good product when, in fact, the process is out of control. This decision is called a *Type II error* and is termed the *consumer's risk*. In statistical control models, we can preset the probabilities of Type I and Type II errors.

Kinds of Control

Figure 12-3 shows that, fundamentally, we control quality by controlling (1) incoming materials, (2) processes at the point of production, and (3) the final performance of products and services, with the maintenance system acting in a supporting role. For some products, quality control of product performance includes the distribution, installation, and use phases.

From the point of view of control methods, we can apply statistical control concepts by sampling a lot of incoming materials to see whether it is acceptable (acceptance sampling) or by sampling the output of a process to keep that process in a state of statistical control (process control).

Acceptance sampling lets us control the level of outgoing quality from an inspection point to ensure that, on the average, no more than some specified percentage of defective items will pass. This procedure assumes that the parts or products already have been produced. We wish to set up procedures and decision rules to ensure that outgoing quality will be as specified or better. In the simplest case of acceptance sampling, we draw a random sample of size n from the total lot N and decide, on the basis of the sample, whether or not to accept the entire lot. If the sample signals a decision to reject the lot, the lot either may be subjected to 100 percent inspection, in which bad parts are sorted out, or may be returned to the original supplier. Parallel acceptance sampling procedures can be used to classify parts as simply good or bad (sampling by attributes) or to make some kind of actual measurement that indicates how good or bad a part is (sampling by variables).

In process control, we monitor the actual ongoing process that makes the units. This allows us to make adjustments and corrections as soon as they are needed, so that bad units in any quantity are never produced. This procedure is a direct application of the statistical control chart, and, as with acceptance sampling, parallel procedures are available for those situations in which sampling is done by attributes and for those in which measurements are made of variables that measure quality characteristics. Figure 12-6 summarizes the classification of statistical control models.

Controlling the Quality of Services

The previous material dealing with industrial quality control has clear objectives on what to measure and control and sophisticated methodology for accomplishing these ends. However, in nonprofit organizations the objectives and outputs seem less well defined and the control methodology relatively crude.

The profit motive provides a focus for all kinds of managerial controls, including

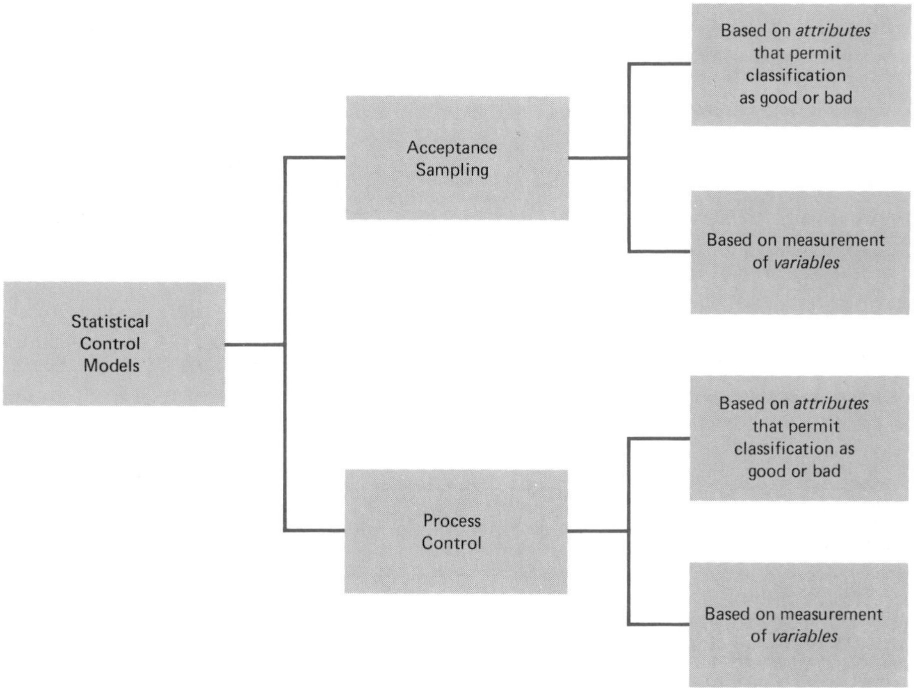

FIGURE 12.6
Classification of statistical control models.

quality. By contrast, nonprofit organizations exist to render service, and their success is judged in those terms. Measuring the quality of services is difficult in part because the attributes of quality are somewhat more diffuse. Is quality health care measured by the death rate, length of hospital stay, or the treatment process used for specific disease? Is the quality of police protection measured by the crime rate, feeling of security by citizens, or indexes of complaints of police excesses? (Note the following anomaly: if the size of the police force is increased, crime rates have been observed to increase, because more of the crimes committed are acted on and reported.) Is the quality of fire service measured by reaction time, the annual extent of fire damage, or some other factor? In partial answer to these questions, we must note that the quality characteristics of most of these kinds of services are multidimensional and often controversial, and reducing quality measurement to something comparable to specific dimensions or chemical composition may be impossible.

Anthony et al. [1972] summarize some of the specific problems affecting managerial control of nonprofit services as follows.

1. *Relationship with the users of services.* In nonprofit organizations, a new "customer" is seen as a problem rather than an opportunity; that individual

places an added strain on existing resources. If the new customer brings added resources, they commonly barely cover incremental costs.

2. *Lack of competition.* Resources for nonprofit organizations are obtained from government appropriations, from endowments, or from fees that recover costs instead of from the competitive marketplace. Thus the necessity to allocate and use resources efficiently is minimized.

3. *Diffused responsibility.* In many nonprofit organizations, the line of responsibility is not a clear single chain of delegation of power.

4. *Politics.* In government organizations, there are strong pressures to make short-run decisions inconsistent with optimum use of resources because of the annual budgeting cycle and the reelection syndrome.

5. *External pressures.* Public review through news media and political parties results in frequently erratic and illogical pressure on managers of public organizations.

6. *Legislative restrictions.* Government organizations must operate within statutes that are much more restrictive than the charters and bylaws of corporations and that often prescribe detailed operating practices that are difficult to change but may not be efficient.

7. *Behavioral factors.* The general "atmosphere" in a nonprofit organization tends to be less conducive to good management control practice than in profit-oriented organizations.

A Framework for Controlling Quality of Services. Adam, Hershauer, and Ruch [1978] have proposed a process for developing standards and measures of quality in services and have applied it in banking. Their approach provides a framework and process for creating a unique set of measures for each service within each organization. It assumes that the knowledge needed to create a measurement system exists within the minds of current systems managers. The group processes involved are designed to coalesce from the managers the definition of the system from a quality point of view, the specific kinds of deviations from accepted norms that affect quality, and ways of measuring each specific deviation.

A shirt laundering service is used as an example to explain their processes to system managers. Shirt laundering involves operations that are simple and well understood. The major steps in this process are as follows:

1. *Define unit operations.* A unit operation is associated with an identifiable change in state in processing that occurs. For example, in the shirt laundering service, the unit operations are (1) receive and tag, (2) wash and clean, (3) dry and press, and (4) package and deliver. The unit operation that results from the process is defined by the participant managers through a carefully designed group process.

2. *Generate key quality deviations.* Deviations represent technical system requirements that are subject to variability and that need to be controlled. Through carefully designed group processes, the managers generate *what* and *where* in the process deviations may occur. A second step in the process is then to distill from the deviations list the key quality deviations. Compared to the industrial quality control systems discussed previously, key quality deviations are equivalent to the dimensions and attributes that need to be controlled because they affect product performance and quality. The key quality deviations are caused by the nature of the items being processed (at entry, in process, or at output) or the nature of the processes used. In the shirt laundering example, the key quality deviations and the unit operation location where they occur (in parenthesis) are as follows:
   a. Number, type, and size of stains in shirts (input, wash, and clean)
   b. Number of buttons missing (input, wash and clean, dry and press)
   c. Item identification missing (receive and tag, wash and clean, dry and press)
   d. Wrong availability (package and deliver)
   e. Days until delivery (wash and clean, dry and press, package and deliver)

3. *Generate measures.* Using the key quality deviations as a basis, the managers developed related measures through group processes. For example, for the deviation "number, type, and size of stains," the measures were:
   a. Customer stain complaints per customer
   b. Worker-hours expended removing stains
   c. Cost of shirt replacements per standard cost for replacements
   d. Number of shirts with a stain per total number of shirts received

4. *Evaluation of measures.* As a basis for finalizing the measures developed in step 3, participants are asked to rate each measure in terms of its value in the outcome of the deviation and to rate the strengths of their convictions of that rating. Only measures whose average rating is above a certain threshold are retained in the measurement system.

Applications in Banking. Adam et al. applied the system for developing quality measures to check processing and to the personnel function in three banks.

For check processing, seven unit operations were defined in two banks as:

1. Receipt of transfer documents

2. Preparation for processing

3. Processing

4. Reconciling and settlement

5. Preparation for dispatch

6. Dispatch

7. Post processing

In bank C, 93 deviations were defined which were reduced to seven key quality deviations. Sixty measures were then defined related to the key deviations. These measures were evaluated by executives in the organization through the rating processes, and a final list of 25 measures were retained. In addition to the more detailed measures, five systemwide measures for check processing were accepted at bank A:

1. Percentage of error-free outgoing cash letters per total worker-hours in check processing

2. Dollar amount of "as of" adjustments per total dollars processed

3. Percentage of error-free outgoing cash letters per cost of 1000 letters

4. Total dollar expense of adjusting key deviations per total dollar expense in check processing

5. Total worker-hours in adjusting key deviations per total worker-hours in check processing

For the personnel function, five unit operations were defined in two banks as:

1. Recruit and employ staff

2. Evaluate employee job assignments

3. Compensate employees: review and adjustment for wages, salaries, and benefits

4. Train

5. Administer personnel services to employees and the public

In bank A, 42 deviations were defined which were reduced to 11 key quality deviations. Forty-three measures were defined related to the key quality deviations. These measures were evaluated by executives through the rating process, and 48 were retained.

Although the process for establishing quality measures proposed by Adam et al. is complex, it reflects the technical-behavioral emphasis of service systems. It is the first comprehensive effort to establish quality measures in service-type operations. The emphasis of the framework on quality measures of service productivity is unique. Most quality control measures in the past have been taken in isolation and do not attempt to relate the measure and quality control effort to input-output values. The validity of the quality control effort in relation to productivity has been the result of a separate judgment that is usually not explicit.

Given the measures of *what* to control for service operations, it is necessary to close the control loop. Standards must be set concerning levels of each of the measures that represent acceptable and unacceptable quality, and corrective action must be taken when measurements signal that the system is out of control. Statistical control procedures similar to those used in industrial quality control may then be used.

Quality Control in Health Care. Mounting costs, complaints, and the enormous increase in public funds allocated to health care have focused attention on the development of formal mechanisms for assessing quality. In 1972, Congress created the Professional Standards Review Organizations (PSROs) as a means of self-regulation in health care. The general thrust of these organizations is to set standards for health care and mechanisms for review to ensure that these standards are maintained in practice. The organizations are intended to be relatively local, and physicians who practice medicine in local medical service areas are to be held responsible for the quality of their practices.

State Boards of Medical Examiners hear cases concerning malpractice. For example, in California during the five-year period from 1970 to 1974, 88 licenses to practice were revoked, and there were 348 suspensions, probations and miscellaneous penalties. There were more than 42,800 active members in the state during the same period. Thus, sanctions and legislation are a means of maintaining broad level control of health care quality. The issue is centered on the tightness of these controls.

Development of Health Care Standards. Current efforts to establish health care standards and control mechanisms focus on an evaluation of outcomes (did the patient die, recover, or recover at what level?). An inquiry into the treatment process used is conducted if the outcomes are outside of standardized limits.

A general model for evaluating the quality of patient care has been developed by Williamson [1971] in this outcomes-process assessment format. In Williamson's format, four factors are evaluated in the following order:

1. The data required to determine the need for care, specific therapy, and prognosis (diagnostic outcomes)

2. The health status of the patients at a given time period following treatment (therapeutic outcomes)

3. The procedures carried out in order to furnish the physician with facts on which to base diagnoses (diagnostic process)

4. The planning, implementation, and evaluation of therapy (therapeutic process)

The heart of the quality control process is in the establishment of standards for outcome criteria. For example, a criterion for death as an outcome might be established by determining the maximum acceptable case fatality rate for patients with a given health problem. "Peer judgment offers a practical method for setting standards." The measurement problems for the four elements in the program are significant; however, the data base is available and can be analyzed.

Definition of Risks and Errors. Although the definitions are not couched in decision theory terms, the medical profession has developed equivalents for Type I and II errors, acceptable quality level, unacceptable quality, and risks. The key definitions are as follows.

False positive, individuals receiving care who did not need it, Type I error.

False negative, individuals needing care who did not receive it, Type II error.

One can then state a maximum false positive for a given health problem as something equivalent to acceptable quality level and the probability of rejecting samples at this level as an equivalent to the producer's risk.

Similarly, one can state a maximum false negative for a given health problem as the limit of bad quality and the probability of accepting this poor quality as the equivalent of the consumer's risk. In fact, both risks are absorbed by the patient (consumer) in the medical system, since a false positive results in a patient's receiving unneeded care that is billed and could be injurious, and a false negative results in a patient's not receiving needed care.

Process Study and Action. In the Williamson control format, measured findings with established criteria reveal whether detailed study of the medical care process is indicated. He suggests that a 95 percent confidence interval covering the measured findings be used. For example, if the maximum acceptable case fatality rate were set at 5 percent, a measured rate of 10 percent with confidence limits of 4 to 15 percent would not be sufficiently different from the criterion to warrant process study. The result of process study, when it is invoked, might bring the standards into question or might result in the alteration of health care procedures.

Example 1. A study was made of urinary tract infections diagnosed in a community hospital in the Midwest involving over 6000 consecutive admissions. Criteria were independently established by group judgment, taking into account the sensitivity and specificity of methods for detecting urinary tract infections and the seriousness of implications to the patient of a "missed diagnosis" or a "misdiagnosis."

The maximum acceptable percentage of false negatives was set at 15 percent and of false positives at 20 percent.

Measurement of actual outcomes revealed that 265 of the 6145 consecutive admissions probably required urinary tract care; however, 187 of these patients did not receive this care from the regular hospital staff, resulting in a false negative rate of over 70 percent. Of 110 patients thought by the hospital staff to have urinary tract infections, 32 had negative urine tests results, producing a false positive rate of 29 percent. Process study in this example was indicated because maximum acceptable criteria for both false negative and false positive diagnoses were exceeded by measured findings [Gonella et al., 1970].

Example 2. Another study of urinary tract infections used the same criteria of maximum acceptable outcomes established for Example 1. Measurement of outcomes was accomplished by an independent study team who examined 133 consecutive new patients admitted to the medical clinic. Over three months later the patients' charts were examined, and recorded results were compared with the findings of the study team. Of 18 patients requiring urinary tract care, 10 (56 percent) were missed by the clinic staff. There were no false positives [Williamson, 1971].

Example 3. A study of heart failure was conducted in a city hospital in the East that was interested in applying the Williamson quality control strategy to the study of patients in the emergency room. The first sample consisted of 113 consecutive admissions suspected of having acute coronary artery disease. Criteria were based on staff judgment and set at 5 percent as the maximum acceptable levels for both false negatives and false positives in diagnostic results. Measurement of diagnostic outcomes indicated a false negative rate of 3 percent and a false positive rate of 0 percent. Process study was not indicated, since the criteria were not exceeded.

Example 4. A study assessing the health status of patients at the end of a one-year follow-up (therapeutic outcomes) was made of 75 patients among the 113 consecutive emergency room patients who were suspected of having an acute coronary occlusion. The criterion of maximum acceptable fatality rate set by the medical staff by peer judgment techniques was 30 percent. Measurement of outcomes on one-year follow-up revealed that 31 percent had died. Process study was not indicated, since the 95 percent confidence limits were not exceeded [Williamson, 1971].

The examples indicate that setting standards for desired outcomes, monitoring performance with comparisons to standard, and taking corrective action in the traditional control loop conception may have valid use in health care systems. While the idea of setting standards for recovery or death rates may be shocking to some, it is realistic in decision theory terms and may result in improvements of care quality and some basis for the control of institutions providing substandard care. In fact, the health care quality control example is most encouraging, since it shows that rigorous concepts of quality control in services may be possible.

THE MAINTENANCE FUNCTION

Quality control procedures are designed to track characteristics of quality and to take action to maintain quality within limits. In some instances the action called for may be equipment maintenance. The maintenance function then acts in a supporting role to keep equipment operating effectively to maintain quality standards, as well as to maintain the quantitative and cost standards of output.

There are alternate policies that may be appropriate, depending on the situation and the relative costs. First, is routine preventive maintenance economical, or will it be less costly to wait for breakdowns to occur and repair the equipment? Are there guidelines that may indicate when preventive maintenance is likely to be economical? What service level for repair is appropriate when breakdowns do occur? How large should maintenance crews be to balance the costs of downtime versus the crew costs?

In addition, there are longer-range decisions regarding the possible overhaul or replacement of a machine.

The decision concerning the appropriate level of preventive maintenance rests on the balance of costs, as indicated in Figure 12-7. Managers will want to select that policy which minimizes the sum of preventive maintenance plus repair costs.

Curve *a* in Figure 12-7 represents the increase in costs that result from higher levels of preventive maintenance. These costs increase because increased level means that we replace parts before they fail more often, and/or we replace more components when preventive maintenance is performed. In addition, there may be more frequent lubrication and adjustment schedules for higher levels of preventive maintenance. Curve *b* of Figure 12-7 represents the declining cost of breakdown and repair as the level of preventive maintenance increases. These costs represent the cost of repair plus downtime costs that result from a breakdown. With higher levels of preventive maintenance, we should anticipate fewer actual breakdowns.

The total incremental cost curve is the sum of curves *a* and *b*. The optimal policy regarding the level of preventive maintenance is defined by the minimum of that curve.

There is a combination of costs that leads to the decision not to use preventive maintenance. Suppose that the breakdown and repair costs did not decline as the level of preventive maintenance increased or declined more slowly than preventive costs increased. Then preventive maintenance would not be justified, since the minimum total cost occurs with no preventive maintenance. The optimal policy would then be to simply repair the machine when breakdowns occurred.

In order to develop a framework for preventive maintenance policy, we need basic data concerning breakdowns.

Breakdown Time Distributions

Breakdown time distribution data are basic to the formulation of any general policies concerning maintenance. Breakdown time distributions show the frequency with

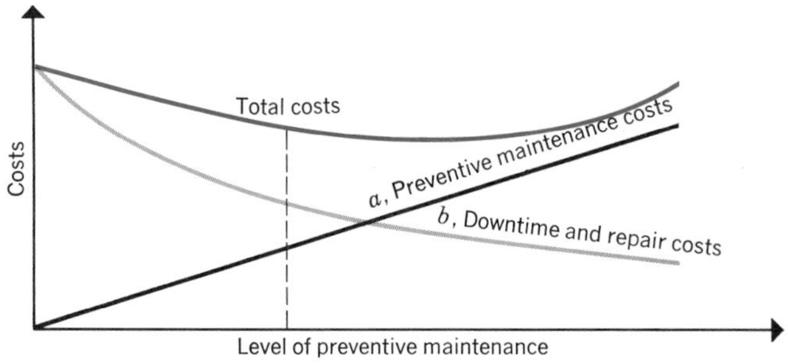

FIGURE 12-7
Balance of costs defining an optimal preventive maintenance policy.

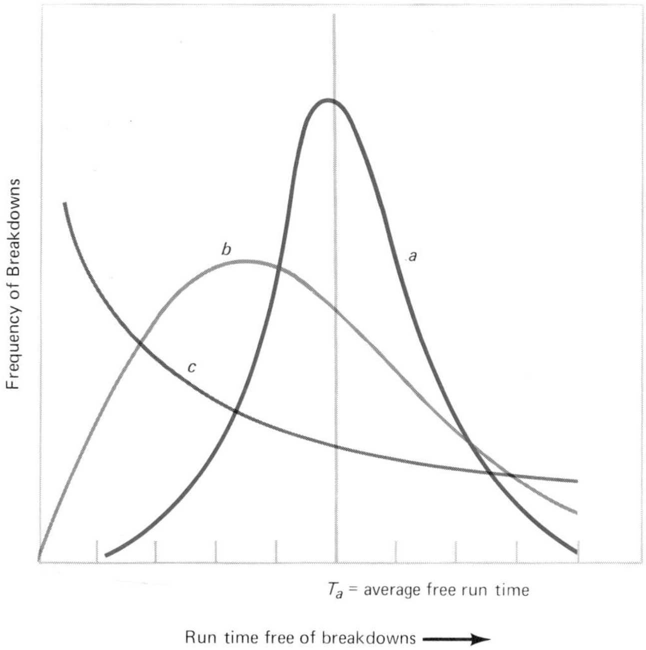

FIGURE 12-8
Frequency distribution of run time free of breakdowns representing three degrees of variability in free run time.

which machines have maintenance-free performance for a given number of operating hours. Ordinarily, they are shown as distributions of the fraction of breakdowns that exceed a given run time. Breakdown time distributions are developed from distributions of run time free of breakdowns, as shown in Figure 12-8.

Figure 12-9 shows three breakdown time distributions. These distributions take different shapes, depending on the nature of the equipment with which we are dealing. For example, a simple machine with a few moving parts would tend to break down at nearly constant intervals following the last repair. That is, it would exhibit minimum variability in breakdown time distributions. Curve *a* of Figure 12-8 would be fairly typical of such a situation. A large percentage of the breakdowns occur near the average breakdown time, T_a, and only a few occur at the extremes.

In a more complex machine with many parts, each part would have a failure distribution. When all these parts were grouped together in a single distribution of the breakdown time of the machine for any reason, we would expect to find greater variability. The machine could break down for any one of a number of reasons. Some breakdowns occur shortly after the last repair, or at any time. Therefore, for the same average breakdown time T_a, we would find much wider variability of breakdown time, as in curve *b* of Figure 12-8.

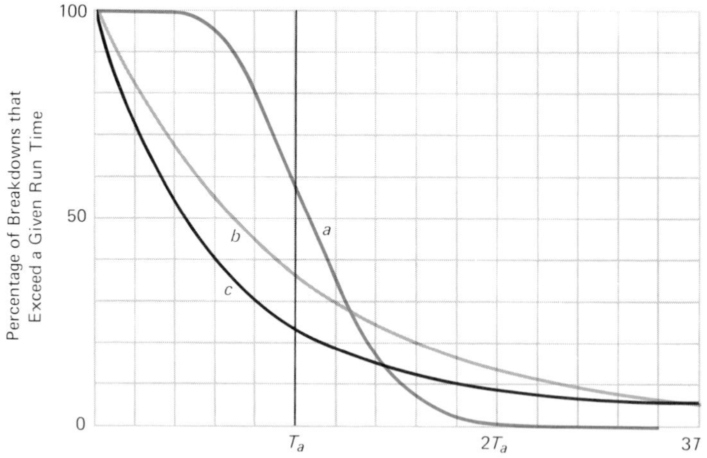

FIGURE 12-9
Breakdown time distributions. Curve *a* exhibits low variability from the average breakdown time T_a. Curve *b* is the negative exponential distribution and exhibits medium variability. Curve *c* exhibits high variability; vertical line shows constant breakdown times.
SOURCE: C. M. Morse, *Queues, Inventories, and Maintenance.* John Wiley, New York, 1958.

To complete the picture of representative breakdown time distributions, curve *c* is representative of distributions with the same average breakdown time T_a, but with wider variability. A large proportion of the breakdowns with a distribution such as curve *c* occurs just after repair; on the other hand, machines may have a long running life after repair. Curve *c* may be typical of machines that require "ticklish" adjustments. If the adjustments are made just right, the machinery may run for a long time; if not, readjustment and repair may be necessary almost immediately.

In models for maintenance, we normally deal with distributions of the percentage of breakdowns that exceed a given run time, as shown in Figure 12-9. They are merely transformations of the distributions of free run time typified by those in Figure 12-8. By examining Figure 12-9, we see that almost 60 percent of the breakdowns exceeded the average breakdown time T_a and that very few of the breakdowns occurred after $2T_a$.

In practice, actual breakdown time distributions often can be approximated by standard distributions, three of which are shown in Figure 12-9. Curve *b* is the negative exponential distribution.

Preventive Maintenance

Assume a preventive maintenance policy for a single machine that provides for an inspection and perhaps replacement of parts after the machine has been running for a fixed time, called the *preventive maintenance period.* The maintenance crew takes an average time, T_m, to accomplish the preventive maintenance. This is the *preventive maintenance cycle.* A certain proportion of the breakdowns will occur before the fixed cycle has been completed. For these cases, the maintenance crew will repair the machine, taking an average time, T_s, for the repair. This is the *repair cycle.* These two patterns of maintenance are diagrammed in Figure 12-10. The probability of occurrence of the two different cycles depends on the specific breakdown time distribution of the machine and the length of the standard preventive maintenance period. If the distribution has low variability and the standard period is perhaps only 80 percent of the average run time without breakdowns, T_a, actual breakdown would occur rather infrequently, and most cycles would be preventive maintenance cycles. If the distribution were more variable for the same standard preventive maintenance period, more actual breakdowns would occur before the end of the standard period. Shortening the standard period would result in fewer actual breakdowns, and lengthening it would have the opposite effect for any distribution.

Assuming that either a preventive maintenance or a repair puts the machine in shape for a running time of equal probable length, the percentage of machine running time depends on the ratio of the standard maintenance period and the average run time, T_a, for the breakdown time distribution. Figure 12-11 shows the relationship between the percentage of time that the machine is working and the ratio of the standard period to average run time, T_a, for the three distributions of breakdown times shown in Figure 12-9. In general, when the standard period is short (say less than 50 percent of T_a), the machine is working only a small fraction of the time. This is because the machine is down so often owing to preventive maintenance. As the standard period is lengthened, more actual breakdowns occur that require repair. For curves *b* and *c*, this improves the fraction of time during which the machine is running because the combination of preventive maintenance time and repair time produces a smaller total downtime.

Curve *a*, however, contains an optimum preventive maintenance period, which maximizes the percentage of machine working time. What is different about curve *a*? It is based on the low variability breakdown time distribution from Figure 12-9. For curve *a*, lengthening the maintenance period beyond about 70 percent of T_a reduces the fraction of machine working time because actual machine breakdowns are more

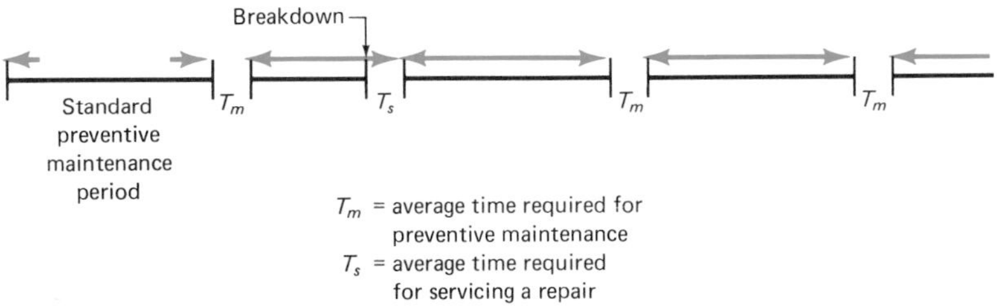

FIGURE 12-10
Illustrative record of machine run time, preventive maintenance time T_m, and service time for actual repairs T_s.

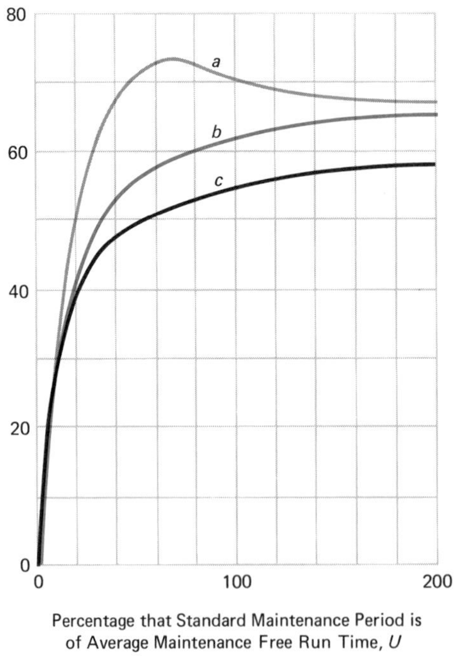

FIGURE 12-11
Percentage of time a machine is working for the three distributions of breakdown time shown in Figure 12-9. Preventive maintenance time, T_m, is 20 percent of T_a; repair time is 50 percent of T_a.
SOURCE: C. M. Morse, *Queues, Inventories, and Maintenance*. John Wiley, New York, 1958.

likely. For the more variable distributions of curves *b* and *c*, this is not true, because breakdowns are more likely throughout the distributions of these curves than they are in curve *a*. Comparable curves can be constructed showing the percentage of time that the machine is in a state of preventive maintenance and the percentage of time that the machine is being repaired because of breakdown.

Guides to a Preventive Maintenance Policy

First, preventive maintenance generally is applicable to machines with breakdown time distributions that have low variability, exemplified by curve *a* of Figure 12-9. In general, distributions with less variability than the exponential, curve *b*, are in this category because low variability means that we can predict with fair precision when the majority of breakdowns will occur. A standard preventive maintenance period can then be set that anticipates breakdowns fairly well.

Equally important, however, is the relation of preventive maintenance time to repair time. If it takes just as long to perform a preventive maintenance as it does to repair the machine, there is no advantage in preventive maintenance, since the amount of time that the machine can work is reduced by the amount of time that it is shut down for repairs. In this situation, the machine will spend a minimum amount of time being down for maintenance if we simply wait until it breaks down.

The effect of downtime costs can modify these conclusions. Suppose that we are dealing with a machine in a production line. If the machine breaks down, the entire line may be shut down, and very high idle labor costs will result. In this situation, preventive maintenance is more desirable than repair *if* the preventive maintenance can take place during second or third shifts, vacations, or lunch hours, when the line normally is down anyway. This is true even when $T_m \geq T_s$. The determination of the standard preventive maintenance period would require a different, but similar, analysis in which the percentage of machine working time is expressed as a function of repair time only, since preventive maintenance takes place outside or normal work time.

An optimal solution would minimize the total of downtime costs, preventive maintenance costs, and repair costs. The effect of the downtime costs would be to justify shorter standard preventive maintenance periods and to justify making repairs more quickly (at higher cost) when they do occur. There are many situations, however, in which extra personnel on a repair job would not speed it up. In such cases, total downtime might be shortened by overtime on multiple shifts and weekends, with higher costs. Optimal solutions would specify the standard preventive maintenance period, the machine idle time, and the repair crew idle time, striking a balance between downtime costs and maintenance costs.

Overhaul and Replacement

In maintaining system reliability, sometimes more drastic maintenance actions are economical. These decisions renew machines through overhauls or replace them when obsolete. Overhaul and replacement decisions can be related to the capital and operating costs (including maintenance) of the equipment. Figure 12-12 shows that while the operating costs are temporarily improved through preventive maintenance,

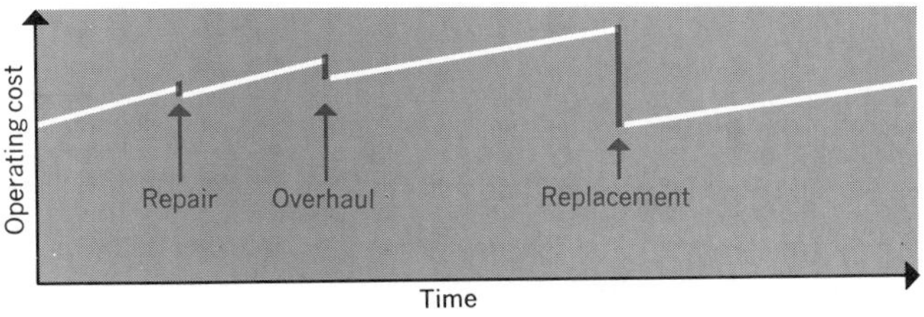

FIGURE 12-12
Operating cost increase versus time, with temporary improvements resulting from repair, overhaul, and replacement.

TABLE 12-1 **Present Values of Repair and Overhaul Alternatives for a Machine (Assume that salvage value is zero. Interest rate = 10 percent)**

(1) Year	(2) Present Value Factor for Future Single Payments[a]	(3) Repair Costs	(4) Present Value of Repair Costs, Col. 2 × Col. 3	(5) Overhaul Costs	(6) Present Value of Overhaul Costs, Col. 2 × Col. 5
Initial	1.000	$500	$500	$1500	$1500
1	0.909	2000	1818	1800	1636
2	0.826	2500	2065	2000	1652
3	0.751	3000	2253	2300	1727
			$6636		$6515

[a] Present value factors from Table H-1 in Appendix H.

repair, and overhaul, there is a gradual cost increase until replacement is finally justified.

Repair Versus Overhaul. The decisions concerning the choice between repair and major overhaul normally occur at the time of a breakdown. Many organizations also have regular schedules for overhaul. For example, trucking companies may schedule major engine overhauls after a given number of miles of operation. These major preventive maintenance actions are meant to anticipate breakdowns and the occurrence of downtime at inconvenient times and perhaps to minimize downtime costs.

Since renewals through overhaul involve future costs, these values must be discounted. For example, suppose that a machine breakdown has just occurred. It will cost $500 to repair the equipment, after which the annual operating costs are expected to be $2000, $2500, and $3000 for the next three years, at which time replacement is planned. If a major overhaul is performed now, the cost will be $1500, with operating costs of only $1800, $2000, and $2300 in the following three years, with the replacement decision probably postponed. Let us first examine just the next three years of cost to see if the overhaul is justified in that time frame. The two alternatives are compared in Table 12-1 by discounting all future costs to present values, using a 10 percent interest rate.* In this instance, the present value of overhaul is lower and tentatively would be the more economical strategy.

Replacement Decisions. If the choice is only between overhaul and replacement, the foregoing analysis may be adequate. However, the replacement alternatives lurk in the background and really need to be considered as a part of a sequential decision strategy. The possible sequences could include repair, overhaul, perhaps a second overhaul, replacement, repair, overhaul, and so on.

An Example. Suppose that a machine is used in a productive system and that it is usually overhauled after two years of operation, or replaced. The present machine

* Present value methods are reviewed in Appendix A, "Capital Costs and Investment Criteria."

was purchased two years ago, and a decision must now be made concerning overhaul or possible replacement. The machine costs \$9000 installed, and annual operating costs (including maintenance) are \$2000 the first year and \$3000 the second year. The machine can be overhauled for \$5000, but operating costs for the next two years will be \$2800 and \$4000 for the first overhaul and \$3500 and \$5000 for the second overhaul.

In deciding whether to overhaul or replace at this time, we should consider the available alternate sequences of decisions. For example, we can overhaul at this time or replace. For each of these possible decisions, we have the same options two years hence, and so on. Figure 12-13 shows the simple decision tree structure.

In order to compare the alternatives, the future costs are discounted to present value, using the present value methods reviewed in Appendix A. The calculations are summarized in Table 12-2 for the four sequences indicated in the decision tree of Figure 12-13. The four alternate strategies are:

1. Replace now and in two years (R-R)
2. Replace now and overhaul in two years (R-OH)
3. Overhaul now and replace in two years (OH-R)
4. Overhaul now and again in two years (OH-OH)

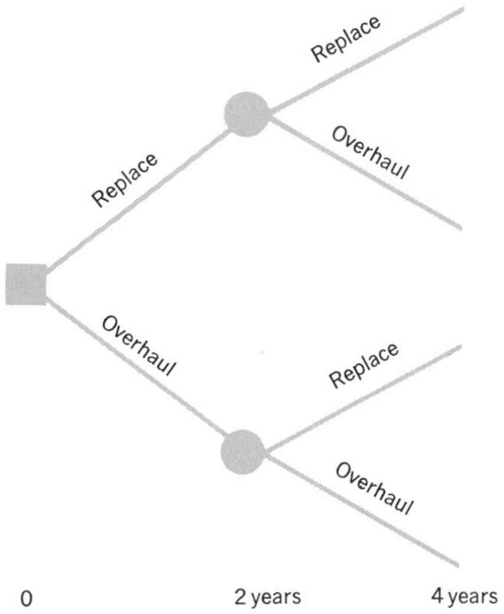

FIGURE 12-13
Decision tree for the overhaul-replacement example.

TABLE 12-2 **Present Values for Four Alternate Strategies Involving Overhaul and Replacement (Assume that salvage values are zero. Interest rate = 10 percent)**

(1) Year	(2) Present Value Factor for Future Single Payments[a]	(3) Costs for R-R Sequence	(4) Present Value of R-R, Col. 2 × Col. 3	(5) Costs for R-OH Sequence	(6) Present Value of R-OH, Col. 2 × Col. 5	(7) Costs for OH-R Sequence	(8) Present Value of OH-R, Col. 2 × Col. 7	(9) Costs for OH-OH Sequence	(10) Present Value of OH-OH, Col. 2 × Col. 9
Initial	1.000	$9,000.00	$9,000.00	$9,000.00	$9,000.00	$5,000.00	$5,000.00	$5,000.00	$5,000.00
1	0.909	2,000.00	1,818.00	2,000.00	1,818.00	2,800.00	2,545.20	2,800.00	2,545.20
2	0.826	3,000.00	2,478.00	3,000.00	2,478.00	4,000.00	3,304.00	4,000.00	3,304.00
Replace or overhaul at end of 2nd year	0.826	9,000.00	7,434.00	5,000.00	4,130.00	9,000.00	7,434.00	5,000.00	4,130.00
3	0.751	2,000.00	1,502.00	2,800.00	2,102.80	2,000.00	1,502.00	3,500.00	2,628.50
4	0.683	3,000.00	2,049.00	4,000.00	2,732.00	3,000.00	2,049.00	5,000.00	3,415.00
			$24,281.00		$22,260.80		$21,834.20		$21,022.70

[a] Present value factors from Table H-1 in Appendix H.

The four-year present value totals in Table 12-2 indicate that the best strategy for this example is to overhaul each two years (OH-OH) and that the next best strategy is to overhaul now and replace in two years (OH-R). This is true in spite of the rapidly mounting operating costs.

Since operating costs do increase so rapidly, perhaps it will be worthwhile to see what happens if we adopt a six-year planning horizon. If a third overhaul is scheduled, the next two years' operating costs will be $4500 and $5500. On the other hand, if we add an overhaul cycle to the third alternative, it will enjoy the relatively low operating cost of a first overhaul. Adding a third cycle to each of the four alternatives results in the following strategies (we are ignoring the additional sequences created by a third branching):

1. R-R-R
2. R-OH-R
3. OH-R-OH
4. OH-OH-OH

Table 12-3 summarizes the calculations for the third cycle, the present values for the first four years, and the six-year present value totals. Alternative 3 (OH-R-OH) is now the lowest-cost strategy. This change in result demonstrates the importance of choosing a horizon that fairly represents all the alternatives. If a fourth second-year cycle were added to the evaluation, it might seem that strategies 3 and 4 are the same but reversed in sequence. But they are not the same because we start from an existing situation with a two-year old machine. Strategy 2 places two replacements in sequence, while strategy 3 alternates overhauls and replacements.

This example assumes replacement with an identical machine, but it is often true that alternate machines will have rather different capital and operating costs characteristics. New machine designs often have improvements owing to mechanization or automation that reduce labor and maintenance costs, and these cost advantages would affect replacement decisions.

IMPLICATIONS FOR THE MANAGER

The general concepts of system reliability are important for managers to understand. When productive systems involve a network of activities with many required sequences, it will be difficult to maintain the reliability of the system as a whole. This system unreliability would be true even though each individual operation might be 99 percent reliable. Managers can improve the reliability by providing parallel capabilities and slack capacity, although these remedies may be expensive.

The most important techniques available to managers to sustain reliability are through quality control and equipment maintenance systems. The quality control system functions as a primary control loop, with the maintenance system providing reliability in the longer term through a secondary control loop.

Quality control begins in the preproduction planning phases of an enterprise, where policies regarding market strategies are developed. Quality standards are then developed out of the iterative process of product/service design and productive system design. The productive system must be designed so that it is capable of producing the quality level set at reasonable cost.

TABLE 12-3 **Present Values for Third Cycle, and Six-Year Totals for Four Alternate Strategies Involving Overhaul and Replacement (Assume that salvage values are zero. Interest rate = 10 percent)**

(1) Year	(2) Present Value Factor for Single Payments[a]	(3) Costs for R-R Sequence	(4) Present Value of R-R, Col. 2 × Col. 3	(5) Costs for R-OH Sequence	(6) Present Value of R-OH, Col. 2 × Col. 5	(7) Costs for OH-R Sequence	(8) Present Value of OH-R, Col. 2 × Col. 7	(9) Costs for OH-OH Sequence	(10) Present Value of OH-OH, Col. 2 × Col. 9
Replace or overhaul at end of 4th year	0.683	$9,000.00	$6,147.00	$9,000.00	$6,147.00	$5,000.00	$3,415.00	$5,000.00	$3,415.00
5	0.621	2,000.00	1,242.00	2,000.00	1,242.00	2,800.00	1,738.80	4,500.00	2,794.50
6	0.564	3,000.00	1,692.00	3,000.00	1,692.00	4,000.00	2,256.00	5,500.00	3,102.00
			$9,081.00		$9,081.00		$7,409.80		$9,311.50
First 4 years from Table 12-2			24,281.00		22,260.80		21,834.20		21,022.70
6-year totals			$33,362.00		$31,341.80		$29,244.00		$30,334.20

[a] Present value factors from Table H-1 in Appendix H.

Monitoring quality levels of output is necessarily a sampling process, since the entire population of output is seldom available for screening. The techniques of statistical quality control are often valid and cost-effective mechanisms for managers to employ.

Quality control of services is difficult for a variety of reasons related to the unique character of services. Attempts are being made to establish a framework for all kinds of control. Legislation has placed great emphasis on quality control in health care systems, and self-regulation experiments are now developing. Nevertheless, quality control techniques in service operations are underdeveloped, partly because rigorous standards for quality of services are not available.

Managers often regard the maintenance function as ancillary to operations, ignoring its crucial role in supporting the reliability system. It is important to understand when preventive maintenance is likely to be appropriate. Analysis of breakdown time distributions provides guidelines to the development of preventive maintenance policies. In general, these policies are appropriate when breakdown time distributions exhibit low variability and when the average time for preventive maintenance is less than the average repair time following breakdown. Also, when downtime costs are large, preventive maintenance is preferable to repair if it can be performed when the facilities are normally down anyway.

The maintenance function extends into decisions involving major overhaul and replacement. These kinds of decisions involve a longer time horizon and the proper handling of capital costs. Present value techniques for the evaluation of alternate strategies may be used, and the strategies need to consider sequences of decisions involving repair, overhaul, and replacement.

REVIEW QUESTIONS AND PROBLEMS

1. If the *reliability* of a productive system means its capability to maintain quality, schedule, and cost standards, which of the following kinds of systems is likely to encounter the greatest quality-reliability problems? Schedule-reliability problems? Why?
   a. Skyscraper construction project
   b. Automobile engine plant
   c. Restaurant
   d. Hospital

2. Compare the reliability of systems with components in series versus components in parallel.

3. How does variability of performance affect system reliability?

4. What are the primary and secondary feedback control loops in the broad reliability system?

5. What measures of quality of output do you think might be important in the following kinds of systems? Be as specific as possible.
   a. Fast food operation, such as McDonalds
   b. Motel
   c. Luxury hotel
   d. Space vehicle manufacture

6. How can you control the aspects of quality that you suggest be measured in your answer to question 5?

7. Define the following terms:
   a. Type I error
   b. Type II error
   c. Producer's risk
   d. Consumer's risk

8. What kinds of control can be exercised in maintaining quality standards?

9. What conditions make acceptance sampling appropriate?

10. What are the criteria of quality in the following?
    a. Banking service
    b. The post office
    c. An institution of higher learning
    d. A hospital
    e. The Internal Revenue Service

11. Discuss some of the problems that affect the managerial control of nonprofit services.

12. In controlling the quality of health care, what is the meaning of a *false positive?* A *false negative?*

13. In the Williamson control model for health services, under what conditions do we examine the treatment processes that are used?

14. What kinds of costs are associated with machine breakdown?

15. Discuss the general methods by which the reliability of productive systems can be maintained.

16. What is a breakdown time distribution?

17. Discuss the types of situations of machine breakdown that are typified by curves *a*, *b*, and *c*, respectively, in Figure 12-8.

TABLE 12-4 **Operating Costs for Problem 22**

Year	New Machine	First Overhaul	Second Overhaul	Third Overhaul	Fourth Overhaul
1	$1000	$1100	$1300	$1700	$2300
2	1100	1300	1700	2300	3200

18. What are the general conditions for which preventive maintenance is appropriate?

19. If it takes just as long to perform a preventive maintenance operation as it does a repair, is there an advantage to preventive maintenance? How can high downtime costs modify this?

20. Rationalize why the operating costs for equipment should increase with time, as indicated in Figure 12-12.

21. If the decision tree of Figure 12-13 were developed through decisions made at the end of four years, how many alternate sequences result? Enumerate them.

22. A company is considering whether to overhaul or replace a machine. The machine was purchased four years ago and was overhauled two years ago.

    A new machine costs $2000, and an overhaul costs $500 and lasts two years. Experience indicates that annual operating costs increase with time owing to increased maintenance charges. Table 12-4 shows operating costs for new and overhauled machines.

    Analyze the situation and indicate what decision should be made. Assume that machines have no salvage value at any time and that the cost of capital is 15 percent per year.

REFERENCES

Adam, E. E., J. C. Hershauer, and W. A. Ruch. *Measuring the Quality Dimension of Service Productivity*. National Science Foundation No. APR 76-07140, University of Missouri—Arizona State University, 1978.

Anthony, R. H., J. Deardon, and R. F. Vancil. *Management Control Systems*. Richard D. Irwin, Homewood, Ill., 1972.

Barlow, R., and L. Hunter. "Optimum Preventive Maintenance Policies," *Operations Research*, *8*(1), 1960, pp. 90–100.

Bennigson, L. A., and A. I. Bennigson. "Product Liability: Manufacturers Beware!" *Harvard Business Review*, May-June 1974.

Boere, N. J. "Air Canada Saves with Aircraft Maintenance Scheduling," *Interfaces*, 7(3), May 1977, pp. 1–13.

Bovaird, R. L. "Characteristics of Optimal Maintenance Policies," *Management Science*, *7*(3), April 1961, pp. 238–254.

Buffa, E. S., and J. S. Dyer. *Management Science/Operations Research: Model Formulation and Solution Methods.* John Wiley, New York, 1977.

Cunningham, C. E., and W. Cox. *Applied Maintainability Engineering.* John Wiley, New York, 1972.

Dodge, H. F., and H. G. Romig. *Sampling Inspection Tables* (2nd ed.). John Wiley, New York, 1959.

Duncan, A. J. *Quality Control and Industrial Statistics* (4th ed.). Richard D. Irwin, Homewood, Ill., 1974.

Eginton, W. W. "Minimizing Product Liability Exposure," *Quality Control*, January 1973.

Enrick, N. L. *Quality Control* (5th ed.). Industrial Press, New York, 1966.

Gilbert, J. O. W. *A Manager's Guide to Quality and Reliability.* John Wiley, New York, 1968.

Goldman, A. S., and T. B. Slattery. *Maintainability: A Major Element of System Effectiveness.* John Wiley, New York, 1964.

Gonella, J. A., M. J. Goran, and J. W. Williamson, et al. "The Evaluation of Patient Care," *Journal of the American Medical Association, 214*(8), 1970, pp. 2040–2043.

Gradon, F. *Maintenance Engineering: Organization and Management.* John Wiley, New York, 1973.

Grant, E. L., and R. S. Leavenworth. *Statistical Quality Control* (4th ed.). McGraw-Hill, New York, 1972.

Hardy, S. T., and L. J. Krajewski. "A Simulation of Interactive Maintenance Decisions," *Decision Sciences, 6*(1), 1975, pp. 92–105.

Harris, D. H., and F. B. Chaney. *Human Factors in Quality Assurance.* John Wiley, New York, 1969.

Jardine, A. K. S. "Decision Making in Maintenance," *Chartered Mechanical Engineer, 21*(1), January 1974, pp. 35–39.

Jardine, A. K. S. *Maintenance, Replacement and Reliability.* John Wiley, New York, 1973.

Juran, J. M., and F. M. Gryna. *Quality Planning and Analysis: From Product Development Through Usage.* McGraw-Hill, New York, 1970.

Kirkpatrick, E. G. *Quality Control for Managers and Engineers.* John Wiley, New York, 1970.

Lusser, R. "The Notorious Unreliability of Complex Equipment," *Astronautics, 3*(2), February 1958, pp. 74–78.

Meister, D., and G. F. Rabideau. *Human Factors Evaluation in System Development.* John Wiley, New York, 1965.

Morse, P. M. *Queues, Inventories, and Maintenance.* John Wiley, New York, 1958.

Peck, L. G., and R. N. Hazelwood. *Finite Queuing Tables.* John Wiley, New York, 1958.

Smith, C. S. *Quality and Reliability: An Integrated Approach.* Pitman, New York, 1969.

Turban, E. "The Use of Mathematical Models in Plant Maintenance," *Management Science, 13*(6), February 1967, pp. 342–358.

Vance, L. L., and J. Neter. *Statistical Sampling for Auditors and Accountants.* John Wiley, New York, 1956.

Williamson, J. W. "Evaluating Quality of Patient Care," *Journal of the American Medical Association, 218*(4), October 25, 1971, pp. 564–569.

PART

FOUR

DESIGN OF PRODUCTIVE SYSTEMS

WHEN ONE THINKS OF A PRODUCTIVE SYSTEM IN THE PHYSIcal sense, there are processes and equipment, jobs, and a facility to house them. Relating these elements in a way that results in an efficient productive system that fits in with the long-term goals of the enterprise is the subject of the two chapters of Part Four.

Figure IV-1 is a diagram that relates the major elements of long-term strategic decisions to the system design. The process diagrammed should be thought of as being dynamic. It must take account of changing consumer preferences; changing demand quantity, mix, and location; changing technological innovations in products *and* processes; and changing social and cultural values of employees.

A good design will be one in which a balance is achieved between the competing criteria and values. Some of the criteria are economic, related to markets and demand; some are behavioral, related to consumer behavior and to job satisfaction; and some are technological, involving products and processes.

Assume we are starting with a new idea for a useful product or service, as we did in Chapter 1 in discussing the BURGER enterprise. Rereading the pages regarding the beginning and evolution of BURGER may be of value at this point, since it provides a narrative that fits into the general framework of Figure IV-1.

Taking BURGER as an example, we started with an attempt to make market predictions and long-range plans (blocks 1 and 2) for an organization to provide a service of certain defined characteristics in the food service market (block 3). As we sharpened our concept of the business (particularly the nature of the food service provided and the food quality), we made broad studies of the size and location of markets and the size and location of branches. These plans became important inputs to blocks 3 and 4, which involve the design of the food service and its impact on the design of facilities.

Recall that at one point in the evolution of BURGER, we made plans to design a modular branch as well as a system of branches. The inputs to the design of the productive system were from blocks 4, 5, 7, and 8. The forecasts (block 5) provided data useful in determining a peak and average capacity, but the output rates were affected by the processes used and the work flow layout (block 7) and the way we decided to put the work teams together (block 8).

The first design was based largely on our experience with the most recent branch, but preliminary cost figures suggested that both labor and material costs were too high. The first alternatives were to relocate some processes at the central warehouse, because they could be done on a larger scale for all branches, with greater mechanization and automatic material handling. This change simplified operations at each branch and made possible the dropping of the production line concept and the adoption of a rotating team approach to staffing the branches. Alternatives were considered that involved the possible alteration of the service offered in order to accommodate some of the needs of the productive system (the interaction between blocks 4 and 6).

With the background of systems concepts, we see feedback loops in Figure IV-1. The productive system design process involves a feedback of information to an analysis and design of the service offered and to the long-range plans that in turn affect

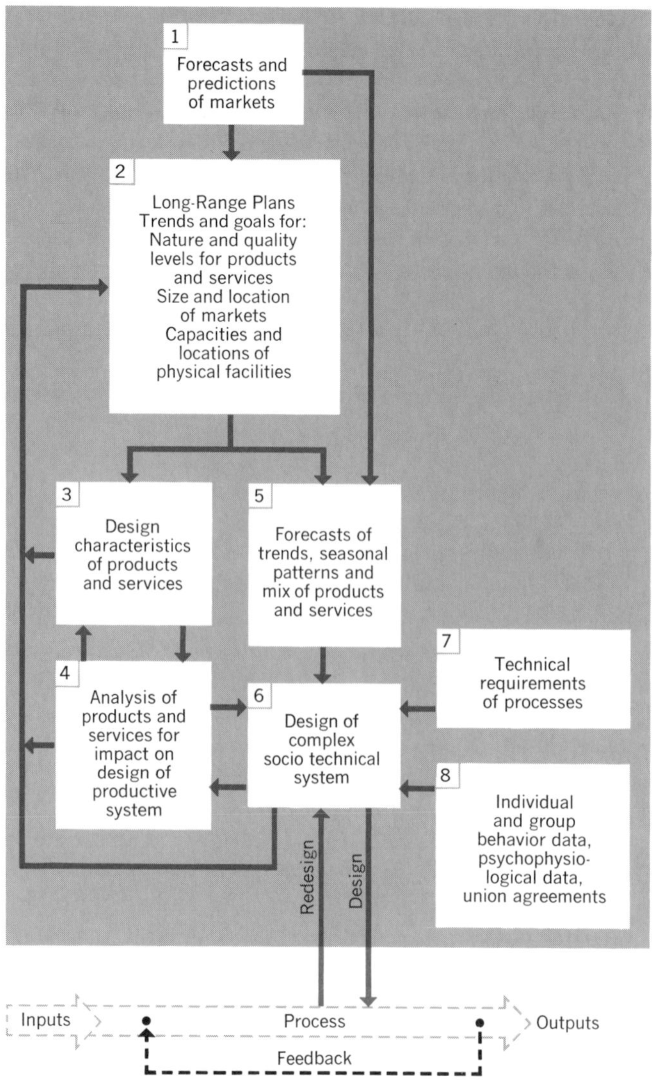

FIGURE IV-1
Process and facilities design module.

capacities and location. In an ongoing situation, there is a continuing information feedback and interaction between all the elements according to the structure of Figure IV-1. The process is dynamic and interactive. The changes are often small ones, but on occasion there may be massive changes, for example, radical product design changes or the abandonment of old facilities and the planning of new ones.

CHAPTER 13 Process and Job Design

THERE ARE STRATEGIC DECISIONS INVOLVING THE DESIGN OF products/services, and a location for the system, that are an integral part of the productive system design. The core of that productive system, however, is the complex of technology and people. We have noted that the entire design process is interdependent and that the products/services and locations are also partially influenced by the productive process, and vice versa. We will concentrate our attention in this chapter on the blending of technology and people as a part of a productive system. Because these productive systems are such a blend, they are often called "sociotechnical systems." In the chapter that follows, we will consider some of the special design problems encountered in integrating the process-job and flow systems.

The result of the process is to develop a rationale for the organization of the work to be done and to relate it to machines and technology. The physical integration of these factors is the facility layout. Whether or not the layout permits an effective design from points of view other than work flow and physical efficiency depends on how technology and people are molded into a system.

It is at this stage that we consider the alternatives of division of labor versus broad-spectrum jobs. In the past, process planning (i.e., technology and layout) has been thought of as the independent variable and people and job designs as the dependent variables. In that framework, job designs were results of process or technology planning. Currently developing concepts and practices consider the two components jointly to produce designs that satisfy the needs of both.

CRITERIA AND VALUES

From the time of the Industrial Revolution to the present, the main pressures influencing the design of processes and jobs have been productivity improvement and economic optimization. Adam Smith stated the advantages of division of labor as the guiding principle, and managers have applied this principle progressively over time. The specialization principle has been applied throughout industry and is currently being applied in service and nonmanufacturing systems.

The specific technology involved determines the extent to which division of labor *can* be pursued and that market size determines how far it *will* be pursued. Present-day technology has indeed fostered specialization, and the size of markets for many products and services is enormous. In short, from the time of Adam Smith to the present, the dominant criterion has been the economic one. It was assumed that finely divided jobs were better because they would increase productivity and that other factors were correlated with productivity. Incentive pay schemes were used to maintain workers' motivation.

Beginning in the early 1930s, however, another criterion was proposed as a counterbalance—job satisfaction. Studies indicated that workers responded to other factors in the work situation. In the late 1940s, the value of the job satisfaction criterion developed from a morale-building program at IBM [Walker, 1950]. The term "job enlargement'' was coined to describe the process of reversing the trend toward specialization. Practical applications of job enlargement were written up in the literature, describing improvements in productivity and quality levels resulting from jobs of broader scope.

Davis [1971a], in commenting on job satisfaction research, points to several *values* held by the organizations that applied the specialization criterion. He says that there was a widely held belief that workers could be viewed as an operating unit; that they could be adjusted and changed by training and incentives to suit the needs of the organization; that workers were viewed as spare parts and were therefore interchangeable in work assignments; that labor was thought of as a commodity to be bought and sold; that materialism in its narrow sense of achieving material comfort justified the means required to achieve it; and that many managers regarded jobs as isolated events in the lives of individuals—a noncareer.

Davis is also critical, however, of most job enlargement and job satisfaction studies because they almost always accept the technology as given and merely attempt to maximize satisfaction within the constraints of technology. As a basis for process planning and job design, Davis concludes that during the industrial era, technology has dominantly determined job content, even in most of the applications taking account of job satisfaction as a criterion.

PROCESS-JOB CONCEPTS

There are constraints imposed by technology that limit the possible arrangements of processes and jobs, and there are constraints imposed by job satisfaction and social system needs. The circle marked ''technological constraints'' in Figure 13-1 indicates that all job designs within the circle represent feasible solutions from a technological point of view, and all points outside the circle are infeasible. Similarly, the circle marked ''social system constraints'' indicates that all job designs within that circle represent feasible solutions from the socio-job satisfaction point of view. Within the shaded area of overlap between the two circles, we have a solution space that meets the constraints of both technology and the social system. The shaded area defines the only solutions that can be regarded as feasible in joint terms. Our objective, then, is to consider jointly the economic and social system variables. We must find the best solution to process and job designs within the feasible shaded solution space. Since optimization is an unclear process in job design, we seek solutions that are acceptable.

Looking within the joint feasible solution space, Scoville [1969] developed models that examine job breadth from the points of view of managers and workers. Figure 13-2 shows a graphic form of the managers' model.

The tasks and duties required can be shuffled in many ways to form the continuum of narrow versus broad job designs. For example, in auto assembly, jobs can be finely divided as with conventional auto assembly lines, or at the opposite extreme, one worker or a team could assemble the entire vehicle. This kind of job enlargement

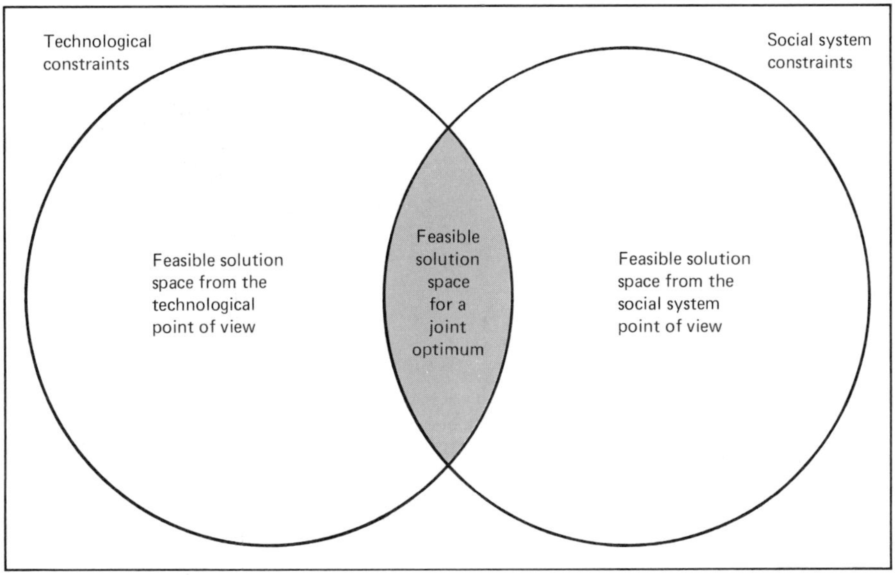

FIGURE 13-1
Feasible solution spaces for technological and social systems, joint solution space.

can be termed "horizontal." Ultimately, a vertical enlargement could be envisioned where jobs incorporate varying degrees of quality control, maintenance, repair, supply, and even supervisory functions.

The curves in Figure 13-2 are rationalized as follows (Scoville argues for the shape of the curves rather than for any specific numerical solution):

1. The wage costs curve reflects low productivity for both very narrow and very broad jobs, with a maximum somewhere in the middle range. Training costs go up as the scope of jobs increase, while turnover costs are most important for narrow jobs. Thus the wage costs curve declines to a minimum and increases thereafter as job breadth increases.

2. Material, scrap, and quality control costs are high with narrow jobs owing to lack of motivation. Penalties are also high for broad jobs because the advantages of division of labor are lost.

3. Supervisory costs decline with broader jobs because that function is progressively incorporated with jobs through job enlargement.

4. Capital costs per worker rise on the assumption that capital-labor cost ratios more than offset the inventory cost of goods in process.

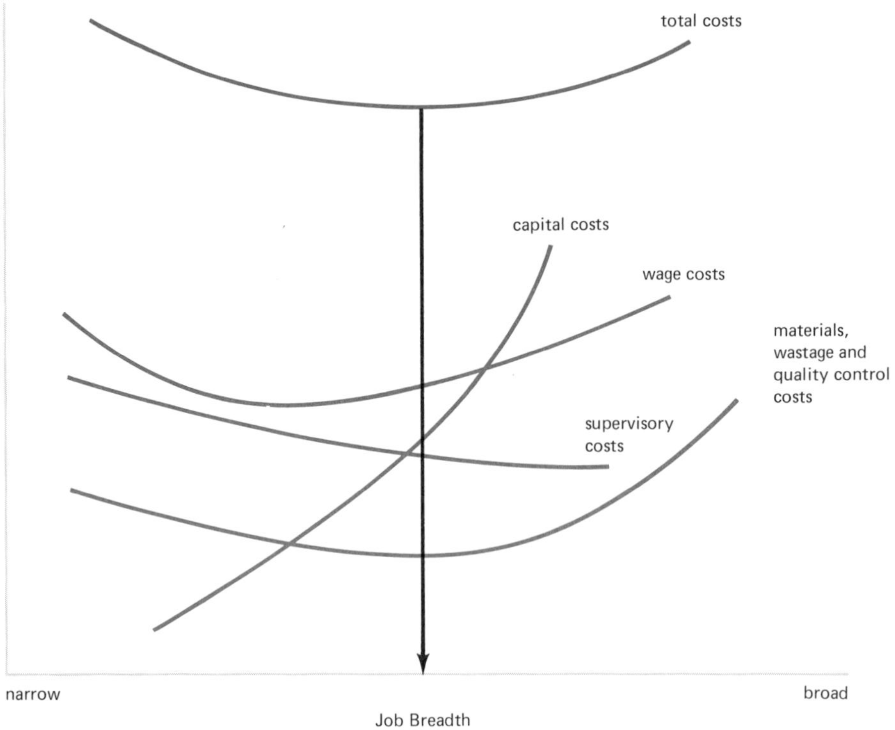

FIGURE 13-2
Manager's model of job breadth.
SOURCE: *J. G. Scoville, "A Theory of Jobs and Training,"* Industrial Relations, 9, *1969, pp. 36–53.*

The total cost curve in Figure 13-2 is the sum of the individual cost component curves and reflects an optimum where managers would choose to operate. As with all cost allocation problems where each pure strategy involves a cost, the joint optimum must be somewhere between the extremes. The pure strategies cannot represent the optimum, since the best solution necessarily results from a balance of costs.

Scoville's model of job breadth from the point of view of workers is shown graphically in Figure 13-3. The wage-productivity and employment probabilities curves are multiplied to produce the discounted expected earnings curve. If one then subtracts the worker-borne training costs, a net economic benefits curve is obtained, which has a maximum near the broad end of the spectrum.

Thus, Scoville's models indicate that there are optimum job designs from an economic point of view. However, since the factors that enter the two models are different, it is unlikely that managers and workers could agree on how work would be organized.

We have a rationale for process and job design in an equilibrium model, where the balance of forces between labor and management are likely to produce an organization of work somewhere within the joint feasible region of Figure 13-1.

TECHNOLOGICAL VIEW OF PROCESS PLANNING AND JOB DESIGN

Although the general methods we will describe were developed in manufacturing systems, they have been adapted and widely used in many other institutions, such as offices, banks, and hospitals.

Figure 13-4 shows the overall development of process plans in a manufacturing situation. Process planning takes as its input the drawings or other specifications that indicate *what* is to be made and the forecasts, orders, or contracts that indicate the *quantity* to be made. The drawings are then analyzed to determine the overall scope of the project. If it is a complex assembled product, considerable effort may go into "exploding" the product into its components of parts and subassemblies. This overall planning may take the form of special drawings that show the relationship of parts, cutaway models, and assembly drawings. Preliminary decisions are made concerning some assembly groupings to determine which parts to make and buy, as well as the general level of tooling expenditures. Then for each part, a detailed routing through the system is developed. Technical knowledge is required concerning processes, machines and their capabilities, costs, and production economics. Ordinarily, a range of processing alternatives is available. The selection may be influenced strongly by the projected volume and stability of product design.

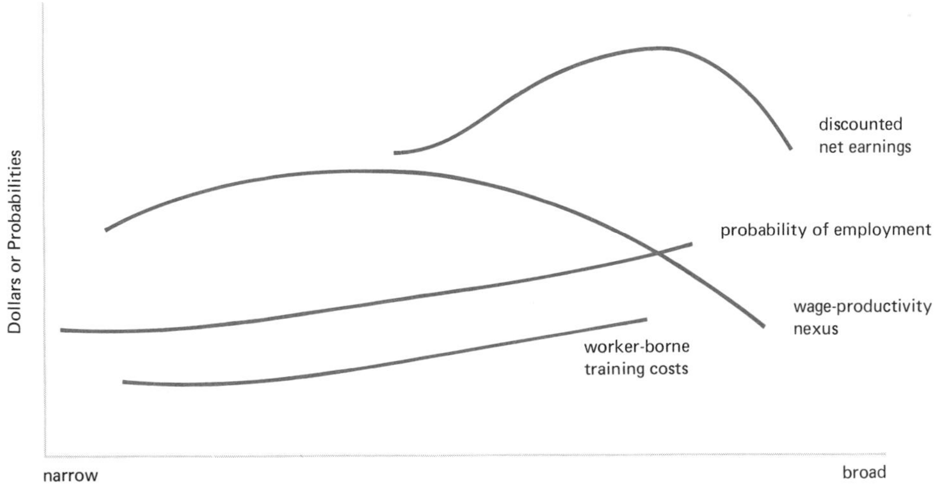

FIGURE 13-3
Worker's model of job breadth.
SOURCE: *J. G. Scoville, "A Theory of Jobs and Training,"* Industrial Relations, 9, *1969, pp. 36–53.*

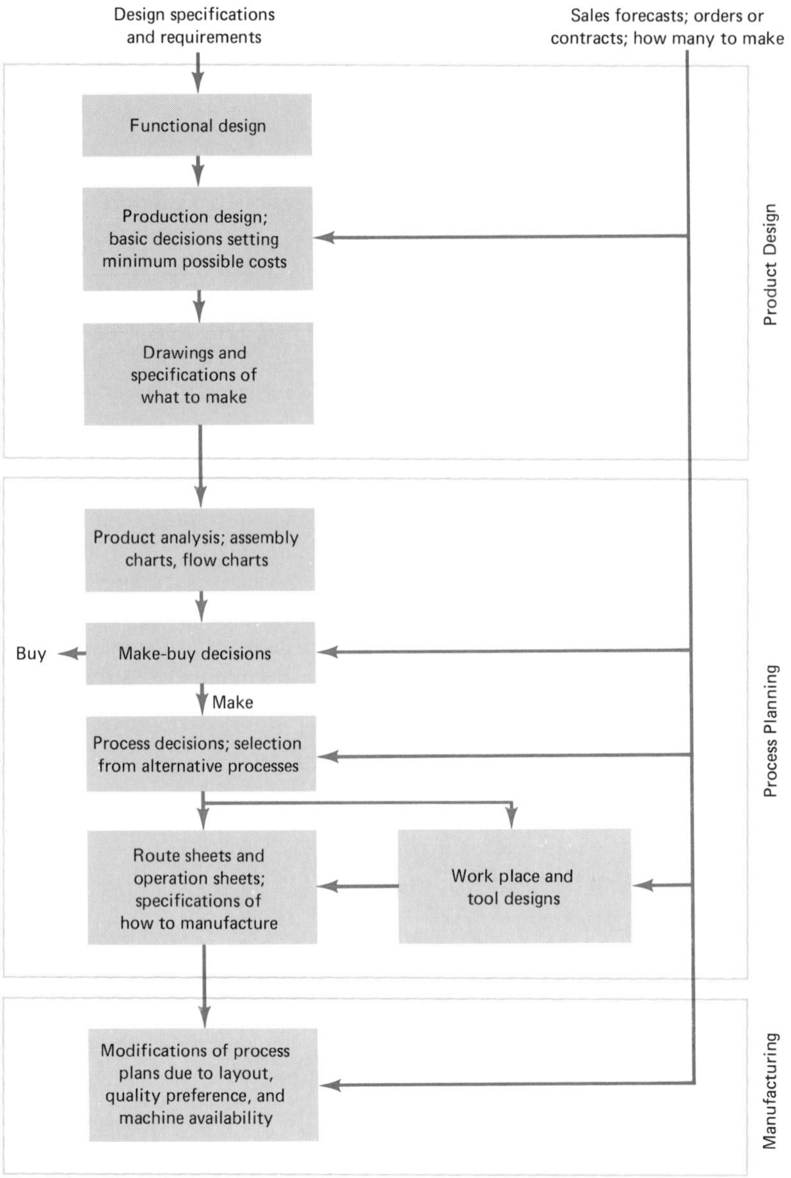

FIGURE 13-4
Development of processing plans.

Product Analysis

The product to be manufactured is analyzed from a technological point of view to determine the processes required.

Assembly Charts. Schematic and graphic models are commonly developed to help visualize the flow of material and the relationship of parts. The assembly chart can be useful in making preliminary plans regarding subassemblies and appropriate general methods of manufacture. For example, where in the process might production lines be appropriate?

Operation Process Charts. Assuming that the product is already engineered, we have complete drawings and specifications of the parts, their dimensions and tolerances, and materials to be used. The engineering drawings specify locations, sizes, and tolerances for holes to be drilled and surfaces to be finished for each part. With this information, the most economical equipment, processes and sequences of processes can be specified.

An operation process chart is then a summary of all required operations and inspections. It is a general plan for manufacture. Although the focus of such charts is on the technological processing required, it is obvious that the jobs to be performed by humans have also been specified. Some discretion in the makeup of jobs still exists, especially in the assembly phase; however, it is clear that "technology is in the saddle."

Analysis of Human–Machine Relationships

Given the product analysis and the required technological processing, individual job designs become the focus. Concepts and methods used have developed over a long period beginning with the scientific management era. The professional designers of processes and jobs have been industrial engineers in industry. In the post-World War II period, psychologists and physiologists have contributed concepts and methods concerning the role of humans in systems, reinforcing the technological view.

Humans Versus Machines. In the technologists' view, human beings have certain physiological, psychological, and sociological characteristics that define their capabilities and limitations in the work situation. These characteristics are thought of, not as fixed quantities, but as distributions that reflect individual variation.

In performing work, human functions are envisioned in three general classifications:

1. Receiving information through the various sense organs.

2. Making decisions based on information received and information stored in the memory of an individual.

3. Taking action based on decisions. In some instances the decision phase may be virtually automatic because of learned responses, as in a highly repetitive task.

In others, the decisions may involve an order of reasoning and the result may be complex.

Note that the general structure of a closed-loop automated system is parallel in concept. Wherein lies the difference? Are automated machines like humans? Yes, in this model of humans in the system, machines and humans are alike in certain important respects. Both have sensors, stored information, comparators, decision makers, effectors, and feedback loops. The difference is in the human being's tremendous range of capabilities and in the limitations imposed by physiological and sociological characteristics. Thus, machines are much more specialized in the kinds and range of tasks they can perform. Machines perform tasks as faithful servants reacting mainly to physical factors. Humans, however, react to their psychological and sociological environments as well as to the physical environment.

Although there are few guides to the allocation of tasks to humans and machines on other than an economic basis, a subjective list of the kinds of tasks most appropriate for humans and for machines is given by McCormick [1970].

Conceptual Framework for Human–Machine Systems

We have noted that humans and machines can be thought of as performing in similar functions in work tasks, although they each have comparative advantages. The functions they perform are represented in Figure 13-5 and are generally comparable to those of the closed-loop feedback system.

Information is received by the *sensing function.* Sensing by humans is accomplished through the sense organs. Machine sensing can parallel human sensing through electronic or mechanical devices. Machine sensing is usually much more specific or single purpose in nature than the broadly capable human senses.

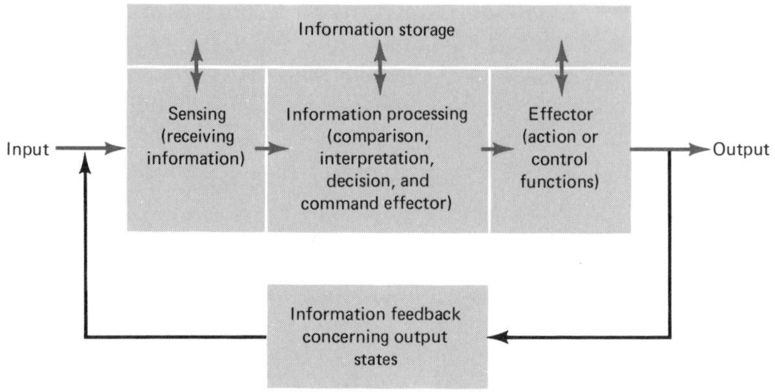

FIGURE 13-5
Functions performed by worker or machine components of worker-machine systems.

Information storage for humans is provided by memory or by access to records. Machine information storage can be obtained by magnetic tape or disk, punched cards, and cams and templets.

The function of *information processing and decision making* is to take sensed and stored information and produce a decision. The processing could be as simple as a choice between two alternatives, depending on input data, or very complex involving deduction, analysis, or computing to produce a decision for which a command is issued to the effector.

The effector or *action function* occurs as a result of decisions and command and may involve the triggering of control mechanisms by humans or machines or a communication of decisions. Control mechanisms would, in turn, cause something physical to happen, such as moving the hands or arms, starting a motor, or increasing or decreasing the depth of a cut on a machine tool.

Input and output are related to the raw material or the thing being processed. The output represents some transformation of the input in line with our previous discussions of systems. The processes themselves may be of any type; chemical, processes to change shape or form, assembly, transport, or clerical.

Information feedback concerning the output states is an essential ingredient, because it provides the basis for control. Feedback operates to control the simplest hand motion through the senses and the nervous system. For machines, feedback concerning the output states provides the basis for machine adjustment. Automatic machines couple the feedback information directly so that adjustments are automatic (closed-loop automation). When machine adjustments are only periodic, based on information feedback, the loop is still closed but not on a continuous and automatic basis.

Types of Human–Machine Systems

We will use the module of the functions performed by humans or machines shown in Figure 13-5 to discuss the basic structure for three typical systems: manual, semiautomatic, and mechanized or automatic systems.

Manual systems involve humans with only mechanical aids or hand tools. The human supplies the power required and acts as controller of the process; the tools and mechanical aids help multiply human efforts. The basic module of Figure 13-5 describes the functions, where the human directly transforms input to output. In addition, we must envision the manual system as operating in some working environment that may have an impact on the human and the output.

Semiautomatic systems involve the human mainly as a controller of the process. The human interacts with the machine by sensing information about the process, interpreting it, and using a set of controls. Power is normally supplied by the machine. There are combinations of the manual and semiautomatic systems, where humans supply some of the system power, perhaps in loading the machine. Common examples of semiautomatic systems are the machine tools frequently used in the mechanical industries.

Automatic systems presumably do not need a human, since all the functions of sensing, information processing and decision making, and action are performed by the machine. Such a system would need to be fully programmed to sense and take required action for all possible contingencies. Automation at such a level is not economically justified, even if the machines could be designed. Therefore, the human functions as a monitor to help control the process. In this role, humans periodically or continuously maintain surveillance over the process through displays that indicate the state of the crucial parameters of the process.

In analyzing manual activity and human–machine cycles, various types of graphic models are used, such as operator charts. These graphic models are used as design aids in conjunction with the principles of motion economy to develop superior designs from the point of view of the effective use of humans. Also, most of these methods require an estimate of how long it takes for the worker to perform the activities involved. A special field called "work measurement" has developed to meet this need in business and industry. Appendix F deals with the methods of work measurement.

Job Design in Relation to Layout

As discussed in the Part Two Introduction, there are two basic types of physical systems, intermittent and continuous. While many actual systems are combinations of these two extremes, the two types illustrate the nature of jobs that result. Figure 13-6 shows the general layout patterns.

Intermittent systems are associated with a functional layout pattern, as shown in Figure 13-6*a*. With functional layout, departments are generic in type, with all the operations of a given type being done within that department. For example, Figure 13-6*a* represents a machine shop, and the Grinding Department performs all kinds of metal work involving grinding. The department has a range of grinding equipment, so it has relatively broad capability within the skill category. There are many examples of other systems where functional layout is employed, for example, hospitals, municipal offices, and so on.

The job designs that are associated with functionally laid out systems tend to be relatively broad, though specialized. For example, mechanics in the Grinding Department can usually perform a wide variety of such work. Normally they are trained to operate several of the machines, because this breadth offers flexibility. Similarly, an X-ray Department in a hospital does a broad range of x-ray work. The variety of work in both situations stems both from the more general nature of operations in the departments and from the fact that each job order that arrives may be slightly different. Such functional systems require employees who can perform within a skill category. They are specialized, but it is more closely related to craft specialties. There is repetition in the work in the sense that an experienced x-ray technician has probably performed all the common types of x-rays previously, but there is a continuing mix of job orders that lends variety to the work.

Continuous systems are associated with line-type operations, as illustrated by Figure 13-6*b*. The flow is organized entirely around the product being produced. Each

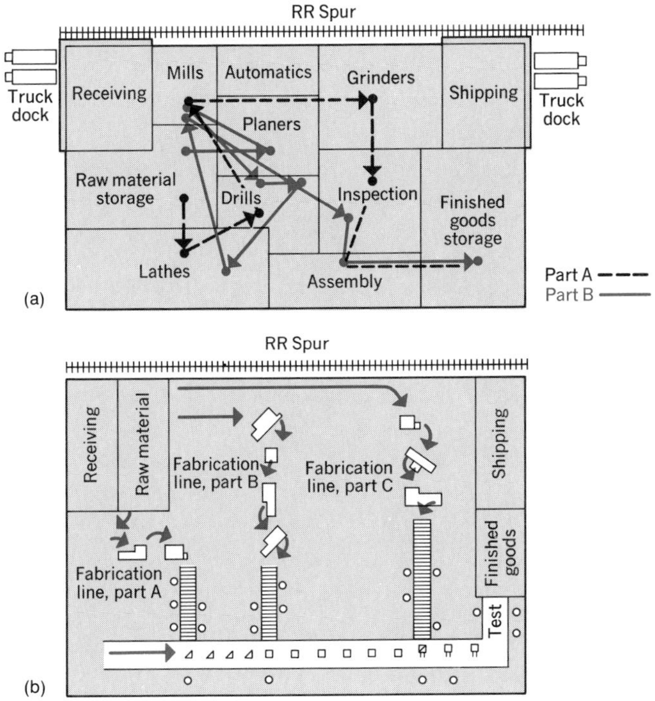

FIGURE 13-6
(a) Functional layout pattern for intermittent flow, and *(b)*, product or line layout for continuous flow.

operation being performed must fit into the flow in a highly integrated manner. The operations are highly repetitive. Each employee performs just a few elements of the work that in themselves may seem unrelated because of the "balance" among operations required to obtain smooth flow. The balance among operations is critical to making the system function as an efficient high-output system. The higher the output requirements, the more likely it is that the system will be balanced with a small cycle time that restricts the content of each individual job.

The jobs that result from the line balance process are likely to contain small work elements that are difficult to relate to end products. To appreciate the situation, the process of assembly line balancing needs to be understood. The actual processes are rather sophisticated, but a simple example will show how jobs are constructed from the work required to assemble a simple toy automobile.

An Example. Figure 13-7 shows a wooden toy car; the parts are named and numbered. By examining the toy car, we can see the sequence restrictions that must be observed in its assembly. For example, the hubcaps must be installed on the wheels prior to subsequent assembly steps to ensure that the wood axle is not broken

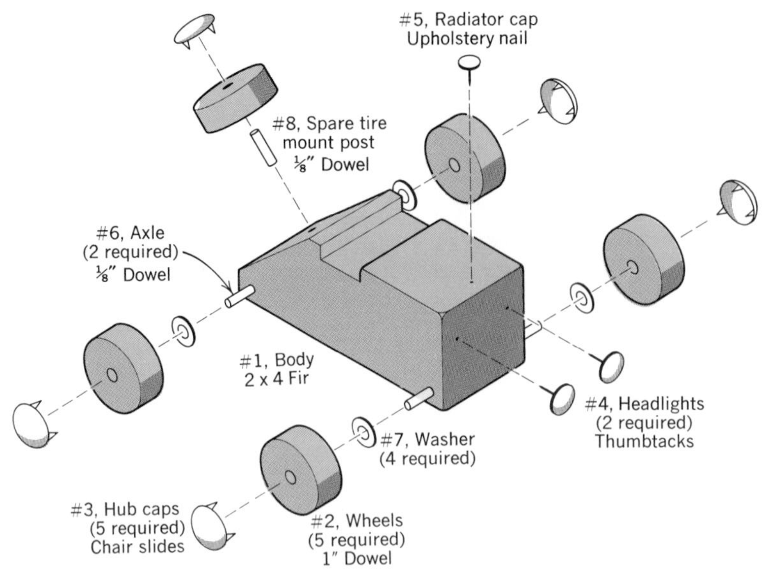

FIGURE 13-7
Toy car with parts and assembly details.

as a result of impact. Finally, the wheels cannot be assembled until the axle has been inserted in the car body.

These sequences must be observed, because the toy car cannot be assembled correctly in any other way. On the other hand, it makes no difference whether the headlights are assembled before or after the wheels are assembled. Similarly, it makes no difference whether the front or rear wheels are assembled first.

These task sequence restrictions are summarized in Table 13-1. In general, the assembly tasks listed in Table 13-1 are broken down into the smallest whole activity. For each task, we note in the right-hand column the task or tasks that must immediately precede it. Tasks *a* and *e* can take any sequence because no tasks need precede them. However, task *b* (install right headlight) must be preceded by task *a* (position body to conveyor). Only the immediate predecessor tasks are listed to avoid redundancy.

The precedence restrictions in Table 13-1 are summarized in the diagram in Figure 13-8. Now we can proceed with the grouping of tasks to obtain balance. But balance at what level? What is to be the capacity of the line? This is an important point and one that makes the balance problem difficult. If there were no capacity restrictions, the problem would be simple; one could take the lowest common multiple approach. But capacity would then be specified by balance rather than by market considerations.

For illustrative purposes, assume that we must balance the line for an 11-second cycle. A completed unit would be produced by the line every 11 seconds. To meet this capacity requirement, no station could be assigned more than 11 seconds' worth of the tasks shown in Figure 13-8. The total of all task times is 53 seconds. Therefore, with an 11-second cycle, five stations is the minimum possible. Any solution that required more than five stations would increase direct labor costs. Figure 13-9 shows a solution that yields five stations.

While this example is simple, it illustrates the concepts of assembly line balancing. As the product becomes more complex, the relationships between tasks that end up within jobs may be more disjointed. As the enterprise is rewarded for its success by public acceptance of its product, volume increases and lines are rebalanced for smaller cycle times to achieve higher output, further restricting the content of jobs.

THE SOCIOTECHNICAL VIEW OF PROCESS PLANNING AND JOB DESIGN

In the sociotechnical view, the concepts and methods of the technological view are mechanistic, humans being thought of as machines or, worse, as just links in ma-

TABLE 13-1

List of Assembly Tasks Showing Performance Times and Sequence Restrictions for the Toy Car Assembly

Task	Task Description	Performance Time, Seconds	Task Must Follow Task Listed Below
a	Position body to conveyor	2	—
b	Install right headlight	3	*a*
c	Install left headlight	3	*a*
d	Install radiator cap	3	*a*
e-1	Install hubcap on spare tire	4	—
e-2	Install hubcap on wheel	4	—
e-3	Install hubcap on wheel	4	—
e-4	Install hubcap on wheel	4	—
e-5	Install hubcap on wheel	4	—
f	Assemble spare tire post on spare tire	2	*e*-1
g	Assemble spare tire subassembly to body	2	*a,f*
h-1	Press axle on wheel	2	*e*-2
h-2	Press axle on wheel	2	*e*-3
i-1	Assemble washer to wheel-axle subassembly	1	*h*-1
i-2	Assemble washer to wheel-axle subassembly	1	*h*-2
j-1	Assemble front wheel-axle-washer subassembly to body	2	*a*, *i*-1
j-2	Assemble back wheel-axle-washer subassembly to body	2	*a*, *i*-2
k-1	Assemble washer to front axle	1	*j*-1
k-2	Assemble washer to back axle	1	*j*-2
l-1	Press second wheel on front axle	2	*e*-4, *k*-1
l-2	Press second wheel on back axle	2	*e*-5, *k*-2
m	"Drive" car off conveyor	2	*b,c,d,g,l*

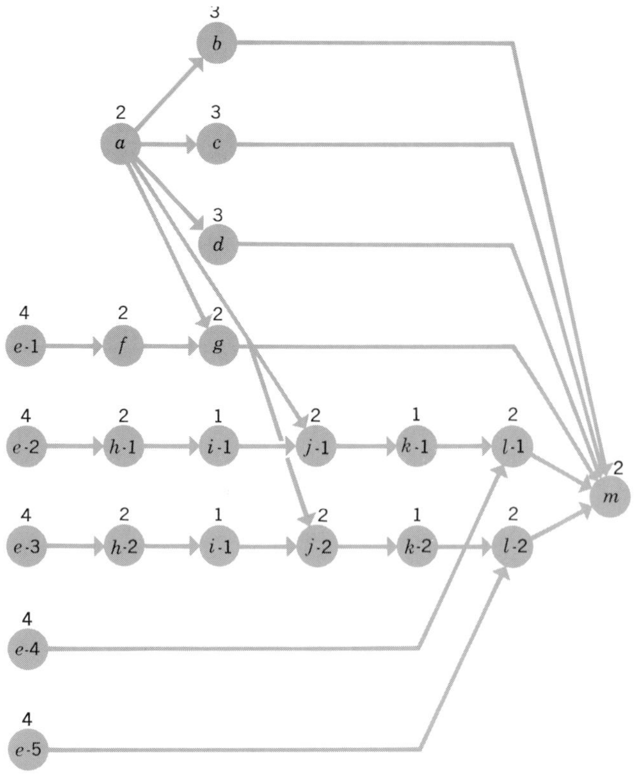

FIGURE 13-8
Precedence diagram representing the sequence requirements shown in Table 13-1 for the toy car assembly example. Numbers indicate the task performance times.

chines. This view states that, without question, the central focus of the technological approach is technology itself and that technology is taken as given without exploring the full range of the possible alternatives.

Briefly, sociotechnical theory rests on two essential premises. The first is that there is a joint system operating, a sociotechnical system, and that joint optimization is appropriate, as indicated by the Venn diagram of Figure 13-1. The second premise is that every sociotechnical system is embedded in an environment. The environment is influenced by a culture and its values and by a set of generally accepted practices, where there are certain roles for organizations, groups, and people.

Englestad [1972] developed a set of psychological job requirements. These requirements are a translation of empirical evidence that suggests that workers prefer tasks of a substantial degree of wholeness, where the individual has control over the materials and the processes involved:

1. The need for the content of a job to be reasonably demanding in terms other than sheer endurance, yet provide at least a minimum of variety (not necessarily novelty).

2. The need for being able to learn on the job (which implies standards and knowledge of results) and to go on learning. Again, there is a question of neither too much nor too little.

3. The need for some minimum area of decision.

4. The need of some minimum degree of social support and recognition in the work place.

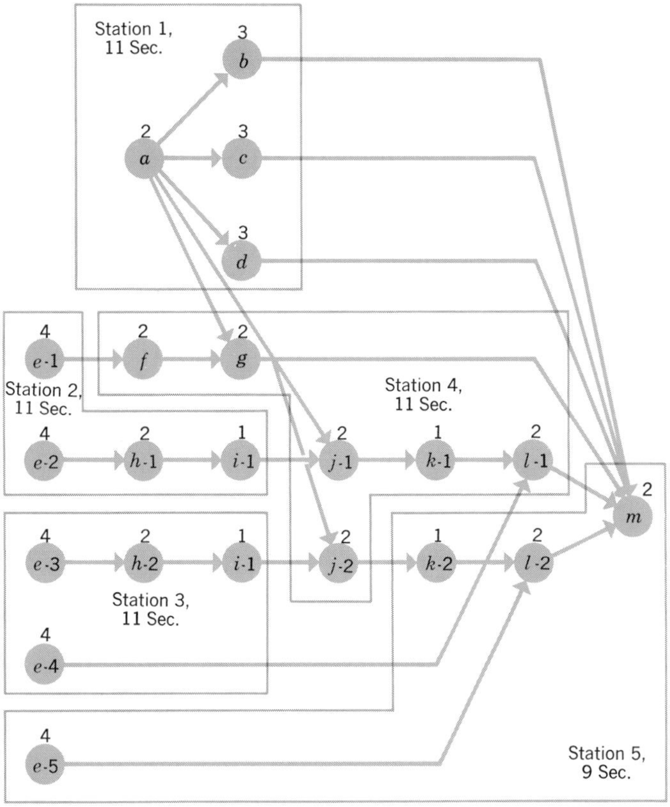

FIGURE 13-9
A solution to the toy car assembly line balancing example. Cycle time = 11 seconds; idle time = 2 seconds in Station 5.

5. The need to be able to relate what one does and what one produces to one's social life.

6. The need to feel that the job leads to some sort of desirable future.

Thus the principles of process and job design may be summarized as the application of the concept of joint optimization between technology and social system values and the ideal of wholeness and self-control. These are the principles that guide the organization of work in specific situations, resulting in the establishment of job content. The determination of job methods in the sociotechnical view results from a concept of the semiautonomous work group who, by and large, create their own work methods.

Job Enlargement and Enrichment

Since the IBM experience with job enlargement in 1950, a number of case and research studies have indicated that benefits can result from jobs of broader scope. One survey attempted to correlate degree of job interest with the number of operations performed. They asked 180 assembly line workers this question: "Would you say your job was interesting?" The results are tabulated in Table 13-2. There are significant differences between the interest-noninterest categories in relation to the number of operations performed. This lends credence to the hypothesis that jobs of broader scope lend interest and satisfaction to the worker.

Quantitative Evidence of the Benefits of Job Enlargement

In most of the case study applications, there was no careful experimental design to ensure that the results reported were measurements of the effects of job enlargement. Control groups normally were not used to make it possible to contrast true effects.

TABLE 13-2 **Number of Operations Performed Correlated with Degree of Job Interest, Present Job**

Operations Performed	Very or Fairly Interesting	Not Very or Not at All Interesting	Total
1	19	38	57
2-5	28	36	64
5 or more	41	18	59
	88	92	180

SOURCE. C. R. Walker and R. H. Guest, *The Man on the Assembly Line*, Harvard University Press, Cambridge, Mass., 1952.

During changeover, many other factors could have entered the situation; we are not always sure that the results did not stem from other factors in the job situation. Therefore, the following carefully controlled study is of interest.

A set of experiments was carried out in the manufacturing department of a West Coast company manufacturing a hospital appliance; the company was unionized. The department that was the subject of the study manufactured a line of similar products and had been the subject of a careful and detailed methods engineering study. At the beginning of the experiment, the product was made on an assembly line, using carefully specified, minutely subdivided operations on which 29 of the department's 35 members worked. Other workers were concerned with preparing and removing material from the line, inspection, and supply. A similar department in the company was used as a control group to permit monitoring of the presence of plantwide changes that might affect employee attitudes, practices, and performance.

The experiment was divided into four phases, as indicated in Table 13-3. The line job design was the original setup with the conveyor pacing all work. The group job design was identical to the line job design except that the conveyor was eliminated as a pacing device. The two individual job designs both involved job enlargement. Workers performed all nine operations at their own work stations, including inspection and obtaining their own supplies.

The results, which are summarized in Figures 13-10 and 13-11, show considerable differences in output and quality for the different job designs. Figure 13-10 shows that the group job design was somewhat poorer than the line job design, producing an average daily productivity index of only 89 percent of the line job design. Also note that the standard deviation of indexes of productivity was over twice as large as that for the line job design. The pacing effect of the conveyor apparently contributed substantially to the maintenance of output. The individual job designs resulted in good productivity levels, but they did not come up to the average output of the original line job design levels, having, respectively, 91.7 and 95.3 percent of the line job design output.

The quality levels for the original line job design were regarded as very good. However, these quality levels improved with the removal of the conveyor pacing. When the responsibility for quality was placed in the hands of the workers, quality levels rose even higher. The individual job designs produced only one-fourth the number of kinked assemblies as the line job design, as shown in Figure 13-11.

The conclusions of the study were as follows. The individual job design:

1. Brought an improvement in quality, although quality levels were high originally.

2. Increased the flexibility of the production process.

3. Permitted identification of individuals having deficiencies in productivity and quality.

4. Reduced service functions in the department, for example, material delivery and inspection.

TABLE 13-3

Experimental Conditions for Four Job Designs

	Purpose	Criteria	Locations	Workers	Total Number of Days	Number Days Worker Assigned	Production Method
A. Line job design	Obtain reference base of job design where separate tasks performed on rotated basis by workers	Quantity, quality, some measures of attitude and satisfaction	Main department	29	26	26	Workers rotate among nine stations on belt conveyor, performing minute specified operations at pace of conveyor
B. Group job design	Eliminate conveyor pacing effects; other conditions primarily the same	Quantity, quality	Adjacent room	29	14	2	Workers rotate among nine individual stations using batch method
C. Individual job design No. 1	Give workers experience on experimental job design where assembly tasks are performed by workers	Quantity, quality	Adjacent room	29	16	2	Workers perform all nine operations at own stations, plus inspection and getting own supplies
D. Individual job design No. 2	Obtain measure on experimental job design	Quantity, quality, some measures of attitude and satisfaction	Main department	21	27	6	Same as C above

SOURCE. A. R. N. Marks, "An Investigation of Modifications of Job Design in an Industrial Situation and Their Effects on Measures of Economic Productivity." Unpublished PhD dissertation, University of California, Berkeley, November 1954.

5. Developed a more favorable attitude toward individual responsibility, individual work rate, effort expenditure, distribution of work load, and making whole units. After experience with the individual job design, workers disliked the lack of personal responsibility characteristics of the line job design.

Job Methods Design and Semiautonomous Work Groups

It has always been assumed that professional job designers should be able to design superior methods of work. After all, they have knowledge of work flow, of human

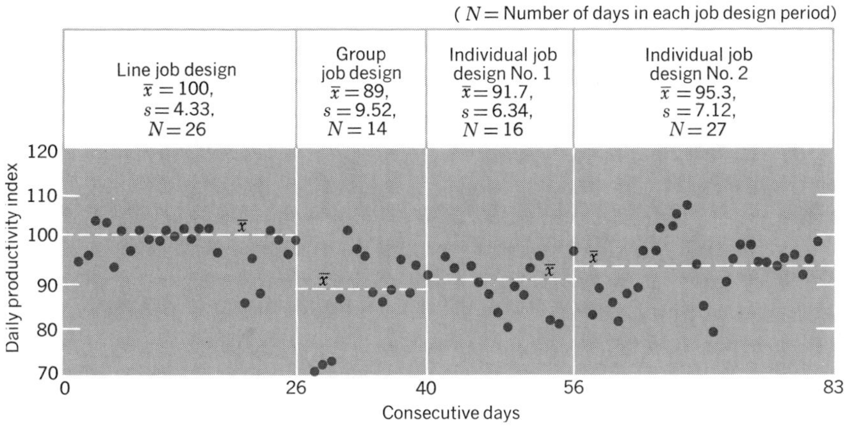

FIGURE 13-10
Average daily productivity indexes for the four job designs.
SOURCE: *A. R. N. Marks, "An Investigation of Modifications of Job Design in an Industrial Situation and Their Effects on Measures of Economic Productivity." Unpublished PhD dissertation, University of California, Berkeley, November 1954, p. 32.*

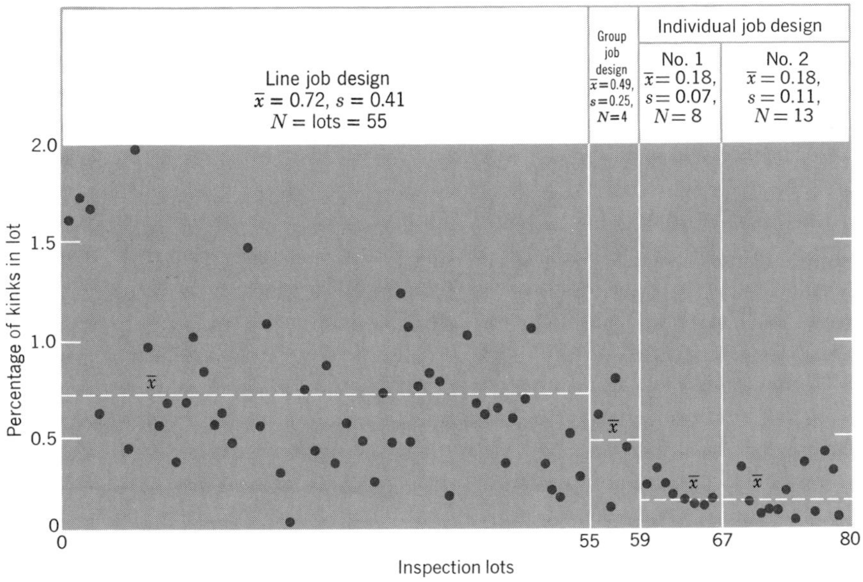

FIGURE 13-11
Percentage of kinked assemblies in consecutive lots for four job designs.
SOURCE: *A. R. N. Marks, "An Investigation of Modifications of Job Design in an Industrial Situation and Their Effects on Measures of Economic Productivity." Unpublished PhD dissertation, University of California, Berkeley, November 1954, p. 87.*

psychological and physiological capabilities, of machines, and of tool design. The questioning of this accepted view began with the recognition of the human problems of resistance to change. We have long been aware of the problems of introducing change in an existing situation. It may well be an oversimplification, but the way around the problem has been to involve the people affected by the change in the process of the change.

Many case studies have shown that involvement seemed to produce a low-threat situation, whereas the more traditional role for the introduction for the change, through the authority system, was a high-threat technique. High-threat techniques produce resistance to change that can manifest itself through noncooperation, poor motivation, and even active opposition or sabotage. A superior job design from the point of view of work flow, tools, and psychophysiological data can be masked by the effects produced through resistance to change.

The low-threat approach tends to minimize opposition, and under certain circumstances, a work group can be a driving force for promoting change. These circumstances involve full participation in the decisions that affect the work group. In its full meaning, this includes the design of the procedures and methods by which the work of the group is carried out.

Coal Mining in England. Trist and co-workers report a study of coal mines in England, which took place over a long period of time with periodic reporting in the literature [Trist and Bamforth, 1951; Emery and Trist, 1960; Trist et al., 1963]. The study involved two alternate ways of organizing work in a coal mining operation, where the technology and other conditions were the same. The conventional approach had been to divide the labor force of 41 workers into 14 segregated task groups that had specialized tasks to perform.

In the conventional organization, activities were divided into seven specialized tasks, each carried out by a different group. Mine output depended on the completion of a working cycle that consisted of preparing an area for coal extraction, using machinery to dig the coal out of the face, and removing the coal with the aid of conveyors. Each of the tasks had to be completed in sequence and on schedule over three working shifts. On each of the shifts, one or more task groups performed their work, provided that the preceding tasks had been completed. The "filling" tasks for coal removal were the most onerous and were frequently not completed, resulting in a delay in the work cycle and a reduction of the output. Each worker was paid an incentive, without reference to the other tasks of workers in the work groups. The result of the conventional system was the development of isolated work groups, each with its own customs, agreements with management, and pay arrangements focused in its own interest. Coordination between workers and groups on different shifts and control of work had to be provided entirely by management.

In the new composite design, all the required roles were internally allocated to members by the work group itself. The basic objective of the new design was to maintain continuity for achieving a complete work cycle that commonly extended over more than one shift. Integration of the objectives of the crews in the three shifts was aided by setting goals for the performance of the entire cycle and making payments to the group as a whole, plus an incentive payment for output. The wage payment scheme placed responsibility on the group as a whole for all operations, generating the need for individuals to interrelate. The nature of the working relations and payment scheme led to the spontaneous development of interchangeability of workers according to need. This in turn required the development of multiskilled workers and an extremely flexible work force.

Average productivity increased from the range of 67-78 percent of potential to 95 percent, and absentee rates improved considerably.

There are many other case studies reporting on results obtained from various aspects of job enlargement and semiautonomous work groups: Harwood Manufacturing Company, by Cartwright and Zander [1960]; home appliance manufacturing, Connant and Kilbridge [1965]; pharmaceutical appliances, Davis and Canter [1956]; maintenance craftsmen in a chemical plant, Davis and Werling [1960]; The paper and pulp industry in Norway, Englestad [1972]; textile weaving in India, Rice [1953]; petroleum refining, Susman [1972] and Taylor [1972]; and IBM Corporation, Walker [1950].

IMPLICATIONS FOR THE MANAGER

Many of the problems of the operations manager stem from employees' dissatisfaction with the nature of their jobs. Labor disputes erupt over issues referred to as working conditions and require endless hours of negotiations for settlement. Even when the issues are not stated in terms of job design, job satisfaction, and dehumanization at the work place, it is often felt that these are the real issues. If the structure of processes, jobs, and layout took account of the worker's model of job breadth, psychological job requirements, and job enrichment principles, perhaps the operations manager would face fewer labor disputes resulting from these causes.

While alterations in the process and job design structures can be made in existing organizations, managers have the greatest opportunity when new facilities are being planned. It is during these opportune times that fundamental alternatives to the organization of work can be considered most easily.

The manager needs to know about the results of studies of alternate job design structures, the fundamental conflict between the managers' and the workers' models of job breadth, psychological job requirements, and the success of semiautonomous work groups in designing their own work methods. If the process and job design structure and the physical layout have conflict built into them, the resulting operating problems may be severe. The moral, "ye shall reap as ye have sown," will be of little consolation.

A manager needs to have a clear understanding of the appropriateness of different layout flow patterns. Aside from the job design aspects, product or line layout is appropriate only under certain conditions. The volume of activity necessary to justify specialized line operations is relatively large and requires a standardized product or service—variations are difficult to incorporate into line systems. The conflict between specialization as a concept and job satisfaction is a difficult one for managers. Specialization has many benefits to society and to an individual enterprise. There seems to be little question that specialized line-type operations are easier to plan for, easier to schedule, easier to control, require less space per unit of capacity, and so on. Yet specialized jobs have created a work environment that raises social responsibility issues for managers.

It seems unlikely that large-scale productive systems can or should be converted from line operations. On the other hand, managers need to raise questions about feasible alternatives that may exist within the joint space diagrammed in Figure 13-1. Experience has shown that there are often solutions to process-job design problems that satisfy both sets of criteria if a genuine effort is made to find them.

REVIEW QUESTIONS

1. In Scoville's view, what are the factors that limit the extent of division of labor in practice?
2. In Davis' view, what are the values held by organizations that applied the division of labor criterion throughout the industrial era?
3. Taking an assembly line as an example, what are the factors that determine the nature and content of jobs that result?
4. What is the sociotechnical systems approach to job design?
5. What is the nature of the manager's model of job breadth? Of the worker's model of job breadth? Do they result in the same kinds of job design?
6. What is the technological view of process planning and job design?
7. What are the tools of product analysis? How are job designs affected by their use?
8. In the technologist's view, what are humans' functions in performing work? In which kinds of functions do humans have superiority over machines and vice versa?
9. What is role of humans in the models for manual, semiautomatic, and automatic systems?
10. What is the sociotechnologist's view of process planning and job design?
11. What are Englestad's psychological job requirements? Are they compatible with the technological view of job design?

12. Given the principles of division of labor and counterbalancing concepts of job enlargement, is there any way of deriving some middle ground concept representing optimal job design?

13. What is a semiautonomous work group? What evidence supports the superiority of this approach? The inferiority?

SITUATIONS

14. Following is a situation that occurred in the Hovey and Beard Company, as reported by J. V. Clark.*

This company manufactured a line of wooden toys. One part of the process involved spray painting partially assembled toys, after which the toys were hung on moving hooks that carried them through a drying oven. The operation, staffed entirely by women, was plagued with absenteeism, high turnover, and low morale. Each woman at her paint booth would take a toy from the tray beside her, position it in a fixture, and spray on the color according to the required pattern. She then would release the toy and hang it on the conveyor hook. The rate at which the hooks moved had been calculated so that each woman, once fully trained, would be able to hang a painted toy on each hook before it passed beyond her reach.

The women who worked in the paint room were on a group incentive plan that tied their earnings to the production of the entire group. Since the operation was new, they received a learning allowance that decreased by regular amounts each month. The learning allowance was scheduled to fall to zero in six months, since it was expected that the women could meet standard output or more by that time. By the second month of the training period, trouble had developed. The women had progressed more slowly than had been anticipated, and it appeared that their production level would stabilize somewhat below the planned level. Some women complained about the speed that was expected of them, and a few of them quit. There was evidence of resistance to the new situation.

Through the counsel of a consultant, the supervisor finally decided to bring the women together for more general discussions of working conditions. After two meetings in which relations between the work group and the supervisor were somewhat improved, a third meeting produced the suggestion that control of the conveyor speed be turned over to the work group. The women explained that they felt that they could keep up with the speed of the conveyor but that they could not work at that pace all day long. They wished to be able to adjust the speed of the belt, depending on how they felt.

After consultation, the supervisor had a control marked, "low, medium, and fast" installed at the booth of the group leader, who could adjust the speed of

* From J. V. Clark, "A Healthy Organization," *California Management Review, 4,* 1962.

the conveyor anywhere between the lower and upper limits that had been set. The women were delighted and spent many lunch hours deciding how the speed should be varied from hour to hour throughout the day. Within a week, a pattern had emerged: the first half-hour of the shift was run on what the women called "medium speed" (a dial setting slightly above the point marked "medium"). The next two and one-half hours were run at high speed, and the half-hour before lunch and the half-hour after lunch were run at low speed. The rest of the afternoon was run at high speed, with the exception of the last 45 minutes of the shift, which were run at medium speed.

In view of the women's report of satisfaction and ease in their work, it is interesting to note that the original speed was slightly below medium on the dial of the new control. The average speed at which the women were running the belt was on the high side of the dial. Few, if any, empty hooks entered the drying oven, and inspection showed no increase of rejects from the paint room. Production increased, and within three weeks the women were operating at 30 to 50 percent above the level that had been expected according to the original design.

Evaluate the experience of the Hovey and Beard Company as it reflects on job design, human relationships, and the supervisor's role. How would you react as the supervisor to the situation where workers determine how the work will be performed? If you were designing the spray painting setup, would you design it differently?

15. The following situation is drawn from an article by G. Woolsey.*

A Canadian manufacturer of heavy equipment had installed an automated warehouse as a base for worldwide supply of spare parts for their equipment. The automated system involved filling orders for spare parts that would ultimately be packed and shipped. Orders were coded by a clerk, indicating the item and its location, using a keyboard and a cathode ray tube (CRT) information display. In the warehouse, forklift operators also had a small CRT on the truck that listed the necessary items and locations to complete an order.

The trucks were routed to the correct position in the warehouse by a computer, the trucks being guided by wires in the floor. On arrival at the correct location, the truck raised the operator automatically to the proper level, so that the item could be picked out of the bin. The truck operator then pressed a button to indicate that the particular item had been obtained. The computer then routed the truck and operator to the next station to pick another item. When all the items for an order had been filled, the computer routed the truck back to the packing and shipping location. The process was then repeated for the next order.

The only control that operators had over their trucks was a "kill" button that could stop them in case of an impending collision. This never happened, since the computer program knew the position of all the trucks and programmed

* From G. Woolsey, "Two Digressions on Systems Analysis: Optimum Warehousing and Disappearing Orange Juice," *Interfaces*, 7(2), February 1977, pp. 17-20.

them around each other. If the operator stepped off the truck, a "dead man" control stopped the truck.

The automated system had replaced a conventional order picking system in which forklift truck operators performed essentially the same process manually.

On installation, the automated system ran smoothly with virtually zero errors in picking items. Then, however, errors increased to new record levels, absenteeism became a problem, and there were some examples of what appeared to be sabotage. The warehouse supervisor was particularly outraged because the company had agreed to a massive across-the-board pay increase for operators in order to get the union to accept the new computer system. The union had argued that the operators now worked with a computer and that the job was more complex.

What is the problem? Should the multimillion dollar automated system be scrapped? Should the jobs be redesigned? What do you propose?

REFERENCES

Barnes, R. M. *Motion and Time Study: Design and Measurement of Work* (6th ed.). John Wiley, New York, 1968.

Cartwright, D., and A. Zander, editors. *Group Dynamics: Research and Theory*. Row, Peterson, Evanson, Ill., 1960.

Chapanis, A. *Man-Machine Engineering*. Wadsworth, Belmont, Calif., 1965.

Conant, E. H., and M. D. Kilbridge. "An Interdisciplinary Analysis of Job Enlargement: Technology, Costs and Behavioral Implications," *Industrial and Labor Relations Review, 18,* October 1965, p. 377.

Davis, L. E. "Job Satisfaction Research: The Post-Industrial View," *Industrial Relations, 10,* 1971, pp. 176-193. Also in L. E. Davis and J. C. Taylor, editors. *Design of Jobs*. Penguin Books, Middlesex, England, 1972.(a)

Davis, L. E. "The Coming Crisis for Production Management," *International Journal of Production Research, 9,* 1971, pp. 65-82. Also in L. E. Davis and J. C. Taylor, editors. *Design of Jobs*. Penguin Books, Middlesex, England, 1972.(b)

Davis, L. E., and R. R. Canter. "Job Design Research," *The Journal of Industrial Engineering, 12*(6), November-December 1956.

Davis, L. E., and A. B. Cherns, editors. *Quality of Working Life: Problems, Prospects and State of the Art.* Vol. I; *Cases,* Vol. II. Free Press, Glencoe, Ill., 1975.

Davis, L. E., and J. C. Taylor, editors. *Design of Jobs*. Penguin Books, Middlesex, England, 1972.

Davis, L. E., and R. Werling. "Job Design Factors," *Occupational Psychology, 24*(2), 1960, pp. 109-132.

Elliot, J. D. "Increasing Office Productivity Through Job Enlargement," in *The Human Side of the Office Manager's Job,* Office Management Series No. 134, American Management Association, New York, 1953, pp. 5-15.

Emery, F. E., and E. L. Trist. "Socio-Technical Systems," in *Management Science,*

Models and Techniques, edited by C. W. Churchman and M. Verhulst. Pergamon Press, London, 1960.

Englestad, P. H. "Socio-Technical Approach to Problems of Process Control," in *Design of Jobs,* edited by L. E. Davis and J. C. Taylor, Penguin Books, Middlesex, England, 1972.

Fogel, L. J. *Biotechnology: Concepts and Applications.* Prentice-Hall, Englewood Cliffs, N.J., 1963.

Guest, R. H. "Job Enlargement: A Revolution in Job Design," *Personnel Administration, 20*, March–April 1957, pp. 13-15.

Marks, A. R. N. "An Investigation of Modifications of Job Design in an Industrial Situation and Their Effects on Measures of Economic Productivity." Unpublished PhD dissertation, University of California, Berkeley, November 1954.

McCormick, E. J. *Human Factors Engineering* (3rd ed.). McGraw-Hill, New York, 1970.

Niebel, B. W. and A. B. Draper. *Product Design and Process Engineering.* McGraw-Hill, New York, 1974.

Rice, A. K. "Productivity and Social Organization in an Indian Weaving Shed," *Human Relations, 6,* November 1953, p. 297.

Scoville, J. G. "A Theory of Jobs and Training," *Industrial Relations, 9*, 1969, pp. 36–53. Also in L. E. Davis, and J. C. Taylor, editors. *Design of Jobs.* Penguin Books, Middlesex, England, 1972.

Susman, G. I. "The Impact of Automation on Work Group Autonomy," *Human Relations, 23*, 1970, pp. 567–577. Also in L. E. Davis and J. C. Taylor, editors. *Design of Jobs.* Penguin Books, Middlesex, England, 1972.

Taylor, J. C. "Experiments in Work System Design: Economic and Human Results," Part I and Part II, *Personnel Review, 6*(3), *6*(4), Summer and Autumn, 1977.

Taylor, J. C. "Some Effects of Technology in Organizational Change," *Human Relations, 24*, 1971, pp. 105–123. Also in L. E. Davis, and J. C. Taylor, editors. *Design of Jobs.* Penguin Books, Middlesex, England, 1972.

Trist, E. L. and K. W. Bamforth, "Some Social and Psychological Consequences of the Longwall Method of Coal Mining," *Human Relations, 4,* February 1951, pp. 3–38.

Trist, E. L., G. W. Higgin, H. Murray, and A. B. Pollack. *Organizational Choice.* Tavistock Publications, London, 1963.

Turner, A. N., and P. R. Lawrence. *Industrial Jobs and the Worker.* Harvard Business School, Boston, 1965.

Walker, C. R. "The Problem of the Repetitive Job." *Harvard Business Review, 28*(3), May 1950, pp. 54–58.

Walker, C. R., and R. H. Guest. *The Man on the Assembly Line.* Harvard University Press, Cambridge, Mass., 1952.

Woodson, W. E. *Human Engineering Guide for Equipment Designers* (2nd ed.). University of California Press, Berkeley, 1966.

CHAPTER 14 Facility Layout

PROCESS AND JOB DESIGN PLANS MUST BE RELATED IN THE PHYSical sense, whatever the basic conception for the organization of work. Facility layout is the integrating phase of the design of a productive system. It is at this stage that we must provide for all the factors of process and job design plans, capacities, space for equipment and personnel, methods for moving materials and supplies through the system, auxiliary and support services, and flexibility for change and future expansion.

Normally, there will be financial constraints, and physical restrictions concerning the site for existing structures. Some of the physical restrictions can have an impact on the intended job designs. All of the various factors we have mentioned tend to interact, creating a very complex process of developing physical layout plans. There is no general overall theory that makes it possible to relate the multitude of factors that influence layout designs. Instead, the development of a good layout is the result of a sequence of major decisions on such questions as location, basic organization of work and work flow, and design capacity. These decisions are followed by decisions concerning the selection and placement of equipment, space allocations, and flow patterns.

THE SERVICE FACILITY MODULE

The service facility module shown in Figure 14-1 represents the assembly of resources that produces some step in the overall process. It involves the personnel, machines, tools, supplies, and energy required. The personnel might be represented by a single individual or by a cooperative team. The machine might be as simple as a lever, holding device, or a pencil, or as complex as a computer. The work going on within the service facility may range from simple repetitive tasks to complex nonrepetitive tasks. Work "arrives" at the service facility by some process—the demand for the service—and is processed by the center. Variable rates of arrival and service are certainly the most common.

Figure 14-2 shows some of the ways that service facilities could be combined in serial fashion, in parallel to provide the needed capacity for a given process or in flexible networks where each service facility performs a generic function. Very complex systems may combine the combinations.

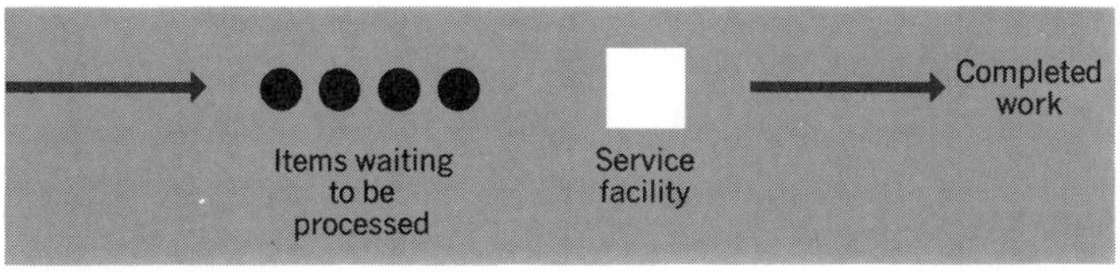

FIGURE 14-1
The service facility module.

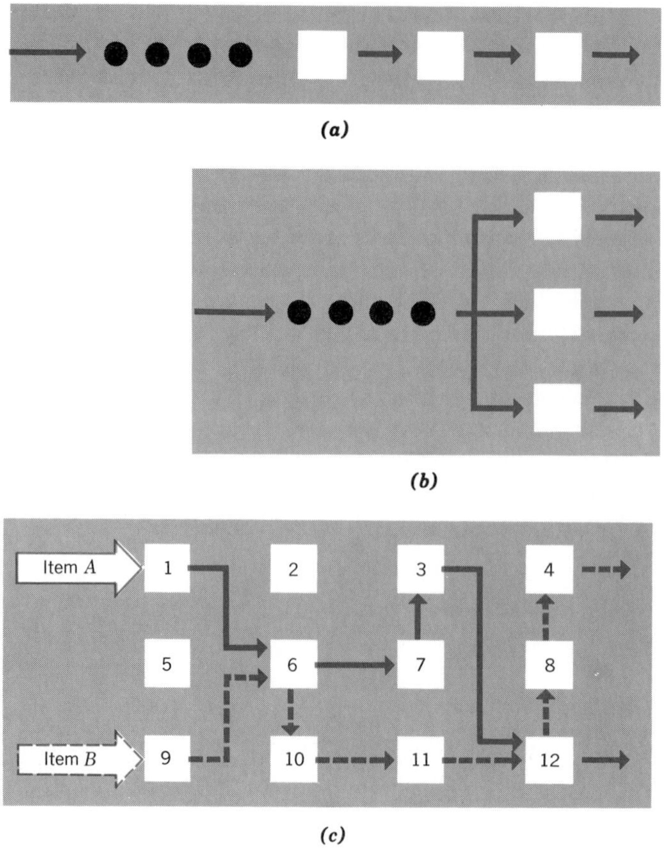

FIGURE 14-2
Combinations of service facility modules. *(a)* Serial combination of service facility modules; *(b)* parallel combination of modules to provide needed capacity for a given process; *(c)* networks of service facility modules that can be combined in flexible sequences.

The examples of each of the classic set of combinations are numerous and occur in all kinds of productive systems. First, the simple module may sometimes be the entire system, such as the one-operator barber or beauty shop and other small enterprises. The serial combination of service centers is represented by industrial production lines, a cafeteria line, or the registration at an enrollment process at colleges and universities. Parallel combinations of modules to provide needed capacity for a given process are common in the multiple check-out counters of supermarkets, multiple tellers windows in banks, larger barber and beauty shops, and all kinds of generic processing centers. Networks of functional service facilities are common in manufacturing and jobbing machine shops, hospitals, and some large office operations where activities are grouped by function.

The service facility module, with its arrival and service rate processes, is an important concept discussed in Appendix D in connection with waiting line models.

BASIC LAYOUT TYPES

The basic layout types that occur are based on the way work flows through the system. When the combinations of service modules is serial, as in Figure 14-2*a*, the layout of physical facilities normally follows the *line* sequence. In manufacturing, such a sequence is usually related to a particular product, and the layout type is called "product" or "line" layout. The flow through such systems is direct; it approaches continuous flow in some automated systems and achieves continuous flow in some chemical processes. Systems that are laid out as lines or by product are also termed "continuous systems." Balance between the capacities of the individual service facilities is an extremely important problem when facilities are arranged in production line fashion.

When service facilities that have a generic or functional capability are combined in flexible networks, as in Figure 14-2*c*, the system is termed "intermittent" because of the nature of the demand on the system. Work flows through the system in batches or special orders and commonly waits for a time before being processed at service facilities required by the products. The utilization of facilities is intermittent because each order may require different processes. The layout of facilities follows a pattern based on the best set of interrelationships between the functional service centers and is called "functional" or "process" layout. Parallel combinations of service modules, as in Figure 14-2*b*, might be used to provide needed capacity at any stage of either type of layout. The names of the service centers that result in functional layout are commonly taken from the name of the process. Therefore, we find lathe departments, milling departments, and assembly departments in manufacturing; x-ray, surgery, and physical therapy departments in hospitals; and accounts receivable, accounts payable, and billing in offices. An important problem in generating good functional layouts is to determine the best pattern of interrelationships between functional service facilities. Diagrams illustrating both product and functional layout were shown in the Part II Introduction (Figure II-1), and in the previous chapter as Figure 13-6.

In manufacturing, the basic layout patterns seldom occur in their pure forms. Fabrications operations are most commonly arranged in functional departments in an effort to achieve good utilization of expensive equipment. Assembly lines are common, but *fixed-position* assembly is common with very large products, such as aircraft and marine diesels, ships, and construction projects.

THE CAPACITY DECISION

The capacity decision for a productive system can be of extreme importance, especially if we are dealing with capital-intensive systems. In most cases, some commitment of assets is involved, as well as a strategy for producing for the market or to meet the demand for a service.

Demand Variations and Sources of Capacity

Demand for products and services may vary in one or more important ways, and these variations make the capacity decision more difficult. For example, if demand is seasonal, should capacity be set for the peak demand or for some lower level? If one can build inventories during the slack season to be moved during the peak demand season, then a smaller physical plant may be feasible. On the other hand, if the system produces a service or a noninventoriable product, this option may not be available. The aggregate scheduling strategies discussed in Chapter 6 have a direct impact on the physical capacity decision.

The possible variations in demand may be seasonal, monthly, weekly, or daily and involve trend. However, our strategies for production may involve several sources of capacity that we can plan to use in ways that may minimize investment in facilities. The sources of capacity might be regular output, changes in the size of the labor force, use of overtime, and seasonal inventories.

Table 14-1 shows general strategies available for capital-intensive versus labor-intensive systems and for situations when seasonal inventories can be used and when they cannot. Of course, sharp dichotomies do not exist between the capital- and labor-intensive system classifications, and many specific situations may use some of the strategies of both. For example, capital-intensive systems also require labor and may be able to use overtime work to some extent.

TABLE 14-1

Nature of Strategies for Use of Sources of Capacity for Capital- and Labor-Intensive Systems When Seasonal Inventories Are Available and When They Are Not

	Seasonal Inventories Available as a Short-term Source of Capacity	Inventories *Not* Available as a Source of Capacity
Capital-intensive systems	Design capacity near an average annual demand level, using seasonal inventories to smooth output. Intensive use of expensive facilities may reduce investment required.	Smooth peak demand through compensatory sales promotion. Intensive use of expensive facilities may reduce investment required. If demand variations are very short term (daily or weekly), try to schedule "arrivals" as in hospitals.
Labor-intensive systems	Use casual labor, overtime work, and inventories in combination to meet peak demand.	Use casual labor, and overtime work in combination to meet peak demand.

Future Capacity

Do we build a capacity that matches present demand experience, or do we attempt to build for some forecast level of 1, 5, or 10 years hence? Can we afford to build for more capacity than is needed currently? We must remember that successive units of capacity are not equally expensive, because capacity is bought in "chunks." Moreover, it is commonly true that at any one level, there will be idle capacity in certain equipment classifications. Therefore, to move to the next level does not entail the purchase of equipment items where we already have idle capacity. Where capacity is built to match some forecast of future needs, it is common to buy equipment for current needs only and provide space for additional equipment when it is needed. Thus, planning is for a future capacity, but extra overhead is carried for the building space only. As the capacity is needed, machines and work places can be integrated into the system without the need for extensive redesign.

The question then really relates to the provision of extra space to match a future forecast. Some additional space can probably be justified when we consider that space added later is more expensive. More important than the cost of additional space itself are redesign costs that would be required to integrate the new space into the productive system. If the new space is not integrated into the system, we would pay extra costs daily in the form of congestion and handling costs. These extra costs of adding future space must be balanced against the incremental costs of building more space now and carrying it as added overhead.

Translating Capacity into Workable Units

We need a unit of capacity that is meaningful for the system and can be translated into requirements for personnel, machines, and space. In manufacturing systems, some aggregate unit of capacity per time period is chosen, such as tons of steel, tons of dog food, number of automobiles, or machine-hours available. In service systems, the units might be numbers of beds for a hospital, numbers of students for a school, numbers of tables or seats or meal servings in a restaurant or mass food operation, or numbers of rooms in a hotel. In merchandising operations, the units might be square feet of selling or display area.

In manufacturing systems capacity allowances are made for certain losses. Through the *facility efficiency factor* we recognize that some capacity cannot be used because of scheduling delays, machine breakdowns, preventive maintenance, and so on. Through the *scrap factor* we recognize that some capacity is used to produce unacceptable product.

FUNCTIONAL LAYOUT FOR INTERMITTENT SYSTEMS

In functional layout, processing units are organized by function on the assumption that certain skills and expertise are available in that service facility. Therefore, an x-ray department of a hospital normally offers such a broad range of skills and knowledge that a physician can call for virtually any kind of x-ray.

Many examples of intermittent systems can be found in practice, for instance, in manufacturing, hospital and medical clinics, large offices, municipal services, and libraries. In every situation, the work is organized according to the function that is performed. The machine shop industry is one of the most common examples, and the name and much of our knowledge of intermittent systems results from the study of such manufacturing systems. Table 14-2 gives a summary of typical departments or service centers that occur in several generic types of intermittent systems.

In all the generic types of intermittent systems, the item being processed (part, product, information, person) normally goes through a processing sequence, but the work to be done and the sequence of processing vary. At each service center, the specification of what is to be accomplished determines the details of processing and the time required. For each service center, we have the general conditions of a waiting line (queuing) system, with variable arrival and processing rates. The detailed structure of the activities within the service center could follow any of the four basic structures of waiting line situations shown in Figure D-1 (Appendix D). When we view an intermittent system as a whole, we can visualize it as a network of queues with variable paths or routes through the system, depending on the details of processing requirements. An important problem in the design of such systems is the relative location of the service centers or departments.

TABLE 14-2

Typical Departments or Service Centers for Various Generic Types of Intermittent Systems

Generic System	Typical Departments or Service Centers
Machine shop	Receive, Stores, Drill, Lathe, Mill, Grind, Heat-Treat, Inspection, Assembly, Ship
Hospital	Receiving, Emergency, Wards, Intensive-Care, Maternity, Surgery, Laboratory, X-ray, Administration, Cashier, etc.
Medical clinic	Initial Processing, External Examination, Eye, Ear, Nose and Throat, X-ray and Fluoroscope, Blood Tests, Electrocardiagraph and Electroencephalograph, Laboratory, Dental, Final Processing.
Engineering office	Filing, Blueprint, Product Support, Structural Design, Electrical Design, Hydraulic Design, Production Liaison, Detailing and Checking, Secretarial Pool.
Municipal offices	Police Dept., Jail, Court, Judge's Chambers, License Bureau, Treasurer's Office, Welfare Office, Health Dept., Public Works and Sanitation, Engineer's Office, Recreation Dept., Mayor's Office, Town Council Chambers.

Decision to Organize Facilities by Function

To obtain reasonable utilization of personnel and equipment in intermittent flow situations, we assemble the skills and machines to perform a given function in one place and then route the items being processed to the functional centers. If we tried to specialize according to the processing requirements of each type of order in production line fashion, we would have to duplicate many kinds of skills and equipment. The utilization for each order might be very low unless the volume for that type were very large. Thus, when flexibility is the basic system requirement, a functional arrangement is likely to be the most economical. The flexibility required may be of several types: flexibility of routes through the system; flexibility in the volume of each order; and flexibility in the processing requirements of the item being processed.

Other advantages of the functional design become apparent when it is compared to the continuous flow or production line concept. The jobs that result from an intermittent organization are likely to be broader in scope and require more job knowledge. One is an expert in some field of work, whether it is heat-treating, medical laboratory work, structural design, or city welfare. Even though the functional mode implies a degree of specialization, it is specialization within a field of activity, and the variety within that field can be considerable. A pride in workmanship has been traditional in this form of organization of work by trades, crafts, and relatively broad specialties. Job satisfaction criteria seem easier to meet in these situations than when specialization results in highly repetitive activities.

The Relative Location of Facilities Problem

Once we decide to organize productive facilities on a functional basis, the central design problem is to relate the individual functional units. In a machine shop, should the lathe department be located adjacent to the mill department? In a hospital, should the emergency room be located adjacent to intensive care? In an engineering office, should product support be located adjacent to electrical design? In municipal offices, should the welfare and health department offices be adjacent to each other? The locations will depend on the need for one facility to be adjacent relative to the need for other pairs of facilities to be adjacent. We must allocate locations based on relative gains and losses for alternatives and seek to minimize some measure of the cost of having facilities nonadjacent.

Criteria. We always are attempting to measure the interdepartmental interactions required by the nature of the system. How much business is carried on between departments, and how do we measure it? In manufacturing systems, material must be handled from department to department; in offices, people walk between locations to do business and communicate; and in hospitals, patients must be moved, and nurses and other personnel must walk from one location to another. Table 14-3 summarizes criteria for four systems.

By their very nature, functional systems have no fixed path. We must aggregate for all paths and seek a combination of relative locations that minimizes the criterion.

TABLE 14-3

Criteria for Determining the Relative Location of Facilities in Intermittent Systems

Generic System	Criterion
Manufacturing	Interdepartment material handling cost
Hospital	Personnel walking cost between departments
Medical clinic	Walking time of patients between departments
Offices	Personnel walking cost between areas and equipment, or face-to-face contacts between individuals

While this location combination may be poor for some paths through the system, in the aggregate it will be the best set of locations.

Complexity of the Relative Location Problem. Figure 14-3 shows, in schematic form, six process areas arranged on a grid. If any of the six departments can be located in any of the six alternate locations, there are 6! = 720 possible arrangements, of which 90 are different in terms of their effects on the cost of interdepartmental transactions. For this trivial problem, we could consider enumerating the different location combinations, comparing aggregate costs, and selecting the combination with minimum cost. However, the number of combinations to evaluate goes up rapidly as we increase the number of departments. For just 9 departments on a 3×3 grid, we have more than 45,000 combinations. For 20 departments arranged on a 4×5 grid, we have 608×10^{15} combinations. Therefore, we must rule out enumeration as a practical approach.

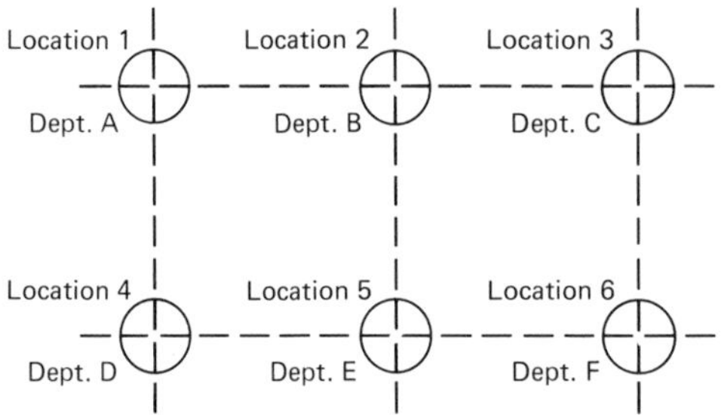

FIGURE 14-3
There are 6! = 720 arrangements of the six process areas in the six locations of the grid.

TABLE 14-4 **Load Summary: Sample of Number of Trips per Month Between All Combinations of Departments for an Industrial Medical Clinic**

From Departments		To: 1 Initial Processing	2 Eye Examination	3 Ear, Nose, and Throat	4 X-ray and Fluoroscope	5 Blood Tests	6 Blood Pressure Check	7 Respiratory Check	8 Electrographic	9 Laboratory	10 Dental Examination	11 Final Processing
Initial processing	1		600									
Eye examination	2			400	100			100				
Ear, nose, and throat	3				350	50						
X-ray and fluoroscope	4						100	450				
Blood tests	5							50				
Blood pressure check	6				100					150	100	
Respiratory check	7						50		450	100		
Electrographic	8						200			250		
Laboratory	9										500	
Dental examination	10											600
Final processing	11											

Operation Sequence Analysis

An early graphical approach to the problem led to a powerful computerized model, which we will also examine. Operation sequence analysis maintains close contact with the nature of the problem and is useful for relatively small problems.

For example, a private industrial clinic performs services under contract to a number of business and industrial firms. These services include medical examinations for new employees, as well as annual physical examinations for all employees. The clinic performs eight types of examination sequences, depending on the details of the individual contracts. Many people who have been examined have complained about excessive walking because of the clinic layout. The director of the clinic wants to know what a good solution would look like.

Table 14-4 represents a one-month sample of the flow between the 11 departments, aggregating over all types of examinations. Table 14-4 is called a load summary and summarizes the flow among all combinations of departments.

The graphical approach to finding a solution places the information contained in the load summary in an equivalent schematic diagram, in which circles represent the functional service centers. Connecting lines are labeled to indicate the intensity of travel or transactions between centers. Figure 14-4 is a first solution and is obtained merely by placing the work centers on a grid, following the logic of the pattern indicated by Table 14-4. The initial solution may be improved by inspecting the effect of changes in location. When an advantageous change is found, the diagram is altered. For example, in Figure 14-4, work center 4 has a total of 300 trips to or from work centers that are not adjacent, that is, 2 and 6. If 4 is moved to the location between 2 and 6, all loads to and from 4 become adjacent.

Further inspection shows that 200 nonadjacent trips occur between work centers 6 and 8. Is an advantageous shift possible? Yes: by moving 9 down and placing 8 in the position vacated by 9, the number of nonadjacent trips is reduced from 200 to 100. Figure 14-5 shows the diagram with the changes incorporated.

Further inspection reveals no further obvious advantageous shifts in location. Figure 14-5, has a $2 \times 100 = 200$ trip-distance rating. For larger problems, the grid distance becomes an important part of the measure of effectiveness, because work centers might be separated by two, three, or four grid units. Figure 14-5 represents a good solution because most of the work centers are adjacent to the other work centers involving interdepartmental flow.

The Block Diagram. The block diagram is developed by substituting estimated areas for the small circles in the schematic diagram. Initially, this can be done with

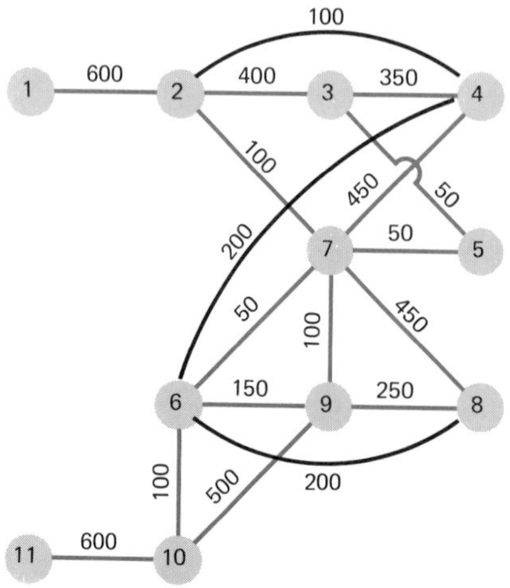

FIGURE 14-4
Initial graphic solution developed from load summary of Table 14-4.
SOURCE: *Figures 14-4 to 14-7 are from E. S. Buffa, "Sequence Analysis for Functional Layouts,"* Journal of Industrial Engineering, *6(2), March-April 1955.*

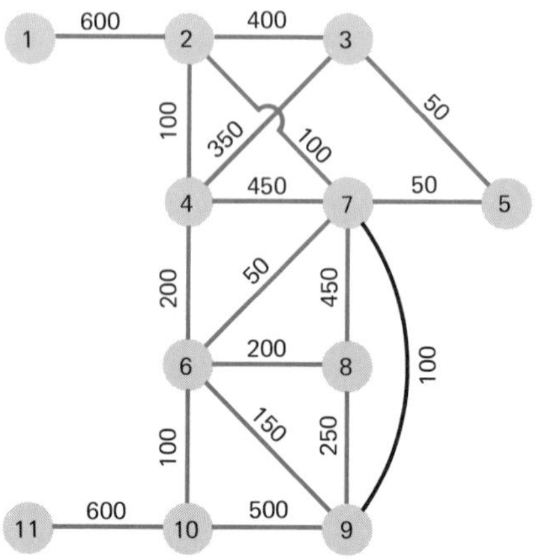

FIGURE 14-5
Schematic diagram incorporating changes suggested by Figure 14-4.

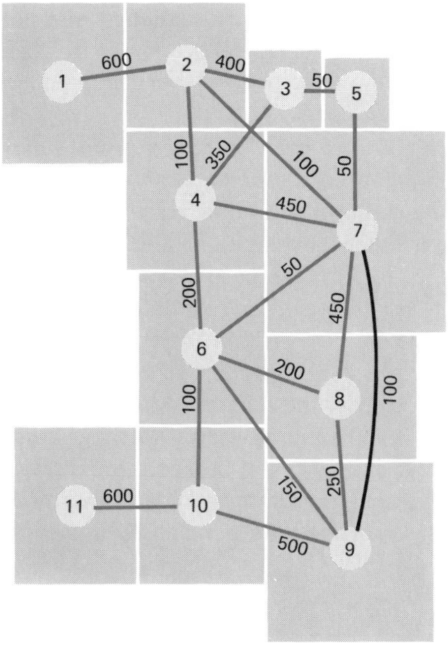

FIGURE 14-6
Initial block diagram. Estimated areas are substituted for circles in the schematic diagram of Figure 14-5.

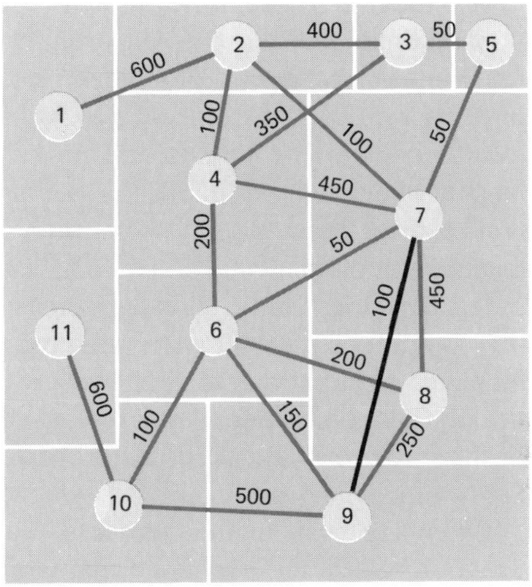

FIGURE 14-7
Block diagram that takes account of rectangular building shape.

block templates to find an arrangement that is compatible with both the flow pattern of the schematic diagram and the various size requirements for departments. Figure 14-6 shows such an initial block diagram. Although the essential character of the schematic diagram is retained, Figure 14-6 obviously does not yet represent a practical solution. A slight variation of the shapes of departments will enable us to fit the system into a rectangular configuration and meet possible shape and dimension restrictions that may be imposed by the site, an existing building, or the desired configuration of a new building. Figure 14-7 shows such a block diagram.

The final block diagram represented by Figure 14-7 becomes an input to detailed layout. The detailed layout phase would undoubtedly require minor shifts in space allocation and shape, but the basic relationships would be retained.

CRAFT (Computerized Relative Allocation of Facilities)

The effectiveness of the graphical approach depends on the analyst's insight, and as the number of activity centers increases, the technique breaks down rapidly. Practical problems in facility location often involve 20 or more activity centers, and this number is already at the limit for feasible use of the operation sequence analysis technique. To overcome this limitation, a computerized relative allocation of facilities technique (CRAFT) was developed, which easily handles up to 40 activity centers and has other important advantages.

The CRAFT Program. The CRAFT program takes as input data matrixes of interdepartmental flow and interdepartmental transaction cost, together with a representation of a block layout. The block layout that is fed in may be the existing layout or any arbitrary starting solution if a new facility is being developed. The program calculates department centers and an estimate of total interaction cost for the input layout. The governing heuristic algorithm then asks: What change in interaction cost would result if locations of departments were exchanged? Within the computer, the locations of the activity centers are exchanged, and the interaction costs are recomputed. Whether the result is an increase or a decrease, the difference is recorded in the computer's memory. The program then asks the same question for other combinations of departments, and again records cost differences. It proceeds in this way through all combinations of exchanges. The present algorithm involves either two- or three-way exchanges. Block [in progress] has extended the principle to four- and five-way exchanges.

When the cost differences for all such combinations have been computed, the program selects the exchange that would result in the largest reduction, makes the exchange in locations on the block diagram, and then prints out the new block layout, the new total interaction cost, the cost reduction just effected, and the departments involved in the exchange. The basic procedure is then repeated, generating a second and a third improved block layout and so on. Finally, when the procedure indicates that no further location cost-reducing exchanges can be made, the final block layout is printed out. This becomes the basis for a detailed template layout of the facility. Typical computer output for the block layout is shown in Figure 14-8. Departmental areas are identified by letter code, and each character represents a given number of square feet.

The program has the capacity for handling 40 activity centers. Any departmental location can be fixed down simply by specifying in the instructions that the department (or departments) is not a candidate for exchange. This feature has great practical importance, because existing layouts cannot be completely rearranged. Fixed locations may develop when costly heavy equipment has been installed, or the location of receiving or shipping facilities may be determined by the location of roads or railroad spurs. Finally, the locations of some work groups often make it desirable to treat them as fixed points in the layout. A new development also makes it possible to include relocation costs as well as interaction costs in the computer program for relayout. [Hicks and Cowan, 1976].

Industrial Application of CRAFT. Computerized facilities layout programs have been most widely applied in industry. However, they have also been used in service and nonmanufacturing applications. To the author's knowledge, the CRAFT program itself has been used in four aircraft plant applications, two of the largest automobile companies, two computer manufacturers, a pharmaceutical manufacturer, a meat packer, a precision machine shop, a movie studio, and a hospital. Since the program

	1	2	3	4	5	6	7	8	9	10	11	12	13	14	15
1	P	P	P	P	P	V	V	V	V	V	V	V	V	V	V
2	P				P	V									V
3	S	P	P	P	P	V									V
4	S	S	S	S	S	V									V
5	S			S	S	V	V	V	V	V	V	V	V	V	V
6	S	S	S	T	T	B	B	B	B	M	I	I	I	N	N
7	T	T	T		T	B			B	I			I	N	N
8	T	T			T	B				B	I		I	N	N
9	R	R	T		T	B	B	B	B	B	I		I	N	N
10	R	R	T		T	C	C	C	C	C	I	I	N		N
11	R	R	T		T	D	D	D	D	D	D	D	N		N
12	G	G	T		T	D						D	N	N	N
13	G	G	T		T	D						D	L	L	L
14	G	G	T	T	T	D					D	L			L
15	G	Q	Q	Q	Q	D					D	L			L
16	G	Q	Q	F	F	D	D	D	D	D	D	L			L
17	F	F	F		F	U	U	U	U	U	U	L	L	L	L
18	F				F	U					U	H	H	H	L
19	F	F	F		F	U					U	H	E	E	E
20	J	J	J	F	F	U	U	U	U	U	U	E			E
21	J	J	J	A	A	K	K	K	K	K	K	E			E
22	A	A	A		A	K					K	E			E
23	A	A	A	A	A	K	K	K	K	K	E	E	E	E	E
24	O	O	O	O	O	O	O	O	O	O	O	O	O	O	O
25	O														O
26	O														O
27	O	O													O
28	O		O	O	O	O	O	O	O	O	O	O	O	O	O

TOTAL COST 900.93 EST COST REDUCTION 3.38 MOVEA Q MOVEB G MOVEC

FIGURE 14-8
Best block layout for alternative 4, dispatch with information system—using specialized material handlers.
SOURCE: *E. S. Buffa, G. C. Armour, and T. E. Vollmann, "Allocating Facilities with CRAFT,"* Harvard Business Review, *March-April 1964.*

is freely available and has been circulated widely, many other applications undoubtedly have been made.

A Precision Manufacturer in the Aerospace Industry. A particularly interesting application was made in the machine shop of a precision manufacturer in the aerospace industry. The shop occupied about 42,000 square feet of floor area, and the majority of orders were for small precision parts in low quantities. Because of the small physical size of orders, most material handling among departments was accomplished by the machinists themselves, carrying orders to and from a central holding and dispatch area (Department K) in tote pans. The current layout had grown around the central holding and dispatch department.

In addition to the layout itself, one important question that management wished to evaluate was the validity of having material flow through the central holding area. Although the use of the central holding area offered a small plant the advantage of close control, it was obvious that the physical flow of material to and from the holding area entailed incremental material handling costs. The alternate policy was to dispatch orders directly from department to department, using an information system for control.

Another operating policy that management wished to evaluate was the use of machinists to accomplish the majority of interdepartmental material handling. Originally, when the plant was small, the machinists were used for this purpose because distances were short and it was felt that these short walks gave machinists a break from usual routines. As time passed, however, the plant was enlarged, and management felt that the use of specialized material handlers should be evaluated.

There were four basic conditions for which the determination of the best layout and the associated material handling costs were desired:

1. Current practice: material flow through Department K. Machinists used for handling.

2. Material flow through Department K. Using material handlers.

3. Dispatch using information system. Machinists used for handling.

4. Dispatch using information system. Using material handlers.

For purposes of analysis, 22 plant areas were designated as department centers. To determine interdepartmental flow matrix, it was necessary to analyze approximately 1600 shop orders (approximately an eight-week sample) on which were indicated the routing required to fabricate the parts. A similar matrix for the four conditions was developed to show the material handling cost in dollars per 100 feet of movement for combinations of departments for which flow occurred.

The CRAFT program was then used to evaluate both layout configuration and the four alternate policies for order dispatching and material handling. Table 14-5 summarizes the results. The far-right column indicates the percentage reduction in the cost of each alternative compared to the cost of the existing layout and current

TABLE 14-5 **Material Handling Costs for Five Layouts**

Alternatives	Total 8-Week Material Handling Cost	Percentage Reduction from Existing Layout
Existing layout—current policies	$3294.98	—
Cost for best layout under following conditions:		
1. Current practice: Material flow through Department *K*—machinists used for handling	2645.08	20%
2. Material flow through Department *K*—using material handlers	2402.97	27%
3. Dispatch utilizing information system—machinists used for handling	1186.89	64%
4. Dispatch utilizing information system—using specialized material handlers	900.93	73%

SOURCE: E. S. Buffa, G. C. Armour, and T. E. Vollmann, "Allocating Facilities with CRAFT," *Harvard Business Review*, March-April 1964.

policies. Alternative 4, which uses an information system for dispatch and control and material handlers, results in a 73 percent reduction. When the eight-week figures are placed on an annual basis, the cost reduction potential is approximately $15,500. If the company demands a 10 percent return on plant investments of this kind and imposes a severe three-year payback for capital recovery and return, it could afford to spend over $38,000 for the plant relocations and alterations necessary for re-layout. Figure 14-8 shows the computer printout for alternative 4.

Improvement in material handling cost is possible through all four policies. Alternative 1 indicates that a 20 percent material handling cost reduction results from a layout improvement with existing policies. An additional 7 percent improvement results from using material handlers. The use of an information system apparently leads to the largest improvement. It is important to note, however, that we are dealing with an integrated system and that interacting effects may have occurred among layout configuration, material handling method, and dispatch and control system.

Office Layout. The same general concepts about the relative location of facilities apply to the layout of large offices with functional groupings, although the measure of effective relative location may change. A number of applications of the CRAFT program in office situations have been reported.

Vollmann, Nugent, and Zartler [1968] applied a variant of the program to the office layout of an oil company in the southwestern United States. The locations of 27 persons and 7 significant pieces of equipment were involved in an office occupying 5600 square feet. The criterion to be minimized was the aggregate of the product of cost-weighted trips by employees by the distance walked. Thus the cost of a highly paid executive walking between centers has a greater significance than lower paid employees when weighted in the criterion. A large number of alternatives were evaluated, including those involving personal preferences for location of certain individuals (e.g., given that a certain individual merits a corner office by virtue of organizational position, what should be the relative location of other individuals?).

Applications to Macro Systems. In larger-scale systems, the same general point of view on the relative location of facilities applies. Many manufacturing plants are enormous in scale, with several buildings covering many acres. The material handling cost between units located in separate buildings can be very large, justifying careful study of relative locations.

One such application took place at a movie studio [Buffa, Armour, and Vollmann, 1974], where the major functional departments—wardrobe, furniture storage, nursery (for dressing a stage with plant material), sound stages, and so on—were interconnected by a material handling system. Material handling was accomplished through a fleet of trucks, and each trip involved truck and driver costs.

In dressing a stage before shooting a scene, the director must specify the kinds of furnishings and other props desired. Because of the esthetic aspects required for this procedure, many trips may be needed to obtain exactly the right effects. After the scene has been shot, all the materials must be redistributed to their storage places for later reuse. In such a situation the material handling costs are immense, and the best relative location of various departments strongly depends on these costs. The CRAFT program indicated that it was possible to reduce one studio's material handling cost by $240,000 per year, through relocation of many of the activities involved. Since relocation on such a scale is expensive, many alternatives were computed, fixing the location of certain facilities (e.g., sound stages) in their present locations and determining the best locations for movable functions.

In another study of the facilities of a large, complex aerospace manufacturer, these same concepts were valuable in relocating departments within the system. The study focused on the relocation of special laboratory service (which had been decentralized) and raised the question of whether or not to centralize—and, if so, which location of several possibilities was most desirable. The results indicated that centralizing the service in one of the available locations was 14 percent less costly than the second best location.

Assumptions and Limitations of CRAFT. CRAFT is a macro layout model that emphasizes the large differences among alternatives. It assumes that a good final layout must be based on a block diagram that permits low-cost aggregate flow. In taking the aggregate view, some of its assumptions should be understood so that results can be interpreted properly.

One of CRAFT's major assumptions is that flow among departments or work centers occurs between the centroids of the departments. When the department shape is approximately square, this is a fairly good assumption. On the other hand, when departments are long narrow rectangles or irregular shapes, the assumption is not a good one. Also, sometimes the minimizing algorithm will produce odd department shapes. The analyst must regard the CRAFT output as an input to further analysis. Department shapes often are regularized anyway to fit in with building limitations (e.g., columns) and to lay out aisles and other spaces in a practical manner.

The nature of the criterion used also involves significant assumptions. Whatever the criterion used—material handling costs, walking cost, and so on—the assumption is that the criterion varies linearly with distance. In addition, the very nature of layout studies places faith in past data on flow and its composition, although the layout itself is made to accommodate future flow patterns. Studies indicate, in some situations at least, that flow patterns are fairly stable over time. However, if they are dynamic, a layout based on CRAFT or any other technique will become obsolete. Finally, in some situations, handling costs may be joined with several products and can only be allocated arbitrarily.

Comparative Effectiveness of Computer Layout Aids. Other approaches to the relative location problem have been developed, some of which have been computerized. Hillier and Connors [1966] developed a quadratic assignment algorithm. In 1967, two computerized approaches were developed that use a rating system to approximate the relative need for pairs of departments to be adjacent or close to each other. ADELP was developed by Seehof and Evans [1967], and CORELAP was developed by Lee and Moore [1967]. Both programs use a scale that assigns numerical weights to the following descriptors: absolutely essential, essential, important, optional importance, unimportant, undesirable, and same department. Independent comparative studies have been carried out by Denholm and Brooks [1970], Ritzman [1972], Nugent, Vollmann, and Ruml [1968], and Zoller and Adendorff [1972].

Comparison with Visually Based Methods. Computer-based methods such as CRAFT have an advantage where the flow through the system does not have dominance, that is, a dominant sequential flow. When flow is dominant, the relative location of departments becomes obvious by inspection, or other visually based methods such as operation sequence analysis.

Project Systems and Fixed-Position Assembly

Large-scale projects are common in today's economy. In terms of facility utilization, they are intermittent systems, and their physical layout deserves a special comment. Examples of familiar large-scale projects are huge aerospace projects such as the Polaris missile, aircraft assembly, ship building, or large construction projects. Some of these kinds of projects involve basically conventional manufacturing systems to produce components that go into the project itself. The systems that manufacture these components will be typical of either intermittent or continuous manufacturing systems.

The heart of the project concept, however, lies in the assembly process, which is usually done on a fixed-position basis, either by necessity (as with buildings, dams, and bridges) or for reasons of economy (as with missiles, aircraft, ships, and other very large projects). In these fixed-position assembly situations, the equivalent of functional centers are commonly arranged around the unit being constructed in "staging areas." Some of the staging areas will be storage locations where the material or components will wait until needed in the process. Other areas may involve some degree of fabrication or prefabrication before final assembly on the unit.

Proximity of staging areas to the major unit may depend on frequency of use and travel time between the staging area and the unit. For example, in constructing a skyscraper, the heating and air-conditioning unit need only be installed once; however, forming lumber and reinforcing steel will be used continuously throughout the rough construction stages. The general concept of the computer location algorithms apply to project systems as well as to the kinds of situations already discussed. Since the project is commonly a one-time system, however, formal location procedures have not been used in practice, although the potential for use is valid.

PRODUCT LAYOUT FOR CONTINUOUS SYSTEMS

Decision to Organize as a Continuous System

The managerial decision to organize the work on a product or line basis is a significant one. Some important requirements should be met, and there are some consequences from the attitudes of the work force that should be carefully weighed before implementing the decision.

If we are to organize as a continuous system, the following requirements should be met:

1. Volume adequate for reasonable equipment utilization
2. Reasonably stable product demand
3. Product standardization
4. Part interchangeability
5. Continuous supply of material

Continuous system concepts have found their greatest field of application in assembly rather than in fabrication. A moment's reflection makes it obvious why this is true. Machine tools commonly have fixed machine cycles; this factor makes it difficult to achieve balance between successive operations. The result is poor equipment utilization and relatively high costs. In assembly operations where the work is more likely to be manual, balance is much easier to obtain because the total job can be divided into minute elements. If station 10 is too short while 16 is too long, part of the work of station 16 probably can be transferred to 10 (perhaps the tightening of a single bolt). Since very little equipment is involved at each station anyway, utilization of equipment may not be of great importance.

When the conditions for continuous systems are met, significant advantages can result. The production cycle is speeded up because materials approach continuous movement. Since very little manual handling is required, the cost of material handling is low. In-process inventories are lower compared with batch processing because of the relatively fast manufacturing cycle. Because aisles are not used for material movement and in-process storage space is minimized, less total floor space is commonly required than for an equivalent functional system, even though more individual pieces of equipment may be required. Finally, the control of the flow of work (production control) is greatly simplified for continuous systems because routes become direct and mechanical. No detailed scheduling of work to individual work places and machines is required, since each operation is an integral part of the line. Scheduling the line as a whole automatically schedules the component operations.

Given the decision for designing a continuous system, the major problems are (1) deciding on a production rate or cycle time, and (2) subdividing the work so that smooth flow can result. The subdivided activities need to be balanced so that each station has an equivalent capacity.

The Line Balance Problem

Balance refers to the equality of capacity or output of each of the successive operations in the sequence of a line. If all are equal, we have perfect balance and expect smooth flow. If they are unequal, the maximum possible output for the line as a whole will be dictated by the slowest operation in the sequence. This slow, or bottleneck, operation restricts the flow on the line much as a half-closed valve restricts the flow of water, even though the pipes in the system may be capable of carrying twice as much water. Thus, when imbalance exists in a line, we have wasted capacity in all operations except the bottleneck operation.

Production Lines as Queuing Systems. One of four basic structures of waiting line models is the single channel, multiple phase case.* This structure is a sequence of several of our basic input-transformation-output modules in series, with the output of one phase or stage being the input to the next. If the line is not mechanically paced, the concept of such a structure parallels the actual flow that takes place in a production line.

The important problem is to balance the stages so that each has the same amount of work to do. Actually, however, we are dealing with work-time distributions, and therefore an interplay between arrivals and service times at each stage will exist, as in any queuing system. This means that there will be some instances when the waiting line is empty and the stage is idle. Therefore, the actual operation of a line complicates the problem of achieving balance. We often try to minimize the probability of having any stage idle by flooding the system with in-process inventory so that work usually will be waiting at each stage.

To date, waiting line models have not been a fruitful approach in solving the line balance problem, even though it gives useful insights into how a line functions. Most of the work on line balance, however, has simplified the basic problem, assuming the

* See Appendix D for a discussion of waiting lines. Figure D-1c shows the structure of the single channel, multiple phase case.

deterministic case where the service times are constant values and work is always available at each stage.

Deterministic Form of Line Balancing. The conceptual framework for the deterministic form of the line balance problem was discussed in Chapter 13. Although the toy car example presented there was very simple, it illustrated how combinations of tasks could be made to form stations in order to achieve balance.

Practical Balancing Methods. These general concepts of line balancing have been implemented through a number of practical methods for the large-scale problems in industry. Perhaps of the greatest interest for large-scale problems are COMSOAL, a Computer Method for Sequencing Operations for Assembly Lines, developed by Arcus [1966]; and a computer model called MALB, developed by Dar-El [1973], based on an improvement of the Ranked Positional Weight Technique.

Auxiliary Balancing Techniques. Other techniques are available to obtain balance in both the design and operation stages. If one station requires greater time than the others, a careful study of the activities may result in a reduction of time. Compensations for imbalance can be accomplished by assigning fast operators to the limiting operations. When some very fast operations cannot be combined into a single station (as might be true with machine operations), material banking before and after the fast operations may be required. These fast operations would be operated during only a small portion of the day; the machine, the operator, or both, could be used for other purposes.

Balancing Fabrication Lines. Conceptually, there is no difference in the balancing procedures for assembly and fabrication lines. However, the fixed machine cycles found in fabrication operations considerably limit freedom in achieving balance. It is often impractical to divide a machine operation into two or more suboperations in order to level out the time requirements at each station. This situation partially accounts for the fact that fabrication lines generally are economical only for very high volumes, since good balance is likely to be achieved only at high levels of output.

Production Line Design and Job Satisfaction

As noted in Chapter 13, the production line concept epitomizes division of labor and specialization and is the target of considerable criticism. The general line balance techniques do not seem to leave a door open for discussion of job design alternatives. The process starts with the determination of a cycle time that meets the output rate needs and progresses toward the generation of stations (jobs) made up of tasks that meet the needs of the cycle and do not violate the technological constraints. Job satisfaction as a criterion seems to have no role in this process. The issue is: Are there alternatives that allow us to address the question of the degree of fractionation of jobs while retaining the production line concept?

Multiple Stations and Parallel Lines. Given a capacity or production rate requirement, we can meet that requirement by a single line with cycle time c, or two parallel lines with cycle time $2c$, or three parallel lines with cycle time $3c$, and so forth. Buxey

[1974] has developed extended line balance programs that enable us to use multiple stations. As the number of parallel lines increases, the scope of jobs increases, and finally, we have complete horizontal enlargement. The point is that the alternatives do exist, even with the line organization of work.

In addition to increasing job scope, there are a number of advantages that result from the parallel line–multiple station concept. First, from the line balance point of view, balance may be easier to achieve because the larger cycle time offers a greater likelihood of attaining a good fit with low residual idle time. This is particularly true when some of the task times are nearly equal to the single-line cycle time. Furthermore, a multiple line design increases flexibility of operations enormously. Output gradations are available. That is, one can have one, two, three, or more lines operating or not operating, and one can work overtime or undertime with all the line combinations. There are fewer dependent operations; for example, if there is a difficulty with an operation in line 1, it may not affect line 2. A machine breakdown in line 1 need not stop the operation of line 2.

From a human organization viewpoint, the parallel line–multiple station concept has all the advantages of horizontal job enlargement. Work groups can be smaller and more cohesive. A team spirit may be engendered by competition between line teams on the bases of output, quality, safety, and other dimensions.

IMPLICATIONS FOR THE MANAGER

Since intermittent systems can process a variety of work, routes through the system must vary, and the resulting problem of physical design and arrangement of work centers can be very complex. Thus the relative location of departments sets the design for the system in an overall sense. The objective of the physical arrangement is to minimize some appropriate measure of the interaction among work centers. This may be material handling cost in manufacturing systems; cost of personnel walking between locations in hospitals, clinics, and offices; or simply the need to have face-to-face contact, as in some office situations.

Operation sequence analysis is a simple, graphic technique for handling small problems. Larger problems may require more powerful computer-aided methods, such as CRAFT. As inputs, the CRAFT program requires (1) an initial spatial array and (2) matrixes of the volume of interaction and interaction cost per volume unit, per unit distance. The algorithm functions by computing the changes in interaction cost that would result if locations were exchanged, making exchanges progressively for those switches with the greatest cost reduction potential. The program may be of value for both new design and redesign of existing facilities.

Large-scale projects are intermittent systems in which the assembly process is usually accomplished in a fixed position. The "relative location" concept for locating staging areas around the unit being constructed is valid but has not been used.

Continuous system design is centered in line balancing, where the problems of system capacity and the capacity of individual operations and their balance are resolved. While the queuing model is conceptually valid in helping us understand what happens in the flow of items in a production line, it provides little help in solving the line balance problem. Job satisfaction is an important element in the decision to design a line system, and parallel lines offer some opportunity to broaden job scope while retaining the advantages of line operations.

REVIEW QUESTIONS AND PROBLEMS

1. In planning for a facility layout, the design capacity is an important decision, since capital investment requirements largely are determined by the decision, and the basic strategy for output scheduling also is set. Under what conditions should we design for peak demand needs? For average demand needs?

2. What are the alternate strategies for providing for future capacity in a facility layout?

3. What are the general economic considerations that result in the use of multiple shifts?

4. Define the terms "facility efficiency factor" and "scrap factor." What are their functions in computing capacity needs?

5. Define the terms "process layout" and "product layout."

6. Under what conditions would process layout be appropriate?

7. What is the relative location of facilities problem?

8. Consider the situation of the private industrial clinic used as an example for operation sequence analysis. Recall that the clinic performs eight types of examination sequences, depending on the details of individual contracts. Since there have been many complaints about excessive walking by those who have been examined, why not organize the entire clinic on the basis of the eight types of examination sequences in production-line fashion?

9. Table 14-3 summarizes types of criteria that might be used to determine the relative location of facilities in functional systems. Can you think of situations where criteria other than those listed might be important?

10. In the operation sequence analysis technique, what is the criterion for deciding whether or not departments should be located adjacent to one another?

11. In the operation sequence analysis technique, how do we translate the schematic diagram into a block diagram?

12. Explain the nature and functioning of the CRAFT algorithm. By what criterion would the algorithm exchange the location of departments in a factory layout? An office layout? A municipal office complex?

13. Explain the assumptions and limitations of the CRAFT layout model.

14. What is fixed position assembly? How does it relate to large-scale project systems?

15. What kinds of material handling systems do you think might be appropriate for process layouts?

16. In the decision to organize for continuous flow, why are each of the following factors requirements that should be met?

    a. Volume adequate for reasonable equipment utilization
    b. Reasonably stable product demand
    c. Product standardization
    d. Part interchangeability
    e. Continuous supply of material

17. Define the meaning of each of the five requirements listed in question 16; that is, what is their meaning with respect to the decision to organize for continuous flow?

18. Define the nature of the line balance problem.

19. Describe the line balance problem as a queuing model. What simplifications are made in the deterministic form of line balancing?

20. How is *fine tuning* of line balance solutions achieved after installation?

21. How is it possible to introduce the question of appropriate job breadth into the process for designing and balancing production lines?

22. Assume a simple situation in which there are four departments to be arranged on a grid, as in Figure 14-3. Because of symmetry, most of the possible arrangements are not really different in their effect on aggregate interaction cost. There are 4! = 24 possible arrangements; however, only three of them are fundamentally different. Enumerate these three arrangements.

23. How do unequal areas change the answer to problem 22? Suppose, for example, that area requirements are:

Department	*Area*
A	$2 \times 4 = 8$
B	$2 \times 2 = 4$
C	$1 \times 2 = 2$
D	$1 \times 2 = 2$
	16

a. How many different arrangements are now possible? Enumerate them.
b. How many different arrangements are possible if the areas are as follows:

Department	*Area*
A	$1 \times 4 = 4$
B	$1 \times 4 = 4$
C	$1 \times 2 = 2$
D	$1 \times 2 = 2$
	12

24. Consider the schematic diagram shown in Figure 14-5. Can you improve it?

25. A manufacturing concern has four departments, and the flow between combinations of departments is as follows:

	To			
From	A	B	C	D
A		2		2
B	2		4	
C		3		1
D	2		1	

a. Using the operation sequence analysis technique, how should the departments be arranged?
b. Now suppose that the area requirements are as follows:
Department A—3600 sq. ft.
Department B—2400 sq. ft.
Department C—2400 sq. ft.
Department D—1600 sq. ft.
Sketch the block diagram based on your answer in part a.

TABLE 14-6 **Operations Sequence Summary for Problem 26**

Machine or Work Center	Area, Square Feet	Work Center Number	Part A	Part B	Part C	Part D	Part E	Part F	Part G
Saw	50	1		2	2				2
Centering	100	2		4	3				3
Milling machines	500	3	5	9	5	5		4	4
Lathes	600	4		5,7	7		5	10	5
Drills	300	5	8	3	4	11	7		6
Arbor press	100	6					11		7
Grinders	200	7		12	12		6		8
Shapers	200	8	9			3			9
Heat treat	150	9	11	4					10
Paint	100	10						11	11
Assembly bench	100	11	12	13	13	13	13	13	12
Inspection	50	12	13	11	11				13
Pack	100	13							
Production Summary									
Pieces per month			500	500	1600	1200	400	800	400
Pieces per load			2	100	40	40	100	100	2
Loads per month			250	5	40	30	4	8	200

c. Now assume that the four departments are located in two separate buildings that are 100 feet apart. The two buildings have floor areas of $60 \times 100 = 6000$ square feet, housing departments A and C, and $40 \times 100 = 4000$ square feet, housing departments B and D, respectively. What should be the space allocation to the four departments if the material handling costs are as follows:

From	To A	To B	To C	To D
A		1	1	2
B	1		1	2
C	1	1		2
D	2	2	2	

26. An organization does job machining and assembly and wishes to redesign its production facilities so that the relative location of departments reflects in a somewhat better way the average flow of parts through the plant.

Table 14-6 shows an operation sequence summary for a sample of seven parts, with approximate area requirements for each of the 13 work centers. The

TABLE 14-7 **Number of Face-to-Face Contacts per Month in an Engineering Office for Problem 27**

	To									
From	(A) Filing	(B) Supervision	(C) Blueprint	(D) Product Support	(E) Structural Design	(F) Electrical Design	(G) Hydraulic Design	(H) Production	(I) Detailing and Checking	(J) Secretarial Pool
A Filing		15				5		10		15
B Supervision	20		25	40	100	90	80	160	85	60
C Blueprint										
D Product Support	10	15					20	280		10
E Structural Design	50	20	600			40			340	50
F Electrical Design			475					160	270	60
G Hydraulic Design	10		460	20				140	320	45
H Production Liaison	20			200	160	190	240			680
I Detailing and Checking		210	690	40	190	240	80			20
J Secretarial Pool		25						15		

TABLE 14-8 **Area Requirements and Average Wage Rates for Ten Groups in a Large Engineering Office**

	Group	Area	Average Hourly Wage
A	Filing	20 × 15 = 300 sq. ft.	$2.25
B	Supervision	30 × 15 = 450 sq. ft.	5.00
C	Blueprinting	40 × 15 = 600 sq. ft.	2.10
D	Product support	25 × 20 = 500 sq. ft.	2.70
E	Structural design	65 × 25 = 1625 sq. ft.	4.50
F	Electrical design	25 × 35 = 875 sq. ft.	4.50
G	Hydraulic design	45 × 30 = 1350 sq. ft.	4.50
H	Production liaison	20 × 70 = 1400 sq. ft.	2.70
I	Detailing and checking	70 × 25 = 1750 sq. ft.	3.60
J	Secretarial pool	70 × 15 = 1050 sq. ft.	2.40
		90 × 110 = 9900 sq. ft.	

numbers in the columns headed by each of the parts indicate the number of the work center to which the part goes next. Just below the sequence summary is a summary of production per month and the number of parts handled at one time through the shop for each part.

a. Develop a load summary showing the number of loads per month going between all combinations of work centers.
b. Develop a schematic layout that minimizes nonadjacent loads.
c. Develop a block diagram that reflects the approximate area requirements and results in an overall rectangular shape.

27. A layout study was made in the engineering office of a large aerospace manufacturer. The study initially focused on the flow of work through the system, but meaningful cost data were difficult to generate on this basis.

The search for realistic measures of effectiveness was finally narrowed down to the relative location of people in the organization as required by their face-to-face contacts with others in the organization. It was decided to collect data on face-to-face contacts initiated by each person for a one-month period. These data were accumulated in the form of Table 14-7.

The entire department was divided into 10 groups, as indicated in Table 14-7. Each cell value indicates the number of face-to-face contacts initiated by that group; for example, 15 from A to B. Table 14-8 summarizes the area requirements for each group, as well as the average hourly wage paid in each group. Figure 14-9 shows the present block layout.

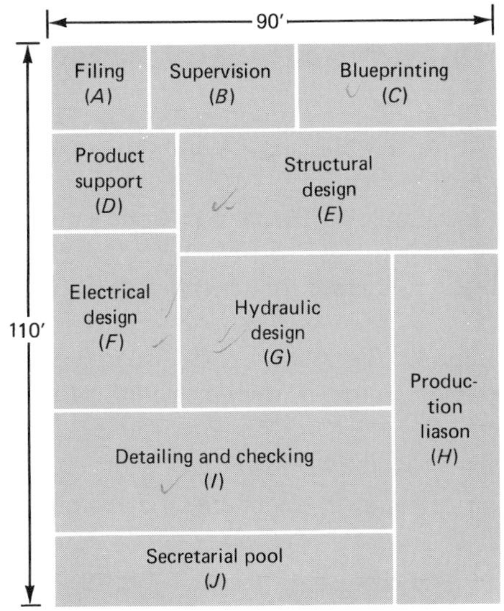

FIGURE 14-9
Existing block layout for ten groups within a large engineering office.

TABLE 14-9

Tasks, Performance Times, and Precedence Requirements for Problem 28

Task	Performance Time, Minutes	Task Must Follow Task Listed Below
a	4	—
b	3	*a*
c	5	*b*
d	2	—
e	4	*c*
f	6	*d*
g	2	—
h	3	*dg*
j	5	*h*
k	2	—
l	3	*k*
m	4	*l*

Prepare a new block layout within the constraints of the overall size and dimensions of the layout shown in Figure 14-9.

28. Table 14-9 shows a list of assembly tasks with sequence restrictions and performance times. Construct a precedence diagram for the assembly. Balance the line for an output rate of 10 units per hour.

REFERENCES

Arcus, A. L. "COMSOAL: A Computer Method for Sequencing Operations for Assembly Lines," *International Journal of Production Research, 4*(4), 1966.

Armour, G. C., and E. S. Buffa. "A Heuristic Algorithm and Simulation Approach to Relative Location of Facilities," *Management Science, 9*(1), 1963, pp. 294–309.

Atkinson, G. A., and R. J. Phillips. "Hospital Design: Factors Influencing the Choice of Shape," *Architects Journal,* April 1964.

Block, T. E. "A Note on 'Comparison of Computer Algorithms and Visual Based Methods for Plant Layout' by M. Scriabin and R. C. Vergin," *Management Science, 24*(2), October 1977, pp. 235–238.

Block, T. E. "PLOP—Plant Layout Optimization Procedure," in progress.

Buffa, E. S. "Communication to the Editor on a Paper by Scriabin and Vergin," *Management Science, 23*, September 1976.

Buffa, E. S. "Sequence Analysis for Functional Layouts," *Journal of Industrial Engineering, 6*(2), March–April 1955.

Buffa, E. S., G. C. Armour, and T. E. Vollmann. "Allocating Facilities with CRAFT," *Harvard Business Review, 42*(2), March–April 1964, pp. 136–159.

Buffa, E. S., and J. G. Miller. *Production-Inventory Systems: Planning and Control* (3rd ed.). Richard D. Irwin, Homewood, Ill., 1979.

Buxey, G. M. "Assembly Line Balancing with Multiple Stations," *Management Science, 20*(6), February 1974, pp. 1010–1021.

Coleman, D. R. "Plant Layout: Computers Versus Humans," *Management Science, 24*(1), September 1977, pp. 107–112.

Dar-El, E. M. "Solving Large Single-Model Assembly Line Problems—a Comparative Study," *AIIE Transactions, 7*, 1975, pp. 302–310.

Dar-El, E. M. (Mansoor). "MALB—a Heuristic Technique for Balancing Large-Scale Single-Model Assembly Lines," *AIIE Transactions, 5*(4), December 1973.

Dar-El, E. M., and R. F. Cother. "Assembly Line Sequencing for Model Mix, *International Journal for Production Research, 13*, 1975, pp. 463–477.

Denholm, D. H., and G. H. Brooks. "A Comparison of Three Computer Assisted Plant Layout Techniques," *Proceedings, American Institute of Industrial Engineers,* 21st Annual Conference and Convention, Cleveland, 1970.

Francis, R. L., and J. A. White. *Facility Layout and Location: An Analytical Approach.* Prentice-Hall, Englewood Cliffs, N. J., 1974.

Hicks, P. E., and T. E. Cowan. "CRAFT-M for Layout Rearrangement," *Industrial Engineering, 8*, May 1976, pp. 30–35.

Hillier, F. S., and M. M. Connors. "Quadratic Assignment Problem Algorithms and the Location of Indivisible Facilities," *Management Science, 13*(1), September 1966, pp. 42–57.

Ignall, E. J. "A Review of Assembly Line Balancing," *Journal of Industrial Engineering, 16*, 1965.

Kilbridge, M. D. "The Balance Delay Problem," *Management Science, 8*(1), October 1961.

Kilbridge, M. D., and L. Wester. "A Heuristic Method of Assembly Line Balancing," *Journal of Industrial Engineering, 12*(4), July–August 1961.

Kilbridge, M. D., and L. Wester. "A Review of Analytical Systems of Line Balancing," *Operations Research, 10*(5), September–October 1962.

Lee, R. C., and J. M. Moore. "COORELAP—Computerized Relationship Layout Planning," *Journal of Industrial Engineering, 18*(3), March 1967, pp. 195–200.

Lew, P., and P. M. Brown. "Evaluation and Modification of CRAFT for an Architectural Methodology," in *Emerging Methods in Environmental Design and Planning,* edited by G. T. Moore. MIT Press, Cambridge, Mass., 1970.

Mansoor, E. M. (Dar-El). "Assembly Line Balancing—an Improvement on the Ranked Positional Weight Technique," *Journal of Industrial Engineering, 15*(5), March–April 1964.

Mastor, A. A. "An Experimental Investigation and Comparative Evaluation of Production Line Balancing Techniques," *Management Science, 16*(11), July 1970, pp. 728–746.

Moore, J. M. *Plant Layout and Design.* Macmillan, New York, 1962.

Muther, R. *Practical Plant Layout* McGraw-Hill, New York, 1956.

Muther, R., and K. McPherson. "Four Approaches to Computerized Layout Planning," *Industrial Engineering, 2*, 1970, pp. 39–42.

Nugent, C. E., T. E. Vollmann, and J. Ruml. "An Experimental Comparison of Techniques for the Assignment of Facilities to Locations," *Operations Research, 16*(1), January–February 1968, pp. 150–173.

Reed, R. *Plant Location, Layout and Maintenance.* Richard D. Irwin, Homewood, Ill., 1967.

Reeve, N. R., and W. H. Thomas. "Balancing Stochastic Assembly Lines," *AIIE Transactions, 5*(3), September 1973.

Ritzman, L. P. "The Efficiency of Computer Algorithms for Plant Layout," *Management Science, 18*(5), January 1972, pp. 240–248.

Scriabin, M., and R. C. Vergin. "Comparisons of Computer Algorithms and Visual Based Methods for Plant Layout," *Management Science, 22*, October 1975, pp. 172–181.

Seehof, J. M. and W. O. Evans. "Automated Layout Design Program," *Journal of Industrial Engineering, 18*(12), December 1967, pp. 690–695.

Vollmann, T. E. "*An Investigation of the Bases for the Relative Location of Facilities.*" Unpublished PhD dissertation, UCLA, 1964.

Vollmann, T. E., and E. S. Buffa, "The Facilities Layout Problem in Perspective," *Management Science, 12*(10), June 1966, pp. 450–468.

Vollmann, T. E., C. E. Nugent, and R. L. Zartler. "A Computerized Model for Office Layout," *Journal of Industrial Engineering, 19*(7), July 1968, pp. 321–327.

Wester, L., and M. D. Kilbridge. "Heuristic Line Balancing: A Case," *Journal of Industrial Engineering, 13*(3), May 1962.

Zoller, K., and K. Adendorff. "Layout Planning by Computer Simulation," *AIIE Transactions, 4*(2), June 1972, pp. 116–125.

PART

FIVE

SYNTHESIS AND CONCLUSION

CHAPTER 15 The Operating Manager

THROUGHOUT THE BOOK, WE HAVE TALKED A GREAT DEAL ABOUT operating systems, their problems, and the nature of analysis and decisions in production/operations management. But operating systems are managed by people who make the judgments.

The managers who run the shops, services, or projects have important and exciting positions. They are "in the line" in organizational terms. From a social point of view, they contribute *directly* to the creation of goods and services—they are not leeches on society. It is a good feeling!

The operating management positions themselves involve the development of strategies as well as the day-to-day management of resources and people. The decisions made are crucial to the success of the organization. They determine whether or not the products and services are of appropriate quality and cost and whether they are delivered to the consumers in timely fashion. Image makers can build on product attributes that exist, but only the operating managers can build the base on which the image can be erected.

THE OPERATING MANAGER AND STRATEGY

There are basically four functional types of managers in an enterprise: finance, marketing, operations, and personnel. There are strategic implications in all of these functions. The long-term financial viability of a firm can be affected by the financing mechanisms chosen. The labor and personnel policies can also have a bearing on the long-term viability. Poor policies can "derail" an otherwise successful venture, and sound policies can build important aspects of human capital. Marketing can promote the fundamental values that exist and sense the directions that may be fruitful, showing the way for the future.

But the nexus of strategic planning is in operations—the other three support the fundamental function of creating goods and services. The operating manager presides over the planning issues that set the course for quality, cost, and delivery of products and services. Operating managers must select an adroit combination of strategies involving processes and product designs for production.

A strategy must be selected that joins the available producing technologies with the nature of the market. Choices can be made that emphasize low cost and availability of products or flexibility and quality. The traditional industrial image of low cost and high production can be precisely the wrong strategy for many situations, as Mr. Ford finally realized with his Model T. In selecting joint strategies, single-minded drives toward extreme positions of the strategy map of Figure II-2 need careful consideration.

Successful operating managers may be rather contemplative, even academic, in their approaches to these strategic issues. High-technology line production systems are good images, but the manager's job is to assess the situation in relation to the market and retain an appropriate degree of flexibility.

The strategic implications of capacity and location planning are enormous. If you overbuild, you load costs of products and services with unneeded overhead. If you underbuild, you may miss market opportunities. If you build capacity in the wrong places, you may increase distribution costs or miss market opportunities. Technological innovations in either products or processes can render physical facilities obsolete. Alternate strategies for adding capacity can have a marked effect on costs and flexibility of operation. Long planning and execution times can compound already fuzzy conclusions on the best strategies. But these strategic issues are a fascinating part of the operating manager's position.

THE "FIRING LINE"

The operating manager is in the vortex of action. Customers and clients want products and services; they complain about quality and availability. Sales personnel make promises that are kept by the operating manager, if they are kept. Suppliers create problems with faulty materials, delays, and shortages. Equipment breaks down. Workers go on strike. Within this madhouse of "action," the operating manager is somehow supposed to produce the products and services in the quantities needed, available when needed, and at a controlled cost and quality.

Surely the operating manager's position must be one of putting out fires. Each of the problems must be dealt with, and sometimes emergencies must be faced. But the firefighting approach to managing operations is seldom a winning philosophy. If for no other reason, operating managers cannot physically survive for long. How then can operatings managers cope?

Managers cope by creating rational systems for handling the great bulk of problems. They do not attempt to reinvent the wheel for each problem that comes along. We have examined many of these systems in this book. They deal with forecasting and scheduling systems and a variety of controls to be sure that the systems are continuing to function properly.

The operating manager's decisions invariably seem to seek balance. There is a price for everything, and you cannot usually get something for nothing. "Free cheese is in mousetraps!" The decision problems reflect a recognition of these realities, and managers are constantly balancing one cost against another. Sometimes the costs are obvious, and we try to construct models that formalize them. Then managers can use the models to help price the nonquantitative aspects of decisions, as well as using them in the decision-making process.

While staff specialists are busy dissecting problems and trying to analyze them, operating managers must try to see relationships and integrate the results. Managers are probably the true systems philosophers. They must continually think in wholistic terms. They must anticipate interacting effects and the unintended effects of decisions. Managers must take to heart Lord Acton's admonition, "You can't change just one thing." There will be unintended effects, and a system-oriented manager will anticipate them with counterbalancing actions.

THE FUTURE OF PRODUCTION/OPERATIONS MANAGEMENT

The impact of pollution and natural resource limits and drain, the overbearing increase in population, and our changing values seem destined to change industrialized society. Predicting the form of the changes may require a crystal ball, and we will not attempt it. However, the impact of these events on Production/Operations Management is somewhat more clear.

First, the importance of well-managed productive systems will increase rather than decrease as even greater importance is placed on the efficient use of resources. In line with the external pressures, more emphasis will be placed on analytical approaches to problems within the framework of systems analysis.

At the same time, we will become masters of technology in productive systems, and jobs will be designed in ways that use the unique capability of both humans and machines appropriately. This latter process will face a severe test if the predictions of doom are correct. Because, if the collapse of modern technology and a breakdown of industrial society should actually occur, we will once again find ourselves doing work more appropriate for machines in order to survive.

PART

SIX

APPENDIXES

APPENDIX A Capital Costs and Investment Criteria

CAPITAL COSTS AFFECT DECISION PROBLEMS IN PRODUCTION/operations management whenever a physical asset or expenditure is involved that provides a continuing benefit or return. From an accounting point of view, the original capital expenditure must be recovered through the mechanism of depreciation and must be deducted from income as an expense of doing business. The number of years over which the asset is depreciated and the allocation of the total amount to each of these years (i.e., whether depreciation is straight-line or some accelerated rate) represent alternative strategies that are directed toward tax policy. We must remember that all of these depreciation terms and allocations are arbitrary and have not been designed from the point of view of cost data for decision making.

A SIMPLE EXAMPLE

We have just installed a piece of equipment that performs a highly specialized operation. The installation was a custom job that fit our particular situation. Because of the specialized nature of the equipment and the custom installation, the equipment has no salvage value.

The useful physical life of the equipment can be prolonged almost indefinitely by maintenance and repair, so it is difficult to say yet what the life of the equipment will be. Since the equipment has no salvage value, the entire $10,000 seems to go down some sort of economic sink the minute the equipment becomes ours. We say that the $10,000 is a *sunk cost*, meaning simply that it is gone forever, regardless of what we may list as the "book value" of the equipment. Since the $10,000 is sunk, it is completely irrelevant to any future decisions because no future decision can affect it.

The cost of owning the equipment is simply $10,000. We hope to spread this total over a period of time so that the *average* annual cost of ownership will not be too great. In five years, the average cost of owning the equipment is only $10,000/5 = $2000 per year. In 10 years, the average annual cost is down to $1000 per year. Regardless of how long we keep the equipment, this cost of owning is irrelevant because it is a past cost. The only future costs that we will incur for this equipment are the costs of operating and maintaining it. Once the machine is installed, these are the only costs that are subject to managerial control by future decisions.

Assume that two workers are required to operate the equipment at $4000 per year per worker. Maintenance costs are expected to be $2000 the first year, thereafter increasing by $200 per year. Figure A-1 shows the total of these costs in relation to time. We assume that there are no other pertinent costs. The original cost of the equipment is sunk, and the only future costs are the operating and maintenance costs. Therefore, we are in a position to see under what conditions we would consider this setup to be obsolete and would replace it. We would replace it any time that we could find a functionally equivalent setup that could offer a total average annual cost of owning plus operation and maintenance that fell below the cost curve of Figure A-1.

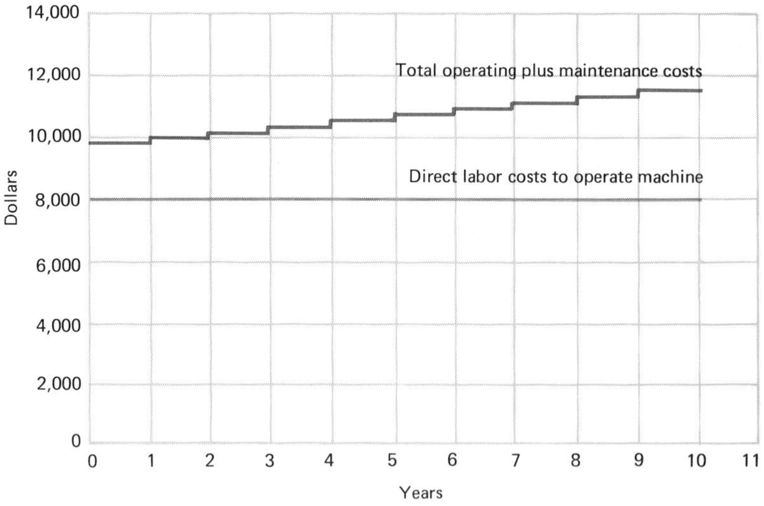

FIGURE A-1
Operating and maintenance costs, showing rising costs of maintenance.

Assume that during the *fourth* year of operation of the present setup a new equipment design is developed. This *new equipment design* has an important advantage over the old one. Because it is more automatic, it can be operated by one person. The more automatic features require an additional $1000 maintenance effort, however. Therefore, the first-year operating and maintenance cost is expected to be $4000 operating labor plus $3000 maintenance. This total of $7000 is expected to increase by $200 per year as before, because of mounting maintenance charges. The improved design costs $12,000, installed. Now we want to see whether the average annual total of the costs of owning plus operating and maintaining the improved equipment design are less than the $10,600 current annual expense (fourth year) for the present setup. While we ignore the original sunk cost of the present setup, we cannot ignore the installed cost of the improved design. It is still a future cost.

The average annual cost of owning (capital cost) and the annual operating and maintenance costs for the improved equipment design are plotted in Figure A-2. The total cost curve is developed simply by adding the cost of owning to the operating and maintenance costs for each year. The total cost is very high during the first two years, it reaches a minimum during the eighth year, and then it begins to rise again. It is high in the early years because the annual average costs of owning the equipment are very high during those years. It begins to rise again after the eighth year because of the influence of rising maintenance costs.

For the present setup, the total incremental cost for the fourth year is $10,600, a sum that we expect will become larger in future years. However, the total average annual cost for the proposed setup will be less than $10,600 after its fourth year. It seems clear that a decision to replace the present setup is needed.

What criteria for comparison are we using here? We are looking at the best possible future cost performance for both the present and proposed setups. The best cost performance possible for the present design is $10,600, which is this year's operating and maintenance cost. The best cost performance possible for the proposed design is achieved if it is held in service for eight years; its total cost would average only $9900 per year for the entire eight-year span. As long as the best performance of the proposed design (represented by the minimum of the total cost curve) is less than the best performance of the present design, it would be economical to make the switch. We must temper this statement by recognizing important intangible values and the accuracy of the cost estimates.

OPPORTUNITY COSTS

Suppose that we are discussing an asset that is used for more general purposes, such as an over-the-road semitrailer truck. Assume that we own such a truck and that the question is, How much will it cost us to *own* this truck for one more year? These costs of owning, or capital costs, cannot be derived from the organization's ordinary accounting records. The cost of owning it for one more year depends on its current

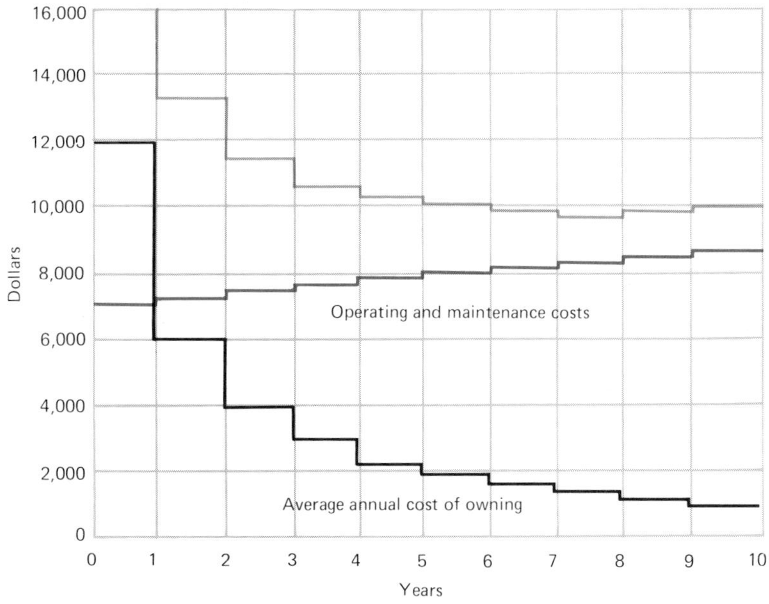

FIGURE A-2
Year-by-year average costs for a proposed replacement machine costing $12,000 initially and having no salvage.

value. If the truck can be sold on the secondhand market for $5000, this is a measure of its economic value. Since it has value, we have two basic alternatives: we can sell it for $5000 or we can retain it. If we sell, the $5000 can earn interest or a return on an alternate investment. If we keep the truck, we forego the return, which then becomes an *opportunity cost* of holding the truck one more year. Similarly, if we keep the truck, it will be worth less a year from now, so there is a second opportunity cost, measured by the fall in salvage value during the year.

The loss of opportunity to earn a return and the loss of salvage value during the year are the costs of continued ownership. They are opportunity costs rather than costs paid out. Nevertheless, they can be quite significant in comparing alternatives that require different amounts of investment. There is one more possible component of capital cost for the next year if the truck is retained—the cost of possible renewals or "capital additions." We are not thinking of ordinary maintenance but of major overhauls such as a new engine or an engine overhaul that extends the physical life for some time. In summary, the capital costs, or costs of owning the truck for one more year, are as follows:

1. Opportunity costs:
   a. Interest on opening salvage value
   b. Loss in salvage value during the year

2. Capital additions or renewals required to keep the truck running for at least an additional year

By assuming a schedule of salvage values, we can compute the year-by-year capital costs for an asset. This is done in Table A-1 for a truck that cost $10,000 intially and for which the salvage schedule is indicated. The final result is the projected capital

TABLE A-1 **Year-by-Year Capital Costs for a Semitrailer Truck, Given a Salvage Schedule (Interest at 10%)**

Year	Year-End Salvage Value	Fall in Salvage Value during Year	Interest on Opening Salvage Value	Capital Cost, Sum of Fall in Value, and Interest
New	$10,000	—	—	—
1	8,300	$1,700	$1,000	$2,700
2	6,900	1,400	830	2,230
3	5,700	1,200	690	1,890
4	4,700	1,000	570	1,570
5	3,900	800	470	1,270
6	3,200	700	390	1,090
7	2,700	500	320	820
8	2,300	400	270	670
9	1,950	350	230	580
10	1,650	300	195	495

TABLE A-2 **Comparison of Capital Costs for Two Machines Costing $10,000 Initially but with Different Salvage Schedules (Interest at 10%)**

Machine 1				Machine 2			
Year-End Salvage Value	Fall in Value During Year	Interest at 10% on Opening Value	Capital Cost	Year-End Salvage Value	Fall in Value During Year	Interest at 10% on Opening Value	Capital Cost
$10,000	—	—	—	$10,000	—	—	—
8,330	$1,670	$1,000	$2,670	7,150	$2,850	$1,000	$3,850
6,940	1,390	833	2,223	5,100	2,050	715	2,765
5,780	1,160	694	1,854	3,640	1,460	510	1,970
4,820	960	578	1,538	2,600	1,040	364	1,404
4,020	800	482	1,282	1,860	740	260	1,000
3,350	670	402	1,072	1,330	530	186	716
2,790	560	335	895	950	380	133	513
2,320	470	279	749	680	270	95	365
1,930	390	232	622	485	195	68	263
1,610	320	193	513	345	140	49	189

cost that is incurred for each year . If we determine the way in which operating and maintenance costs increase as the truck ages, we can plot a set of curves of yearly costs. The combined capital plus operating and maintenance cost curve will have a minimum point. This minimum of the combined cost curve defines the best cost performance year in the life of the equipment. Beyond that year, the effect of rising maintenance costs more than counterbalances the declining capital costs. Note that such a plot is different from Figure A-2, in which we plotted annual average costs instead of yearly costs.

OBSOLESCENCE AND ECONOMIC LIFE

By definition, when a machine is obsolete, an alternate machine or system exists that is more economical to own and operate. The existence of the new machine does not cause any increase in the cost of operating and maintaining the present machine. Those costs already are determined by the design, installation, and condition of the present machine. However, the existence of the new machine causes the salvage value of the present setup to fall, inducing an increased capital cost. For assets in technologically dynamic classifications, the salvage value schedule falls rapidly in anticipation of typical obsolescence rates. Economic lives are very short. On the other hand, when the rate of innovation is relatively slow, salvage values hold up fairly well.

Table A-2 compares year-by-year capital costs for two machines that initially cost $10,000 but that have different salvage schedules. The value of machine 1 holds the best; machine 2 has more severe obsolescence reflected in its salvage schedule. The result is that capital costs in the initial years are greater for machine 2 than for machine 1. The average capital costs for the first five years are:

Machine 1	\$1913
Machine 2	\$2198

Therefore, if the schedules of operating expenses for the two machines were identical, machine 1 would seem more desirable. However, since the timing of the capital costs is different for the two machines, we adjust all figures to their equivalent present values.

PRESENT VALUES

Since money has a time value, future expenditures and opportunity costs will have different present values. Since money can earn interest, \$1000 in hand now is equivalent to \$1100 a year from now if the present sum can earn interest at 10 percent. Similarly, if we must wait a year to receive \$1000 that is due now, we should expect not \$1000 a year from now but \$1100. When the time spans involved are extended, the appropriate interest is compounded, and its effect becomes much larger. The timing of payments and receipts can make an important difference in the value of various alternatives.

We know that if a principal sum P is invested at interest rate i, it will yield a future total sum S in n years hence, if all the earnings are retained and compounded. Therefore, P in the present is entirely equivalent to S in the future by virtue of the compound amount factor. That is,

$$S = P(1 + i)^n \tag{1}$$

where $(1 + i)^n$ = the compound amount factor for interest rate i and n years.

Similarly, we can solve for P to determine the present worth of a sum to be paid n years hence:

$$P = \frac{S}{(1 = i)^n} = S \times PV_{sp} \tag{2}$$

where PV_{sp} = the present value of a single payment S to be made n years hence with interest rate i. Therefore, if we were to receive a payment of \$10,000 in 10 years, we should be willing to accept a smaller but equivalent sum now. If interest at 10 percent were considered fair and adequate, that smaller but equivalent sum would be

$$P = 10{,}000 \times 0.3855 = \$3855$$

since

$$\frac{1}{(1 + 0.10)^{10}} = PV_{sp} = 0.3855$$

Now let us return to the example of the two machines. The capital costs for each machine occur by different schedules because of different salvage values. If all future values were adjusted to the present as a common base time, we could compare the totals to see which investment alternative was advantageous. We have done this in Table A-3, where we have assumed an operating cost schedule in column 2, determined combined operating and capital costs in columns 5 and 6, and listed present values in columns 8 and 9. The present value of the entire stream of expenditures and opportunity costs is $32,405 for machine 1. The net difference in present values for the two machines is shown at the bottom of Table A-3. Since the operating cost schedule was identical for both machines, the difference reflects differences in the present worth of capital costs. Obviously, the method allows for different operating cost schedules as well.

There are some difficulties with the methods just described. First, we have assumed that the schedule of salvage values is known, which is not usually true. Second, at some point in the life of the machines it becomes economical to replace them with identical models. Therefore, a chain of identical machines should be considered for comparative purposes; the machine is replaced in the year in which operating and capital costs are exactly equal to the interest on the present worth of all future costs. The essence of this statement is that we are seeking a balance between this year's costs (operating and capital costs) and opportunity income from disposal (interest on

TABLE A-3
Present Value of Capital and Operating Costs for the Two Machines from Table A-2 [Schedule of operating costs is the same for both machines (interest at 10%)]

		Capital Costs (from Table A-2)		Combined Operating and Capital Costs		Present Worth Factor for Year Indicated[b]	Present Worth of Combined Costs for Year Indicated	
Year (1)	Operating Cost (2)	Machine 1 (3)	Machine 2 (4)	Machine 1 (5)	Machine 2 (6)	(7)	Machine 1 (8)	Machine 2 (9)
1	$3,000	$2,670	$3,850	$5,670	$6,850	0.909	$5,154	$6,227
2	3,200	2,223	2,765	5,423	5,965	0.826	4,490	4,939
3	3,400	1,854	1,970	5,254	5,370	0.751	3,946	4,033
4	3,600	1,538	1,404	5,138	5,004	0.683	3,509	3,418
5	3,800	1,282	1,000	5,082	4,800	0.621	3,156	2,981
6	4,000	1,072	716	5,072	4,716	0.565	2,866	2,665
7	4,200	895	513	5,095	4,713	0.513	2,614	2,418
8	4,400	749	365	5,149	4,765	0.467	2,405	2,225
9	4,600	622	273	5,222	4,873	0.424	2,214	2,066
10	4,800	513	189	5,313	4,989	0.386	2,051	1,926
Totals							$32,405	$32,898

Machine 1, present worth of all future values is total of column (8) less present worth of tenth-year salvage value, i.e., $32,405—1610[a] × 0.386 = 32,405 − 621 = $31,784.
Machine 2, $32,898 − 345[a] × 0.386 = 32,898 − 133 = 32,765.

[a] Tenth-year salvage values from Table A-2.

[b] From Table H-1 (Appendix H)

the present worth of all future costs). When the opportunity income from disposal is the greater of the two, replacement with the identical machine is called for. Most common criteria for comparing alternate capital investments circumvent these problems by (1) assuming an economic life and (2) assuming some standard schedule for the decline in value of the asset. We will now consider some of these criteria.

COMMON CRITERIA FOR COMPARING ECONOMIC ALTERNATIVES

Some of the common criteria used to evaluate proposals for capital expenditures and compare alternatives involving capital assets are (1) present values, (2) average investment, (3) rate of return, and (4) payoff period.

Present Value Criterion

Present value methods for comparing alternatives take the sum of present values of all future out-of-pocket expenditures and credits over the economic life of the asset. This figure is compared for each alternative. If differences in revenue also are involved, their present values also must be accounted for. Table H-1 in Appendix H gives the present values for single future payments or credits; Table H-2 gives present values for annuities for various years and interest rates. An *annuity* is a sum that is received or paid annually. The factors in Table H-2 convert the entire series of annual sums to present values for various interest rates and years. We will use the notation PV_a for the present value factor of an annuity.

An Example. Suppose we are considering a machine that costs $15,000, installed. We estimate that the economic life of the machine is eight years, at which time its salvage value is expected to be about $3000. For simplicity's sake, we take the average operating and maintenance costs to be $5000 per year. At 10 percent interest, the present value of the expenditures and credits is

Initial investment	$\$15{,}000 \times PV_{sp} = 15{,}000 \times 1{,}000 =$	15,000
Annual operating and maintenance costs	$5{,}000 \times PV_a = 5{,}000 \times 5.335 =$	26,675
		41,675
Less credit of present value of salvage to be received in eight years	$3{,}000 \times PV_{sp} = 3{,}000 \times 0.467 =$	1,401
		$40,274

The net total of $40,274 is the present value of the expenditures and credits over the eight-year expected life of the machine. The initial investment is already at present value; that is, $PV_{sp} = 1$. The annual costs of operation and maintenance are an eight-year annuity, so the entire stream of annual costs can be adjusted to present value by the multiplication of PV_a from Table H-2. Finally, the present value of the salvage is deducted. This total could be compared with comparable figures for other alternatives over the same eight-year period. Suppose that another alternate machine is estimated to have a different economic life (perhaps four years). Then, to make the present value totals comparable, we compare two cycles of the four-year machine with one cycle of the eight-year machine. If the operating and maintenance costs increased as the machine aged, the present value of the expenditure in each year would be determined separately by PV_{sp}.

Average Investment Criterion

Average investment methods estimate an average annual cost of owning plus operating and maintaining an asset. The average annual capital costs are approximated by average salvage loss plus interest on the average investment, assuming that the decline in value of the asset is on a uniform or straight-line basis. Figure A-3 shows the assumed structure for the decline in value of an asset and the calculation of the average investment. Thus, if a machine cost \$10,000 and was estimated to have a 10-year life with a salvage value of \$1000, the capital costs are approximated by

$$\text{Average annual salvage loss} = \frac{10{,}000 - 1{,}000}{10} = \qquad \$900$$

Annual interest on average investment, at 10 percent

$$= \frac{(10{,}000 + 1{,}000)0.10}{2} = 5{,}500 \times 0.10 = \qquad \underline{\$550}$$

$$\text{Average annual capital cost} = \$900 + \$550 = \qquad \$1450$$

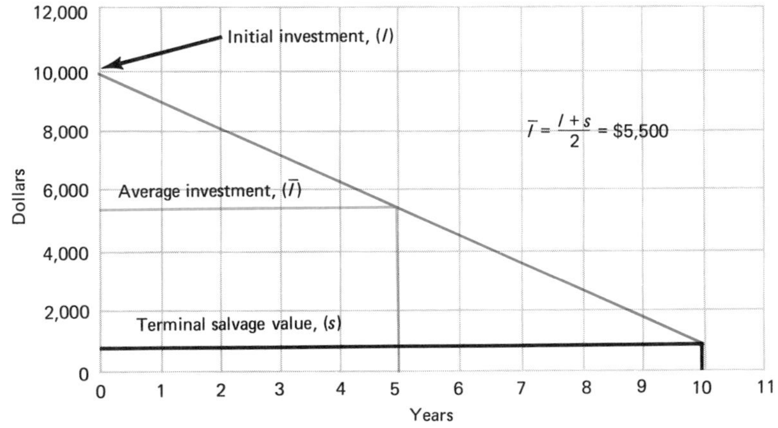

FIGURE A-3
Relationship of initial investment, terminal salvage value and average investment for "average investment methods."

If operating and maintenance costs were estimated to average \$12,500 per year over the 10-year machine life, the total average annual cost for comparison would be 1,450 + 12,500 = \$13,950. Differences in annual revenue among alternatives can be accounted for in the operating costs. Comparable calculations for alternate machines would form the basis for comparison. For a strictly economic comparison, the alternative presenting the lowest average annual cost would be selected.

Rate of Return Criterion

One common method of evaluating new projects or comparing alternate courses of action is to calculate a rate of return, which is then judged for adequacy. Usually, no attempts are made to consider interest costs, so the resulting figure is referred to as the "unadjusted" rate of return (i.e., unadjusted for interest values). It is computed as follows:

$$\text{Unadjusted rate of return} = \frac{100(\text{net monetary operating advantage} - \text{amortization})}{\text{average investment}}$$

The net monetary advantage reflects the algebraic sum of incremental costs of operation and maintenance and possible differences in revenue. If the rate computed is a "before-tax" rate, then the amortization

$$\frac{\text{incremental investment}}{\text{economic life}}$$

is subtracted, and the result is divided by average investment and multiplied by 100 to obtain a percentage return. If an "after-tax" rate is sought, the net increase in income taxes due to the project is subtracted from the net monetary advantage, and the balance of the calculation is as before. Obviously, the adequacy of a given rate of return changes drastically if it is being judged as an after-tax return.

An Example. Assume that new methods have been proposed for the line assembly of a product, each assembly being completed by one individual. The new methods require the purchase and installation of conveyors and fixtures that cost \$50,000 installed, including the costs of re-layout. The new line assembly methods require five fewer assemblers. After the increased maintenance and power costs are added, the net monetary operating advantage is estimated as \$20,000 per year. Economic life is estimated at five years. The unadjusted before-tax return is

$$\frac{20{,}000 - \dfrac{50{,}000}{5}}{\dfrac{50{,}000}{2}} \times 100 = 40 \text{ percent}$$

The after-tax return requires that incremental taxes be deducted. Incremental taxable income will be the operating advantage less increased allowable tax depreciation. Assuming straight-line depreciation and an allowed depreciation term of eight years, incremental taxable income is \$20,000 less \$50,000/8, or \$20,000 − \$6,250 = \$13,750. Assuming an income tax rate of 50 percent, the incremental tax due to the project is \$6875. The after-tax return is therefore

$$\frac{20{,}000 - 6{,}875 - 10{,}000}{25{,}000} \times 100 = \frac{3{,}125 \times 100}{25{,}000} = 12.5 \text{ percent}$$

Whether or not either the before- or after-tax rates calculated in this example are adequate must be judged in relation to the risk involved in the particular venture and the returns possible through alternate uses of the capital.

Payoff Periods

The payoff period is the time required for an investment to "pay for itself" through the net operating advantage that would result from its installation. It is calculated as follows:

$$\text{Payoff period in years} = \frac{\text{net investment}}{\text{net annual operating advantage after taxes}}$$

The payoff period for the conveyor installation that we discussed previously is

$$\frac{\$50{,}000}{\$13{,}125} = 3.8 \text{ years}$$

It is the period of time for the net after-tax advantage to equal exactly the net total amount invested. Presumably, after that period, "it is all gravy"; the \$13,125 per year is profit, since the invested amount has been recovered. If the economic life of the equipment is five years and 10 percent is regarded as an appropriate rate of after-tax return for the project, what *should* the payoff period be? Obviously, the period for both capital recovery and return is the five-year economic life. The period that recovers capital only, but also allows enough time in the economic life to provide the return, will be somewhat shorter and will depend on the required rate of return. The payoff period is another interpretation that can be given to the present value factors for annuities, PV_a, given in Table H-2 (Appendix H). As an example, for an economic life of five years and a return rate of 10 percent, $PV_a = 3.791$ from Table H-2. This indicates that capital recovery takes place in 3.791 years. The equivalent of 10 percent compound interest takes place in 5.000 − 3.791 = 1.209 years. Therefore, any of the PV_a values in Table H-2 for a given economic life in years and a given return rate indicate the shorter period in years required to return the investment; they give the payoff period directly.

The proper procedure would be to estimate economic life and to determine the applicable return rate. Determine from the present value tables the payoff period associated with these conditions. Then compute the actual payoff period of the project in question and compare it with the standard period from the tables. If the computed period is less than, or equal to, the standard period, the project meets the payoff and risk requirements that are imposed. If the computed value is greater than the table value, the project would earn less than the required rate.

REVIEW QUESTIONS AND PROBLEMS

1. A trucking firm owns a five-year old truck that it is considering replacing. The truck can be sold for $5000, and Blue Book values indicate that this salvage value would be $4000 one year from now. It also appears that the trucker would need to spend $500 on a transmission overhaul if the truck were to be retained. What are the trucker's projected capital costs for next year? Interest is at 10 percent.

2. What is the present value of the salvage of a machine that can be sold 10 years hence for $2500? Interest is at 10 percent.

3. What is the future value in 25 years of a bond that earns interest at 10 percent and has a present value of $10,000?

4. What interest rate would a $10,000 bond have to earn to be worth $50,000 in 10 years?

5. At 8 percent interest, how many years will it take money to double itself?

6. What is the present value of an income stream of $1500 for 15 years at 10 percent interest?

7. What is the value of an annuity of $2000 per year for 10 years at the end of its life? Interest is at 10 percent.

8. The proud owner of a new automobile states that she intends to keep her car for only two years in order to minimize repair costs, which she feels should be near zero during the initial period. She paid $4000 for the car new, and Blue Book value schedules suggest that it will be worth only $2000 two years hence. She normally drives 10,000 miles per year, and she estimates that her cost of operation is $0.10 per mile. What are her projected capital costs for the first two years, if interest of 6 percent represents a reasonable alternate investment for her?

9. Suppose that we are considering the installation of a small computer to accomplish internal tasks of payroll computation, invoicing, and other routine accounting. The purchase price is quoted as $300,000 and the salvage value five years later is expected to be $100,000. The operating costs are expected to be

$100,000 per year, mainly for personnel to program, operate, and maintain the computer. What is the present value of the costs to own and operate the computer over its five-year economic life? The value of money in the organization is 15 percent.

10. Using the average investment criterion as a basis for comparison, at what annual lease cost would we break even for the data on the computer in problem 9? If the computer is leased, maintenance is furnished, reducing annual operating costs to $70,000.

11. An aggressive marketer of a new office copier has made its machine available for sale as well as lease. The idea of buying a copying machine seems revolutionary, but less so when we examine our present costs, which come to $6500 per year for lease plus per copy charges of 2 cents per page. If we own a machine, the cost of paper and maintenance is projected to be $1500 per year. The new copier costs $10,000, installed, and is assumed to have an economic life of five years and a salvage value of $2000 (assume 50,000 pages per year).
    a. What is the projected unadjusted rate of return if we install the copier?
    b. If incremental taxes for the project are $1000, what is the adjusted rate of return?

12. What is the actual payoff period for the office copier project discussed in problem 11? If interest is 10 percent, what should the minimum payoff period be to make the investment economically sound? Does the office copier project meet the payoff standard?

REFERENCES

Anthony, R. N., and G. A. Welsch. *Fundamentals of Management Accounting* (Rev. ed.). Richard D Irwin, Homewood, Ill., 1977.

Grant, E. L., and W. G. Ireson. *Principles of Engineering Economy* (5th ed.). Ronald Press, New York, 1970.

Reisman, A. *Managerial and Engineering Economics* Allyn & Bacon, Boston, 1971.

Terborgh, G. *Business Investment Policy.* Machinery and Allied Products Institute, Washington, D.C., 1958.

Thuesen, H. G., W. J. Fabrycky, and G. J. Thuesen. *Engineering Economy* (5th ed.). Prentice-Hall, Englewood Cliffs, N.J., 1977.

APPENDIX B Linear Programming

LINEAR OPTIMIZATION MODELS ARE CHARACTERIZED BY LINEAR mathematical expressions. In addition, they are deterministic in nature, that is, they do not take account of risk and uncertainty; the parameters of the model are assumed to be known with certainty. Finally, linear programming is used to allocate some limited or scarce resource so that we can make decisions to optimize (either minimize or maximize) a stated objective function.

ELEMENTS OF THE MODEL BUILDING PROCESS

To develop a linear optimization model, we use the following process:

1. Define the decision variables.

2. Define the objective function, Z, a linear equation involving the decision variables, which identifies the objective in the problem-solving effort. This equation predicts the effects on the objective of choosing different values of the decision variables.

2. Define the constraints as linear expressions involving the decision variables. The constraints specify restrictions on the decisions. Alternatives can be generated by selecting values for the decision variables that satisfy these constraints.

FORMULATION OF A TWO-PRODUCT MODEL

Assume a simple situation for a plant that manufactures washing machines and dryers. The major manufacturing departments are the stamping department, the motor and transmission department, and the final assembly lines for the washer and dryer. The stamping department fabricates a large number of metal parts of both products, and the motor and transmission department produces the power units for both products. Monthly departmental capacities are:

Stamping department	10,000 washers or	10,000 dryers
Motor and transmission department	16,000 washers or	7,000 dryers
Washer assembly department	9,000 washers	—
Dryer assembly department	—	5,000 dryers

The same facilities for the two products are used for stamping and for motor and transmission. The stamping department can produce parts for 10,000 washers per month or 10,000 dryers per month, as well as combination amounts of washers and dryers. We have a similar situation with the motor and transmission department, but the final assembly lines are separate for the two products. The unit contribution to profit and overhead is \$90 for washers and \$100 for dryers.

First, set up the restrictions of the problem. We will denote the number of washers

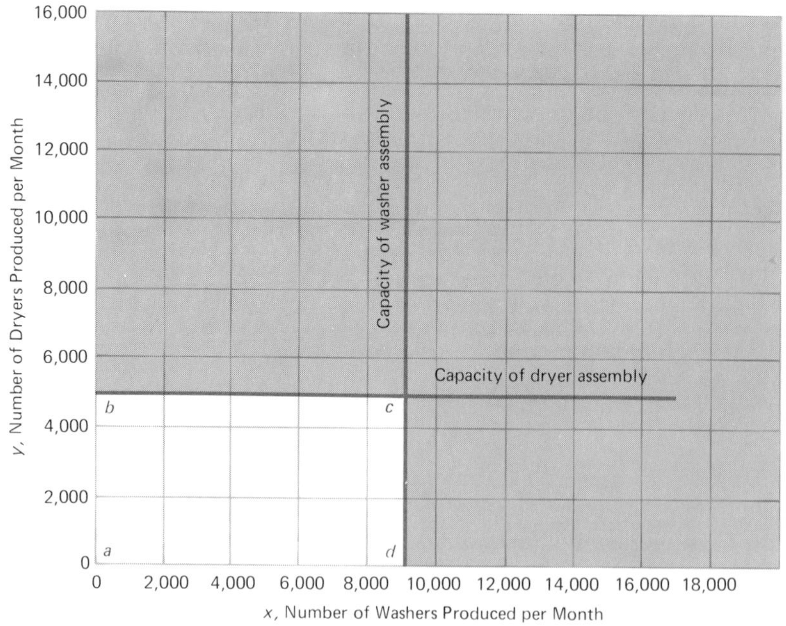

FIGURE B-1
Graphic illustration of limitations imposed by the assembly line capacity of washers and dryers.

by x and the number of dryers by y. The simplest restrictions are those imposed by the two final assembly lines. They indicate that the number of washers per month must be less than or equal to 9000 and that the number of dryers must be less than or equal to 5000. These are plotted in Figure B-1. The shaded areas indicate those parts of the graph that are eliminated as feasible solutions to the problem because of assembly line capacity limitations. Any solution to the problem must be a combination of a number of washers and dryers that falls within area *abcd*.

The other restrictions to the problem are only slightly different. The stamping department capacity limits production to 10,000 washers per month, 10,000 dryers per month, or any comparable combination of washer and dryer production. Satisfactory combinations are 8000 washers and 2000 dryers, 5000 each, and so on. This restriction is shown in Figure B-2 as a straight line that restricts further the space allowing feasible solutions.

Figure B-2 also shows the limitations imposed by the capacities of the motor and transmission department, represented by another straight line that goes through the two points ($x = 0$, $y = 7000$) and ($x = 16{,}000$, $y = 0$). Combinations of washer and dryer production that are equivalent fall on this straight line; for example, washers, 8000; dryers, 3500. The combination of washer and dryer production that we are seeking lies within the area *abefgd*. All other combinations have been eliminated by the stated restrictions of the problem.

There are additional restrictions that are implied by the logic of the problem; that is, there cannot be negative production. All values of x and y must be greater than or equal to zero.

With this general background, let us now formulate the washer-dryer problem as a linear optimization model.

Definition of Decision Variables

Since some of the facilities are shared and each product's costs and profits are different, we must decide how to utilize the available capacities in the most profitable way. Dryers seemingly are more profitable. However, when the manager tried producing the maximum amount of dryers within market limitations, using the balance of his capacity to produce washers, it was found that such an allocation of departmental capacities resulted in poor profit performance. The manager now feels that some appropriate balance between the two products is best, the objective being to determine the monthly production rates for each product.

The decision variables are

x = the number of washers to be produced per month
y = the number of dryers to be produced per month

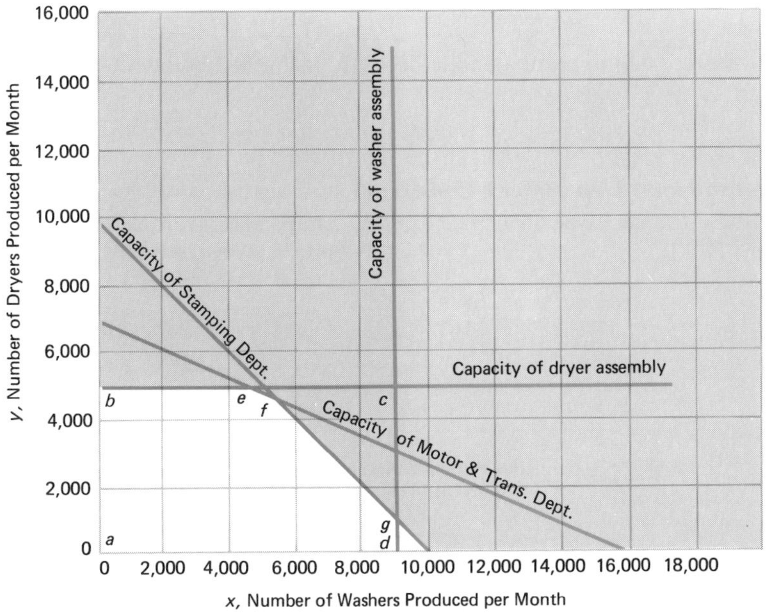

FIGURE B-2
Remaining limitations imposed by capacities of the stamping department and motor and transmission department. The area enclosed by *abefgd* includes all feasible solutions to the problem.

Definition of the Objective Function

The physical plant and basic organization exists and represents the fixed costs of the enterprise. Since we know that these costs are irrelevant to the production scheduling decision, we ignore them. The manager, however, has obtained price and variable cost information and has computed the contribution to profit and overhead per unit, as shown in Table B-1. The objective is to maximize profit, and the contribution rates are related linearly to this objective. Therefore, the objective function to maximize is the sum of the total contribution from washers x, $(90x)$, plus the total contribution from dryers y, $(100y)$, or

$$\text{Maximize } Z = 90x + 100y$$

Definition of Constraints

In general, the constraints were defined as the departmental capacities and were illustrated graphically in Figures B-1 and B-2. However, we must state these constraints in mathematical form.

Stamping Department Constraint. The equation for the line describing the stamping department capacities that is shown in Figure B-2 takes the form $y = mx + b$, where m is the slope of the line and b is the y intercept. Since the slope is -1 and the y intercept is 10,000, the equation is

$$y = -x + 10{,}000$$

Since we wish to express all the combinations of x and y that fall below the line, the constraint is stated as

$$x + y \leq 10{,}000$$

Motor and Transmission Department Constraint. In a similar way, the equation for the line describing the capacities for the motor and transmission department that is shown in Figure B-2 is

$$y = 7/16x + 7000$$

TABLE B-1

Sales Prices, Variable Costs, and Contributions for Washers and Dryers

	Sales price, P	Variable costs, v	Contribution to profit and overhead, $C = p - v$
Washers, x	\$400	\$310	\$90
Dryers, y	450	350	100

Transposing, reducing the fraction to a decimal, and expressed as a constraint:

$$0.4375x + y \leq 7000$$

Assembly Line Constraints. The mathematical statements for the assembly line constraints are simple, since the washer assembly line can produce no more than 9000 washers and the dryer assembly line has a capacity of 5000 dryers. Therefore,

$$x \leq 9000$$
$$y \leq 5000$$

Minimum Production Constraints. The minimum production for each product is zero; therefore,

$$x \leq 0$$
$$y \leq 0$$

The Linear Optimization Model

We now can summarize a statement of the linear optimization model for the two-product washer-dryer company in the standard linear programming format, as follows:

Maximize $Z = 90x + 100y$
Subject to:

$x + y \leq 10{,}000$	(Stamping department)
$0.4375x + y \leq 7{,}000$	(Motor and transmission department)
$x \leq 1{,}000$	(Washer assembly)
$y \leq 5{,}000$	(Dryer assembly)
$x, y \geq 0$	(Minimum production)

SOLUTION AND INTERPRETATION FOR THE WASHER-DRYER MODEL

We will assume that we have a mechanism for solving linear optimization models when they are formulated in the standard format just shown. Indeed, linear programming computing codes commonly are available in both an interactive mode (using a time-share terminal) and a batch mode (for large-scale linear programming problem solutions). To use either kind of computing program to solve linear optimization models, we must present the problem to the "black box" in the precise form required. This input format usually is more user-oriented in interactive time-share systems, and we will use one of these kinds of programs to illustrate the solution to our washer-dryer problem.*

* See J. W. Buckley, M. R. Nagarai, B. L. Sharp, and J. W. Schenck, *Management Problem-Solving with APL*. John Wiley, New York, 1974.

Many computer programs for linear programming are available, and the instructions for each individual program are unique to that program, the documentation of which indicates exactly how to provide input. The form of the computer output may vary from program to program, but the content is similar. (The computer input for the interactive program used is shown in Figure B-3*a*.)

Computer Output

Figure B-3*b* shows the solution output. First, the terminal prints the optimum value of the objective function, \$946,666.667. In other words, it states that $Z = 946{,}667$ in the objective function for an optimal solution.

Next, the terminal prints the values of the variables in the optimal solution. Note that scientific notation is used; that is, the value of each variable is followed by "E" and some number. This means that the number preceding the E is to be multiplied by that number of 10's. For example, E1 means multiply by 10, E2 by 100, and so forth. E0 indicates that the multiplier is 1.

Consider only the first two variables listed in the solution, which we have named WASH and DRY. The solution states that their optimal values are 5333.3 and 4666.7, respectively. This is point *f* in Figure B-2, the point where the capacity constraint lines

```
      LPENTER
ENTER THE NAME OF THIS PROJECT WASHER-DRYER PROBLEM
MAXIMIZE OR MINIMIZE:  MA
OBJECTIVE FUNCTION:Z=90WASH+100 DRY
ENTER CONSTRAINT EQUATIONS, (STRIKE JUST A CARRIAGE RETURN TO STOP INPUT)
[001]  WASH+DRY ≤ 10000
[002]  .4375WASH+DRY ≤ 7000
[003]  WASH ≤ 9000
[004]  DRY ≤ 5000
```

(a)

```
      LPRUN

                    WASHER-DRYER PROBLEM

THE OPTIMAL VALUE OF THE OBJECTIVE FUNCTION IS: 946666.667

                    THE VARIABLES IN THE SOLUTION ARE

      VARIABLE      WASH    AT LEVEL       5.3333E3
                    DRY                    4.6667E3
                    SLK3                   3.6667E3
                    SLK4                   3.3333E2
```

(b)

FIGURE B-3
(*a*) Computer input, and (*b*) computer solution to the washer-dryer problem.

for the stamping department and the motor and transmission department intersect. Using the solution values for WASH and DRY, we insert them in the objective function and compute Z:

$$Z = 90 \times 5333.3 + 100 \times 4666.7 = 946{,}667$$

This checks with the optimal value of Z given by the computer solution.

Check one further bit of logic: if the solution to our problem is at the intersection of the two capacity constraint equations, then we should be able to solve the equations for the two lines simultaneously to determine those values of WASH and DRY that are common to the equations. First, use the equation for the stamping department constraint and solve for y, the number of dryers:

$$y = 10{,}000 - x \tag{1}$$

Then substitute this value of y in the constraint equation for the motor and transmission department:

$$y = 7000 - 7/16x \tag{2}$$

Solving simulataneously:

$$10{,}000 - x = 7000 - 7/16x$$
$$9/16x = 3000$$
$$x = 5333.3 \text{ washers}$$

Using $x = 5333.3$ in Equation 1:

$$y = 10{,}000 - 5333.3 = 4666.7 \text{ dryers}$$

Of course, this agrees with the approximate values that we could read from the graph of Figure B-2, as well as with the computer solution shown in Figure B-3.

We also can verify the solution graphically by plotting various values of the objective function on the graph of constraints. At a total contribution of \$450,000, for example, the objective function is

$$90x + 100y = 450{,}000$$

when

$$x = 0 \qquad y = 4500$$

and when

$$y = 0 \qquad x = 5000$$

The line passing through these two points is shown as the \$450,000 line in Figure B-4. This line defines all combinations of washer and dryer production that yield a total contribution of 450,000. Now choose a larger value of contribution, perhaps \$630,000. The \$630,000 line plotted in Figure B-4 defines all combinations of washer and dryer production that yield this total contribution. Note that this line is parallel to the \$450,000 line. If we again increase contribution to \$900,000 and plot its line in Figure B-4, we see that we will be limited in the size of total contribution by the

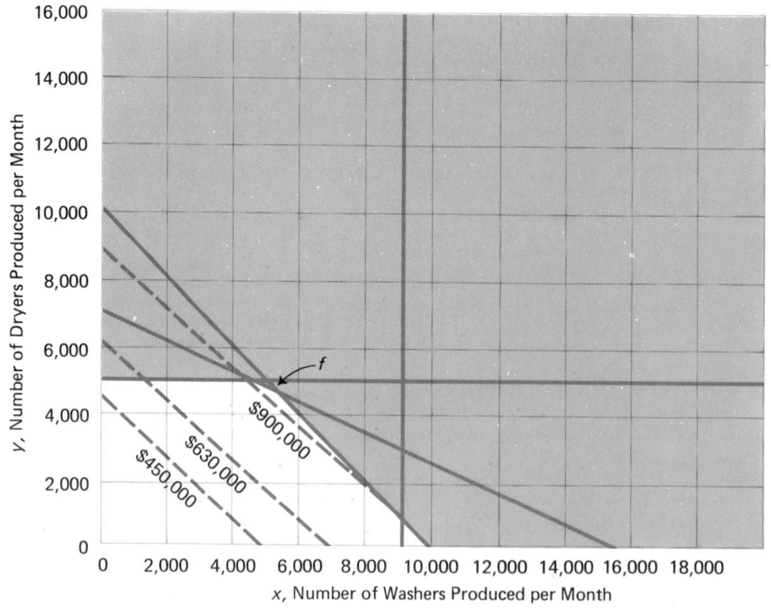

FIGURE B-4
Contribution lines plotted to show the effect of larger and larger contributions. The maximum possible contribution within the area of permitted solutions is defined by a line through point *f*.

point *f*, which defines the combination of washer and dryer production that produces the largest possible contribution with the space of feasible solutions. As we noted before, point *f* is the intersection of the two lines that define stamping capacity and motor and transmission capacity, and the solution is determined by the simultaneous solution of the equations for these two lines.

Slack Variables

As noted, the solution is at point *f*, where the two constraint equations for stamping capacity and motor and transmission capacity intersect. Another way to interpret this fact is that this solution completely utilizes the capacity of these two departments—there is no residual slack in their capacities. This is important because any of the other feasible solutions in the polygon *abefgd* of Figure B-2 would have involved some slack capacity in one or both of these two departments. If there had been slack capacity for either of these two departments in the optimum solution, that fact would have been indicated in the computer output for the optimum solution.

Note that the computer gave us the value of variables for which we did not ask explicitly, SLK3 and SLK4. These are the slack values for constraints 3 and 4, the assembly line constraints. Constraint 3, WASH $\leq$ 9000 was the washer assembly line capacity. The solution shows us that if we were producing according to the optimum solution, there would be slack capacity of 9000 − 5333.3 = 3666.7 units in the washer assembly line. Similarly, the value of SLK4 indicates that there is slack capacity of 5000 − 4666.7 = 333.3 units in the dryer assembly line.

These interpretations of the optimum solution to the washer-dryer production problem are rather simple. The important point is that equivalent interpretations of more complex problems are a straightforward extension of these ideas. The solution will state the combination of variables that optimizes the objective function. Some, but not all, of the constraints will be the controlling ones, and there will be slack in some of the resources (i.e., they will not all be fully utilized). In our example, the slack was in the use of the assembly lines for the two products. However, if the capacity of the dryer assembly line had been only 4000 units, it would have become one of the controlling ("tight") constraints (as may be seen from either Figure B-2 or B-4), and there would have been some slack capacity in the motor and transmission department.

SIMPLEX FORMULATION AND SOLUTION

We will develop the basic simplex method of solution by means of a simplified problem. A manufacturer has two products, I and II, both of which are made in two steps by machines A and B. The process times per hundred for the two products on the two machines are as follows (setup times are negligible):

Product	*Machine A*	*Machine B*
I	4 hours	5 hours
II	5 hours	2 hours

For the coming period, machine A has available 100 hours and B has available 80 hours.

The contribution for product I is \$10 per 100 units; for product II, \$5 per 100 units. The manufacturer is in a market upswing and can sell as much as can be produced of both products for the immediate period ahead. We wish to determine how much of products I and II should be produced to maximize contribution.

Formulation of the Problem

Let x_I be the amount in 100's of product I, and x_{II} be the amount in 100's of product II that will be produced. We are limited only by the available hours on machines A and B, so we know that the sum of the times spent producing the two products on the two machines cannot exceed 100 hours for machine A and 80 hours for machine B. We can express this symbolically as

$$\text{Machine A} \quad 4x_I + 5x_{II} \leq 100 \qquad (3)$$

$$\text{Machine B} \quad 5x_I + 2x_{II} \leq 80 \qquad (4)$$

Since the total contribution, which we want to be as large as possible, depends only on the amounts of the two products produced, we can state:

$$\text{Maximize } Z = 10x_{\text{I}} = 5x_{\text{II}} \tag{5}$$

This is the *objective function*. We want to find the combination of values for x_{I} and x_{II} that fits into the restrictions imposed by the manufacturing times for the two products and the total available time (as expressed by Equations 3 and 4), and that, in addition, makes the total contribution (as expressed by Equation 5) a maximum.

We want to recognize that it is possible that machines A and B could have idle time. If W_{A} is the idle time on machine A and W_{B} the idle time on machine B, Equations 3 and 4 become

$$4x_{\text{I}} + 5x_{\text{II}} + W_{\text{A}} = 100 \tag{6}$$

$$5x_{\text{I}} + 2x_{\text{II}} + W_{\text{B}} = 80 \tag{7}$$

Initial Solution

Rearrange Equations 6 and 7 by placing the variables x_{I}, x_{II}, W_{A}, and W_{B} at the heads of columns, with only the coefficients below in columns. This is done in Table B-2.

We have two equations and four unknowns. A maximum of two of these variables can have positive values, and at least two of them must be zero, if we solve the equations simultaneously. This provides the key to establishing an initial solution, which we then may use as a basis for improvement. If we start out by assuming that x_{I} and x_{II} both are zero, it is easy to see by Equations 6 and 7 that W_{A} must be 100 and W_{B} must be 80. Admittedly, this solution is a very poor solution because it says that all the available time is idle, but it fits the problem equations and allows us to start.

We place this initial trivial solution in the tableau, as shown in Table B-2. This form of a linear optimization model is called the "simplex tableau." Note that we have placed the coefficients of the objective function above the array in Table B-3. The stub identifies the variables in the solution and shows their values. To the right in the stub, we have the contribution rates of the variables in the objective function. Neither W_{A} nor W_{B} is in the objective function, so zeros are placed in this column. This makes good sense; if the two machines are completely idle, the contribution is zero. The value of the objective function at this point is

$$10(0) + 5(0) + (0)W_{\text{A}} + (0)W_{\text{B}} = 0$$

TABLE B-2

Problem Equations in Tableau Form

x_{I}	x_{II}	W_A	W_B	
4	5	1	0	100
5	2	0	0	80

TABLE B-3

Initial Solution Shown in Tableau

10	5	0	0 ← Coefficients of objective function added			
x_I	x_{II}	W_A	W_B			
4	5	1	0	100	W_A	0
5	2	0	1	80	W_B	0

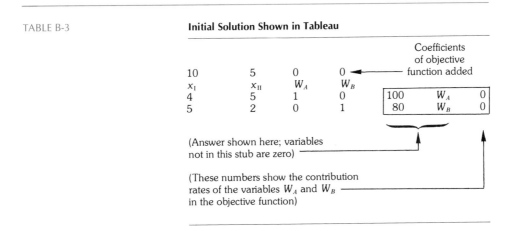

(Answer shown here; variables not in this stub are zero)

(These numbers show the contribution rates of the variables W_A and W_B in the objective function)

Figure B-5 shows the nomenclature of the various parts of the tableau. The solution stub always will contain three columns: the body and identity will vary in size, depending on the particular problem.

To improve the initial solution, we must have a measure of the potential improvement that would be made in the objective function by bringing some of the variables into the solution that currently are zero, instead of the slack variables currently in the solution. We will develop an *index row*, which will be placed just below the present initial tableau. These index numbers will appear under the constant column, the body, and the identity. They are calculated from the formula:

Index number = (number in objective row at head of column) $-\sum$ (numbers in column) $\times$ (corresponding number in objective column)

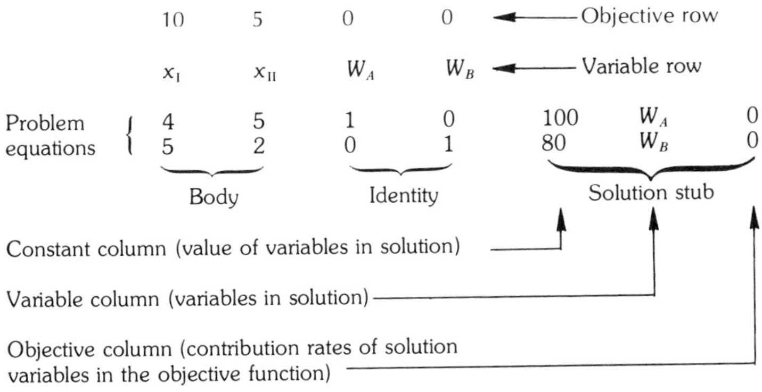

Figure B-5
Nomenclature of the simplex tableau

For our problem, the index row numbers are:

(1) Index number for constant column
$= 0 - (100 \times 0 + 80 \times 0) = 0$

(2) Index number for first column of body
$= 10 - (4 \times 0 + 5 \times 0) = 10$

(3) Index number for second column of body
$= 5 - (5 \times 0 + 2 \times 0) = 5$

(4) Index number for first column of identity
$= 0 - (1 \times 0 + 0 \times 0) = 0$

(5) Index number for second column of identity
$= 0 - (0 \times 0 + 1 \times 0) = 0$

We now place the index numbers in the initial simplex tableau (see Table B-4). This was a trivial step in that we merely copied the objective row coefficients and inserted the value of the objective function, 0. This trivial transformation occurs only when the objective column contains all zeros, that is, when the variables in the solution all have the value of zero.

The larger the positive number in the index row, the greater the potential improvement per unit of the new variable to be introduced. If all the numbers under the body and identity of the index row were zero or negative, no further improvement could be obtained, which would indicate that the solution presented in the stub was optimal.

We see from Table B-4 that the column headed by the variable x_I has the greatest improvement potential, so we select it as the *key column*. This selection means that the variable x_I will be introduced into the solution in favor of W_A or W_B. To determine whether x_I will replace W_A or W_B, we must select a *key row*. To do this, *we divide each number in the constant column by the corresponding positive nonzero number in the key column*. The resulting quotients are compared, and the *key row* is selected

TABLE B-4 **Initial Simplex Tableau Showing Index Row**

	10	5	0	0			
	x_I	x_{II}	W_A	W_B			
	−4	5	1	0	100	W_A	0
	5	2	0	1	80	W_B	0
Index row →	10	5	0	0	0		

TABLE B-5 **Initial Simplex Tableau with Key Column, Row, and Number Identified**

	10	5	0	0			
	x_I	x_{II}	W_A	W_B			
	4	5	1	0	100	W_A	0
Key row →	5	2	0	1	80	W_B	0
	10	5	0	0	0		

Key number (5 in the key row) — Key column (x_I)

as the row yielding the smallest nonnegative quotient. For our problem, the quotients are

$$\text{First row } \frac{100}{4} = 25$$

$$\text{Second row } \frac{80}{5} = 16 \text{ (key row)}$$

The smallest nonnegative quotient is 16, computed for the second row, and it becomes the key row. The number that is common to both the key column and the key row is designated the *key number*. Table B-5 shows the tableau, with the key column, row, and number identified.

Let us pause for a moment to examine our rationale. We know that x_I will be introduced into the solution, and we wish to know its maximum value consistent with both problem equations, assuming that none of the variables can take on negative values. In the first equation, x_I would be the largest possible when x_{II} and W_A were zero; that is,

$$4x_I + 5x_{II} + W_A = 100$$

or

$$x_I = \frac{100}{4} = 25$$

In the second equation, x_I would be the largest possible when x_{II} and W_B were zero; that is,

$$5x_I + 2x_{II} + W_B = 80$$

or

$$x_I = \frac{80}{5} = 16$$

TABLE B-6 **Simplex Tableu with Main Row of New Table**

1st tableau	10	5	0	0			
	x_I	x_{II}	W_A	W_B			
	4	5	1	0	100	W_A	0
	5	2	0	1	80	W_B	0
	10	5	0	0	0		
2nd tableau							
	1	$^2/_5$	0	$^1/_5$	16		Main row

Thus, the second equation is the one that limits x_I. It can be no larger than 16, and this fact dictates the selection of the second row as the key row.

Once we have selected the key column and row, we can prepare a new tableau representing an improved solution. The first step in developing the new tableau is to calculate the coefficients for the *main row*. This main row appears in the same relative position in the new tableau as the key row in the preceding one. It is computed by dividing the coefficients of the key row by the key number. Table B-6 shows this development. The variable and its objective number from the head of the key column, x_I and 10, are placed in the stub of the main row, replacing W_B and 0 from the previous tableau. The balance of the objective and variable columns in the stub is copied from the previous tableau; the new tableau, developed to this point, now appears as Table B-7.

Now all the remaining coefficients in the new tableau, including the constant col-

TABLE B-7

Simplex Tableau with Variable and Objective Columns of Stub Completed

1st tableau	10	5	0	0			
	x_I	x_{II}	W_A	W_B			
	4	5	1	0	100	W_A	0
	5	2	0	1	80	W_B	0
	10	5	0	0	0		
2nd tableau							
						W_A	0
	1	$^2/_5$	0	$^1/_5$	16	x_I	10

umn, the body, the identity, and the index row, can be calculated to complete the tableau by the following formula:

$$\text{New number} = \text{old number} - \frac{\begin{pmatrix}\text{corresponding}\\ \text{number of}\\ \text{key row}\end{pmatrix} \times \begin{pmatrix}\text{corresponding}\\ \text{number of}\\ \text{key column}\end{pmatrix}}{\text{key number}}$$

(1) First row constant column

$$\text{New number} = 100 - \frac{80 \times 4}{5} = 36$$

(2) First row, first column of body

$$\text{New number} = 4 - \frac{5 \times 4}{5} = 0$$

(3) Index row, constant column

$$\text{New number} = 0 - \frac{80 \times (10)}{5} = -160$$

The remaining coefficients can be calculated in the same way and the completed improved solution is shown in Table B-8, with a new index row that shows any new possibilities for improvement.

The solution at this stage is

$$\begin{aligned} x_I &= 16 \\ x_{II} &= 0 \\ W_A &= 36 \\ W_b &= 0 \end{aligned}$$

TABLE B-8

Simplex Tableau with First Iteration Completed

1st tableau	10	5	0	0			
	x_I	x_{II}	W_A	W_B			
	4	5	1	0	100	W_A	0
	5	2	0	1	80	W_B	0
	10	5	0	0	0		
2nd tableau							
	0	$^{17}/_5$	1	$^{-4}/_5$	36	W_A	0
	1	$^{2}/_5$	0	$^{1}/_5$	16	x_I	10
	0	1	0	2	−160		

The values in the constant column of the stub indicate that x_{II} and W_B are zero, since they are not in the stub at all. The value of the objective function for this solution is \$160, which is given in the constant column, index row. (A negative value for the objective function in the constant column indicates a positive contribution.)

However, Table B-8 shows that the solution still can be improved, since a (+1) appears in the index row under the variable x_{II}. Since it is the only positive number in the index row, it is selected as the key column for the next iteration. The key row is selected in the same way as before. The two quotients are

First row $$\frac{36}{17/5} = \frac{180}{17} = 10.59$$

Second row $$\frac{16}{2/5} = 40$$

The first row has the smallest nonnegative quotient, so it is selected as the key row. A new main row is calculated, as before, by dividing the coefficients in the main row by the key number. The new variable, x_{II}, and its objective number are entered in the stub, and the new numbers in the body, identity, and index row are computed as before. The remaining variable and its objective number are copied from the preceding iteration tableau; Table B-9 shows the new solution.

The new solution in Table B-9 is optimal, since no further improvement is indicated in the index row. The values of the variables for the optimal solution are

$$\begin{aligned} x_I &= 200/17 \\ x_{II} &= 180/17 \\ W_A &= 0 \\ W_B &= 0 \end{aligned}$$

If these values are inserted in Equations 6 and 7, we find that they check exactly. The solution indicates the products I and II should be produced in the amounts shown by x_I and x_{II} (in 100's) to produce a maximum contribution of $C = 2900/17 =$ \$170.59. For each iteration, the value of the objective function is shown in the constant column, index row. It was \$160 for the second solution and $2900/17 =$ \$170.59 for the optimal solution. The solution is unique; that is, no other combination of x_I and x_{II} will yield a contribution figure that is as high as this.

Procedure Summary

1. *Formulate the problem as a linear optimization model.*

2. *Develop the initial simplex tableau*, including the initial trivial solution and the index row numbers. The index row numbers in the initial tableau are calculated by the formula:

TABLE B-9 **Simplex Tableau, Second and Final Iteration Completed**

1st tableau	10	5	0	0			
	x_I	x_{II}	W_A	W_B			
	4	5	1	0	100	W_A	0
	5	2	0	1	80	W_B	0
	10	5	0	0	0		
2nd tableau							
	0	$^{17}/_5$	1	$^{-4}/_5$	36	W_A	0
	1	$^2/_5$	0	$^1/_5$	16	x_I	10
	0	1	0	2	−160		
3rd tableau							
	0	1	$^5/_{17}$	$^{-4}/_{17}$	$^{180}/_{17}$	x_{II}	5
	1	0	$^{-2}/_{17}$	$^5/_{17}$	$^{200}/_{17}$	x_I	10
	0	0	$^5/_{17}$	$^{30}/_{17}$	$^{-2900}/_{17}$		

$$\text{Index number} = \left(\begin{array}{c}\text{number in}\\\text{objective}\\\text{row at head}\\\text{of column}\end{array}\right) - \sum \left(\begin{array}{c}\text{numbers}\\\text{in}\\\text{column}\end{array}\right) \times \left(\begin{array}{c}\text{corresponding}\\\text{number in}\\\text{objective}\\\text{column}\end{array}\right)$$

3. *Select the key column,* the column with the largest positive index number in the body or the identity.

4. *Select the key row,* the row with the smallest nonnegative quotient, which is obtained by dividing each number of the constant column by the corresponding positive, nonzero number in the key column.

5. *The key number* is at the intersection of the key row and key column.

6. *Develop the main row of the new table.*

$$\text{Main row} = \frac{\text{key row of preceding table}}{\text{key number}}$$

   The main row appears in the new tableau in the same relative position as the key row of the preceding table.

7. *Develop the balance of the new tableau.*
   a. The variable and its objective number at the head of the key column are entered in the stub of the new tableau to the right of the main row, replacing the variable and objective number from the key row of the preceding tableau.

b. The remainder of the variable and objective columns are reproduced in the new tableau exactly as they were in the preceding tableeau.
c. The balance of the coefficients for the new tableau are calculated by the formula:

$$\text{New number} = \text{old number} - \frac{\begin{pmatrix}\text{corresponding}\\ \text{number of}\\ \text{key row}\end{pmatrix} \times \begin{pmatrix}\text{corresponding}\\ \text{number of}\\ \text{key column}\end{pmatrix}}{\text{key number}}$$

8. *Repeat (iterate) steps 3 through 7c* until all the index numbers (not including the constant column) are negative or zero. An optimal solution then results.

The interpretation of the resulting optimum solution is as follows: the solution appears in the stub. The variables shown in the variable column have values that are shown in the corresponding rows of the constant column. The value of the objective function is shown in the constant column, index row. All variables not shown in the stub are equal to zero.

REVIEW QUESTIONS AND PROBLEMS

1. Outline the model building process that is used to develop a linear optimization model.

2. A chemical manufacturer produces two products, x and y. Each product is manufactured by a two-step process that involves blending and mixing in machine A, and packaging on machine B. The two products complement each other, since the same production facilities can be used for both products, thus achieving better utilization of these facilities.

   Since the facilities are shared and each product's costs and profits are different, we must figure out how to utilize the available machine time in the most profitable way. Chemical x seemingly is more profitable. However, when the manager tried producing the maximum amount of chemical x within market limitations, using the balance of machine time to produce chemical y, she found that such an allocation of machine time resulted in poor profit performance. She now feels that some appropriate balance between the two products is best and wishes to determine the production rates for each product per two-week period.

   What are the decision variables for a linear optimization model that is designed to solve the chemical manufacturer's problem?

3. Referring to the problem stated in question 2, the manager has obtained the following price and cost information on the two chemicals.

	Sales Price	Variable Costs
Chemical x	\$360	\$290
Chemical y	\$460	\$400

The manager wishes to maximize profit in her decision.
What is the objective function for the chemical manufacturer's problem?

4. Further data for the chemical manufacturer's problem stated in questions 2 and 3 relate to the process time for the two products on the mixing machine (machine A) and the packaging machine (machine B), as follows:

Product	Machine A	Machine B
x	4 hours	4 hours
y	5 hours	2 hours

For the upcoming two-week planning period, machine A has available 90 hours and machine B has available 80 hours. Furthermore, market forecasts indicate that the maximum sale of chemical x is 18 units and of y is 20 units.
Define the constraints for a linear optimization model for the chemical production problem.

5. Make a complete mathematical statement for the linear optimization model developed in questions 2, 3, and 4.

6. Plot the constraints for the linear optimization model, stated in question 5, on a graph. Define the feasible solution space.

7. Using the same graph of the problem constraints developed in question 6, plot objective function lines for $Z = \$1200$, and $Z = \$1300$.

8. What is the function of slack variables in the simplex solution to linear optimization models?

9. What is shown in each of the following in the simplex tableau?
   a. Objective row
   b. Variable row
   c. Objective column
   d. Variable column
   e. Constant column

10. Outline the procedure for the simplex solution of linear optimization models.

PROBLEMS

11. Once upon a time, Lucretia Borgia invited 50 enemies to dinner. The pièce de résistance was to be poison. In those crude days, only two poisons were on the market, poison X and poison Y. Before preparing the menu, however, the remarkably talented young lady considered some of the restrictions placed on her scheme:
    a. If she used more than one-half pound of poison, the guests would detect it and refuse to eat.
    b. Lucretia's own private witch, a medieval version of the modern planning staff, once propounded some magic numbers for her in the following doggerel:

       One Y and X two,
       If less than half,
       Then woe to you.

    c. Poison X will kill 75 people per pound, and poison Y will kill 200 people per pound.
    d. Poison X costs 100 solid gold pieces per pound, and Poison Y costs 400 solid gold pieces per pound.

    After devising a menu to cover up the taste of the poison, Lucretia found that she was very short of solid gold pieces. In fact, unless she were very careful, she would not be able to have another scheduled poisoning orgy that month. So she called in her alchemist, a very learned man, and told him about her problem. The alchemist had little experience in solving problems of this type, but he was able to translate the four restrictions into mathematical statements.

    (1) $X + Y \leq 1/2$
    (2) $2X + Y \geq 1/2$
    (3) $75X + 200Y \geq 50$
    (4) $100X + 400Y = \text{cost}$

    Assist the alchemist in solving this problem, using graphic methods. The penalty for failure will be an invitation to next month's dinner.

12. A company makes four products x_1, x_2, x_3, and x_4, which flow through four departments: drill, mill, lathe, and assembly. The hours of department time required by each of the products per unit are:

	Drill	Mill	Lathe	Assembly
x_1	3	0	3	4
x_2	7	2	4	6
x_3	4	4	0	5
x_4	0	6	5	3

The unit contributions of the four products and hours of availability in the four departments are:

Product	*Contribution*
x_1	$ 9
x_2	18
x_3	14
x_4	11

Department	*Hours Available*
Drill	70
Mill	80
Lathe	90
Assembly	100

Formulate the problem as a linear optimization model to determine optimal product mix.

13. A company makes five products, x_1, x_2, x_3, x_4, and x_5, which flow through five departments: blanking, forming, straightening, brazing, and assembly. The following processing times are required:

	Time Required				
Product	Blanking (hours)	Forming (hours)	Straightening (hours)	Brazing (hours)	Assembly (hours)
x_1	1	1	2	0	1
x_2	1	0	½	1	2
x_3	2	3	0	½	1
x_4	0	2	2	1	2
x_5	½	1	½	2	1

The contribution of each product is as follows:

Product	*Contribution*
x_1	$ 8 per unit
x_2	9 per unit
x_3	13 per unit
x_4	15 per unit
x_5	7 per unit

Time available in the various departments and the estimated incremental cost of idle time are:

Department	Time Available, Hours	Estimated Incremental Cost of Idle-Time per Hour
Blanking	115	$12
Forming	100	8
Straightening	140	3.50
Brazing	90	30
Assembly	110	12

Formulate the problem in the simplex tableau.

REFERENCES

Bierman, H., C. P. Bonini, and W. H. Hausman. *Quantitative Analysis for Business Decisions* (5th ed.). Richard D. Irwin, Homewood, Ill., 1977.

Buckley, J. W., M. R. Nagarai, B. L. Sharp, and J. W. Schenck. *Management Problem-Solving with APL*. John Wiley, New York, 1974.

Buffa, E. S., and J. S. Dyer. *Management Science/Operations Research: Model Formulation and Solution Methods*. John Wiley, New York, 1977.

Levin, R. I., and C. A. Kirkpatrick. *Quantitative Approaches to Management* (3rd ed.). Prentice-Hall, Englewood Cliffs, N.J., 1975.

Thierauf, R. J., and R. C. Klekamp. *Decision Making through Operations Research* (2nd ed.). John Wiley, New York, 1975.

APPENDIX C Linear Programming—Distribution Methods

EXAMPLE PROBLEM

Consider the distribution situation of the Pet Food Company. There are three factories located in Chicago, Houston, and New York that produce some identical products. There are five major distribution points that serve various market areas in Atlanta, Buffalo, Cleveland, Denver, and Los Angeles. The three factories have capacities that determine the availability of product, and the market demand in the five major areas determines the requirements to be met. The problem is to allocate available product at the three factory locations to the five distribution points so that demand is met and distribution cost is minimized for the system.

Data for our illustrative problem are shown in Table C-1. There are 46,000 cases per week available at the Chicago plant, 20,000 at Houston, and 34,000 at New York, or a total of 100,000 cases per week of pet food. Similarly, demands in the five market areas are indicated in the bottom row of the table and also total 100,000 cases per week. Equality of availability and demand is not a necessary requirement for solution.

These figures of units available and required are commonly termed the "rim conditions." Table C-1 also shows the distribution costs per 1000 cases for all combinations of factories and distribution points. These figures are shown in the small boxes; for example, the distribution cost between the Chicago plant and the Altanta distribution point is $18 per 1000 cases. For convenience in notation the plants are labeled A, B, and C, and the distribution points V, W, X, Y, and Z. Table C-1 is called the "transportation or distribution table."

TABLE C-1 **Transportation Table for the Pet Food Company. Quantities of Product Available at Factories and Required at Distribution Points, and Distribution Costs per Thousand Cases**

From Factories / To Distr. Points	Atlanta (V)	Buffalo (W)	Cleveland (X)	Denver (Y)	Los Angeles (Z)	Available from Factories, 1000's
Chicago (A)	18	16	12	28	54	46
Houston (B)	24	40	36	30	42	20
New York (C)	22	12	16	48	44	34
Required at distribution points, 1000's	27	16	18	10	29	100

An Initial Solution

We will establish an initial solution in an arbitrary way, ignoring the distribution costs. Beginning in the upper left-hand corner of the transportation table (called the northwest corner) note that A has 46 (1000) cases available and V needs 27 (1000). We assign the 27 from A to V. (See Table C-2, where circled numbers represent assigned product, e.g., 27 in box AV means 27,000 cases are to go from A to V.) We have not used up A's supply, so we move to the right under column W and assign the maximum possible to the route AW, 16. We still have not used up A's supply, so we move to the right under column X and assign the balance of A's supply, 3, to X.

Examining the requirements for X, we note that it has a total requirement of 18, so we drop down to row B and assign the balance of X's requirements, 15, from B's supply of 20. We then move to the right again and assign the balance of B's supply to Y, 5. We continue in this way, stair-stepping down the table, until all the arbitrary assignments have been made as in Table C-2. The distribution costs, $2984, are calculated below Table C-2. Note that we have seven squares with assignments (n rows + m columns - 1) and eight open squares without assignments. This requirement for $m + n - 1$ assignments avoids degeneracy, to be discussed later. Table C-2 is the northwest corner initial solution.

Methods for obtaining better starting soluions, and simplifications will be discussed later in this appendix.

TABLE C-2 **Northwest Corner Initial Solution**

From Factories \ To Distr. Points	Atlanta (V)	Buffalo (W)	Cleveland (X)	Denver (Y)	Los Angeles (Z)	Available from Factories, 1000's
Chicago (A)	[18] (27)	[16] (16)	[12] (3)	[28]	[54]	46
Houston (B)	[24]	[40]	[36] (15)	[30] (5)	[42]	20
New York (C)	[22]	[12]	[16]	[48] (5)	[44] (29)	34
Required at distribution points, 1000's	27	16	18	10	29	100

Total distribution cost:

AV,	27×18	=	486
AW,	16×16	=	256
AX,	3×12	=	36
BX,	15×36	=	540
BY,	5×30	=	150
CY,	5×48	=	240
CZ,	29×44	=	1276
			$2984

Test for Optimality

Since the northwest corner solution in Table C-2 was established arbitrarily, it is not likely to be the best possible solution. Therefore, we need to develop a method for examining each of the open squares in the transportation table to determine if improvements can be made in total distribution costs by shifting some of the units to be shipped to these routes. In making the shifts, we must be sure that any new solution satisfies the supply and demand restrictions shown in the rim conditions of the transportation table.

In evaluating each open square, the following steps are used:

Step 1
Determine a closed path, starting at the open square being evaluated and "stepping" from squares with assignments back to the original open square. Right-angle turns in this path are permitted only at squares with assignments and at the original open square. Since only the squares at the turning points are considered to be on the closed path, both open and assigned squares may be skipped over.

Step 2
Beginning at the square being evaluated, assign a plus sign and then alternate minus and plus signs at the assigned squares on the corners of the paths.

Step 3
Add the unit costs in the squares with plus signs, and subtract the unit costs in the squares with minus signs. If we are minimizing costs, the result is the net change in cost per unit from the changes made in the assignments. If we are maximizing profits, the result is the net change in profit.

Step 4
Repeat steps 1, 2, and 3 for each open square in the transportation table.

Steps 1 and 2 involve the assignment of a single unit to the open square and then adjusting the shipments in the squares with assignments until all the rim conditions are satisfied.

Step 3 simply calculates the cost (or contribution to profit) that would result from such a modification in the assignments. If we are minimizing costs, and if the net changes are all greater than or equal to zero for all open squares, we have found an optimal solution. If we are maximizing profits, and all net changes are less than or equal to zero for all open squares, we have found an optimal solution.

The application of these steps to open square BV is shown in Table C-3. In this case, the closed path forms a simple rectangle. The net change in total distribution cost resulting from shifting one unit to route BV is

$$24 - 18 + 12 - 36 = -18$$

TABLE C-3 **Evaluation of Square BV for Possible Improvement. The Change in Total Distribution Cost Resulting from Shifting One Unit to Route BV is: 24 − 18 + 12 − 36 = − 18**

From Factories \ To Distr. Points	Atlanta (V)	Buffalo (W)	Cleveland (X)	Denver (Y)	Los Angeles (Z)	Available from Factories, 1000's
Chicago (A)	[18] (−) **27**	[16] **16**	[12] (+) **3**	[28] +22	[54] +52	46
Houston (B)	[24] (+) −18	[40] 0	[36] **15** (−)	[30] **5**	[42] +16	20
New York (C)	[22] −38	[12] −46	[16] −38	[48] **5**	[44] **29**	34
Required at distribution points, 1000's	27	16	18	10	29	100

TABLE C-4 **Evaluation of Square CV for Possible Improvement. The Change in Total Distribution Cost Resulting from Shifting One Unit to Route CV is: 22 − 18 + 12 − 36 + 30 − 48 = −38**

From Factories \ To Distr. Points	Atlanta (V)	Buffalo (W)	Cleveland (X)	Denver (Y)	Los Angeles (Z)	Available from Factories, 1000's
Chicago (A)	[18] (−) **27**	[16] **16**	[12] (+) **3**	[28]	[54]	46
Houston (B)	[24]	[40]	[36] (−) **15**	[30] **5** (+)	[42]	20
New York (C)	[22] (+) −38	[12]	[16]	[48] **5** (−)	[44] **29**	34
Required at distribution points, 1000's	27	16	18	10	29	100

This value is entered in the bottom left-hand corner of the square BV in Table C-3. The evaluation of open squares does not always follow the simple rectangular path. For example, the closed path for evaluating square CV is shown in Table C-4, and the net change in total distribution cost resulting from shifting one unit to route CV is

$$22 - 18 + 12 - 36 + 30 - 48 = -38$$

The net changes in total distribution cost for all the open squares are shown in Table C-3, in the lower left-hand corner of each open square.

Improving the Solution

Since each negative net change indicates the amount by which the total distribution cost will decrease if one unit were shifted to the route of that cell, we will be guided by these evaluations as indexes of potential improvement. Notice in Table C-3 that four of the open squares indicate potential improvement, three indicate that costs would increase if units were shifted to those routes, and one indicates that costs would not change if units were shifted to its route. Which change should be made first in determining a new improved solution? One reasonable rule for small problems and hand solutions is to select the square with the most negative index number when we are minimizing costs. Therefore, we choose square CW.

To improve the solution, we carry out the following steps for square CW:

Step 1

Identify again the closed path for the chosen open square, and assign the plus and minus signs around the path as before. Determine the minimum number of units assigned to a square on this path that is marked with a minus sign.

Step 2

Add this number to the open square and to all other squares on the path marked with a plus sign. Subtract this number from the squares on the path marked with a minus sign. This step is a simple accounting procedure for observing the restrictions of the rim conditions.

The closed path for open square CW is +CW −AW +AX −BX +BY −CY. The minimum number of units in a square with a minus sign is five in CY. Thus, we add five units to squares CW, AX, and BY, and subtract five units from squares AW, BX, and CY. This reassignment of units generates a new solution, as shown in Table C-5.

Taking Table C-5 as the current solution, we repeat the process, reevaluating all open squares, as shown in Table C-5. Note that there are now only two open squares with negative index numbers, indicating potential improvement. Following our previous procedure, open square BZ indicates that the largest improvement could be made by shifting assignments to that route.

An Optimal Solution

The process is continued until all open squares show no further improvement. At this point, an optimal solution has been obtained and is shown in Table C-6, where all

TABLE C-5 **New Solution Resulting After Shifting Five Units to Square CW. The Number of Units That Could Be Shifted to CW Was Limited to Five by Square CY in the Previous Solution**

From Factories \ To Distr. Points	Atlanta (V)	Buffalo (W)	Cleveland (X)	Denver (Y)	Los Angeles (Z)	Available from Factories, 1000's
Chicago (A)	[18] (27)	[16] (11)	[12] (8)	[28] +22	[54] +6	46
Houston (B)	[24] −18	[40] 0	[36] (10)	[30] (10)	[42] −30	20
New York (C)	[22] +8	[12] (5)	[16] +8	[48] +46	[44] (29)	34
Required at distribution points, 1000's	27	16	18	10	29	100

the index numbers for open squares are either positive or zero. The total distribution cost required by the optimal solution is \$2446, which is \$538 less than the original northwest corner solution. Although this 18 percent improvement is impressive, we should note that we started with a rather poor initial solution. Better starting solutions can be obtained for simple problems by making initial assignments to the most promising routes while observing the restrictions of the rim conditions and being sure that there are exactly $n + m - 1$ assignments. Solutions with $n + m - 1$ assignments

TABLE C-6 **An Optimal Solution. Evaluation of All Open Squares in This Table Results in No Further Improvement. Total Distribution Cost = \$2446**

From Factories \ To Distr. Points	Atlanta (V)	Buffalo (W)	Cleveland (X)	Denver (Y)	Los Angeles (Z)	Available from Factories, 1000's
Chicago (A)	[18] (27)	[16] +6	[12] (18)	[28] (1)	[54] +14	46
Houston (B)	[24] +2	[40] +30	[36] +22	[30] (9)	[42] (11)	20
New York (C)	[22] 0	[12] (16)	[16] 0	[48] +16	[44] (18)	34
Required at distribution points, 1000's	27	16	18	10	29	100

are called basic solutions, solutions with fewer assignments are called degenerate, and solutions with more than $n + m - 1$ assignments are called nonbasic solutions.

Alternate Optimal Solutions

The fact that open squares CV and CX in Table C-6 have zero evaluations is important and gives us flexibility in determining the final plan of action. These zero evaluations allow us to generate other solutions that have the same total distribution costs as the initial optimal solution generated in Table C-6.

As an example, since open square CV has a zero evaluation, we may make the shifts in allocations indicated by its closed path and generate the alternate basic optimal solution shown in Table C-7. Still another alternate optimum solution could be generated by shifting assignments to square CX.

We are not yet finished, because we can generate literally dozens of other optimal solutions from each of the basic solutions. In generating the alternate basic optimum solution in Table C-7, we shifted 9000 units to open square CV, which had a zero evaluation. It was not necessary for us to shift the entire 9000 units, however. We could have shifted only 5000 units, or 4000, or 3000, or any fractional amount of the 9000 units. All of the basic alternate optimum solutions could also be varied in this way, producing nonbasic alternate optimum solutions.

Where we have optimum solutions containing open squares with zero evaluations, we have great flexibility in distribution at minimum cost. Where fractional units are permitted, we have in fact a tremendous number of alternate optimal solutions. This may often make it possible to satisfy subjective factors in the problem and still retain minimum distribution costs.

TABLE C-7 **Alternate Basic Optimum Solution**

From Factories / To Distr. Points	Atlanta (V)	Buffalo (W)	Cleveland (X)	Denver (Y)	Los Angeles (Z)	Available from Factories, 1000's
Chicago (A)	[18] (18)	[16] +8	[12] (18)	[28] (10)	[54] +14	46
Houston (B)	[24] +4	[40] +30	[36] +16	[30] 0	[42] (20)	20
New York (C)	[22] (9)	[12] (16)	[16] 0	[48] +16	[44] (9)	34
Required at distribution points, 1000's	27	16	18	10	29	100

UNEQUAL SUPPLY AND DEMAND

Now suppose that supply exceeds demand, as shown in Table C-8. The total available supply is still 100,000 cases; however, the aggregate demand at the five distribution points totals only 95,000 cases. This situation can be handled in the problem by creating a dummy distribution point to receive the extra 5000 cases. The nonexistent distribution point is assigned zero distribution costs as shown, since the product will never be shipped. The optimal solution then assigns 95,000 of the 100,000 available units in the most economical way to the five real distribution points and assigns the balance to the dummy distribution point. Table C-8 shows an optimal distribution plan for this situation.

When demand exceeds supply, we can resort to a modification of the same technique. Create a dummy factory to take up the slack. Again, zero distribution costs are assigned to the dummy factory, since the product will never be shipped. The solution then assigns the available product to the distribution points in the most economical way, indicating which distribution points should receive "short" shipments, so that the total distribution costs are minimized.

TRANSSHIPMENT

It often happens that direct routes between all origins and destinations are not used. Major routes may exist through distribution centers on the way to smaller receipt points. The transshipment model is a modification of the distribution model that allows transshipment points. See Buffa and Dyer [1977] for methods.

TABLE C-8 **Distribution Table with Supply Exceeding Demand: Optimum Solution**

From Factories \ To Distr. Points	Atlanta (V)	Buffalo (W)	Cleveland (X)	Denver (Y)	Los Angeles (Z)	Dummy	Available from Factories, 1000's
Chicago (A)	[18] (25)	[16] 8	[12] (11)	[28] (10)	[54] 14	[0] 4	46
Houston (B)	[24] 4	[40] 30	[36] 22	[30] 0	[42] (20)	[0] 2	20
New York (C)	[22] 0	[12] (15)	[16] (7)	[48] 16	[44] (7)	[0] (5)	34
Required at distribution points, 1000's	25	15	18	10	27	5	100

TABLE C-9 **Evaluation of Square AX Produces Degeneracy**

From \ To	V	W	X	Y	Z	Available 1000s ↓
A	42 (−) 6	42 13	44 (+)	40	44	19
B	34 (+) 6	42	40 6 (−)	46 16	48	28
C	46	44	42	48 1	46 24	25
Required 1000s →	12	13	6	17	24	72

DEGENERACY IN DISTRIBUTION PROBLEMS

Another aspect of the mechanics of developing a solution is the condition known as *degeneracy*. Degeneracy occurs in distribution problems when, in shifting assignments to take advantage of a potential improvement, more than one of the existing assignments go to zero. Degeneracy also can occur in an initial solution that does not meet the $m + n - 1$ requirement for the number of allocations. Examination of the problem in Table C-9 shows that degeneracy is about to happen. This problem was set up in the usual way, and an initial northwest corner solution was established. The open squares were evaluated column by column, as before, and changes in assignments were made when they indicated potential improvement.

In Table C-9, we are evaluating square AX by the closed path pattern. Potential improvement is indicated, since a unit of allocation reduces transportation costs by \$4 per thousand. We wish to press this advantage to the maximum by shifting as much as possible to AX. We are limited, however, by both squares AV and BX, each of which has an allocation of 6000 units assigned to it. When the shift in assignment is made, both AV and BX go to zero. This is shown in Table C-10. We now have only six allocations instead of the seven we had before, and we do not meet the restriction on the method of solution that we stated earlier: that the number of allocations must be $m + n - 1$. The practical effect of this is that several of the open squares, namely AV, BW, CW, BX, CX, AY, and AZ, cannot be evaluated in the usual way because a closed path cannot be established for them.

TABLE C-10 **Problem Now Degenerate; Squares AV, BW, CW, BX, CX, AY, and AZ Cannot be Evaluated**

From \ To	V	W	X	Y	Z	Available 1000s ↓
A	42	42 (13)	44 (6)	40	44	19
B	34 (12)	42	40	46 (16)	48	28
C	46	44	42	48 (1)	46 (24)	25
Required 1000s →	12	13	6	17	24	72

TABLE C-11 **Degeneracy Resolved by Use of the ϵ Allocation**

From \ To	V	W	X	Y	Z	Available 1000s ↓
A	42 (ϵ)	42 (13)	44 (6)	40	44	19
B	34 (12)	42	40	46 (16)	48	28
C	46	44	42	48 (1)	46 (24)	25
Required 1000s →	12	13	6	17	24	72

We can resolve the degeneracy, however, by regarding one of the two squares in which allocations have disappeared as an allocated square with an extremely small allocation, which we will call an ϵ allocation. This is illustrated in Table C-11. Conceptually, we will regard the ϵ allocation as infinitesimally small, so that it does not affect the totals indicated in the rim. The ϵ allocation, however, does make it possible to meet the $m + n - 1$ restriction on the number of allocations so that evaluation paths may be established for all open squares. The ϵ allocation is simply manipulated as though it were no different from the other allocations.

If, in subsequent manipulations, the ϵ allocation square is the one that limits shifts in assignments, it simply is shifted to the square being evaluated, and the usual procedures are then continued. This is illustrated in Table C-12, where we are attempting to evaluate square AZ by the closed path shown. A potential improvement of \$8 per 1000 units is indicated, but the limiting allocation at a negative square is the ϵ allocation. The net effect of adding and subtracting the ϵ allocation around the closed path is to move the ϵ allocation from square AV to square AZ. The procedure then is continued as before, until an optimal solution is obtained.

As the procedure continues, the ϵ allocation may disappear. This is illustrated in Table C-13, in which we are evaluating the open square CX. Potential improvement of \$4 per 1000 units is indicated, and here we are limited not by the ϵ allocation but by the allocation of 6000 units at AX. In making the adjustments, we add and subtract 6000 units around the closed path according to the signs indicated, and the result is that the ϵ allocation at AZ becomes 6000 units. We now have seven squares with

TABLE C-12 **Shift of ϵ Allocation When it is Limiting**

From \ To	V	W	X	Y	Z	Available 1000s ↓
A	[42] (−) ϵ	[42] 13	[44] 6	[40]	[44] ϵ (+)	19
B	[34] (+) 12	[42]	[40]	[46] (−) 16	[48]	28
C	[46]	[44]	[42]	[48] (+) 1	[46] 24 (−)	25
Required 1000s →	12	13	6	17	24	72

TABLE C-13 **Disappearance of the ε Allocation When It Falls at a Positive Corner of an Evaluation Path**

From \ To	V	W	X	Y	Z	Available 1000s ↓
A	42	42 (13)	44 (−) (6)	40	44 (+) (ε)	19
B	34 (12)	42	40	46 (16)	48	28
C	46	44	42 (+)	48 (1)	46 (24) (−)	25
Required 1000s →	12	13	6	17	24	72

TABLE C-14 **Distribution Table with Initial VAM Row and Column Differences Shown**

	↓ 8	0	2	6	2		
From \ To	V	W	X	Y	Z	Available 1000s ↓	
A	8	8	10	6	10	19	2
B	0	8	6	12	14	28	6
C	12	10	8	14	12	25	2
Required 1000s →	11	13	7	17	24	72	

positive allocations, and the ϵ allocation is no longer needed. In carrying through the solution of larger-scale problems, we may find that degeneracy appears and disappears in the routine solution of a problem or that more than one ϵ allocation exists. Also, optimal solutions may be degenerate.

TECHNIQUES FOR SIMPLIFYING PROBLEM SOLUTION

We can simplify the arithmetic complexity considerably by using two methods. A little thought about the example that we used will convince us that it is the cost differences that are important in determining the optimal allocation, rather than their absolute values. Therefore, we can reduce all costs by a fixed amount, and the resulting allocation will be unchanged. In our illustrative example, we may subtract 12 from all distribution cost values, so that the numbers with which we must work are of such magnitude to allow many evaluations to be accomplished by inspection. Another simplification in the arithmetic may be accomplished by expressing the rim conditions in the simplest terms. For example, in our illustration, we expressed the rim conditions in thousands of units, thus enabling us to work with two-digit numbers only.

Getting an Advantageous Initial Solution

The northwest corner initial solution is not used in practice, since ordinarily it is a rather poor solution that involves a number of steps to develop an optimal solution. Placing the lowest cost cell in the northwest corner gives an advantageous start. The usual procedure is to start with some solution by inspecting the most promising routes and entering allocations that are consistent with the rim conditions. In establishing such an initial solution, the only rules to be observed are that there must be exactly $m + n - 1$ allocations, so that it is possible to evaluate all open squares by the closed path methods.

If the initial solution turns out to be degenerate, it is simple to increase the allocations to the exact number required by resorting to the ϵ allocation. There are a number of short-cut methods that commonly are used, such as row minimum, column minimum, matrix minimum, and VAM. Although they all have merits, we will discuss VAM in some detail because it seems particularly valuable for hand computation of fairly large-scale problems. Of course, computer solutions should be used for large-scale problems.

Vogel's Approximation Method (VAM)

VAM facilitates a very good initial solution, which usually is the optimal solution. The technique is a simple one, and it considerably reduces the amount of work that is required to generate an optimal solution. We will use a new problem to illustrate. Table C-14 shows a distribution table with the distribution costs all reduced by the constant amount, \$34. The steps in determining an initial VAM solution are as follows:

1. *Determine the difference between the two lowest distribution costs for each row and each column.* This has been done in Table C-14, and the figures at the heads of columns and to the right of the rows represent these differences. For example, in column V, the three distribution costs are 8, 0, and 12. The two lowest costs are 8 and 0, and their difference is 8. In row A, the two lowest distribution costs are 6 and 8, or a difference of 2. The other figures at the heads of the columns and to the right of the rows have been determined in a similar way.

2. *Select the row or column with the greatest difference.* For the example, the row or column with the greatest difference is column V, which has a difference of 8.

3. *Assign the largest possible allocation within the restrictions of the rim conditions to the lowest cost square in the row or column selected.* This has been done in Table C-15. Under column V, the lowest cost square is BV, which has a cost of 0, and we have assigned 11 units to that square. The 11-unit assignment is the largest possible because of the restriction imposed by the number required at distribution point V.

TABLE C-15 **First VAM Assignment Satisfies V's Requirement (Row and Column VAM Differences Are Recalculated)**

	↓ 8	0	2	6	2		
From \ To	V	W	X	Y	Z	Available 1000s ↓	
A	[8] X	[8]	[10]	[6]	[10]	19	2
B	[0] (11)	[8]	[6]	[12]	[14]	28	~~6~~2
C	[12] X	[10]	[8]	[14]	[12]	25	2
Required 1000s →	11	13	7	17	24	72	

TABLE C-16 **Second VAM Assignment Satisfies Y's Requirement (Row and Column VAM Differences Are Recalculated)**

	~~8~~	0	2	↓ ~~6~~	2		
From \ To	V	W	X	Y	Z	Available 1000s ↓	
A	[8] X	[8]	[10]	[6] (17)	[10]	19	2
B	[0] (11)	[8]	[6]	[12] X	[14]	28	~~6~~2
C	[12] X	[10]	[8]	[14] X	[12]	25	2
Required 1000s →	11	13	7	17	24	72	

4. *Cross out any row or column that has been completely satisfied by the assignment just made.* For the assignment just made at BV, the requirements for V are entirely satisfied. Therefore, we may cross out the other squares in that column, since we can make no future assignments to them. This is shown in Table C-15.

5. *Recalculate the differences as in step 1, except for rows or columns that have been crossed out.* This has been done in Table C-15, where row B is the only one affected by the assignment just made.

6. *Repeat steps 2 to 5 until all assignments have been made.*
   a. Column Y now exhibits the greatest difference; therefore, we allocate 17 units to AY, since it has the smallest distribution cost in column Y. Since Y's requirements are completely satisfied, the other squares in that column are crossed out. Differences are recalculated. This entire step is shown in Table C-16.
   b. The recalculated differences now show five of the columns and rows with a difference of 2. The lowest cost square in any column or row is BX, which has a cost of 6. We assign 7 units to BX, which completely satisfies the

TABLE C-17 **Third VAM Assignment**

	~~8~~	0	↓ ~~2~~	~~6~~	2		
From \ To	V	W	X	Y	Z	Available 1000s ↓	
A	[8] X	[8]	[10] X	[6] (17)	[10]	19	2
B	[0] (11)	[8]	[6] (7)	[12] X	[14]	28	~~6~~~~2~~6
C	[12] X	[10]	[8] X	[14] X	[12]	25	2
Required 1000s →	11	13	7	17	24	72	

TABLE C-18 **Fourth and Fifth VAM Assignments**

	~~8~~	0	2	~~6~~	2		
From \ To	V	W	X	Y	Z	Available 1000s ↓	
A	[8] X	[8] (2)	[10] X	[6] (17)	[10] X	19	2←
B	[0] (11)	[8] (10)	[6] (7)	[12] X	[14] X	28	~~6~~~~2~~6←
C	[12] X	[10]	[8] X	[14] X	[12]	25	2
Required 1000s →	11	13	7	17	24	72	

TABLE C-19 **Final Assignments at CW and CZ Balance with Rim Restrictions and Yield VAM Initial Solution Which is Optimal**

From \ To	V	W	X	Y	Z	Available 1000s ↓
A	8 X	8 (2)	10 X	6 (17)	10 X	19
B	0 (11)	8 (10)	6 (7)	12 X	14 X	28
C	12 X	10 (1)	8 X	14 X	12 (24)	25
Required 1000s →	11	13	7	17	24	72

requirements at X. Table C-17 shows the allocation of 7 units at BX, the crossing out of the other squares in column X, and the recalculation of cost differences for the remaining rows and columns.

c. Row B now shows a cost difference of 6, and we allocate 10 units to the low cost square BW, as shown in Table C-18. This completes row B. Recalculated cost differences now show that all remaining cost differences in rows and columns are 2. The lowest cost square available is AW, so we allocate 2 units there to complete row A. This step is also shown as part of Table C-18.

d. The last two allocations at CW and CZ are made by inspection of the rim conditions (shown in Table C-19). An evaluation of the open squares in Table C-19 shows that this solution is optimal.

PROBLEMS

1. A company has factories at A, B, and C, which supply warehouses at D, E, F, and G. Monthly factory capacities are 70, 90, and 115, respectively. Monthly

warehouse requirements are 50, 60, 70, and 95, respectively. Unit shipping costs are as follows:

	To			
From	D	E	F	G
A	$17	$20	$13	$12
B	$15	$21	$26	$25
C	$15	$14	$15	$17

Determine the optimum distribution for this company to minimize shipping costs.

2. A company with factories at A, B, and C supplies warehouses at D, E, F, and G. Monthly factory capacities are 20, 30, and 45, respectively. Monthly warehouse requirements are 10, 15, 40, and 30, respectively. Unit shipping costs are as follows:

	To			
From	D	E	F	G
A	$6	$10	$ 6	$ 8
B	$7	$ 9	$ 6	$11
C	$8	$10	$14	$ 6

Determine the optimum distribution for this company to minimize shipping costs.

3. A company has factories at A, B, and C, which supply warehouses D, E, F, and G. Monthly factory capacities are 300, 400, and 500, respectively. Monthly warehouse requirements are 200, 240, 280, and 340, respectively. Unit shipping costs are as follows:

	To			
From	D	E	F	G
A	$7	$ 9	$ 9	$ 6
B	$6	$10	$12	$ 8
C	$9	$ 8	$10	$14

Determine the optimum distribution for this company to minimize shipping costs.

4. A company has factories at A, B, and C, which supply warehouses at D, E, F, and G. Monthly factory capacities are 160, 150, and 190, respectively. Monthly warehouse requirements are 80, 90, 110, and 160, respectively. Unit shipping costs are as follows:

	To			
From	D	E	F	G
A	$42	$48	$38	$37
B	$40	$49	$52	$51
C	$39	$38	$40	$43

Determine the optimum distribution for this company to minimize shipping costs.

5. A company has factories at A, B, C, and D, which supply warehouses at E, F, G, H, and I. Monthly factory capacities are 200, 225, 175, and 350, respectively. Monthly warehouse requirements are 130, 110, 140, 260, and 180, respectively. Unit shipping costs are as follows:

	To				
From	E	F	G	H	I
A	$14	$19	$32	$ 9	$21
B	$15	$10	$18	$ 7	$11
C	$26	$12	$13	$18	$16
D	$11	$22	$14	$14	$18

Determine the optimum distribution for this company to minimize shipping costs. (*Hint:* Use VAM for initial solution.)

6. A company has factories at A, B, and C, which supply warehouses at D, E, F, and G. Monthly factory capacities are 250, 300, and 200, respectively for regular production. If overtime production is utilized, the capacities can be increased to 320, 380, and 210, respectively. Incremental unit overtime costs are $5, $6, and $8 per unit, respectively. The current warehouse requirements are 170, 190, 230, and 180, respectively. Unit shipping costs between the factories and warehouses are:

	To			
From	D	E	F	G
A	$8	$ 9	$10	$11
B	$6	$12	$ 9	$ 7
C	$4	$13	$ 3	$12

Determine the optimum production-distribution for this company to minimize costs.

7. A company with factories at A, B, C, and D supplies warehouses at E, F, G, and H. Monthly factory capacities are 100, 80, 120, and 90, respectively, for regular

production. If overtime production is utilized, the capacities can be increased to 120, 110, 160, and 140, respectively. Incremental unit overtime costs are $5, $2, $3, and $4, respectively. Present incremental profits per unit, excluding shipping costs, are $14, $9, $16, and $27, respectively, for regular production. The current monthly warehouse requirements are 110, 70, 160, and 130, respectively. Unit shipping costs are as follows:

	To			
From	E	F	G	H
A	$3	$4	$5	$7
B	$2	$9	$6	$8
C	$4	$3	$8	$5
D	$6	$5	$4	$6

Determine the optimum production-distribution for this company. (*Hint:* This problem requires that you maximize profits.) What simple change in the procedures makes it possible to maximize rather than minimize?

REFERENCES

Bierman, H., C. P. Bonini, and W. H. Hausman. *Quantitative Analysis for Business Decisions* (5th ed.). Richard D. Irwin, Homewood, Ill., 1977.

Buffa, E. S., and J. S. Dyer. *Management Science/Operations Research: Model Formulation and Solution Methods*. John Wiley, New York, 1977.

Levin, R. L., and C. A. Kirkpatrick. *Quantitative Approaches to Management* (3rd ed.). Prentice-Hall, Englewood Cliffs, N.J., 1975.

Reinfeld, N. V., and W. R. Vogel. *Mathematical Programming* Prentice-Hall, Englewood Cliffs, N.J., 1958.

Thierauf, R. J., and R. C. Klekamp. *Decision Making through Operations Research* (2nd ed.). John Wiley, New York: 1975.

APPENDIX D Waiting Lines

THE ORIGINAL WORK IN WAITING LINE THEORY WAS DONE BY A. K. Erlang, a Danish telephone engineer. Erlang started his work in 1905 in an attempt to determine the effect of fluctuating demand (arrivals) on the utilization of automatic dial equipment. Since the end of World War II, Erlang's work has been extended and applied to a variety of situations.

In all instances, the general input-output module is descriptive, with the time between arrival of individual inputs at the service facility commonly being random. Also, the time for service or processing is a random variable. Table D-1 shows the waiting line model elements for a number of common situations.

STRUCTURE OF WAITING LINE MODELS

There are four basic waiting line structures that describe the general conditions at the service facility. The simplest structure shown in Figure D-1*a* is the basic service facility module. This is called the single-channel, single-phase case. There are many illustrations in practice of the simple module: the cashier at a restaurant, any "single window" operation in a post office or bank, or a one-chair barbershop. If the number of processing stations is increased but still draws on a single waiting line, we have the multiple-channel, single-phase case shown in Figure D-1*b*.

A simple assembly line or a cafeteria line has a number of service facilities in tandem and is called the single-channel, multiple-phase case shown in Figure D-1*c*. Finally, the multiple-channel, multiple-phase case might be illustrated by two or more parallel assembly lines, as shown in Figure D-1*d*. Combinations of any or all of the basic four structures could also exist in networks of queues.

The analytical methods for waiting lines divide into two main categories for any of the basic structures in Figure D-1, depending on the size of the source population of the inputs. When the source population is very large and, in theory at least, the length of the waiting line could grow without fixed limits, the applicable models are termed "infinite." On the other hand, when the arriving unit comes from a source having a fixed upper limit, the applicable models are termed "finite." For example, if we are dealing with the maintenance of a bank of 20 machines, a machine breakdown representing an arrival, the maximum waiting line is 20 machines waiting for service, and a finite model is needed. If, on the other hand, we operated an auto repair shop, the source population of breakdowns is very large, and an infinite model would be appropriate.

There are other variations in waiting line structures that are important in certain applications. For example, the "queue discipline" implied in Figure D-1 is first-come, first-served. Obviously there are many other possibilities involving priority systems. For example, in an outpatient clinic, emergencies and patients with appointments are usually taken ahead of walk-in patients. In job shop scheduling systems, there has been a great deal of experimentation with alternate priority systems, discussed in Chapter 9.

Finally, the nature of the distributions of arrival and service is an important structural characteristic of waiting line models. Some mathematical analysis is available for

TABLE D-1 **Waiting Line Model Elements for Some Commonly Known Situations**

	Unit Arriving	Service or Processing Facility	Service or Process Being Performed
Ships entering a port	Ships	Docks	Unloading and loading
Maintenance and repair of machines	Machine breaks down	Repair crew	Repair machine
Assembly line, not mechanically paced	Parts to be assembled	Individual assembly operations or entire line	Assembly
Doctor's office	Patients	Doctor, staff and facilities	Medical care
Purchase of groceries at a supermarket	Customers with loaded grocery carts	Checkout counter	Tabulation of bill, receipt of payment and bagging of groceries
Auto traffic at an intersection or bridge	Automobiles	Intersection or bridge with control points such as traffic lights or toll booths	Passage through intersection or bridge
Inventory of items in a warehouse	Order for withdrawal	Warehouse	Replenishment of inventory
Machine shop	Job order	Work center	Processing

distributions that follow the Poisson, or Erlang process (with some variations), or when arrivals or service are constant. If distributions are different from those mentioned, or are empirical, simulation (discussed in Appendix E) is likely to be the practical mode of analysis.

INFINITE WAITING LINE MODELS

It will be useful to restrict our thinking to situations involving the first-come, first-served queue discipline and the Poisson distribution of arrivals. We will deal initially with the single-channel, single-phase case (our basic service facility module). Later, we will also discuss the multiple-channel case.

Poisson Arrivals

The Poisson distribution function represents arrival rates in a large number of real-world situations. It is a discrete function, meaning that it deals with whole units of arrivals, and therefore, fractions of people, products, or machines do not have meaning, nor will negative values. The Poisson distribution function is given by

$$f(x) = \frac{\lambda^x e^{-\lambda}}{x!} \quad (1)$$

where λ is the average arrival rate and x is any specific number of arrivals. For example if $\lambda = 4$ per hour, then the probability of $x = 6$ in one hour is

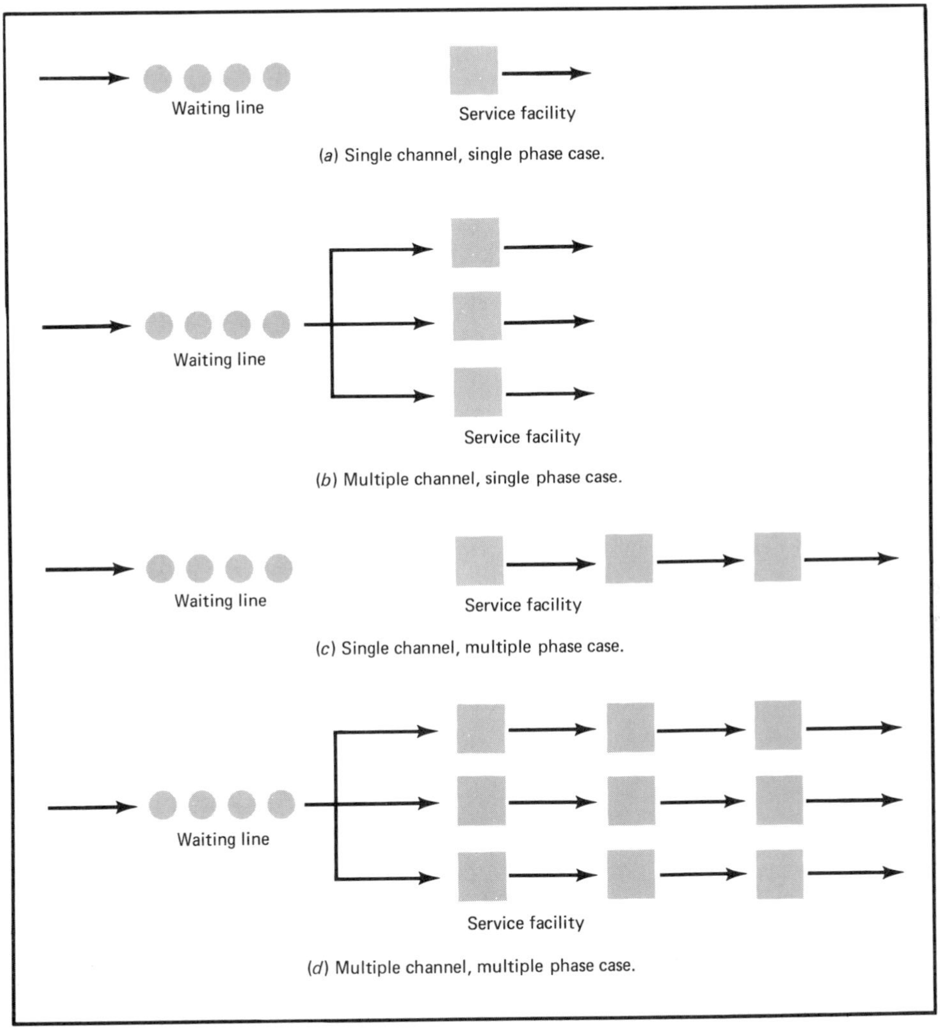

FIGURE D-1
Four basic structures of waiting line situations. *(a)* Single-channel, single phase; *(b)* multiple-channel, single-phase; *(c)* single-channel, multiple-phase; *(d)* multiple-channel, multiple-phase.

$$f(6) = \frac{4^6 e^{-4}}{6!} = \frac{4096 \times 0.0183}{720} = 0.104$$

The Poisson distribution for several values of λ is shown in Figure D-2. The Poisson distribution is typically skewed to the right. The distribution is simple in that the standard deviation is expressed solely in terms of the mean, $\sigma_\lambda = \sqrt{\lambda}$.

There is considerable evidence that the Poisson distribution represents arrival rates in many applications. *However, a negative exponential distribution for interarrival times is a Poisson distribution, when the data are transformed to arrival rates.* Also, many empirical studies have validated the Poisson arrival rate in general industrial operations, traffic flow, and various service operations.

Although we cannot say that all arrival rate distributions are adequately described by the Poisson, we can say that it is usually worth checking to see if it is true, because then a fairly simple analysis may be possible. It is not illogical that arrivals may follow the Poisson distribution when many factors affect arrival time, since the Poisson distribution corresponds to completely random arrivals. This means that each arrival is independent of other arrivals, as well as of any condition of the waiting line. The practical question is whether or not the Poisson distribution is a reasonable approximation to reality. Statistical tests of goodness of fit may be used.

Poisson Arrivals—Service Time Distribution Not Specified

Since Poisson arrivals are common, a useful model is one that depends on Poisson arrivals but accepts any service time distribution. We assume also that the mean service rate is greater than the mean arrival rate; otherwise, the system would be unstable and the waiting line would become infinitely large. The queue discipline is first-come, first-served, and arrivals wait for service; that is, they neither fail to join the line nor leave it because it is too long. Under these conditions the expected length of the waiting line is

$$L_q = \frac{(\lambda\sigma)^2 + (\lambda/\mu)^2}{2(1 - \lambda/\mu)} \tag{2}$$

where L_q is the expected length of the waiting line
λ is the mean arrival rate from a Poisson distribution
μ is the mean service rate
σ is the standard deviation of the distribution of service times

We define the ratio $r = \lambda/\mu$ as the flow intensity, and we will use this ratio as an index for a useful table of values L_q for various numbers of service channels (Table H-3, Appendix H).

The average utilization of the service facility is defined as $\rho = \lambda/M\mu$, where M is the number of service channels ($M = 1$ in the single-channel case that we are discussing.) Since ρ represents the proportion of time that the service facility is in use, it also represents the expected number of individuals or units being served. Then (1

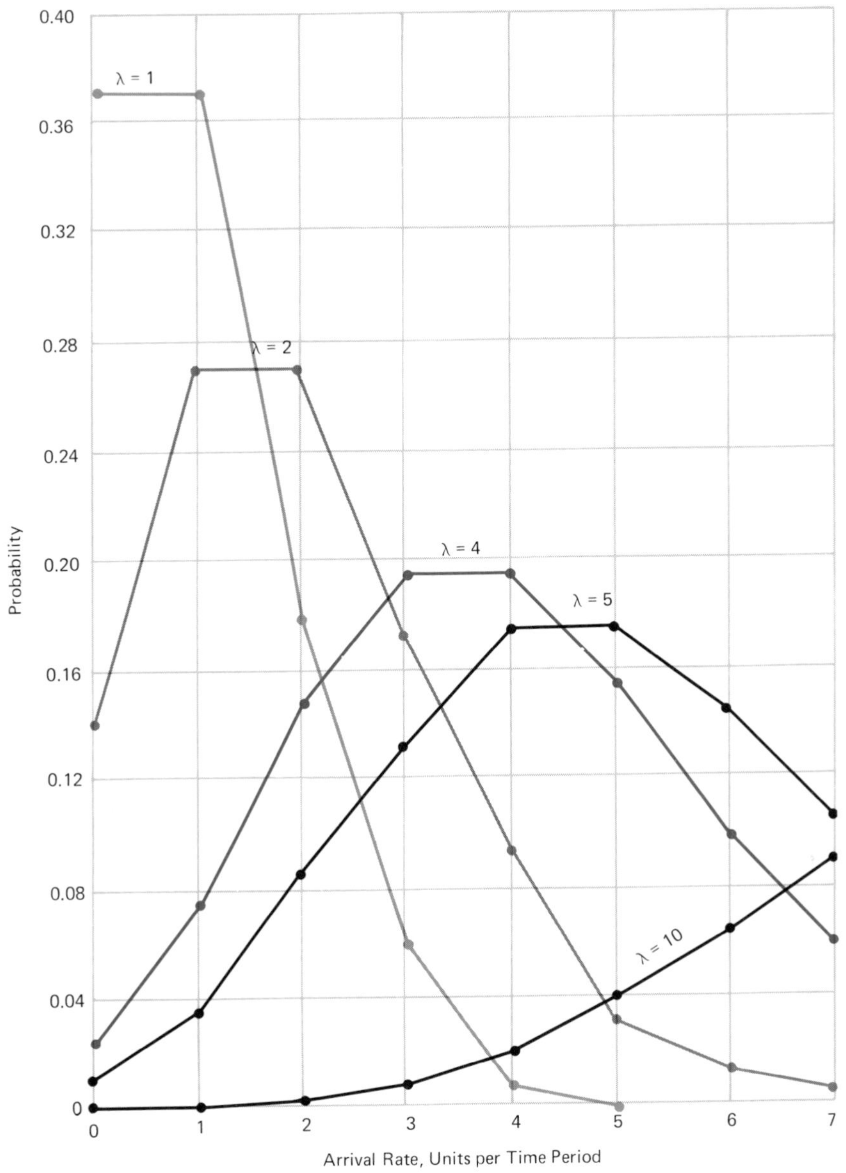

FIGURE D-2
Poisson distributions for several mean arrival rates.

$-\ \rho)$ is the proportion of service facility idle time, when no one is being served. Since λ/μ is the number being served, the total average number in the system, L, is the average number in line plus the average number being served, or

$$L = L_q = + \lambda/\mu = L_q = + r \tag{3}$$

Similar logic leads to the expected waiting time in line W_q, and the time in system including service, W. The reciprocal of the mean arrival rate is the mean time between arrivals $(1/\lambda)$. The product of the mean time between arrivals and the waiting line length is the waiting time. Also, the product of the mean time between arrivals and L (number in line plus the average number being served) is the average time in the system. Therefore, we also have the following relationships:

$$W_q = \frac{L_q}{\lambda} \tag{4}$$

$$W = \frac{L}{\lambda} \tag{5}$$

Equations 3, 4, and 5 are useful relationships. The general procedure would be to compute L_q from Equation 2 and compute the values of L, W_q, and W as needed, given the value of L_q.

An Example. A pharmaceutical manufacturer produces most of its products for inventory in order to give off-the-shelf service to distributors. The factory warehouse maintains one truck dock for supplying small orders on items for which the distributor has run out of stock and needs immediate replenishment. The distributors send their trucks to the factory's finished goods warehouse with authorized orders, which are assembled while the truck waits. Data indicate that there are $\lambda = 7$ trucks per hour arriving on the average and that the average order is filled in six minutes. Since the warehouse is large, the estimate of the standard deviation of service time is $s = 10$ minutes.

Complaints have been received from the distributors that it takes too long to get an order filled, including waiting time plus service time. Calculations yield the following:

$$\lambda = 7/\text{hour}$$

$$\mu = 60/6 = 10/\text{hour},\ s = 1/6 \text{ hours}$$

$$L_q = \frac{(7/6)^2 + (7/10)^2}{2(1 - 7/10)} = 3.085 \text{ trucks waiting}$$

$$L = 3.085 + 0.7 = 3.785 \text{ trucks in the system}$$

$$W_q = 3.085/7 = 0.441 \text{ hours, or } 26.44 \text{ minutes}$$

$$W = 3.785/7 = 0.541 \text{ hours, or } 32.44 \text{ minutes}$$

The calculations provide a verification of the logic in that the average truck waits 26.44 minutes in line plus an average of 6 minutes for service, or 32.44 minutes in the system. Thus, another logical relationship is

$$W = W_q + \frac{1}{\mu}$$

The manufacturer is aware that the expansion of the truck dock would probably solve the problem, but this would require a large capital expenditure, as well as disruption of operations during construction. Instead, the rather large standard deviation of service time is noted.

A first solution is to identify the items subject to emergency orders and to relocate them together in the same general area. Although the manufacturer has doubts about this solution, the effect is estimated to reduce the standard deviation of service time from 10 minutes to 5 minutes. Assuming that the mean service time is not affected, we have an indication of the sensitivity of the system to changes in the variability of the service time. The new values for the queuing statistics are $L_q = 1.384$, $L = 2.084$, $W_q = 11.86$ minutes, and $W = 17.86$ minutes. Waiting time is cut to only 45 percent of the former value, and for the moment at least, the manufacturer seems to have found a way to avoid the capital expenditure.

Before attempting to reorganize the storage system, the manufacturer estimates the effect of having the distributors phone in their orders, so that they can be assembled and waiting in most instances. It is estimated that the service time would be only 3 minutes with a standard deviation of $s = 2$ minutes. The queuing statistics that result are $L_q = 0.419$, $L = 0.769$, $W_q = 3.59$ minutes, and $W = 6.59$ minutes. The manufacturer feels that this is a practical solution that still requires no capital expenditure and reduces the trucker waiting plus service time even more.

Service Time Distributions

Although there is considerable evidence that arrival processes tend to follow the Poisson distribution, service time distributions seem to be much more varied in their nature. It is for this reason that the previous model involving Poisson arrivals and an unspecified service time distribution is so valuable. With Equation 2, one can compute the queuing statistics, knowing only the mean service rate and the standard deviation of service time.

The negative exponential distribution has been one of the prominent models for service time, but there is less evidence that the assumption is valid.

Evidence indicates that in some cases the negative exponential distribution fits. For example, the service time at a tool crib was nearly exponentially distributed. The distribution of local telephone calls not made from a pay station has also been shown to be described by the negative exponential distribution.

Model for Poisson Input and Negative Exponential Service Times

The negative exponential distribution is completely described by its mean value, the standard deviation being equal to the mean. Therefore, we can describe this model as a special case of Equation 2. The mean of the negative exponential distribution is the reciprocal of the mean service rate, that is, $1/\mu$. Therefore, $\sigma = 1/\mu$, and if we substitute this value for the standard deviation in Equation 2 and simplify, we obtain

$$L_q = \frac{\lambda^2}{\mu(\mu - \lambda)} \tag{6}$$

also, the probability of n units in the system is

$$P_n = \left(\frac{\lambda}{\mu}\right)^n \left(1 - \frac{\lambda}{\mu}\right) \tag{7}$$

The other relationships between L_q, L, W_q, and W, expressed by Equations 3, 4, and 5, hold for the negative exponential service time model as well as for the distribution-free case. For the sake of simplicity, many individuals prefer to use Equation 2, inserting the appropriate value of σ to reflect the special case.

We can now check to see the effect of exponential service times on the queuing statistics for the pharmaceutical manufacturer's problem. If we assume that the service time in that situation was represented by a negative exponential distribution, then $\sigma = 1/\mu = 1/10$, and the value of L_q from Equation 2 is 1.633. The other queue statistics are $L = 2.33$, $W_q = 14.0$ minutes, and $W = 20.0$ minutes.

Constant Service Times

Although constant service times are not common in real situations, they may be reasonable in cases where a machine processes arriving items by a fixed time cycle. Also, constant service times represent a lower bound on the value of σ in Equation 2. As such, constant service time is a special case of Equation 2 that can be reduced to a simpler form by inserting $\sigma = 0$. The resulting equation for constant service times is

$$L_q = \frac{\lambda^2}{2\mu(\mu - \lambda)} \tag{8}$$

Again, the other relationships between L_q, L, W_q, and W expressed by Equations 3, 4, and 5 hold for the constant service time distribution case. Again, for comparison, we examine the effect on queue statistics if the pharmaceutical manufacturer could have made service time constant at six minutes. Substituting in Equation 2, we have $L_q = 0.817$, $L = 1.517$, $W_q = 7.0$ minutes, and $W = 13.0$ minutes.

Then Equation 2 is a fairly general model with service time distributions described by the negative exponential, or constant service times, as special cases. Figure D-3

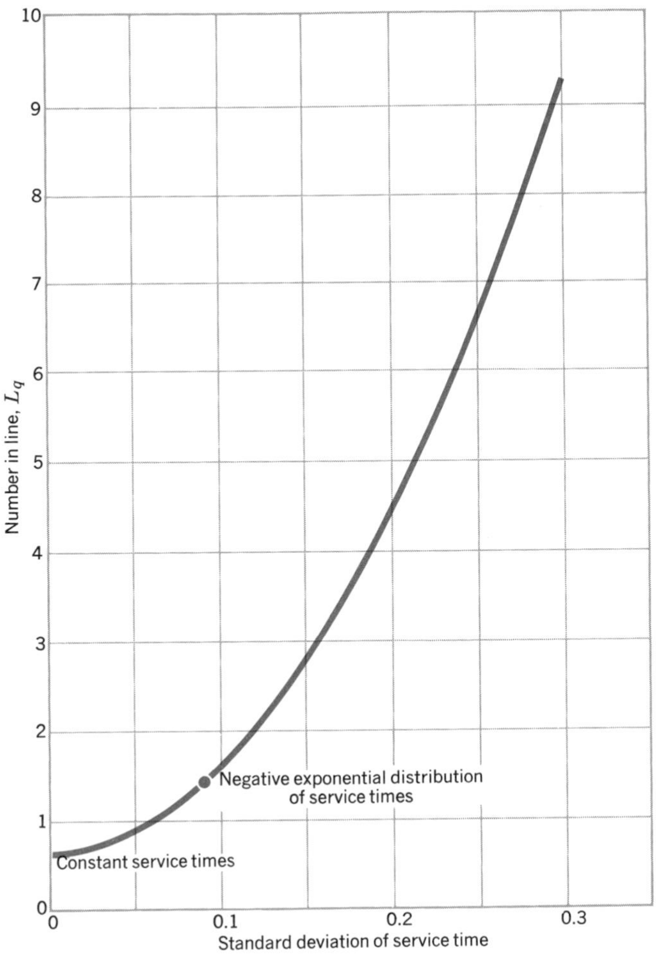

FIGURE D-3
Relation between waiting line length L_q and standard deviation of service time, for a single-channel system.

shows a graph of L_q for various values of the standard deviation, including the values for the negative exponential distribution and constant service times. Values of the standard deviation greater than that for the negative exponential distribution occur and are termed "hyperexponential" distributions. The extreme values shown are not representative of values found in real situations; however, the tail of the curve is shown to indicate how rapidly L_q increases with increased variability in the service time distribution.

Relationship of Queue Length to Utilization

We defined $\lambda/M\mu = \rho$ as the service facility utilization. For the single channel case, if $\lambda = \mu$, then $\rho = 1$, and theoretically the service facility is used 100 percent of the time. However, let us see what happens to the queue length as ρ varies from zero to one. Figure D-4 summarizes the result for Poisson input and negative exponential service time. As ρ approaches unity, the number waiting in line increases rapidly and approaches infinity. You can see that this must be true by examining Equations 2, 6, and 8 for L_q. In all cases, the denominator goes to zero as ρ approaches unity and the value of L_q becomes infinitely large.

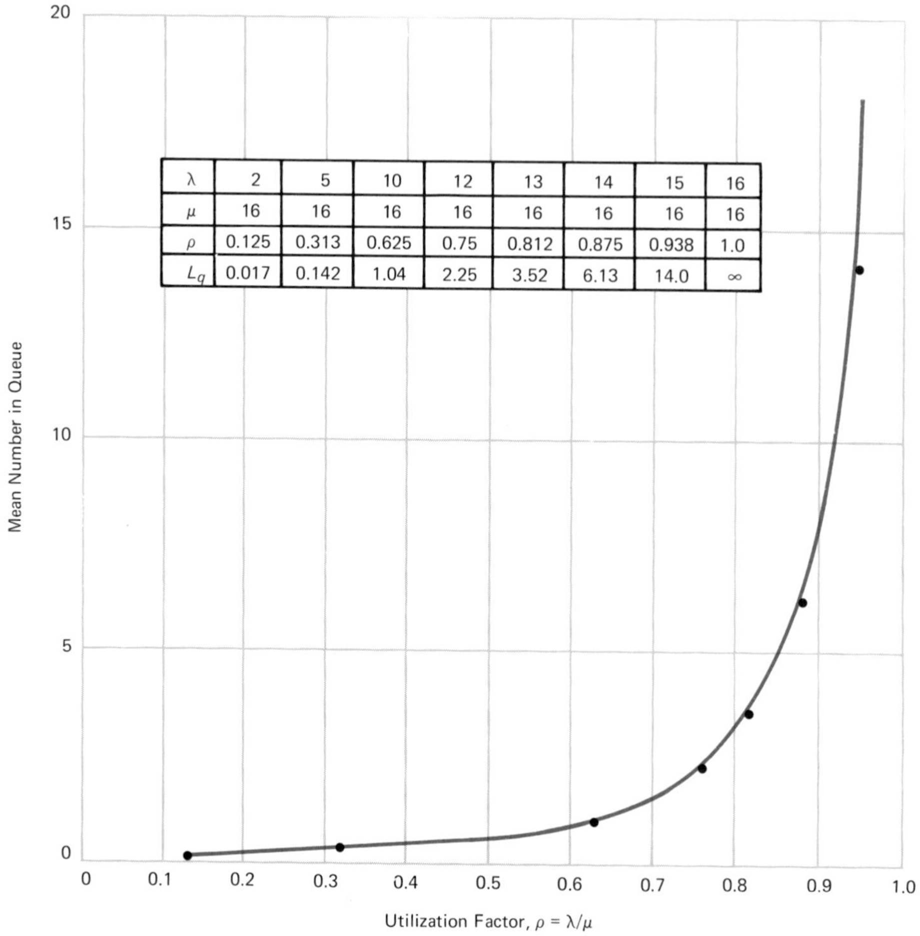

λ	2	5	10	12	13	14	15	16
μ	16	16	16	16	16	16	16	16
ρ	0.125	0.313	0.625	0.75	0.812	0.875	0.938	1.0
L_q	0.017	0.142	1.04	2.25	3.52	6.13	14.0	∞

FIGURE D-4
Relationship of queue length to the utilization factor ρ, for a single-channel system.

We see now that one of the requirements of any practical system is that $\mu \geq \lambda$; otherwise we cannot have a stable system. If units are arriving faster on the average than they can be processed, the waiting line and waiting time will increase continuously, and no steady state can be achieved. This simple fact also indicates that there is a value placed on idle time in the service facility. We must trade off service facility idle time with the value of good service.

Multiple-Channel, Single-Phase Case

In the multiple-channel case, we assume the conditions of Poisson arrivals, exponential service times, and first-come, first-served queue discipline. The effective service rate $M\mu$ must be greater than the arrival rate λ, where M is the number of channels. The facility utilization factor is $\rho = \lambda/M\mu$.

As with the single-channel case, it is first necessary to calculate L_q in order to solve problems. Since the formula for L_q is quite complex for the multiple-channel case, we have computed it for various values of M (the number of service channels), and $r = \lambda/\mu$ in Table H-3, Appendix H.

Given the value of L_q, the other queue statistics can be computed from Equations 3, 4, and 5. Also, the probability that all servers or channels are busy, that is, the probability of a delay, P(Busy), may be computed by inserting the table value of L_q in

$$P(\text{Busy}) = \frac{L_q}{r/(M - r)} \tag{9}$$

As an example, assume that the pharmaceutical manufacturer decides to expand facilities and add a second truck dock. What is the effect on average truck waiting time? Recall the basic data: $\lambda = 7$ per hour, $\mu = 10$ per hour, and $M = 2$. From Table H-3 for $M = 2$ and $r = \lambda/\mu = 7/10 = 0.70$, we find that $L_q = 0.098$ trucks in line. Then, $W_q = L_q/\lambda = 0.098/7 = 0.014$ hours, or 0.84 minutes. Compare these results with the single-channel solution for exponential service time of $W_q =$ 14.0 minutes. Obviously, the second dock would eliminate the truck waiting time problem. Note that overall utilization of the facilities declines from $\rho = \lambda/M\mu = 7/(1 \times 10) = 0.70$ to $\rho = 7/(2 \times 10) = 0.35$.

The Effect of Pooling Facilities

To compare the effects of increasing or decreasing the number of channels, we can simply refer to Table H-3. If for example, $r = 0.9$ for the single-channel case, then L_q is approximately 8 from Table H-3. Adding a second channel reduces the average line length to $L_q = 0.23$. Adding a third channel reduces it to $L_q = 0.03$. The effects are surprisingly large; that is, we can obtain disproportionate gains in line length and waiting time by increasing the number of channels. We can see intuitively that this might be true from Figure D-4, since queue length (and waiting time) begins to increase very rapidly at about $\rho = 0.8.$, for the single channel case. A rather small increase in capacity of the system (decrease in ρ) at these high loads can produce a disproportionate decrease in line length and waiting time.

The effects of pooling facilities may be illustrated by comparing the doubling of capacity within the same service facility with the doubling of capacity through parallel facilities. A common function in machine shops is the tool crib, where mechanics check out special tools to be used in conjunction with the processing of a job order. Suppose that we have an existing tool crib, and we are contemplating doubling its capacity in order to give better service and reduce the time spent by mechanics in obtaining needed tools. The issue is whether the system capacity should be doubled within the present facility by adding a second server, or whether a second parallel facility should be established in another location.

Suppose that the average Poisson arrival rate of mechanics at the tool crib is $\lambda = 0.9$ per minute and that the average exponential service time is 60 seconds, or $\mu = 60/60 = 1.0$ per minute. Then, $r = 0.9/1.0 = 0.9$. This is the base case, and the waiting line statistics are summarized in Table D-2. For the base case, mechanics wait in line an average of 9 minutes and spend a total of 10 minutes in the system, including the 1 minute for service. It is obvious that service needs to be improved.

Now, examine the times for the two proposals for doubling capacity summarized in Table D-2. For the single large tool crib with two servers, all of the waiting line statistics are drastically reduced. On the other hand, doubling capacity through two independent tool cribs also results in large reductions in both waiting time and time in system, but the centralized facility is even better. One large facility provides better service than an equivalent number of smaller facilities. Of course, there are other advantages and disadvantages that the manager must weigh and trade off in making a final decision concerning the centralization or decentralization of facilities.

If we visualize the two decentralized facilities functioning side by side, we can see intuitively why the system time increases. If facility 1 were busy and had mechanics waiting, while at the same time facility 2 happened to be idle, someone from the facility 1 waiting line could be served immediately by facility 2, thereby reducing waiting time for that individual. The two facilities are drawing on one waiting line.

TABLE D-2

Effects of Doubling Capacity Within the Same Tool Crib Versus Parallel Tool Cribs

	Base Case ($M = 1, r = 0.9$)	Capacity Doubled: Within Same Tool Crib, Two Servers ($M = 2, r = 0.9$)	Capacity Doubled: By Adding a Second Tool Crib ($M = 1, r = 0.45$)
L_q, from Table H-3	8.1	0.2285	0.3681
$W_q = L_q/\lambda$	9.0 min	0.2538	0.8180
$L = L_q + r$	9.0	1.1285	0.8180
$W = L/\lambda$	10.0 min	1.2538	1.8180

TABLE D-3

Effects of Doubling Capacity by Doubling Service Rate Versus Adding a Second Service Channel

		Capacity Doubled	
	Base Line ($M = 1, r = 0.9$)	Service Rate Doubled, One Server ($M = 1, r = 0.45$)	Two Servers ($M = 2, r = 0.9$)
L_q from Table H-3	8.1	0.3681	0.2285
$W_q = L_q/\lambda$	9.0 min	0.4090	0.2538 min
$L = L_q + r$	9.0	0.8181	1.1285
$W = L/\lambda$	10.0 min	0.9090 min	1.2538 min

When the two facilities are physically decentralized, the facilities must draw on two independent waiting lines, and the idle capacity of one cannot be used by the waiting mechanics of the other.

Now, let us compare the result obtained in Table D-2 with the doubling of capacity by doubling the service rate. Suppose that an automated system for obtaining tools from storage were available that increased the average service rate from $\mu = 1$ per minute to $\mu = 2$ per minute. The system capacity is then doubled using a single server. Then, $r = 0.9/2 = 0.45$; Table D-3 summarizes the comparative waiting line statistics that result, compared to doubling capacity by adding a second server. Note that waiting time is greater for the single-server system; however, the waiting time is more than offset by faster service rendered by the automated system. Alternatives of increasing capacity within the same facility by adding personnel versus system improvements that increase service rate need to be evaluated carefully.

FINITE WAITING LINE MODELS

Many practical waiting line problems that occur in productive systems have the characteristics of finite waiting line models. This is true whenever the population of machines, workers, or items that may arrive for service is limited to a relatively small, finite number. The result is that we must express arrivals as a unit of the population rather than as an average rate. In the infinite waiting line case, the average length of the waiting line is effectively independent of the number in the arriving population. However, in the finite case, the number of the queue may represent a significant proportion of the arriving population, thereby affecting the probabilities associated with arrivals.

The computations for the resulting mathematical formulations are somewhat more difficult than for the infinite queue case. Fortunately, however, *Finite Queuing Tables* [Peck and Hazelwood, 1958] are available that make problem solving very simple.

Although there is no definite number that we can point to as a dividing line between finite and infinite applications, the finite queuing tables have data for populations from 4 up to 250, and these may be taken as a general guide. We have reproduced these tables for populations of 5, 10, 20, and 30 in Appendix H, Table H-4, to illustrate their use in the solution of finite queuing problems. The tables are based on a finite model for exponential times between arrivals and service times and on a first-come, first-served queue discipline.

Use of the Finite Queuing Tables

The tables are indexed first by N, the size of the population. For each population size, data are classified by X, the service factor (comparable to the utilization factor in infinite queues), and by M, the number of parallel channels. For a given N, X, and M, three factors are listed in the tables: D (the probability of a delay; i.e., if a unit calls for service, the probability that it will have to wait); F (an efficiency factor, used to calculate other important data); and the mean number of the waiting line, L_q. To summarize, we define the factors just expressed, plus those that may be calculated, as follows:

N = population (number of machines, customers, etc.)
μ = mean service rate
λ = mean arrival rate *per population unit*
X = service factor $= \dfrac{\lambda}{\lambda + \mu}$
M = number of service channels
D = probability of delay (probability that if a unit calls for service, it will have to wait)
F = efficiency factor
L_q = mean number in waiting line $= N(1 - F)$
W_q = mean waiting time $= \dfrac{1}{\mu X}\left(\dfrac{1 - F}{F}\right)$
H = mean number of units being serviced $= FNX = L - L_q$
J = mean number of units running $= NF(1 - X)$
$M - H$ = average number of servers idle

The procedure for a given case is as follows:

1. Determine the mean service rate μ and the mean arrival rate λ, based on data or measurements of the system being analyzed.

2. Compute the service factor $X = \lambda/(\lambda + \mu)$.

3. Locate the section of the tables listing data for the population size N.

4. Locate the service factor calculated in step 2 for the given population.

5. Read the values of D, F, and L_q for the number of channels M, interpolating between values of X when necessary.

6. Compute values for W_q, H, and J as required by the nature of the problem.

Example. A hospital ward has 30 beds in one section, and the problem centers on the appropriate level of nursing care. The hospital management believes that patients should have immediate response to a call at least 80 percent of the time because of possible emergencies. The mean time between calls is 95 minutes *per patient,* for the 30 patients. The service time is approximated by a negative exponential distribution, and mean service time is five minutes.

The hospital manager wishes to staff the ward to give service so that 80 percent of the time there will be no delay. Nurses are paid $5 per hour, and one concern involves the cost of idle time at this level of service. Finally, as a base of comparison, the manager wishes to know how much more patients will have to pay for the 80 percent criterion compared with a 50 percent service level for immediate response, which is the current policy.

The solutions to the problems posed by the hospital manager are developed through finite queuing models. A finite model is required because the maximum possible queue is 30 patients waiting for nursing care.

The mean service time is 5 minutes ($\mu = 0.2$ per minute, or 12 per hour), the mean time between calls is 95 minutes per patient ($\lambda = 0.0105$ per minute, or 0.632 per hour), and therefore the service factor is $X = \lambda/(\lambda + \mu) = 0.632/12.632 = 0.05$. Scanning the *finite queuing tables* (Table H-4) under Population $N = 30$, and $X = 0.05$, we seek data for the probability of a delay of $D = 0.20$, since we wish to establish service so that there will be no delay 80 percent of the time. The closest we can come to providing this level of service is with $M = 3$ nurses, and with corresponding data from Table H-4 of $D = 0.208$, $F = 0.994$, and $L_q = 0.18$.

The cost of this level of service is the cost of employing three nurses or $5 \times 3 =$ \$15 per hour, or \$360 per day, assuming day and night care. The average number of calls waiting to be serviced will be $L_q = 0.18$, and the mean waiting time will be

$$W_q = \frac{1}{\mu X}\left(\frac{1-F}{F}\right) = \frac{1}{0.2 \times 0.05}\left(\frac{1-0.994}{0.994}\right) = 0.6 \text{ minutes}$$

The waiting time due to queuing effect is negligible, as intended.

The average number of patients being served will be $H = FNX = 0.994 \times 30 \times 0.05 = 1.49$, and the average number of nurses idle will be $3 - 1.49 = 1.51$. The equivalent value of this idleness is $1.51 \times 5 \times 24 =$ \$181.20 per day.

Finally, the number of nurses needed to provide immediate service 50 percent of the time is $M = 2$ from Table H-4 ($D = 0.571$, $F = 0.963$, and $L_q = 1.11$). The average waiting time under this policy is $W_q = 3.84$ minutes. The average cost to patients of having the one additional nurse to provide the higher level of service is $5 per hour or $120 per day. Divided among 30 patients, the cost is $4 per patient per day.

COSTS AND CAPACITY IN WAITING LINE MODELS

Although many decisions concerning service systems may turn on physical factors of line length, waiting time, and the service facility utilization, system designs will often depend on comparative costs for alternatives. The costs involved are commonly the costs or providing the service versus the waiting time costs. In some instances the waiting time costs are objective, as when the enterprise is employing both the servers and those waiting. The company tool crib facility discussed above is such a case. The company absorbed all the travel time and waiting time costs, as well as the cost of providing the service. In such an instance, a direct cost-minimizing approach can be taken, balancing the waiting costs, or the time in system costs, against the costs of providing the service.

When the arriving units are customers, clients, or patients, the cost of making them wait is less obvious. If they are customers, excessive waiting may cause irritation and loss of goodwill and eventually lost sales. Placing a value on goodwill, however, is not a straightforward exercise. In public service and other monopoly situations, the valuation of waiting cost may be even more tenuous because the individual cannot make alternate choices. In all of these situations where objective costs cannot be balanced, it may be necessary to set a standard for waiting time, adjusting capacity to keep average waiting time below a stated figure.

IMPLICATIONS FOR THE MANAGER

Managers can gain insight into how their systems function through an understanding of waiting line models. The interplay between random arrivals and variable service time produces the condition where clients and customers must wait for service, and yet the system must have idle capacity if it is to provide good service. One of the really important managerial concepts that results from an understanding of waiting line models is that idle capacity has value in these kinds of systems. In designing service systems, managers can balance the costs of providing capacity against the need for good service. The magnitude of the idle capacity needed to provide emergency medical, fire, and police service is quite large, because the acceptable standards for response cannot tolerate a long waiting time.

A manager who understands the nature of waiting line systems is in a position to design policies, procedures, and physical layouts that retain the maximum control over the system. First, is there any way of "managing" the arrival process? If demand on the system can be smoothed, made more predictable, or even scheduled, there will be benefits that result in terms of lower capacity requirements for the same service level and fewer crisis to manage. Appointment systems in doctors' offices, and the hospital admissions system with a call list are examples. Half of the congestion or queuing effect is in the service time variation, but the other half is due to the variable arrival process (compare Equations 6 and 8, or examine Figure D-3). In other words, managers may make important improvements *not* assuming that the arrival process is out of their control.

In some instances, we may be able to get the customers, patients, or clients to do something productive while they are waiting. If we can transfer some of the service activity to the one being serviced, costs may be reduced, and this approach may be one of the few strategies available for improving productivity in service activities. The acceptance of the idea of customers doing part of the work required to serve them has become quite widespread in such operations as self-service markets, gas stations, and cafeterias.

Managers should be aware of the sensitivity of service system performance to variation in the number of parallel channels. Line length, waiting time, and time in the system drop dramatically when parallel channels are added. Also, managers need to understand the important service advantages that result from pooling resources. It is not true that facilities should always be centralized, but a manager should be in a position to evaluate or trade off the service improvement that results from centralization in relation to the other advantages and disadvantages.

The physical flow aspects of queuing systems are not normally complex, but nonetheless need careful thought. First, good design segregates the waiting line from the service and exit flow paths, in particular so that the waiting line and exit path are not in conflict. We have all observed that this flow principle in often violated in practice. Second, in multiple window operations such as in banks and post offices, the formation of a single waiting line takes advantage of the pooling concept, converting the system to a single large facility capable of better service. Today, it is common for banks and post offices to lay out the flow pattern with a single waiting line, but it was only a few years ago that managers of these systems did not take advantage of the this concept.

REVIEW QUESTIONS AND PROBLEMS

1. Classify the following in terms of the four basic waiting line structures:
   a. Assembly line
   b. Large bank—six tellers (one waiting line for each)
   c. Cashier at a restaurant
   d. One-chair barbershop
   e. Cafeteria line
   f. Jobbing machine shop
   g. General hospital
   h. Post office—four windows drawing from one waiting line

2. Define the following terms:
   a. Arrival process
   b. Queue discipline
   c. Infinite waiting line model
   d. Finite waiting line model
   e. Single-phase model

f. Multiple-phase model
g. Single-channel model
h. Multiple-channel model

3. Given a Poisson distribution of arrivals with a mean of $\lambda = 5$ per hour, what is the probability of an arrival of $x = 4$ in one hour? What is the probability of the occurrence of 15 minutes between arrivals?

4. The barber of a one-chair shop finds that sometimes customers are waiting, but sometimes he has nothing to do and can therefore read sports magazines. He prefers to keep a rather steady pace when he is working, hoping to get blocks of time for reading and keeping up on sports. In the hope of improving his situation, he kept records for several weeks and found that an average of one customer per hour comes in for haircuts (Poisson distribution). It takes him an average of 20 minutes per haircut and the standard deviation of his sample of service times is 5 minutes.
   a. What is the average number of customers waiting for service?
   b. What is the average customer waiting time?
   c. How much of the time does the barber have to read sports magazines?
   d. The other barber in town has fallen ill, and our barber's business increases to an average number of two customers per hour. What happens to the average number of customers waiting, the average waiting time, and the time available for reading sports magazines?
   e. After practicing at home on his children, the barber finds that he can reduce both the haircutting time to 10 minutes and the variance to virtually zero by using a bowl and only electric clippers (no scissors). Will this solve the waiting time problem in the interim while the other barber is ill? How would it affect the barber's available reading time?
   f. With the other barber ill, our barber finds that his customers are screaming for better service and that his reading time is available only in small increments. He feels under great pressure but is afraid to try the technological improvements he has developed. How can he solve his problem?

5. A taxicab company has four cabs that operate out of a given taxi stand. Customer arrival rates and service rates are described by the Poisson distribution. The average arrival rate is 10 per hour, and the average service time is 20 minutes. The service time follows a negative exponential distribution.
   a. Calculate the utilization factor.
   b. From Table H-3, determine the mean number of customers waiting.
   c. Determine the mean number of customers in the system.
   d. Calculate the mean waiting time.
   e. Calculate the mean time in the system.
   f. What would be the utilization factor if the number of taxicabs were increased from four to five?

g. What would be the effect of the change in part f on the mean number in the waiting line?
h. What would be the effect of reducing the number of taxicabs from four to three?

6. Why is queuing felt to be an unacceptable method for dealing with overload conditions in medical systems?

7. A stenographer has five persons for whom he performs stenographic services. Arrival rates are adequately represented by the Poisson distribution, and service times follow the negative exponential distribution. The arrival rate is five jobs per hour. The average service time is 10 minutes.
   a. Calculate the mean number in the waiting line.
   b. Calculate the mean waiting time.
   c. Calculate the mean number of units being served.
   d. What is the probability that an individual bringing work to the stenographer will have to wait?

8. In the manufacture of photographic film, there is a specialized process of perforating the edges of the 35-mm film used in movie and still cameras. A bank of 20 such machines is required to meet production requirements. The severe service requirements cause breakdowns that need to be repaired quickly because of high downtime costs. Because of breakdown rates and downtime costs, management is considering the installation of a preventive maintenance program that they hope will improve the situation.

   The present breakdown rate is three per hour per machine, or a time between breakdowns of 20 minutes. The average time for service is only three minutes. The breakdown rate follows a Poisson distribution, and the service times follow a negative exponential distribution. The crew simply repairs the machines in the sequence of breakdown. Machine downtime is estimated to cost $9 per hour, and present repair parts cost an average of $1 per breakdown. Maintenance repairmen are paid $6 per hour. The breakdown rates, service times, and repair parts costs are expected to change with different levels of preventive maintenance, as indicated in Table D-4.

   What repair crew size and level of preventive maintenance should be adopted to minimize costs?

SITUATIONS

9. A large manufacturing concern with a 100-acre plant had a well-established medical facility that was located at the plant offices at the eastern edge of the property. The plant had grown over the years from east to west, and travel time

TABLE D-4

Expected Changes in Breakdown Rates, Service Time, and Cost of Repair Parts for Three Levels of Preventive Maintenance

Level of Preventive Maintenance	Breakdown Rate (%)	Service Time (%)	Cost of Repair Parts (%)
L_1	−30	+20	+ 50
L_2	−40	+35	+ 80
L_3	−50	+75	+120

to the medical facility became so great that the management was considering dividing the facility. The second unit was to be established near the center of the west end of the plant. A study had been made of weighted travel times for the present single-facility system and for the proposed two-facility system. The result indicated that average travel time for the present large medical facility was 20 minutes. There averaged 1080 visits per week, or 360 worker-hours for travel time. The two-facility plan would reduce the average travel time to 10 minutes, or 180 worker-hours per week.

The medical facility was open 40 hours per week, and the number of visits averaged 27 per hour (Poisson distribution). The average service time was 20 minutes (negative exponential distribution). Twelve physicians handled the load, and the proposal was to divide the medical staff, assuming that the load would divide about equally. The physicians were each paid $1100 per week. The average hourly wage of employees coming to the medical facility was $7 per hour. There was substantial opposition to the proposal to decentralize the medical facility. A considerable investment was required in expensive medical equipment that would have to be duplicated in the west-end medical center. In addition, the medical staff did not wish to be divided for professional reasons, since cross-consultation between doctors would tend to be limited within each center.

a. What is the most economical deployment plan, exclusive of the investment required? Is 12 physicians the most appropriate number?
b. Are there other proposals that you feel should be considered?
c. Based on objective and subjective factors, what decisions should be made?

10. The manager of a large bank has the problem of providing teller service for customer demand, which varies somewhat during the business day from 10 A.M. to 4 P.M. She has a total capacity of six windows and can assign unneeded tellers to other useful work about 60 percent of the time. She also wishes to give excellent service, which she defines in terms of customer waiting time as $W_q \leq 2$ minutes. There is controversy about this service standard, however; some feel that the average waiting time should be no longer than one minute, while others feel a four-minute standard would be adequate.

In order to give the best service for any situation, the manager has arranged the layout so that customers form a single waiting line from which the customer at the head of the line goes to the first available teller. The arrival pattern is as follows:

10:00 A.M.–11:30 A.M.	λ = 1.8 customers per minute
11:30 A.M.–1:30 P.M.	λ = 4.8 customers per minute
1:30 P.M.–3:00 P.M.	λ = 3.8 customers per minute
3:00 P.M.–4:00 P.M.	λ = 4.6 customers per minute

The arrival distributions follow the Poisson distribution; that is, the mean value of arrivals varies but always forms a Poisson distribution. The average service time is one minute and the distribution of service times is approximated by the negative exponential distribution. Tellers are paid $4 per hour.

The manager wishes to compare the cost of the several service standard policies. Because of the difficulty with reassigning tellers to useful work in all instances, she is also considering the use of part-time tellers.

What action should be taken by the bank manager?

11. A university must maintain a large and complex physical plant, so that plant maintenance is an important support function. The plant maintenance department maintains a crew of six maintenance mechanics who respond to calls for service from department heads and other authorized personnel on the campus. They respond to a wide variety of calls that range from simple adjustments of room thermostats to actual repair of plant and equipment. In some instances, extensive work involving specialized personnel may be required, and this is scheduled separately.

    During the work day, there are five calls per hour, and the distribution of the call rate is approximated by a Poisson distribution. The average time for service is 60 minutes, including travel time both ways and the time to actually perform the required work and is approximated by the negative exponential distribution. The wage rate of the mechanics is $8 per hour.

    One of the mechanics has just resigned for personal reasons, and the university business manager has refused to replace her because of the budget squeeze. The head of the plant maintenance department is furious and produces a file of complaints from department heads about the slow response to calls for service. The business manager implies that the slow service reflects inefficiency and that it is time for the plant maintenance department to "shape up."

    How many maintenance mechanics are economically justified? What action do you feel should be taken?

12. Since hospital nursing requirements depend on individual patient needs as well as on the number of patients, a method of classifying patients by their relative

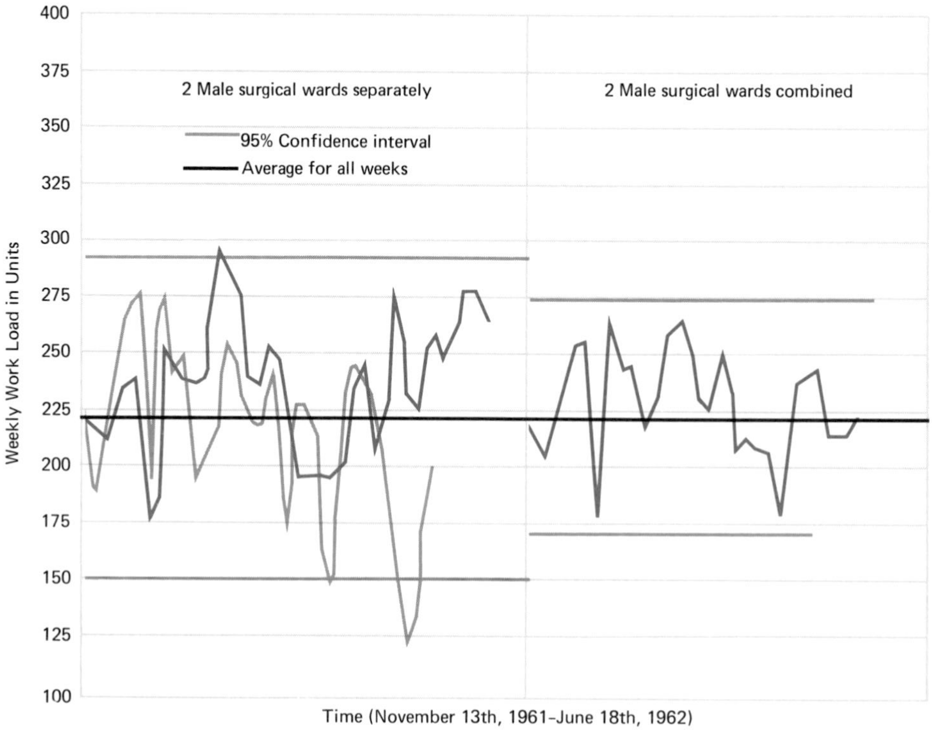

FIGURE D-5
Effect of combining two wards on fluctuation of work-load.
SOURCE: *A. Barr, "Measuring Nursing Care—Operational Research in Nursing," in* Problems and Progress in Medical Care. *Oxford University Press, New York, 1964, pp. 77–90.*

nursing needs was developed by Flagle [1960]. Patients were classified into three groups in terms of nursing care needs: self-care, partial care, and total care. A sample survey then indicated that the distributions of care time were very different for the three categories. The distribution for self-care patients had a mean of one-half hour and a very small standard deviation. The distribution of partial-care patients had a mean of one hour and a somewhat larger standard deviation. Finally, the distribution for total-care patients had a mean of 2.5 hours and a very large standard deviation. Studies of the number of patients in the total-care category in four different wards varied significantly. However, there seemed to be no correlation in work load between the wards.

A similar study of load fluctuation in England also indicated no correlation in weekly load between wards. Figure D-5 indicates these load variations for two wards separately on the left and when combined, on the right. The study indicated that the variance was reduced by about 30 percent when the two wards were combined.

What are the implications of this information concerning work loads when one views the wards as a waiting line system? If you were managing this system, what guidelines for staffing the wards might be indicated by the load information?

REFERENCES

Bleuel, W. H. "Management Science's Impact on Service Strategy," *Interfaces, 6*(1), Part 2, November 1975, pp. 4–12.

Brigham, F. "On a Congestion Problem in an Aircraft Factory," *Operations Research, 3*(4), 1955, pp. 412–428.

Buffa, E. S., and J. S. Dyer. *Management Science/Operations Research: Model Formulation and Solution Methods.* John Wiley, New York, 1977.

Cosmetatos, G. P. "The Value of Queueing Theory—A Case Study," *Interfaces,* 9(3), May 1979, pp. 47–51.

Flagle, C. D. "The Problem of Organization for Hospital Inpatient Care," in *Management Sciences: Models and Techniques,* Vol. 2. Pergamon Press, Paris, 1960, pp. 275–287.

Foote, B. L. "A Queuing Case Study of Drive-In Banking," *Interfaces, 6*(4), August 1976, pp. 31–37.

McKeown, P. G. "An Application of Queueing Analysis to the New York State Child Abuse and Maltreatment Register Reporting System," *Interfaces,* 9(3), May 1979, pp. 20–25.

Morse, P. M. *Queues, Inventories and Maintenance.* John Wiley, New York, 1957.

Nelson, R. T. "An Empirical Study of Arrival, Service Time, and Waiting Time Distributions of a Job Shop Production Process," Research Report No. 60, *Management Sciences Research Project,* UCLA, 1959.

Peck, L. G., and R. N. Hazelwood. *Finite Queuing Tables.* John Wiley, New York, 1958.

Thierauf, R. J., and R. C. Klekamp. *Decision Making Through Operations Research* (2nd. ed.). John Wiley, New York, 1975.

Wagner, H. M. *Principles of Operations Research* (2nd ed.). Prentice-Hall, Englewood Cliffs, N.J., 1975.

APPENDIX E Monte Carlo Simulation

SIMULATED SAMPLING, KNOWING GENERALLY AS MONTE CARLO, makes it possible to introduce into a system data that have the statistical properties of an empirical distribution. If the model involves the flow of orders according to the actual demand distribution experienced, we can simulate the "arrival" of an order by Monte Carlo sampling from the distribution, so that the timing and flow of orders in the simulated system parallels experience. If we are studying the breakdown of a certain machine as a result of bearing failure, we can simulate typical breakdown times through simulated sampling from the distribution of bearing lives.

A COMPUTED EXAMPLE

Suppose we are dealing with the maintenance of a bank of 30 machines, and, initially, we wish to estimate what level of service can be maintained by one mechanic. We have the elements of a waiting line situation, with machine breakdowns representing arrivals, the mechanic being the service facility, and repair time representing service time. If the distributions the time between breakdowns and service times followed the negative exponential distribution, the simplest procedure would be to use the formulas and calculate the average time that a machine waits, the mechanic's idle time, and so forth. We can see by inspection of Figure E-1 and E-2 that the distributions are not similar to the negative exponential, so simulation is an alternative. The procedure is as follows:

1. *Determine the distributions of time between breakdowns and service time*. If these data were not available directly from records, we would have to make a study to determine the distributions. Figures E-1 and E-2 show the distributions of breakdowns and repair times for 73 breakdowns.

2. *Convert the frequency distribution to cumulative probability distributions (see Figures E-3 and E-4)*. This conversion is accomplished by summing the frequencies that are less than or equal to each breakdown or repair time and plotting them. The

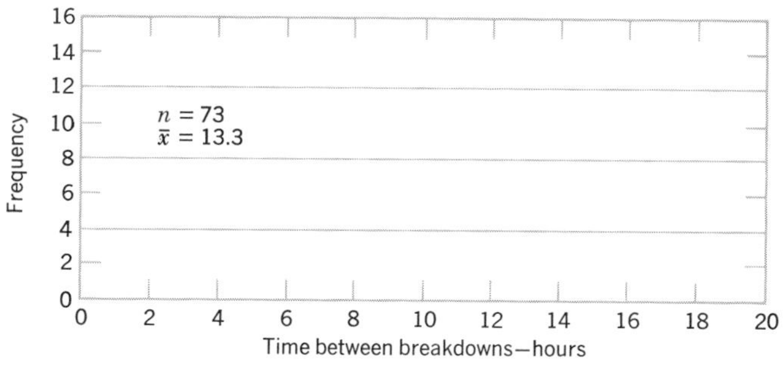

FIGURE E-1
Frequency distribution of the time between breakdowns for 30 machines.

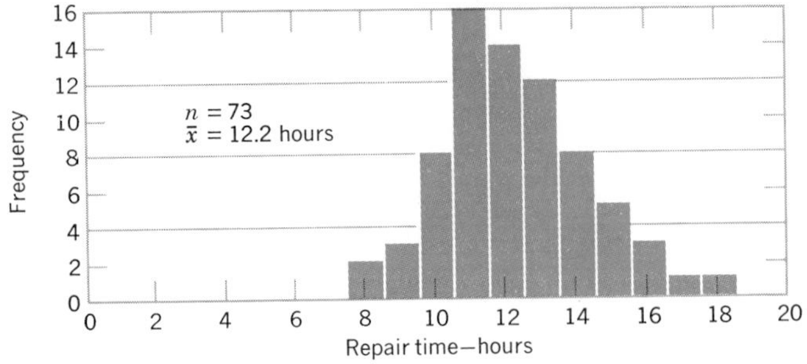

FIGURE E-2
Frequency distribution of the repair time for 73 breakdowns.

cumulative frequencies are then converted to probabilities by assigning the number 1.0 to the maximum value.

As an example, let us take Figure E-1 and convert it to the cumulative distribution of Figure E-3. Beginning at the lowest value for breakdown time, 10 hours, there are four occurrences. Four is plotted on the cumulative chart for the breakdown time of 10 hours. For the breakdown time, 11 hours, there were 10 occurrences, but there were 14 occurrences of 11 hours or less, so the value 14 is plotted for 11 hours. For

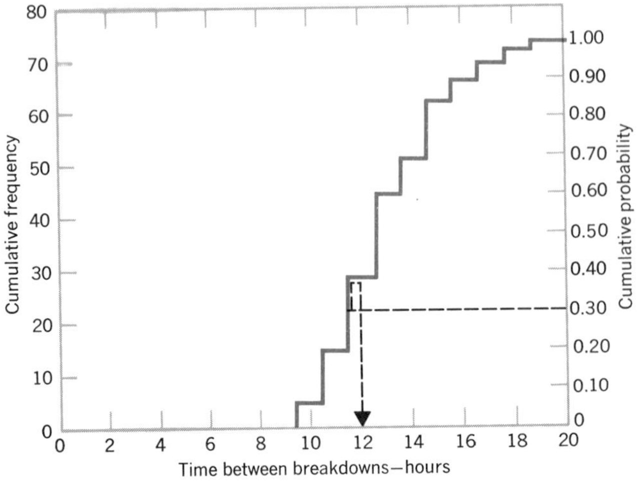

FIGURE E-3
Cumulative distribution of breakdown times.

the breakdown time, 12 hours, there were 14 occurrences recorded, but there were 28 occurrences of breakdowns for 12 hours or less.

Figure E-3 was constructed from Figure E-1 by proceeding in this way. When the cumulative frequency distribution was completed, a cumulative probability scale was constructed on the right of Figure E-3 by assigning the number 1.0 to the maximum value, 73, and dividing the resulting scale into 10 equal parts. This results in a cumulative empirical probability distribution. From Figure E-3, we can say that 100 percent of the breakdown time values were 19 hours or less; 99 percent were 18 hours or less, and so on. Figure E-4 was constructed from Figure E-2 in a comparable way.

3. *Sample at random from the cumulative distributions to determine specific breakdown times and repair times to use in simulating the repair operation.* We do this by selecting numbers between 001 and 100 at random (representing probabilities in percentage). The random numbers could be selected by any random process, such as drawing numbered chips from a box. The easiest way is to use a table of random numbers, such as those included in Table H-5 of Appendix H. (Pick a starting point in the table at random and take two-digit numbers in sequence in that column, for example.)

The random numbers were used to enter the cumulative distributions to obtain time values. An example is shown in Figure E-3. The random number 30 is shown to select for us a breakdown time of 12 hours. We can now see the purpose behind the conversion of the original distribution to a cumulative distribution. Only one

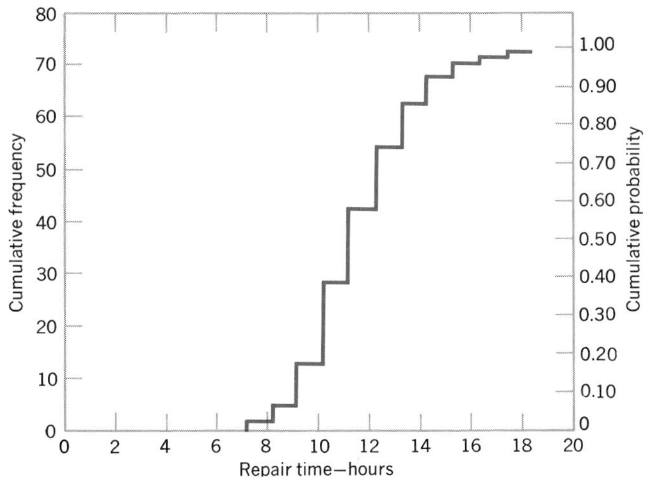

FIGURE E-4
Cumulative distribution of repair times.

breakdown time can now be associated with a given random number. In the original distribution, two values would result because of the bell shape of the curve.

By using random numbers to obtain breakdown time values in this fashion from Figure E-3, we will obtain breakdown time values in proportion to the probability of occurrence indicated by the original frequency distribution. With this framework we can construct a table of random numbers that select certain breakdown times. For example, reading from Figure E-3, the random numbers 6 through 19 result in a breakdown time of 11 hours, and so on. This is the same as saying that 5 percent of the time we would obtain a value of 10 hours, 14 percent of the time we would obtain a breakdown time of 11 hours, and so forth. Table E-1 shows the random number equivalents for Figures E-3 and E-4.

Sampling from either the cumulative distributions of Figures E-3 and E-4 or from Table E-1 will produce breakdown times and repair times in proportion to the original distributions, just as if actual breakdowns and repairs were happening. Table E-2 gives a sample of 20 breakdown and repair times determined in this way.

4. *Simulate the actual operation of breakdowns and repairs.* The structure of the simulation of the repair operation is shown by the flowchart of Figure E-5. This operation involves the selection of a breakdown time and determining whether or not the mechanic is available. If the mechanic is not available, the machine must wait until one is, and the wait time may be computed. If the mechanic is available, the question is, did the mechanic have to wait? If waiting was required, we compute the mechanic's idle time. If the mechanic did not have to wait, we select a repair time and proceed according to the flowchart, repeating the overall process as many times as desired, stopping the procedure when the desired number of cycles has been completed.

The simulation of the repair operation is shown in Table E-3, using the breakdown times and repair times selected by random numbers in Table E-2. We assume that time begins when the first machine breaks down and calculate breakdown time from that point. The repair time required for the first breakdown was 15 hours, and since this is the first occurrence in our record, neither the machine nor the mechanic had to wait. The second breakdown occurred at 18 hours, but the mechanic was available at the end of 15 hours, waiting 3 hours for the next breakdown to occur.

We proceed in this fashion, adding and subtracting, according to the requirements of the simulation model to obtain the record of Table E-3. The summary at the bottom of Table E-3 shows that for the sample of 20 breakdowns, total machine waiting time was 11 hours, and total mechanic's idle time was 26 hours. To obtain a realistic picture we would have to use a much larger sample. Using the same data on breakdown and repair time distributions, 1000 runs using a computer yielded 15.9 percent machine wait time and 7.6 percent mechanic's idle time. Of course, the mechanic is presumably paid for an eight-hour day regardless of the division between idle and service time; however, knowing idle time available may be a guide to the assignment of "fill-in" work.

TABLE E-1

Random Numbers Used to Draw Breakdown Times and Repair Times in Proportion to the Occurrence Probabilities of the Original Distributions

Breakdown Times		Repair Times	
These Random Numbers ↓	→ Select These Breakdown Times ↓	These Random Numbers ↓	→ Select These Repair Times ↓
1–5	10 hours	1–3	8 hours
6–19	11	4–7	9
20–38	12	8–18	10
39–60	13	19–40	11
61–77	14	41–59	12
78–85	15	60–75	13
86–90	16	76–86	14
91–95	17	87–93	15
96–99	18	94–97	16
0=100	19	98–99	17
		0=100	18

TABLE E-2

Simulated Sample of Twenty Breakdown and Repair Times

Breakdown Times		Repair Times	
Random Number	Breakdown Time from Figure E-3	Random Number	Repair Time from Figure E-4
83	15	91	15
97	18	4	9
88	16	72	13
12	11	12	10
22	12	30	11
16	11	32	11
24	12	91	15
64	14	29	11
37	12	33	11
62	14	8	10
52	13	25	11
9	11	74	13
64	14	97	16
74	14	70	13
15	11	15	10
47	13	43	12
86	16	42	12
79	15	25	11
43	13	71	13
35	12	14	10

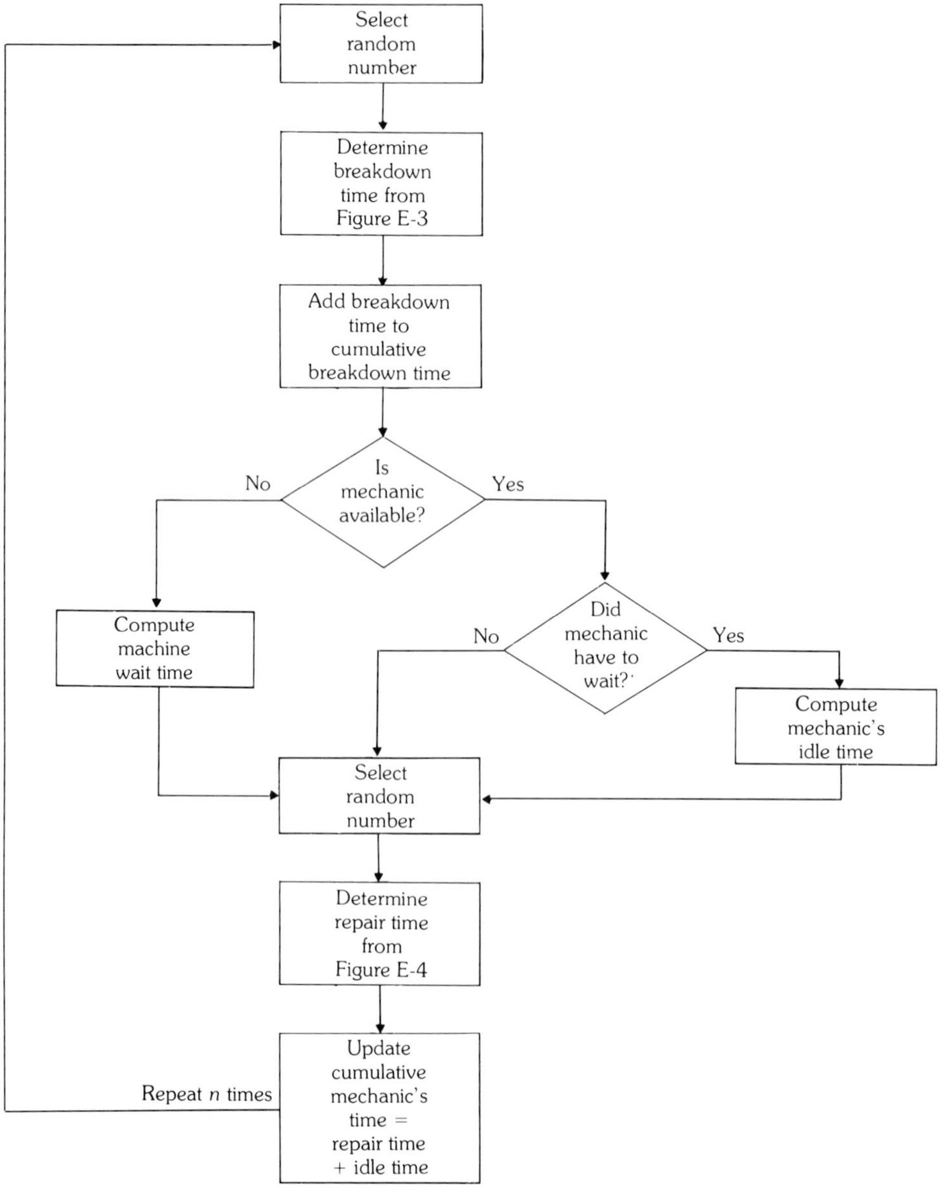

Figure E-5.
Flow chart showing structure of repair simulation.

TABLE E-3 **Simulated Breakdown and Repair for Twenty Breakdowns**

Time of Breakdown	Time Repair Begins	Time Repair Ends	Machine Wait Time	Repair Mechanics Idle Time
0	0	15	0	0
18	18	27	0	3
34	34	47	0	7
45	47	57	2	0
57	57	68	0	0
68	68	79	0	0
80	80	95	0	1
94	95	106	1	0
106	106	117	0	0
120	120	130	0	3
133	133	144	0	3
144	144	157	0	0
158	158	174	0	1
172	174	187	2	0
183	187	197	4	0
196	197	209	1	0
212	212	224	0	3
227	227	238	0	3
240	240	253	0	2
252	253	263	1	0

Total machine wait time = 11 hours
Total mechanic's idle time = 26 hours

PROBLEMS

1. A sample of 100 arrivals of customers at a check-out station of a small store occurs according to the following distribution:

Time Between Arrivals, Minutes	Frequency
0.5	2
1.0	6
1.5	10
2.0	25
2.5	20
3.0	14
3.5	10
4.0	7
4.5	4
5.0	2
	100

A study of the time required to service the customers by adding up the bill, receiving payment, making change, placing packages in bags, and so on yields the following distribution:

Service Time, Minutes	Frequency
0.5	12
1.0	21
1.5	36
2.0	19
2.5	7
3.0	5
	100

a. Convert the distributions to cumulative probability distributions.
b. Using a simulated sample of 20, estimate the average percentage of customer waiting time and the average percentage of idle time of the server.

2. The manager of a drive-in restaurant is attempting to determine how many "car hops" he needs during his peak load period. As a policy, he wishes to offer service such that average customer waiting time does not exceed two minutes.
   a. How many car hops does he need if the arrival and service distributions are as follows and if any car hop can service any customer?
   b. Simulate for various alternate numbers of car hops with a sample of 20 arrivals in each case.

Time Between Successive Arrivals, Minutes	Frequency	Car Hop Service Time, Minutes	Frequency
0.0	10	0.0	0
1.0	35	1.0	5
2.0	25	2.0	20
3.0	15	3.0	40
4.0	10	4.0	35
5.0	5		100
	100		

3. A company maintains a bank of machines that are exposed to severe service, causing bearing failure to be a common maintenance problem. There were three bearings in the machine that caused trouble. The general practice had been to replace bearings when they failed. However, excessive downtime costs raised the question of whether or not a preventive policy was worthwhile. The company wished to evaluate three alternate policies:
   a. The current practice of replacing bearings that fail.
   b. When a bearing fails, replace all three.

c. When a bearing fails, replace that bearing plus other bearings that have been in use 1700 hours or more.

Time and cost data are as follows:

Maintenance mechanics time:	
Replace 1 bearing	5 hours
Replace 2 bearings	6 hours
Replace 3 bearings	7 hours
Maintenance mechanic's wage rate	\$3 per hour
Bearing cost	\$5 each
Downtime costs	\$2 per hour

A record of the actual working lives of 200 bearings results in the following distribution:

Bearing Life, Hours	Frequency
1100	3
1200	10
1300	12
1400	20
1500	27
1600	35
1700	30
1800	25
1900	18
2000	15
2100	4
2200	1
	200

Simulate approximately 20,000 hours of service for each of the three alternate policies.

REFERENCES

Buffa, E. S., and J. S. Dyer. *Management Science/Operations Research: Model Formulation and Solution Methods.* John Wiley, New York, 1977.

Fetter, R. B., and J. D. Thompson. "The Simulation of Hospital Systems," *Operations Research,* 13(5), 1965, pp. 689–711.

Kwak, N. K. P., P. J. Kuzdrall, and H. H. Schmitz. "The GPSS Simulation of Scheduling Policies for Surgical Patients," *Management Science,* 22(9), May 1976, pp. 982–989.

Maisel, H., and G. Gnugnuoli. *Simulation of Discrete Stochastic Systems.* SRA, Chicago, 1972.

Reitman, J. *Computer Simulation Applications.* John Wiley, New York, 1971.

Schriber, T. J. *Simulation Using GPSS.* John Wiley, New York, 1974.

APPENDIX F Work Measurement

PERFORMANCE STANDARDS

Performance standards provide data that are basic to many decision-making problems in production/operations management. The performance standard is of critical importance because labor cost is a predominant factor, influencing many decisions that must be made. For example, decisions to make or buy, to replace equipment, or to select certain manufacturing processes require estimates of labor costs. These decisions necessarily require an estimate of how much output can be expected per unit of time.

Performance standards also provide basic data used in the day-to-day operation of a plant. For example, scheduling or loading machines demands a knowledge of the projected time requirements. For custom manufacture, we must be able to give potential customers a bid price and delivery date. The bid price is ordinarily based on expected costs for labor, materials, and overhead, plus profit. Labor cost is commonly the largest single component in such situations. To estimate labor cost requires an estimate of how long it will take to perform the various operations.

Finally, performance standards provide the basis for labor cost control. By measuring worker performance in comparison to standard performance, indexes can be computed for individual workers, whole departments, divisions, or even plants. These indexes make it possible to compare performance on completely different kinds of jobs. Standard labor costing systems and incentive wage payment systems are based on performance standards.

INFORMAL STANDARDS

Every organization has performance standards of sorts. Even when they seem not to exist formally, supervisors have standards in mind for the various jobs based on their knowledge of the work and past performances. These types of standards are informal. Standards based on supervisor's estimates and past performance data have weaknesses, however. First, in almost all such situations, methods of work performance have not been standardized. Therefore, it is difficult to state what output rate, based on past records, is appropriate, because past performance may have been based on various methods. Since it has been demonstrated that output rates depend heavily on job methods, standards based on past performance records might not be dependable. A second major defect in standards based on estimates and past performance records is that they are likely to be too strongly influenced by the working speeds of the individuals who held the jobs. Were those workers high or low performers?

THE CORE OF THE WORK MEASUREMENT PROBLEM

We wish to set up standards that are applicable to the working population, not just to a few selected people within that population. The standards problem is comparable in some ways to that of designing a lever with the proper mechanical advantage to match the capabilities of workers—but not just any worker: the force required to pull the lever should accommodate perhaps 95 to 99 percent of the population, so that anyone who comes to the job will have the necessary arm strength.

We require a knowledge of the distribution of performance times for the entire working population doing the job. For example, suppose that we have 500 people all doing an identical task; we make sample studies on all of them and plot the data. Figure F-1 shows the results of such a study. The distribution shows that average performance time varies from 0.28 minute to 0.63 minute per piece. If past records reflected data from one or more individuals taken at random from the population of 500, a standard based on their performance might not fit the whole population very well. On the other hand, if we have good data concerning the entire distribution, as in Figure F-1, we can set up standards that probably would be appropriate for everyone. One way to do this is to set the standard so that it accommodates about 95 percent of the population. For Figure F-1, a standard performance time of about 0.48 minute is one that about 95 percent of the individuals exceeded. If we pegged the standard at this level, we would expect that practically all employees on the job should be able to meet or exceed the standard.

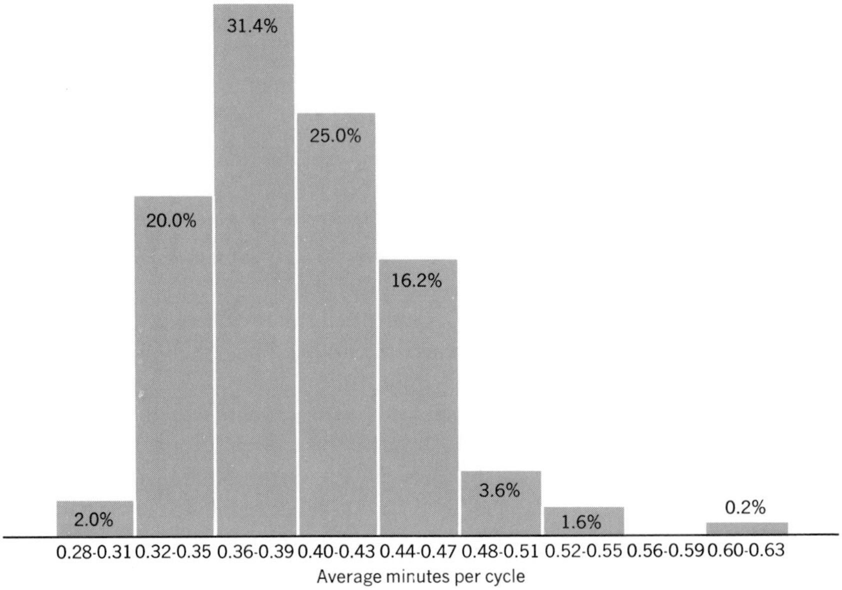

FIGURE F-1
Percentage distribution of the performance of 500 people performing a wood block positioning task. About 5.4 percent of the people averaged 0.48 minutes per cycle or longer.
SOURCE: *Adapted from R. M. Barnes,* Motion and Time Study: Design and Measurement of Work *(6th ed.). John Wiley, New York, 1968.*

Some managers feel that it is not good to quote minimum performance standards such as these for fear that they will encourage relatively poor performance as acceptable. These people prefer to say that the standard performance is about the average of the distribution (0.395 minute for Figure F-1) and expect that most workers will produce about standard, that some will fall below, and that some will exceed standard. Both systems of quoting standards are used, although the practice of quoting *minimum acceptable values* is more common than that of quoting average values.

The distribution of Figure F-1 shows how long it took, on the average, to perform the task. Using the minimum acceptable level as a standard of performance, we will call the actual work time at that level the *normal time*. The normal time for the data of Figure F-1 is 0.48 minute; the total standard time is then

Normal time + standard allowance for personal time
+ allowance for measured delays normal to the job
+ fatigue allowance

We will discuss the several allowances later, but the central question now is, How do we determine normal time in the usual situation when only one or a few workers are on the job? The approach to this problem used in industry is called performance rating.

PERFORMANCE RATING

Performance rating is a critically important part of any formal means of work measurement. To be able to rate accurately requires considerable experience. A pace or performance level is selected as standard. An analyst observes this pace, compares it with various other paces, and learns to judge pace level in percentage of the standard pace. For Figure F-1, we called the cycle time of 0.48 minute "normal," and the pace or rate of output associated with this time is normal pace. A pace of work that is 25 percent faster would require proportionately less time per cycle, or $0.48/1.25 = 0.381$ minute. If a skilled analyst observed a worker performing the task on which Figure F-1 is based and rated performance at 125 percent of normal while simultaneously measuring the actual average performance time as 0.381 minute, 25 percent would be added to observed time to adjust it to the normal level. In this instance, performance rating is perfect, since $0.381 \times 1.25 = 0.48$. Other perfect combinations of rating and actual observed time are 150 percent and 0.32 minute, 175 percent and 0.274 minute, 90 percent and 0.533 minute, and so forth.

In an actual work measurement situation, the analyst does not have the answer beforehand, so actual time taken to do the task and performance rating must be done simultaneously. The normal time is then computed as

$$\text{Normal time} = \text{actual observed time} \times \frac{\text{performance rating}}{100}$$

All formal work measurement systems involve this rating or judgment of working pace, or some equivalent procedure. Alternate methods will be considered later in this appendix.

How Accurate Is Performance Rating?

In the actual work measurement situation, it is necessary to compare a mental image of "normal performance" with what is observed. This rating enters the computation of performance standards as a factor, and the final standard can be no more accurate than the rating. How accurately can experienced people rate? Controlled studies in which films have been rated indicate a standard deviation of 7 to 10 percent. In other words, experienced people probably hold these limits about 68 percent of the time. Therefore, the effect of the element of judgment in current work measurement practice is considerable.

WORK MEASUREMENT SYSTEMS

All practical work measurement systems involve (1) the measurement of actual observed time and (2) the adjustment of observed time to obtain "normal time" by means of performance rating. The alternate systems that we will discuss combine these factors in somewhat different ways.

Stopwatch Methods

By far, the most prevalent approach to work measurement currently used involves a stopwatch time study and simultaneous performance rating of the operation to determine normal time. The general procedure is as follows:

1. Standardize methods for the operation; that is, determine the standard method, specifying work-place layout, tools, sequence of elements, and so on. Record the resulting standard practice.

2. Select for study an operator who is experienced and trained in the standard methods.

3. Determine the elemental structure of the operation for timing purposes. This may involve a breakdown of the operation into elements and the separation of the elements that occur during each cycle from those that occur only periodically or randomly. For example, tool sharpening might be required each 100 cycles to maintain quality limits. Machine adjustments might occur at random intervals.

4. Observe and record the actual time required for the elements, making simultaneous performance ratings.

5. Determine the number of observations required to yield the desired precision of the result based on the sample data obtained in step 4. Obtain more data as required.

6. Compute normal time = average observed actual time × average rating factor/100.

7. Determine allowances for personal time, delays, and fatigue.

8. Determine standard time = normal times for elements + time for allowances.

Breakdown of Elements. Common practice is to divide the total operation into elements rather than to observe the entire cycle as a whole. There are several reasons why this practice is followed:

1. The element breakdown helps to describe the operation in some detail, indicating the step-by-step procedure followed during the study.

2. More information is obtained that may have valuable use for comparing times for like elements on different jobs and for building up a handbook of standard data times for common elements in job families. With standard data for elements, cycle times for new sizes can be forecast without additional study.

3. A worker's performance level may vary in different parts of the cycle. With an element breakdown, different performance ratings can be assigned to different elements where the overall cycle is long enough to permit separate evaluation of performance.

In breaking down an operation into elements, it is common practice to make elements a logical component of the overall cycle, as illustrated in Figure F-2. For example, element 1, "pick up piece and place in jig," is a fairly homogeneous task. Note that element 4, "drill 1/4-inch hole," is the machining element, following the general practice to separate machining time from handling time. Finally, constant elements are usually separated from elements that might vary with size, weight, or some other parameter.

Taking and Recording Data. Figure F-2 is a sample study in which 20 cycles were timed by the continuous method; that is, the stopwatch is allowed to run continuously, being read at the breakpoints between elements. Elapsed times for elements are then obtained by successive subtraction. *Repetitive* or "snap-back" methods of reading the watch are also common. In repetitive timing, the observer reads the watch at the end of each element and snaps the hand back to zero, so that each reading gives the actual time without the necessity of subtraction. Comparative studies indicate that the two methods are equally accurate.

Other data recorded in Figure F-2 identify the part, operation, operator, material, and so on, as well as check data of elapsed time of the study and the number of completed units. The "selected times" represent averages of the element times; the cycle "selected time" is merely the sum of the element averages. A single performance rating of 100 percent was made for the study, and a 5 percent allowance was added to obtain the standard time of 1.17 minutes per piece.

OBSERVATION SHEET

SHEET 1 OF 1 SHEETS	DATE
OPERATION Drill ¼″ Hole	OP. NO. D-20
PART NAME Motor Shaft	PART NO. MS-267
MACHINE NAME Avey	MACH. NO. 2174
OPERATOR'S NAME & NO. S.K. Adams 1347	MALE [√] FEMALE [ ]
EXPERIENCE ON JOB 18 Mo. on Sens. Drill	MATERIAL S.A.E. 2315
FOREMAN H. Miller	DEPT. NO. DL 21

BEGIN	FINISH	ELAPSED	UNITS FINISHED	ACTUAL TIME PER 100	NO. MACHINES OPERATED
10:15	10:38	23	20	115	1

ELEMENTS	SPEED	FEED		1	2	3	4	5	6	7	8	9	10	SELECTED TIME
1. Pick Up Piece and Place in Jig			T	.12	.11	.12	.13	.12	.10	.12	.12	.14	.12	
			R	.12	.29	.39	.54	.66	.77	.92	8.01	14	.32	
2. Tighten Set Screw			T	.13	.12	.12	.14	.11	.12	.12	.13	.12	.11	
			R	.25	.41	.51	.68	.77	.89	7.04	.14	.26	.43	
3. Advance Drill to Work			T	.05	.04	.04	.04	.05	.04	.04	.04	.03	.04	
			R	.30	.45	.55	.72	.82	.93	.08	.18	.29	.47	
4. DRILL ¼″ HOLE	980	H	T	.57	.54	.56	.51	.54	.58	.52	.53	.59	.56	
			R	.87	.99	3.11	4.23	5.36	6.51	.60	.71	.88	11.03	
5. Raise Drill from Hole			T	.04	.03	.03	.03	.03	.03	.03	.03	.04	.03	
			R	.91	2.02	.14	.26	.39	.54	.63	.74	.92	.06	
6. Loosen Set Screw			T	.06	.06	.07	.06	.06	.06	.06	.06	.07	.08	
			R	.97	.08	.21	.32	.45	.60	.69	.80	.99	.14	
7. Remove Piece from Jig			T	.08	.09	.08	.08	.09	.08	.07	.08	.09	.07	
			R	1.05	.17	.29	.40	.54	.68	.76	.88	10.08	.21	
8. Blow Out Chips			T	.13	.10	.12	.14	.13	.12	.13	.12	.12	.11	
			R	.18	.27	.41	.54	.67	.80	.89	9.00	.20	.32	
9.			T											
			R											
10. (1)			T	.12	.11	.13	.14	.12	.12	.11	.13	.12	.12	.12
			R	11.44	.56	.69	.82	.87	17.01	18.09	.21	.31	.42	
11. (2)			T	.12	.14	.12	.11	.12	.10	.13	.15	.12	.11	.12
			R	.56	.70	.81	.93	.99	.11	.22	.36	.43	.53	
12. (3)			T	.04	.04	.04	.03	.04	.04	.04	.04	.04	.04	.04
			R	.60	.74	.85	.96	16.03	.15	.26	.40	.47	.57	
13. (4)			T	.54	.53	.55	.52	.57	.54	.50	.53	.55	.54	.54
			R	12.14	13.27	14.40	15.48	.60	.69	.76	.93	21.02	22.11	
14. (5)			T	.03	.03	.03	.03	.03	.03	.03	.03	.03	.03	.03
			R	.17	.30	.43	.51	.63	.72	.79	.96	.05	.14	
15. (6)			T	.06	.06	.06	.07	.06	.05	.06	.06	.05	.06	.06
			R	.23	.36	.49	.58	.69	.77	.85	20.02	.10	.20	
16. (7)			T	.08	.08	.09	.08	.08	.07	.08	.06	.08	.08	.08
			R	.31	.44	.58	.66	.77	.84	.93	.08	.18	.28	
17. (8)			T	.14	.12	.10	.09	.12	.14	.15	.11	.12	12	.12
			R	.45	.56	.68	.75	.89	.98	19.08	.19	.30	22.40	
18.			T											1.11
			R											

SELECTED TIME	RATING	NORMAL TIME	TOTAL ALLOWANCES	STANDARD TIME
1.11	100%	1.11	5%	1.17

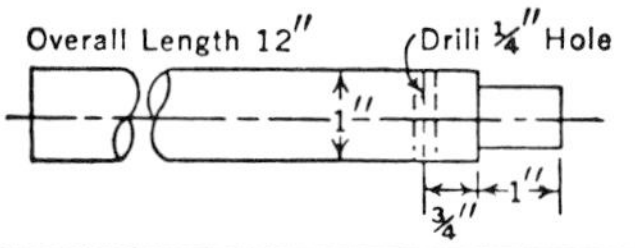

TOOLS, JIGS, GAUGES: Jig No. D-12-33
Use H.S. Drill ¼″ Diam.
Hand Feed
Use Oil - S4

TIMED BY J.B.M.

FIGURE F-2
Stopwatch time study of a drilling operation made by the continuous method.
SOURCE: *R. M. Barnes,* Motion and Time Study: Design and Measurement of Work *(6th ed.). John Wiley, 1968.*

Adequacy of Sample Size. We are attempting to estimate, from the sample times and performance ratings observed, a normal time of performance. The precision desired will determine how many observations will be required. For example, if we wanted to be 95 percent sure that the resulting answer, based on the sample, was within ±5 percent, we would calculate the sample size *n* required from a knowledge of the mean and standard deviation of our sample data. If we wanted greater confidence or closer precision, the sample size would have to be larger.

Figure F-3 is a convenient chart for estimating required sample sizes to maintain a ±5 percent precision in the answer for 95 and 99 percent confidence levels. To use the chart, we merely calculate the mean value of $\bar{x}$ and the standard deviation based on the sample data. The "coefficient of variation" is simply the percentage variation, $100(s_x/\bar{x})$. The chart is entered with the calculated coefficient of variation,

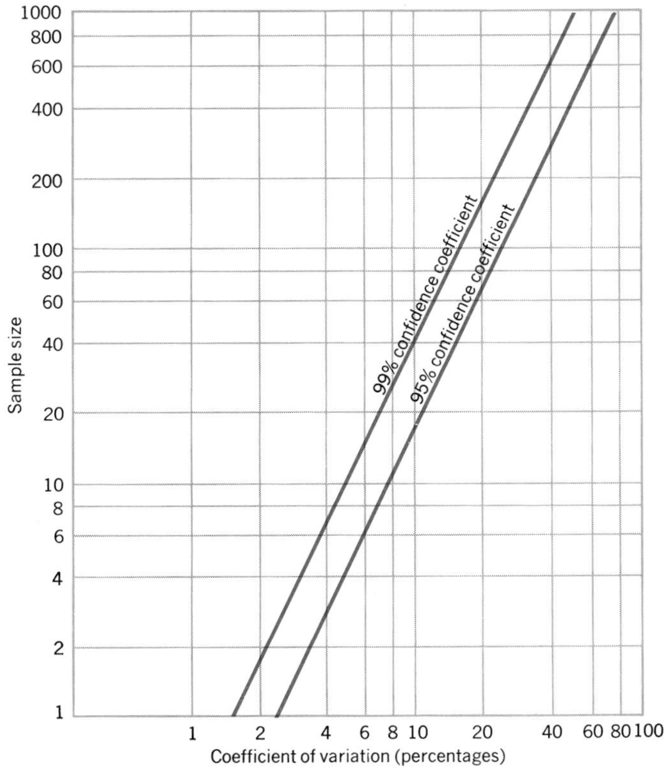

FIGURE F-3
Chart for estimating the sample size required to obtain maximum confidence intervals of ± 5 percent for given coefficient of variation values.
SOURCE: *A. Abruzzi,* Work Measurement. *Columbia University Press, New York, 1952.*

and the sample size is read off for the confidence level desired. The most common confidence level is 95 percent in work measurement.

Let us test the adequacy of the sample taken in the study of Figure F-2. First, was $n = 20$ adequate for estimating the overall cycle within a precision of ±5 percent and a confidence level of 95 percent? Table F-1 shows the calculation of the coefficient of variation for the cycle times as about 5 percent; that is, the standard deviation of 0.057 minute is about 5 percent of the mean cycle time of 1.12 minutes. From Figure F-3, we see that a sample of $n = 4$ would be adequate to maintain a precision within ±5 percent of the correct mean cycle time, 95 percent of the time. For a confidence level of 99 percent, $n = 10$. Our actual sample of 20 was more than adequate.

The reason that the small sample size was adequate is easy to see. The variability of the readings is small in relation to the mean cycle time, so a good estimate of cycle

TABLE F-1 **Cycle Times from Figure F-2, and Calculated Mean Value, Standard Deviation, Coefficient of Variation and Required Sample Sizes from Figure F-3**

Cycle Number	Cycle Time (minutes)	Cycle Time (squared)	Cycle Number	Cycle Time (minutes)	Cycle Time (squared)
1	1.18	1.395	11	1.13	1.280
2	1.09	1.190	12	1.11	1.235
3	1.14	1.300	13	1.12	1.255
4	1.13	1.280	14	1.07	1.145
5	1.13	1.280	15	1.14	1.300
6	1.13	1.280	16	1.09	1.190
7	1.09	1.190	17	1.10	1.215
8	1.11	1.235	18	1.11	1.235
9	1.20	1.440	19	1.11	1.235
10	1.12	1.255	20	1.10	1.215
			Sum	22.40	25.150

$$\bar{x} = \frac{22.40}{20} = 1.12$$

$$s = \sqrt{\frac{\sum x_i^2 - \frac{(\sum x_i)^2}{n}}{n - 1}}$$

$$= \sqrt{\frac{25.150 - \frac{(22.40)^2}{20}}{19}} = 0.057$$

$$\text{Coefficient of variation} = \frac{0.057 \times 100}{1.12} = 5.09\%$$

From Figure F-3:

$n \approx = 4$ @ 95% confidence level.

$n \approx = 10$ @ 99% confidence level.

time is obtained by only a few observations. This is commonly true of operations dominated by a machine cycle. In this case, the actual drill time is almost half of the total cycle, and the machining time itself does not vary much.

If all we wanted was an estimate of cycle time, we could stop at this point. Suppose, however, that we want estimates of each of the average element times to be adequate for future use as elemental standard data. Was the sample size of $n = 20$ adequate for each of these elements? Take element 1 as an example. The mean element time is $\bar{x} = 0.121$ minute, the standard deviation is $s = 0.0097$ minute, and the coefficient of variation is 8 percent. From Figure F-3, we see that we should have taken a sample of $n = 10$ for a 95 percent confidence level and $n = 20$ for a 99 percent confidence level. The reason why a larger sample is needed for element 1 than for the entire cycle is that element 1 is somewhat more variable than is the total cycle (coefficient of variation of 8 percent compared with only 5 percent for the cycle). Therefore, if data on each of the elements are needed, the element for which the largest sample size is indicated, from Figure F-3, dictates the minimum sample size for the study. This procedure ensures the precision and confidence requirements for the limiting element and yields better results than this on all other elements.

Procedures for Ensuring Consistency of Sample Data. A single study always leaves open the question, Were the data representative of usual operating conditions? If a similar study were made on some other day of the week or some other hour of the day, would the results be different? This question suggests the possibility of dividing the total sample into smaller subsamples taken at random times. Then, by setting up control limits based on an initial sample, we can determine if the subsequent data taken are consistent. That is, did all the data come from a common universe?

This situation is comparable to that found in quality control. If a point falls outside of the $\pm 3s$ control limits, we know that the probability is high that some assignable cause of variation is present which has resulted in an abnormally high or low set of sample readings. These assignable causes could be anything that could have an effect on the time of production, such as material variations from standard or changes in tools, work-place, methods of work, or the working environment. As with quality control, we would attempt to determine the nature of these assignable causes and eliminate data where abnormal readings have an explanation.

The general procedure is as follows:*

1. Standardized methods, select operator, and determine elemental breakdown as before.

2. Take an initial sample study.
   a. Compute preliminary estimates of $\bar{x}$ and s.
   b. Determine estimate of total sample needed from Figure F-3.
   c. Set up control limits for balance of the study based on preliminary estimates of $\bar{x}$ and s.

* For detailed procedures with appropriate charts for estimating sample sizes, precision limits, and control limits, see Barnes [1968].

3. Program and execute the balance of the study:
   a. Divide the total sample by the subsample size to find the number of separate subsamples to obtain. Subsample sizes are commonly four to five.
   b. Randomize the time when these subsamples will be taken. A random number table is useful.
   c. At the random times indicated, obtain subsample readings and plot points on a control chart. If points fall outside limits, investigate immediately to determine the cause. Eliminate data from computations for standards where causes can be assigned.
   d. When the study is complete, make a final check to be sure that the precision and confidence level of the result are at least as good as desired.

4. Compute normal time, determine allowances, and compute standard time as before.

Work Sampling

The unique thing about work sampling is that it accomplishes the results of stopwatch study without the need for an accurate timing device. Work sampling was first introduced to industry by L. H. C. Tippett in 1934. However, it has been in common use only since about 1950.

We can illustrate the basic idea of work sampling by a simple example. Suppose we wish to estimate the proportion of time that a worker, or a group of workers, spends working and the proportion of time spent not working. We can do this by long-term time studies in which we measure the work time, the idle time, or both. This would probably take a day or longer, and after measuring we would not be sure that the term of the study covered representative periods of work and idleness.

Instead, suppose that we make a large number of *random* observations in which we simply determine whether the operator is working or idle and tally the result (see Figure F-4). The percentages of the tallies that are recorded in the "working" and "idle" classifications are estimates of the actual percentage of time that the worker was working and idle. Herein lies the fundamental principle behind work sampling: *the number of observations is proportional to the amount of time spent in the working or idle state.* The accuracy of the estimate depends on the number of observations, and we can preset precision limits and confidence levels.

Number of Observations Required. The statistical methods of work sampling depend on the distributions for proportions. Recall that

$$\bar{p} = \frac{x}{n} = \frac{\text{number observed in classification}}{\text{total number of observations}}$$

and

$$s_p = \sqrt{\frac{\bar{p}(1 - \bar{p})}{n}}$$

	Tally	Number	Per cent
Working	𝍸 𝍸 𝍸 𝍸 𝍸 𝍸 𝍸 𝍸 𝍸 𝍸 𝍸 𝍸 𝍸 𝍸 𝍸 𝍸 𝍸 𝍸 𝍸 /	96	88.9
Idle	𝍸 𝍸 //	12	11.1
Total		108	100.0

FIGURE F-4
Work sampling tally of working and idle time.

From these simple formulas for mean proportion and the standard deviation of a proportion, charts and tables have been developed that give directly the number of observations required for a given value of $\bar{p}$, precision limits, and the 95 percent confidence level. Estimates of sample sizes can be obtained from Figure F-5.*

Note that the number of observations required is fairly large. For example, to maintain a precision in the estimate of $\bar{p}$ of ± 1.0 percentage point at 95 percent confidence, 10,000 observations are required if $\bar{p}$ is in the neighborhood of 50 percent; that is, to be 95 percent sure that an estimate of $\bar{p} = 50$ percent is between 49 and 51 percent. About 3600 observations are required to hold an estimated $\bar{p} =$ 10 percent between 9 and 11 percent. Smaller samples are required for looser limits. Although these numbers of observations seem huge, we must remember that the nature of the observation required is merely a recognition of whether or not the employee is working, or possibly a classification of worker activity into various reasons for idleness.

Measuring Delays and Allowances. One common use of work sampling is to determine the percentage of time that workers are actually spending for personal time and delays that are a part of the job. The resulting information could then be used as the basis for the percentage allowances that enter into the calculation of standard time.

Consider as an example the determination of delay and personal allowances in a lathe department of a machine shop. There are 10 workers involved. The delays of which we are speaking are a part of the job such as waiting for tools, materials, and instructions; machine cleanup; securing an inspector's approval; change of jobs; and minor mechanical difficulties. We wish to determine the extent of the delays and how much time workers are spending for personal time. The procedure is as follows:

1. *Design work sampling study.*
   a. Estimate preliminary values for the percentage of time spent in the three categories of work, delay, and personal time from past knowledge, studies, supervisor's estimates, or a preliminary study of the jobs. These preliminary

* More complete information on sample sizes is available in Barnes [1957].

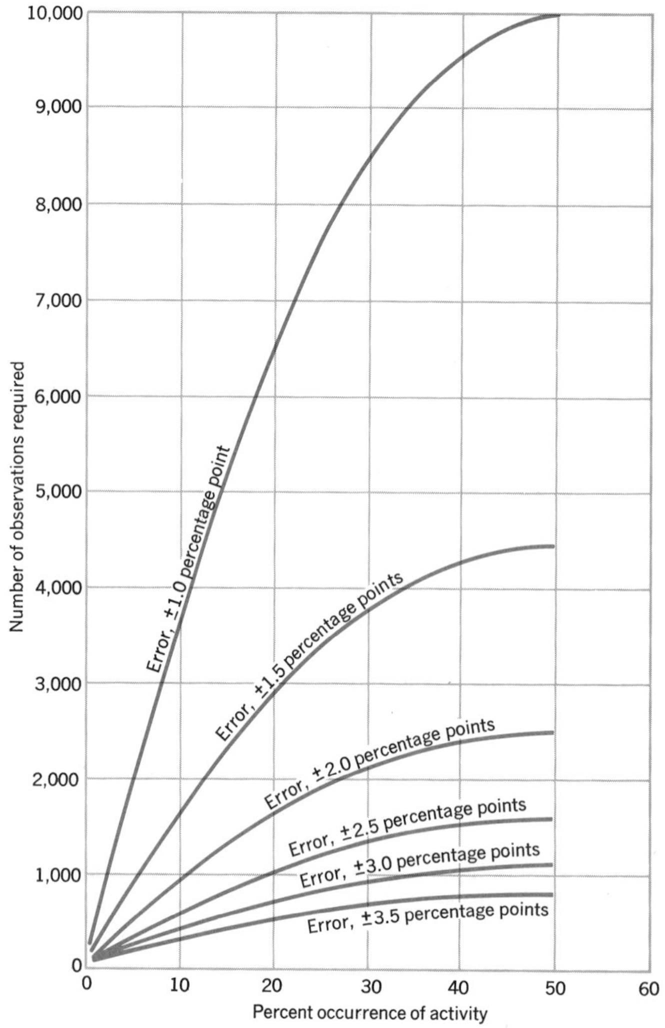

FIGURE F-5
Curves for determining the number of observations required to maintain precision within the percentage points indicated at the 95 percent confidence level.

estimates are necessary to gauge the magnitude of the data-taking phase. Based on a composite of past information and foremen's estimates, our best guesses are:

Work	85 percent
Delay	10 percent
Personal time	5 percent

b. Set desired precision limits of estimates to be obtained. We deicde that ±1.0

percentage point at 95 percent confidence on the delay estimate will be controlling. Thus, if the estimate for delays is actually 10 percent, we want to be 95 percent sure that it is not less than 9 percent or more than 11 percent, with 10 percent being the most probable value.

c. Estimate total the number of readings from Figure F-5. For $\bar{p} = 10$ percent, $N = 3600$ for ± 1.0 percentage point error. Note from Figure F-5 that our precision for personal time of 5 percent would then be slightly better than ± 1.0 percentage point and for working time, slightly worse.

d. Program the total number of readings over the desired time span of the study. We decide that 3600 readings over a 2-week period (10 working days) will cover a representative period. Therefore, we propose to obtain 3600/10 = 360 observations per day. Since there are 10 workers involved, we will obtain 10 observations each time we sample. So we need to program 360/10 = 36 random sampling times each day for 10 days to obtain the total of 3600 readings. The easiest way to select 36 random sampling times is to use a random number table.

e. Plan the physical aspects of the study. This includes an appropriate data sheet, as well as a determination of the physical path, observation points, and the like, so that the results are not biased because workers see the observer coming and change activities accordingly.

2. *Take the data as planned.* Table F-2 shows a summary of the actual data taken in this instance with a breakdown between morning and afternoon observations. The percentages for "work," "delay," and "personal time" have been computed for each half-day and for the total sample.

3. *Recheck precision of results and consistency of data.* A final check of the delay percentage of 9.97 percent shows that the number of readings taken was adequate to maintain the ± 1.0 percentage point precision on the delay time. The consistency of the data could be checked by setting up a control chart for proportions to see if any subsample points fell outside the limits. Other statistical tests comparing morning observations with afternoon observations could also be carried through.

Based on the work sampling study, we could then conclude that the delay part of the work in the lathe department was about 10 percent. We are 95 percent sure that the sampling error has been held to no more than ± 1.0 percentage point, and it is probable that it is less. We have based these conclusions on a study that covered two weeks of time, with any time of the day being equalled likely as a sampling time. The personal time of 5.6 percent is slightly greater than the company standard practice of allowing 5 percent; however, 5 percent is within the probable range of error of estimate.

Determining Production Standards. The previous example showed the use of work sampling to determine percentage allowances for noncyclical elements such as delays and for personal time. Why not carry the idea forward one more step and

TABLE F-2 **Summary of Work Sampling Data for Lathe Department Study**

Date		Total Observations	Work		Delay		Personal	
			Obs.	Percent	Obs.	Percent	Obs.	Percent
10-2	A.M.	190	152	80.0	24	12.6	14	7.4
	P.M.	170	145	85.3	14	8.2	11	6.5
10-3	A.M.	160	144	90.0	10	6.3	6	3.7
	P.M.	200	158	79.0	19	9.5	23	11.5
10-4	A.M.	150	127	84.7	15	10.0	8	5.3
	P.M.	210	182	86.6	23	11.0	5	2.4
10-5	A.M.	180	142	78.9	24	13.3	14	7.8
	P.M.	180	148	82.2	20	11.1	12	6.7
10-6	A.M.	220	189	85.9	24	10.9	7	3.2
	P.M.	140	114	81.4	17	12.1	9	6.5
10-9	A.M.	210	185	88.2	14	6.6	11	5.2
	P.M.	150	135	90.0	9	6.0	6	4.0
10-10	A.M.	190	155	81.6	25	13.2	10	5.2
	P.M.	170	146	85.9	14	8.2	10	5.9
10-11	A.M.	200	166	83.0	22	11.0	12	6.0
	P.M.	160	136	85.0	14	8.8	10	6.2
10-12	A.M.	140	118	84.3	15	10.7	7	5.0
	P.M.	220	185	84.1	25	11.4	10	4.5
10-13	A.M.	210	181	86.2	19	9.1	10	4.7
	P.M.	150	130	86.7	12	8.0	8	5.3
		3600	3038	84.4	359	9.97	203	5.63

utilize the observations on percentage of work time to establish production standards? What additional data do we need? If we know (1) how many pieces were produced during the total time of the study and (2) the performance rate for each observation of work time, we could compute normal time as follows:

$$\text{Normal time} = \frac{\begin{pmatrix}\text{total}\\ \text{time of}\\ \text{study in}\\ \text{minutes}\end{pmatrix} \times \begin{pmatrix}\text{work time}\\ \text{in decimals}\\ \text{from work}\\ \text{sampling study}\end{pmatrix} \times \begin{pmatrix}\text{average}\\ \text{performance}\\ \text{rating in}\\ \text{decimals}\end{pmatrix}}{\text{total number of pieces produced}}$$

Standard time is then computed as before:

$$\text{Standard time} = \text{normal time} + \text{allowances for delays, fatigue, and personal time}$$

We have already seen how the allowances for delays and personal time can be determined from work sampling. Here we see the complete determination of a production standard without the use of a precise timing device. All that was needed was a calendar from which we might calculate the total available time.

Although work sampling can be used in most situations, its most outstanding field of application is in the measurement of noncyclical types of work, where many different tasks are performed but where there is no set pattern of cycle or regularity. In many jobs, the frequency of tasks within the job is based on a random demand function. For example, a storeroom clerk may fill requisitions, unpack and put away stock, deliver material to production departments, clean up the storeroom, and so on. The frequency and time requirements of some of these tasks depend on things outside the control of the clerk. To determine such production standards by stopwatch methods would be difficult or impossible. Work sampling fits this situation ideally because, through its random sampling approach, reliable estimates of time and performance for these randomly occurring tasks can be obtained.

Standard Data Work Measurement Systems

Two kinds of standard data are used: universal data based on minute elements of motion (often called universal or microdata) and standard data for families of jobs (often called macrodata or element standard data).

Universal Standard Data. Universal standard data give time values for fundamental types of motions, so complete cycle times can be synthesized by analyzing the motions required to perform the task. Fundamental time values of this nature can be used as building blocks to forecast the standard time, provided that the time values are properly gathered and that the various minute motion elements required by the tasks are analyzed perfectly.

The result provided by these synthetic standards is an estimate of normal time for the task. Standard time is then determined as before by adding allowances for delay, fatigue, and personal time.

Does performance rating enter into standards developed from universal data? Not for each standard developed, because the analyst simply uses the time value from the table for a given motion. However, performance rating was used to develop the time values that are in the tables. So the rating factor enters the system, but not for each occasion that the data are used.

Many people feel that universal standard data lead to greater consistency of standards, since analysts are not called on to judge working pace in order to develop a standard. This does not mean that judgment is eliminated from the use of universal standard data systems, however. A great deal of judgment is required in selecting the appropriate classifications of motion to use in analyzing an operation. An inexperienced person ordinarily will not be able to perform these selections accurately enough for the purpose of determining production standards.

Using universal standard data as the sole basis for determining production standards is not common. In most cases where data of this kind are used, they are employed in conjunction with some other technique, such as stopwatch study or work sampling. The reason for this methodology seems to be that most organizations feel more comfortable when some actual direct measurement of the work involved has been made.

Standard Data for Job Families. Standard data for job families give normal time values for major elements of jobs (macrostandard data). Also, time values for machine setup and for different manual elements are given, so a normal time for an entirely new job can be constructed by an analysis of blueprints to see what materials are specified, what cuts must be made, how the work piece can be held in the machine, and so forth. Unlike the universal standard data discussed previously, however, the time values for these elements have been based on actual previous stopwatch or other measurement of work within the job family.

In these previous studies, the operations were consistently broken down into common elements until finally a system of data emerged that showed how "normal element time" varied with size, depth of cut, material used, the way the work piece was held in the machine, and the like. At that point the data themselves could be used to estimate production standards without a separate study actually being performed on every different part. Again, although individual performance rating does not enter each application of the standard data, it was used in constructing the data originally. As before, final production standards are determined by adding allowances for delays, fatigue, and personal time to the normal cycle time derived from the standard data. Macrostandard data are in common use, especially in machine shops where distinct job families have a long-standing tradition. The occurrence of this kind of standard data is likely to exist wherever job families exist or when parts of products occur in many sizes or types. Macrostandard data have a large field of application where short runs of custom parts and products occur. In these instances, if we attempt to determine production standards by actual measurement, the order may be completed by the time the production standard has been determined. The result will be of no value unless the identical part is reordered.

ALLOWANCES IN PRODUCTION STANDARDS

Allowances are commonly added to the computed normal time for delay, fatigue, and personal time. Allowances for delay and fatigue depend on the nature of the operation. They may not exist for some activities. The usual approach is to express allowances in percentage of the total available time. Thus, a 10 percent allowance over an 8-hour (480-minute) day is the equivalent of 48 minutes.

Delay Allowances

Delay allowances must be based on actual measurement of the magnitude of the delays. Although stopwatch study can be used, work sampling provides a much more efficient means of obtaining accurate data because delays often occur randomly. Work sampling expresses its measurement of delays directly in terms of percent of the total available time.

Fatigue and Personal Allowances

For some very heavy industrial jobs, an employee might work 20 minutes and rest 20 minutes. This type of very heavy work is not common today, but it occurs often enough for a continuing interest to be maintained on the subject of physical fatigue and rest allowances.

Unfortunately, we still lack an accepted framework for the establishment of rest allowances based on any rational or scientific measurements. In most instances, schedules of *fatigue allowances* for various types of work are used based on general acceptability and are often the subject of agreements between labor and management.

Allowances for *personal time* provide at least a minimum of time that the worker can be away from the job. This personal time allows a break from both the physical and psychological stresses that a job may contain and is, in a sense, a minimum fatigue allowance. The minimum allowance is normally 5 percent of the total available time.

Application of Allowances in Production Standards

The usual interpretation of the meaning of percentage allowances is that they allow a percentage of the total available time. A personal time allowance of 5 percent translates into $0.05 \times 480 = 24$ minutes of personal time in a normal 8-hour day. If the normal time has been measured as 1.20 minutes per piece, then the personal time must be prorated properly to the normal time in computing standard time per piece:

$$\begin{aligned}\text{Standard time} &= \text{normal time} \times \frac{100}{100 - \text{percentage allowance}} \\ &= 1.20 \times \frac{100}{95} = 1.263 \text{ minutes per piece}\end{aligned}$$

If all the allowances for delay, fatigue, and personal time are expressed as percentages of total available time, they can be added together to obtain a single total percentage allowance figure. Then, standard time can be computed from normal time by a single calculation using the preceding formula.

REFERENCES

Barnes, R. M. *Motion and Time Study: Design and Measurement of Work* (6th ed.). John Wiley, New York, 1968.

Barnes, R. M. *Work Sampling* (2nd ed.). John Wiley, New York, 1957.

Nadler, G. *Work Design A Systems Concepts* (Rev. ed.). Richard D. Irwin, Homewood, Ill., 1970.

Niebel, B. W. *Motion and Time Study* (6th ed.). Richard D. Irwin, Homewood, Ill., 1973.

Rice, R. S. "Survey of Work Measurement and Wage Incentives," *Industrial Engineering, 9*(7), July 1977, pp. 18–31.

APPENDIX G Linear Decision Rule Model

BASIC MODEL

The objective function is the sum of all costs over the planning horizon, as indicated in Equation 1. These costs are represented by the four cost functions segregated in Equation 2 as regular payroll, hiring and layoff, overtime, and inventory-connected costs.

The problem is then to minimize costs over N periods, or

$$C_N = \sum_{t=1}^{N} C_t \tag{1}$$

and

$$\begin{aligned}
C_t = {} & [(c_1 W_t) && \text{Regular payroll costs} \\
& + c_2 (W_t - W_{t-1})^2 && \text{Hiring and layoff costs} \\
& + C_3 (P_t - c_4 W_t)^2 + c_5 P_t - c_6 W_t && \text{Overtime costs} \\
& + c_7 (I_t - c_8 - c_9 S_t)^2] && \text{Inventory-connected costs}
\end{aligned} \tag{2}$$

subject to restraints,

$$I_{t-1} + P_t - F_t = I_t \qquad t = 1, 2, \ldots N \tag{3}$$

The total cost for N periods is given by Equation 1 and the monthly cost, C_t, is given by Equation 2. Equation 3 states the relationship between beginning inventory, production, and sales during the month, and ending inventory.

Equations 1, 2, and 3 are general and applicable to a broad range of situations. By estimating the values of the c's a specific factory cost structure can be specified. For a paint factory, Equation 4 is the result:

$$\begin{aligned}
C_N = \sum_{t=1}^{N} \{ & [340 W_t] + [64.3(W_t - W_{t-1})^2] \\
& + [0.20(P_t - 5.67 W_t)^2 + 51.2 P_t - 281 W_t] \\
& + [0.0825(I_t - 320)^2]\}
\end{aligned} \tag{4}$$

OPTIMAL DECISION RULES FOR THE PAINT FACTORY

A solution to Equation 4 was obtained by differentiating with respect to each decision variable. The result for the paint factory is contained in Equation 5 and 6:

$$P_t = \left\{ \begin{array}{l} +0.463\ F_t \\ +0.234\ F_{t+1} \\ +0.111\ F_{t+2} \\ +0.046\ F_{t+3} \\ +0.013\ F_{t+4} \\ -0.002\ F_{t+5} \\ -0.008\ F_{t+6} \\ -0.010\ F_{t+7} \\ -0.009\ F_{t+8} \\ -0.008\ F_{t+9} \\ -0.007\ F_{t+10} \\ -0.005\ F_{t+11} \end{array} \right\} + 0.993\ W_{t-1} + 153 - 0.464 I_{t-1} \qquad (5)$$

$$W_t = 0.743\ W_{t-1} + 2.09 - 0.010 I_{t-1} + \left\{ \begin{array}{l} +0.0101\ F_t \\ +0.0088\ F_{t+1} \\ +0.0071\ F_{t+2} \\ +0.0054\ F_{t+3} \\ +0.0042\ F_{t+4} \\ +0.0031\ F_{t+5} \\ +0.0023\ F_{t+6} \\ +0.0016\ F_{t+7} \\ +0.0012\ F_{t+8} \\ +0.0009\ F_{t+9} \\ +0.0006\ F_{t+10} \\ +0.0005\ F_{t+11} \end{array} \right\} \qquad (6)$$

where P_t = is the number of units of product that should be produced during the forthcoming month t.

W_{t-1} = is the number of employees in the work force at the beginning of the month (end of the previous month).

I_{t-1} = is the number of units of inventory minus the number of units on back order at the beginning of the month.

W_t = is the number of employees that will be required for the current month t. The number of employees that should be hired is therefore $W_t - W_{t-1}$.

F_t = is a forecast of number of units of product that will be ordered for shipment during the current month t.

F_{t+1} = is the same for the next month, $t + 1$, etc.

Equations 5 and 6 would be used at the beginning of each month. Equation 5 determines the aggregate production rate and Equation 6, the aggregate size of the work force.

APPENDIX H Tables

TABLE H-1 ***PV_{sp}*, Present-Value Factors for Future Single Payments**

Years Hence	1%	2%	4%	6%	8%	10%	12%	14%	15%	16%	18%	20%
1	0.990	0.980	0.962	0.943	0.926	0.909	0.893	0.877	0.870	0.862	0.847	0.833
2	0.980	0.961	0.925	0.890	0.857	0.826	0.797	0.769	0.756	0.743	0.718	0.694
3	0.971	0.942	0.889	0.840	0.794	0.751	0.712	0.675	0.658	0.641	0.609	0.579
4	0.691	0.924	0.855	0.792	0.735	0.683	0.636	0.592	0.572	0.552	0.516	0.482
5	0.951	0.906	0.822	0.747	0.681	0.621	0.567	0.519	0.497	0.476	0.437	0.402
6	0.942	0.888	0.790	0.705	0.630	0.564	0.507	0.456	0.432	0.410	0.370	0.335
7	0.933	0.871	0.760	0.665	0.583	0.513	0.452	0.400	0.376	0.354	0.314	0.279
8	0.923	0.853	0.731	0.627	0.540	0.467	0.404	0.351	0.327	0.305	0.266	0.233
9	0.914	0.837	0.703	0.592	0.500	0.424	0.361	0.308	0.284	0.263	0.225	0.194
10	0.905	0.820	0.676	0.558	0.463	0.386	0.322	0.270	0.247	0.227	0.191	0.162
11	0.896	0.804	0.650	0.527	0.429	0.350	0.287	0.237	0.215	0.195	0.162	0.135
12	0.887	0.788	0.625	0.497	0.397	0.319	0.257	0.208	0.187	0.168	0.137	0.112
13	0.879	0.773	0.601	0.469	0.368	0.290	0.229	0.182	0.163	0.145	0.116	0.093
14	0.870	0.758	0.577	0.442	0.340	0.263	0.205	0.160	0.141	0.125	0.099	0.078
15	0.861	0.743	0.555	0.417	0.315	0.239	0.183	0.140	0.123	0.108	0.084	0.065
16	0.853	0.728	0.534	0.394	0.292	0.218	0.163	0.123	0.107	0.093	0.071	0.054
17	0.844	0.714	0.513	0.371	0.270	0.198	0.146	0.108	0.093	0.080	0.060	0.045
18	0.836	0.700	0.494	0.350	0.250	0.180	0.130	0.095	0.081	0.069	0.051	0.038
19	0.828	0.686	0.475	0.331	0.232	0.164	0.116	0.083	0.070	0.060	0.043	0.031
20	0.820	0.673	0.456	0.312	0.215	0.149	0.104	0.073	0.061	0.051	0.037	0.026

TABLE H-2 ***PV_a*, Present-Value Factors for Annuities**

Years (n)	1%	2%	4%	6%	8%	10%	12%	14%	15%	16%	18%	20%
1	0.990	0.980	0.962	0.943	0.926	0.909	0.893	0.877	0.870	0.862	0.847	0.833
2	1.970	1.942	1.886	1.883	1.783	1.736	1.690	1.647	1.626	1.605	1.566	1.528
3	2.941	2.884	2.775	2.673	2.577	2.487	2.402	2.322	2.283	2.246	2.174	2.106
4	3.902	3.808	3.630	3.465	3.312	3.170	3.037	2.914	2.855	2.798	2.690	2.589
5	4.853	4.713	4.452	4.212	3.993	3.791	3.605	3.433	3.352	3.274	3.127	2.991
6	5.795	5.601	5.242	4.917	4.623	4.355	4.111	3.889	3.784	3.685	3.498	3.326
7	6.728	6.472	6.002	5.582	5.206	4.868	4.564	4.288	4.160	4.039	3.812	3.605
8	7.652	7.325	6.733	6.210	5.747	5.335	4.968	4.639	4.487	4.344	4.078	3.837
9	8.566	8.162	7.435	6.802	6.247	5.759	5.328	4.946	4.772	4.607	4.303	4.031
10	9.471	8.983	8.111	7.360	6.710	6.145	5.650	5.126	5.019	4.833	4.494	4.192
11	10.368	9.787	8.760	7.887	7.139	6.495	5.988	5.453	5.234	5.029	4.656	4.327
12	11.255	10.575	9.385	8.384	7.536	6.814	6.194	5.660	5.421	5.197	4.793	4.439
13	12.134	11.343	9.986	8.853	7.904	7.103	6.424	5.842	5.583	5.342	4.910	4.533
14	13.004	12.106	10.563	9.295	8.244	7.367	6.628	6.002	5.724	5.468	5.008	4.611
15	13.865	12.849	11.118	9.712	8.559	7.606	6.811	6.142	5.847	5.575	5.092	4.675
16	14.718	13.578	11.652	10.106	8.851	7.824	6.974	6.265	5.954	5.669	5.162	4.730
17	15.562	14.292	12.166	10.477	9.122	8.022	7.120	6.373	6.047	5.749	5.222	4.775
18	16.398	14.992	12.659	10.828	9.372	8.201	7.250	6.467	6.128	5.818	5.273	4.812
19	17.226	15.678	13.134	11.158	9.604	8.365	7.366	6.550	6.198	5.877	5.316	4.844
20	18.046	16.351	13.590	11.470	9.818	8.514	7.469	6.623	6.259	5.929	5.353	4.870

TABLE H-3 **Values of L_q for $M = 1 - 15$, and Various Values of $r = \lambda/\mu$. Poisson Arrivals, Negative Exponential Service Times**

	Number of Service Channels M														
r	1	2	3	4	5	6	7	8	9	10	11	12	13	14	15
0.10	0.0111														
0.15	0.0264	0.0008													
0.20	0.0500	0.0020													
0.25	0.0833	0.0039													
0.30	0.1285	0.0069													
0.35	0.1884	0.0110													
0.40	0.2666	0.0166													
0.45	0.3681	0.0239	0.0019												
0.50	0.5000	0.0333	0.0030												
0.55	0.6722	0.0449	0.0043												
0.60	0.9000	0.0593	0.0061												
0.65	1.2071	0.0767	0.0084												
0.70	1.6333	0.0976	0.0112												
0.75	2.2500	0.1227	0.0147												
0.80	3.2000	0.1523	0.0189												
0.85	4.8166	0.1873	0.0239	0.0031											
0.90	8.1000	0.2285	0.0300	0.0041											
0.95	18.0500	0.2767	0.0371	0.0053											
1.0		0.3333	0.0454	0.0067											
1.2		0.6748	0.0904	0.0158											
1.4		1.3449	0.1778	0.0324	0.0059										
1.6		2.8444	0.3128	0.0604	0.0121										
1.8		7.6734	0.5320	0.1051	0.0227	0.0047									
2.0			0.8888	0.1739	0.0398	0.0090									
2.2			1.4907	0.2770	0.0659	0.0158									
2.4			2.1261	0.4305	0.1047	0.0266	0.0065								
2.6			4.9322	0.6581	0.1609	0.0426	0.0110								
2.8			12.2724	1.0000	0.2411	0.0659	0.0180								
3.0				1.5282	0.3541	0.0991	0.0282	0.0077							
3.2				2.3856	0.5128	0.1452	0.0427	0.0122							
3.4				3.9060	0.7365	0.2085	0.0631	0.0189							
3.6				7.0893	1.0550	0.2947	0.0912	0.0283	0.0084						
3.8				16.9366	1.5184	0.4114	0.1292	0.0412	0.0127						

TABLE H-3 **Values of L_q for $M = 1 - 15$, and Various Values of $r = \lambda/\mu$. Poisson Arrivals, Negative Exponential Service Times**
Continued

	Number of Service Channels M														
r	1	2	3	4	5	6	7	8	9	10	11	12	13	14	15
4.0					2.2164	0.5694	0.1801	0.0590	0.0189						
4.2					3.3269	0.7837	0.2475	0.0827	0.0273	0.0087					
4.4					5.2675	1.0777	0.3364	0.1142	0.0389	0.0128					
4.6					9.2885	1.4867	0.4532	0.1555	0.0541	0.0184					
4.8					21.6384	2.0708	0.6071	0.2092	0.0742	0.0260					
5.0						2.9375	0.8102	0.2786	0.1006	0.0361	0.0125				
5.2						4.3004	1.0804	0.3680	0.1345	0.0492	0.0175				
5.4						6.6609	1.4441	0.5871	0.1779	0.0663	0.0243	0.0085			
5.6						11.5178	1.9436	0.6313	0.2330	0.0883	0.0330	0.0119			
5.8						26.3726	2.6481	0.8225	0.3032	0.1164	0.0443	0.0164			
6.0							3.6828	1.0707	0.3918	0.1518	0.0590	0.0224			
6.2							5.2979	1.3967	0.5037	0.1964	0.0775	0.0300	0.0113		
6.4							8.0768	1.8040	0.6454	0.2524	0.1008	0.0398	0.0153		
6.6							13.7692	2.4198	0.8247	0.3222	0.1302	0.0523	0.0205		
6.8							31.1270	3.2441	1.0533	0.4090	0.1666	0.0679	0.0271	0.0105	
7.0								4.4471	1.3471	0.5172	0.2119	0.0876	0.0357	0.0141	
7.2								6.3135	1.7288	0.6521	0.2677	0.1119	0.0463	0.0187	
7.4								9.5102	2.2324	0.8202	0.3364	0.1420	0.0595	0.0245	0.0097
7.6								16.0379	2.9113	1.0310	0.4211	0.1789	0.0761	0.0318	0.0129
7.8								35.8956	3.8558	1.2972	0.5250	0.2243	0.0966	0.0410	0.0168
8.0									5.2264	1.6364	0.6530	0.2796	0.1214	0.0522	0.0220
8.2									7.3441	2.0736	0.8109	0.3469	0.1520	0.0663	0.0283
8.4									10.9592	2.6470	1.0060	0.4288	0.1891	0.0834	0.0361
8.6									18.3223	3.4160	1.2484	0.5286	0.2341	0.1043	0.0459
8.8									40.6824	4.4806	1.5524	0.6501	0.2885	0.1298	0.0577
9.0										6.0183	1.9368	0.7980	0.3543	0.1603	0.0723
9.2										8.3869	2.4298	0.9788	0.4333	0.1974	0.0899
9.4										12.4189	3.0732	1.2010	0.5287	0.2419	0.1111
9.6										20.6160	3.9318	1.4752	0.6437	0.2952	0.1367
9.8										45.4769	5.1156	1.8165	0.7827	0.3588	0.1673
10.0											6.8210	2.2465	0.9506	0.4352	0.2040

TABLE H-4 **Finite Queuing Tables**

Population 5 X	M	D	F	L_q	X	M	D	F	L_q	X	M	D	F	L_q
Population 5					.090	2	.044	.998	.010	.170	3	.017	.999	.005
						1	.350	.960	.200		2	.145	.985	.075
.012	1	.048	.999	.005	.095	2	.049	.997	.015		1	.611	.853	.735
.019	1	.076	.998	.010		1	.368	.955	.255	.180	3	.021	.999	.005
.025	1	.100	.997	.015	.100	2	.054	.997	.015		2	.161	.983	.085
.030	1	.120	.996	.020		1	.386	.950	.250		1	.638	.836	.820
.034	1	.135	.995	.025	.105	2	.059	.997	.015	.190	3	.024	.998	.010
.036	1	.143	.994	.030		1	.404	.945	.275		2	.177	.980	.100
.040	1	.159	.993	.035	.110	2	.065	.996	.020		1	.665	.819	.905
.042	1	.167	.992	.045		1	.421	.939	.305	.200	3	.028	.998	.010
.044	1	.175	.991	.045	.115	2	.071	.995	.025		2	.194	.976	.120
.046	1	.183	.990	.050		1	.439	.933	.335		1	.689	.801	.995
.050	1	.198	.989	.055	.120	2	.076	.995	.025	.210	3	.032	.998	.010
.052	1	.206	.988	.060		1	.456	.927	.365		2	.211	.973	.135
.054	1	.214	.987	.065	.125	2	.082	.994	.030		1	.713	.783	1.085
.056	2	.018	.999	.005		1	.473	.920	.400	.220	3	.036	.997	.015
	1	.222	.985	.075	.130	2	.089	.993	.035		2	.229	.969	.155
.058	2	.019	.999	.005		1	.489	.914	.430		1	.735	.765	1.175
	1	.229	.984	.080	.135	2	.095	.993	.035	.230	3	.041	.997	.015
.060	2	.020	.999	.005		1	.505	.907	.465		2	.247	.965	.175
	1	.237	.983	.085	140	2	.102	.992	.040		1	.756	.747	1.265
.062	2	.022	.999	.005		1	.521	.900	.500	.240	3	.046	.996	.020
	1	.245	.982	.090	.145	3	.011	.999	.005		2	.265	.960	.200
.064	2	.023	.999	.005		2	.109	.991	.045		1	.775	.730	1.350
	1	.253	.981	.095		1	.537	.892	.540	.250	3	.052	.995	.025
.066	2	.024	.999	.005	.150	3	.012	.999	.005		2	.284	.955	.225
	1	.260	.979	.105		2	.115	.990	.050		1	.794	.712	1.440
.068	2	.026	.999	.005		1	.553	.885	.575	.260	3	.058	.994	.030
	1	.268	.978	.110	.155	3	.013	.999	.005		2	.303	.950	.250
.070	2	0.27	.999	.005		2	.123	.989	.055		1	.811	.695	1.525
	1	.275	.977	.115		1	.568	.877	.615	.270	3	.064	.994	.030
.075	2	.031	.999	.005	.160	3	.015	.999	.005		2	.323	.944	.280
	1	.294	.973	.135		2	.130	.988	.060		1	.827	.677	1.615
.080	2	.035	.998	.010		1	.582	.869	.655	.280	3	.017	.993	.035
	1	.313	.969	.155	.165	3	.016	.999	.005		2	.342	.938	.310
.085	2	.040	.998	.010		2	.137	.987	.065		1	.842	.661	1.695
	1	.332	.965	.175		1	.597	.861	.695	.290	4	.007	.999	.005

SOURCE: Adapted from L. G. Peck and R. N. Hazelwood, *Finite Queuing Tables*. John Wiley, New York, 1958.

TABLE H-4 **Finite Queuing Tables**
Continued

Population 5–10

X	M	D	F	L_q	X	M	D	F	L_q	X	M	D	F	L_q
	3	.079	.992	.040	.440	4	.037	.996	.020		1	.998	.308	3.460
	2	.362	.932	.340		3	.238	.960	.200	.700	4	.240	.960	.200
	1	.856	.644	1.780		2	.652	.807	.965		3	.678	.815	.925
.300	4	.008	.999	.005		1	.969	.451	2.745		2	.950	.568	2.160
	3	.086	.990	.050	.460	4	.045	.995	.025		1	.999	.286	3.570
	2	.382	.926	.370		3	.266	.953	.235	.750	4	.316	.944	.280
	1	.869	.628	1.860		2	.686	.787	1.065		3	.763	.777	1.115
.310	4	.009	.999	.005		1	.975	.432	2.840		2	.972	.532	2.340
	3	.094	.989	.055	.480	4	.053	.994	.030	.800	4	.410	.924	.380
	2	.402	.919	.405		3	.296	.945	.275		3	.841	.739	1.305
	1	.881	.613	1.935		2	.719	.767	1.165		2	.987	.500	2.500
.320	4	.010	.999	.005		1	.980	.415	2.925	.850	4	.522	.900	.500
	3	.103	.988	.060	.500	4	.063	.992	.040		3	.907	.702	1.490
	2	.422	.912	.440		3	.327	.936	.320		2	.995	.470	2.650
	1	.892	.597	2.015		2	.750	.748	1.260	.900	4	.656	.871	.645
.330	4	.012	.999	.005		1	.985	.399	3.005		3	.957	.666	1.670
	3	.112	.986	.070	.520	4	.073	.991	.045		2	.998	.444	2.780
	2	.442	.904	.480		3	.359	.927	.365	.950	4	.815	.838	.810
	1	.902	.583	2.085		2	.779	.728	1.360		3	.989	.631	1.845
.340	4	.013	.999	.005		1	.988	384	3.080					
	3	.121	.985	.075	.540	4	.085	.989	.055			Population 10		
	2	.462	.896	.520		3	.392	.917	.415					
	1	.911	.569	2.155		2	.806	.708	1.460	.016	1	.144	.997	.03
.360	4	.017	.998	.010		1	.991	.370	3.150	.019	1	.170	.996	.04
	3	.141	.981	.060	.560	4	.098	.986	.070	.021	1	.188	.995	.05
	2	.501	.880	.600		3	.426	.906	.470	.023	1	.206	.994	.06
	1	.927	.542	2.290		2	.831	.689	1.555	.025	1	.224	.993	.07
.380	4	.021	.998	.010		1	.993	.357	3.215	.026	1	.232	.992	.08
	3	.163	.976	.120	.580	4	.113	.984	.080	.028	1	.250	.991	.09
	2	.540	.863	.685		3	.461	.895	.525	.030	1	.268	.990	.10
	1	.941	.516	2.420		2	.854	.670	1.650	.032	2	.033	.999	.01
.400	4	.026	.997	.015		1	.994	.345	3.275		1	.285	.988	.12
	3	.186	.972	.140	.600	4	.130	.981	.095	.034	2	.037	.999	.01
	2	.579	.845	.755		3	.497	.883	.585		1	.302	.986	.14
	1	.952	.493	2.535		2	.875	.652	1.740	.036	2	.041	.999	.01
.420	4	.031	.997	.015		1	.996	.333	3.335		1	.320	.984	.16
	3	.211	.966	.170	.650	4	.170	.972	.140	.038	2	.046	.999	.01
	2	.616	.826	.870		3	.588	.850	.750		1	.337	.982	.18
	1	.961	.471	2.645		2	.918	.608	1.960	.040	2	.050	.999	.01

TABLE H-4 Continued **Finite Queuing Tables**

Population 10

X	M	D	F	L_q	X	M	D	F	L_q	X	M	D	F	L_q
	1	.354	.980	.20	.085	3	.037	.999	.01		2	.437	.947	.53
	1	.371	.978	.22		1	.692	.883	1.17		1	.919	.680	3.20
.042	2	.055	.999	.01		2	.196	.988	.12	.145	4	.032	.999	.01
.044	2	.060	.998	.02	.090	3	.043	.998	.02		3	.144	.990	.10
	1	.388	.975	.25		2	.216	.986	.14		2	.460	.941	.59
.046	2	.065	.998	.02		1	.722	.867	1.33		1	.929	.662	3.38
	1	.404	.973	.27	.095	3	.049	.998	.02	.150	4	.036	.998	.02
.048	2	.071	.998	.02		2	.237	.984	.16		3	.156	.989	.11
	1	.421	.970	.30		1	.750	.850	1.50		2	.483	.935	.65
.050	2	.076	.998	.02	.100	3	.056	.998	.02		1	.939	.644	3.56
	1	.437	.967	.33		2	.258	.981	.19	.155	4	.040	.998	.02
.052	2	.082	.997	.03		1	.776	.832	1.68		3	.169	.987	.13
	1	.454	.963	.37	.105	3	.064	.997	.03		2	.505	.928	.72
.054	2	.088	.997	.03		2	.279	.978	.22		1	.947	.627	3.73
	1	.470	.960	.40		1	.800	.814	1.86	.160	4	.044	.998	.02
.056	2	.094	.997	.03	.110	3	.072	.997	.03		3	.182	.986	.14
	1	.486	.956	.44		2	.301	.974	.26		2	.528	.921	.79
.058	2	.100	.996	.04		1	.922	.795	2.05		1	.954	.610	3.90
	1	.501	.953	.47	.115	3	.081	.996	.04	.165	4	.049	.997	.03
.060	2	.106	.996	.04		2	.324	.971	.29		3	.195	.984	.16
	1	.517	.949	.51		1	.843	.776	2.24		2	.550	.914	.86
.062	2	.113	.996	.04	.120	4	.016	.999	.01		1	.961	.594	4.06
	1	.532	.945	.55		3	.090	.995	.05	.170	4	.054	.997	.03
.064	2	.119	.995	.05		2	.346	.967	.33		3	.209	.982	.18
	1	.547	.940	.60		1	.861	.756	2.44		2	.571	.906	.94
.066	2	.126	.995	.05	.125	4	.019	.999	.01		1	.966	.579	4.21
	1	.562	.936	.64		3	.100	.994	.06	.180	5	.013	.999	.01
.068	3	.020	.999	.01		2	.369	.962	.38		4	.066	.996	.04
	2	.133	.994	.06		1	.878	.737	2.63		3	.238	.978	.22
	1	.577	.931	.69	.130	4	.022	.999	.01		2	.614	.890	1.10
.070	3	.022	.999	.01		3	.110	.994	.06		1	.975	.549	4.51
	2	.140	.994	.06		2	.392	.958	.42	.190	5	.016	.999	.01
	1	.591	.926	.74		1	.893	.718	2.82		4	.078	.995	.05
.075	3	.026	.999	.01	.135	4	.025	.999	.01		3	.269	.973	.27
	2	.158	.992	.08		3	.121	.993	.07		2	.654	.873	1.27
	1	.627	.913	.87		2	.415	.952	.48		1	.982	.522	4.78
.080	3	.031	.999	.01		1	.907	.669	3.01	.200	5	.020	.999	.01
	2	.117	.990	.10	.140	4	.028	.999	.01		4	.092	.994	.06
	1	.660	.899	1.01		3	.132	.991	.09		3	.300	.968	.32

TABLE H-4 Continued **Finite Queuing Tables**

Population 10

X	M	D	F	L_q	X	M	D	F	L_q	X	M	D	F	L_q
	2	.692	.854	1.46		1	.999	.370	6.30		2	.968	.584	4.16
	1	.987	.497	5.03	.280	6	.018	.999	.01	.360	7	.014	.999	.01
.210	5	.025	.999	.01		5	.081	.994	.06		6	.064	.995	.05
	4	.108	.992	.08		4	.252	.972	.28		5	.205	.978	.22
	3	.333	.961	.39		3	.571	.896	1.04		4	.474	.923	.77
	2	.728	.835	1.65		2	.903	.692	3.08		3	.798	.787	2.13
	1	.990	.474	5.26		1	.999	.357	6.43		2	.978	.553	4.47
.220	5	.030	.998	.02	.290	6	.022	.999	.01	.380	7	.019	.999	.01
	4	.124	.990	.10		5	.093	.993	.07		6	.083	.993	.07
	3	.366	.954	.46		4	.278	.968	.32		5	.247	.971	.29
	2	.761	.815	1.85		3	.603	.884	1.16		4	.533	.906	.94
	1	.993	.453	5.47		2	.918	.672	3.28		3	.840	.758	2.42
.230	5	.037	.998	.02		1	.999	.345	6.55		2	.986	.525	4.75
	4	.142	.988	.12	.300	6	.026	.998	.02	.400	7	.026	.998	.02
	3	.400	.947	.53		5	.106	.991	.09		6	.105	.991	.09
	2	.791	.794	2.06		4	.304	.963	.37		5	.292	.963	.37
	1	.995	.434	5.66		3	.635	.872	1.28		4	.591	.887	1.13
.240	5	.044	.997	.03		2	.932	.653	3.47		3	.875	.728	2.72
	4	.162	.986	.14		1	.999	.333	6.67		2	.991	.499	5.01
	3	.434	.938	.62	.310	6	.031	.998	.02	.420	7	.034	.993	.07
	2	.819	.774	2.26		5	.120	.990	.10		6	.130	.987	.13
	1	.996	.416	5.84		4	.331	.957	.43		5	.341	.954	.46
.250	6	.010	.999	.01		3	.666	.858	1.42		4	.646	.866	1.34
	5	.052	.997	.03		2	.943	.635	3.65		3	.905	.700	3.00
	4	.183	.983	.17	.320	6	.036	.998	.02		2	.994	.476	5.24
	3	.469	.929	.71		5	.135	.998	.12	.440	7	.045	.997	.03
	2	.844	.753	2.47		4	.359	.952	.48		6	.160	.984	.16
	1	.997	.400	6.00		3	.695	.845	1.55		5	.392	.943	.57
.260	6	.013	.999	.01		2	.952	.617	3.83		4	.698	.845	1.55
	5	.060	.996	.04	.330	6	.042	.997	.03		3	.928	.672	3.28
	4	.205	.980	.20		5	.151	.986	.14		2	.996	.454	5.46
	3	.503	.919	.81		4	.387	.945	.55	.460	8	.011	.999	.01
	2	.866	.732	2.68		3	.723	.831	1.69		7	.058	.995	.05
	1	.998	.384	6.16		2	.961	.600	4.00		6	.193	.979	.21
.270	6	.015	.999	.01	.340	7	.010	.999	.01		5	.445	.930	.70
	5	.070	.995	.05		6	.049	.997	.03		4	.747	.822	1.78
	4	.228	.976	.24		5	.168	.983	.17		3	.947	.646	3.54
	3	.537	.908	.92		4	.416	.938	.62		2	.998	.435	5.65
	2	.886	.712	2.88		3	.750	.816	1.84	.480	8	.015	.999	.01

TABLE H-4 **Finite Queuing Tables**

Continued

Population 10-20

X	M	D	F	L_q	X	M	D	F	L_q	X	M	D	F	L_q
	7	.074	.994	.06		8	.072	.994	.06		6	.995	.667	3.33
	6	.230	.973	.27		7	.242	.972	.28	.950	9	.630	.938	.62
	5	.499	.916	.84	.600	6	.518	.915	.85		8	.934	.841	1.59
	4	.791	.799	2.01		5	.795	.809	1.91		7	.994	.737	2.63
	3	.961	.621	3.79		4	.953	.663	3.37					
	2	.998	.417	5.83		3	.996	.500	5.00			Population 20		
.500	8	.020	.999	.01	.650	9	.021	.999	.01					
	7	.093	.992	.08		8	.123	.988	.12	.005	1	.095	.999	.02
	6	.271	.966	.34		7	.353	.954	.46	.009	1	.171	.998	.04
	5	.553	.901	.99		6	.651	.878	1.22	.011	1	.208	.997	.06
	4	.830	.775	2.25		5	.882	.759	2.41	.013	1	.246	.996	.08
	3	.972	.598	4.02		4	.980	.614	3.86	.014	1	.265	.995	.10
	2	.999	.400	6.00		3	.999	.461	5.39	.015	1	.283	.994	.12
.520	8	.026	.998	.02	.700	9	.040	.997	.03	.016	1	.302	.993	.14
	7	.115	.989	.11		8	.200	.979	.21	.017	1	.321	.992	.16
	6	.316	.958	.42		7	.484	.929	.71	.018	2	.048	.999	.02
	5	.606	.884	1.16		6	.772	.836	1.64		1	.339	.991	.18
	4	.864	.752	2.48		5	.940	.711	2.89	.019	2	.053	.999	.02
	3	.980	.575	4.25		4	.992	.571	4.29		1	.358	.990	.20
	2	.999	.385	6.15	.750	9	.075	.994	.06	.020	2	.058	.999	.02
.540	8	.034	.997	.03		8	.307	.965	.35		1	.376	.989	.22
	7	.141	.986	.14		7	.626	.897	1.03	.021	2	.064	.999	.02
	6	.363	.949	.51		6	.870	.792	2.08		1	.394	.987	.26
	5	.658	.867	1.33		5	.975	.666	3.34	.022	2	.070	.999	.02
	4	.893	.729	2.71		4	.998	.533	4.67		1	.412	.986	.28
	3	.986	.555	4.45	.800	9	.134	.988	.12	.023	2	.075	.999	.02
.560	8	.044	.996	.04		8	.446	.944	.56		1	.431	.984	.32
	7	.171	.982	.18		7	.763	.859	1.41	.024	2	.082	.999	.02
	6	.413	.939	.61		6	.939	.747	2.53		1	.449	.982	.36
	5	.707	.848	1.52		5	.991	.625	3.75	.025	2	.088	.999	.02
	4	.917	.706	2.94		4	.999	.500	5.00		1	.466	.980	.40
	3	.991	.535	4.65	.850	9	.232	.979	.21	.026	2	.094	.998	.04
.580	8	.057	.995	.05		8	.611	.916	.84		1	.484	.978	.44
	7	.204	.977	.23		7	.879	.818	1.82	.028	2	.108	.998	.04
	6	.465	.927	.73		6	.978	.705	2.95		1	.519	.973	.54
	5	.753	.829	1.71		5	.998	.588	4.12	.030	2	.122	.998	.04
	4	.937	.684	3.16	.900	9	.387	.963	.37		1	.553	.968	.64
	3	.994	.517	4.83		8	.785	.881	1.19	.032	2	.137	.997	.06
.600	9	.010	.999	.01		7	.957	.777	2.23		1	.587	.962	.76

TABLE H-4 **Finite Queuing Tables**
Continued

Population 20

X	M	D	F	L_q	X	M	D	F	L_q	X	M	D	F	L_q
.034	2	.152	.996	.08		2	.392	.978	.44	.095	5	.031	.999	.02
	1	.620	.955	.90		1	.922	.785	4.30		4	.112	.996	.08
.036	2	.168	.996	.08	.062	4	.029	.999	.02		3	.326	.980	.40
	1	.651	.947	1.06		3	.124	.996	.08		2	.733	.896	2.08
.038	3	.036	.999	.02		2	.413	.975	.50		1	.998	.526	9.48
	2	.185	.995	.10		1	.934	.768	4.64	.100	5	.038	.999	.02
	1	.682	.938	1.24	.064	4	.032	.999	.02		4	.131	.995	.10
.040	3	.041	.999	.02		3	.134	.996	.08		3	.363	.975	.50
	2	.202	.994	.12		2	.433	.972	.56		2	.773	.878	2.44
	1	.712	.929	1.42		1	.944	.751	4.98		1	.999	.500	10.00
.042	3	.047	.999	.02	.066	4	.036	.999	.02	.105	5	.046	.999	.02
	2	.219	.993	.14		3	.144	.995	.10		4	.151	.993	.14
	1	.740	.918	1.64		2	.454	.969	.62		3	.400	.970	.60
.044	3	.053	.999	.02		1	.953	.733	5.34		2	.809	.858	2.84
	2	.237	.992	.16	.068	4	.039	.999	.02		1	.999	.476	10.48
	1	.767	.906	1.88		3	.155	.995	.10	.110	5	.055	.998	.04
.046	3	.059	.999	.02		2	.474	.966	.68		4	.172	.992	.16
	2	.255	.991	.18		1	.961	.716	5.68		3	.438	.964	.72
	1	.792	.894	2.12	.070	4	.043	.999	.02		2	.842	.837	3.26
.048	3	.066	.999	.02		3	.165	.994	.12	.115	5	.065	.998	.04
	2	.274	.989	.22		2	.495	.962	.76		4	.195	.990	.20
	1	.815	.881	2.38		1	.967	.699	6.02		3	.476	.958	.84
.050	3	.073	.998	.04	.075	4	.054	.999	.02		2	.870	.816	3.68
	2	.293	.988	.24		3	.194	.992	.16	.120	6	.022	.999	.02
	1	.837	.866	2.68		2	.545	.953	.94		5	.076	.997	.06
.052	3	.080	.998	.04		1	.980	.659	6.82		4	.219	.988	.24
	2	.312	.986	.28	.080	4	.066	.998	.04		3	.514	.950	1.00
	1	.858	.851	2.98		3	.225	.990	.20		2	.895	.793	4.14
.054	3	.088	.998	.04		2	.595	.941	1.18	.125	6	.026	.999	.02
	2	.332	.984	.32		1	.988	.621	7.58		5	.088	.997	.06
	1	.876	.835	3.30	.085	4	.080	.997	.06		4	.245	.986	.28
.056	3	.097	.997	.06		3	.257	.987	.26		3	.552	.942	1.16
	2	.352	.982	.36		2	.643	.928	1.44		2	.916	.770	4.60
	1	.893	.819	3.62		1	.993	.586	8.28	.130	6	.031	.999	.02
.058	3	.105	.997	.06	.090	5	.025	.999	.02		5	.101	.996	.08
	2	.372	.980	.40		4	.095	.997	.06		4	.271	.983	.34
	1	.908	.802	3.96		3	.291	.984	.32		3	.589	.933	1.34
.060	4	.026	.999	.02		2	.689	.913	1.74		2	.934	.748	5.04
	3	.115	.997	.06		1	.996	.554	8.92	.135	6	.037	.999	.02

TABLE H-4 Continued **Finite Queuing Tables**

Population 20

X	M	D	F	L_q	X	M	D	F	L_q	X	M	D	F	L_q
	5	.116	.995	.10		6	.099	.995	.10		8	.054	.998	.04
	4	.299	.980	.04		5	.248	.983	.34		7	.140	.992	.16
	3	.626	.923	1.54		4	.513	.945	1.10		6	.306	.975	.50
	2	.948	.725	5.50		3	.838	.830	3.40		5	.560	.931	1.38
.140	6	.043	.998	.04		2	.993	.587	8.26		4	.834	.828	3.44
	5	.131	.994	.12	.180	7	.044	.998	.04		3	.981	.649	7.02
	4	.328	.976	.48		6	.125	.994	.12	.240	9	.024	.999	.02
	3	.661	.912	1.76		5	.295	.978	.44		8	.068	.997	.06
	2	.960	.703	5.94		4	.575	.930	1.40		7	.168	.989	.22
.145	6	.051	.998	.04		3	.879	.799	4.02		6	.351	.969	.62
	5	.148	.993	.14		2	.996	.555	8.90		5	.613	.917	1.66
	4	.358	.972	.56	.190	8	.018	.999	.02		4	.870	.804	3.92
	3	.695	.900	2.00		7	.058	.998	.04		3	.988	.623	7.54
	2	.969	.682	6.36		6	.154	.991	.18	.250	9	.031	.999	.02
.150	7	.017	.999	.02		5	.345	.971	.58		8	.085	.996	.08
	6	.059	.998	.04		4	.636	.914	1.72		7	.199	.986	.28
	5	.166	.991	.18		3	.913	.768	4.64		6	.398	.961	.78
	4	.388	.968	.64		2	.998	.526	9.48		5	.664	.901	1.98
	3	.728	.887	2.26	.200	8	.025	.999	.02		4	.900	.780	4.40
	2	.976	.661	6.78		7	.074	.997	.06		3	.992	.599	8.02
.155	7	.021	.999	.02		6	.187	.988	.24	.260	9	.039	.998	.04
	6	.068	.997	.06		5	.397	.963	.74		8	.104	.994	.12
	5	.185	.990	.20		4	.693	.895	2.10		7	.233	.983	.34
	4	.419	.963	.74		3	.938	.736	5.28		6	.446	.953	.94
	3	.758	.874	2.52		2	.999	.500	10.00		5	.712	.884	2.32
	2	.982	.641	7.18	.210	8	.033	.999	.02		4	.924	.755	4.90
.160	7	.024	.999	.02		7	.093	.995	.10		3	.995	.576	8.48
	6	.077	.997	.06		6	.223	.985	.30	.270	10	.016	.999	.02
	5	.205	.988	.24		5	.451	.954	.92		9	.049	.998	.04
	4	.450	.947	.86		4	.745	.874	2.52		8	.125	.992	.16
	3	.787	.860	2.80		3	.958	.706	5.88		7	.270	.978	.44
	2	.987	.622	7.56		2	.999	.476	10.48		6	.495	.943	1.14
.165	7	.029	.999	.02	.220	8	.043	.998	.04		5	.757	.867	2.66
	6	.088	.996	.08		7	.115	.994	.12		4	.943	.731	5.38
	5	.226	.986	.28		6	.263	.980	.40		3	.997	.555	8.90
	4	.482	.951	.98		5	.505	.943	1.14	.280	10	.021	.999	.02
	3	.813	.845	3.10		4	.793	.852	2.96		9	.061	.997	.06
	2	.990	.604	7.92		3	.971	.677	6.46		8	.149	.990	.20
.170	7	.033	.999	.02	.230	9	.018	.999	.02		7	.309	.973	.54

TABLE H-4 Continued **Finite Queuing Tables**

Population 20

X	M	D	F	L_q	X	M	D	F	L_q	X	M	D	F	L_q
	6	.544	.932	1.36		8	.309	.973	.54		5	.992	.624	7.52
	5	.797	.848	3.04		7	.529	.935	1.30	.420	13	.019	.999	.02
	4	.958	.708	5.84		6	.766	.862	2.76		12	.055	.997	.06
	3	.998	.536	9.28		5	.933	.748	5.04		11	.131	.991	.18
.290	10	.027	.999	.02		4	.993	.605	7.90		10	.265	.977	.46
	9	.075	.996	.08	.340	11	.029	.999	.02		9	.458	.949	1.02
	8	.176	.988	.24		10	.079	.996	.08		8	.678	.896	2.08
	7	.351	.967	.66		9	.179	.987	.26		7	.863	.815	3.70
	6	.592	.920	1.60		8	.347	.967	.66		6	.965	.711	5.78
	5	.833	.828	3.44		7	.573	.924	1.52		5	.996	.595	8.10
	4	.970	.685	6.30		6	.802	.846	3.08	.440	13	.029	.999	.02
	3	.999	.517	9.66		5	.949	.729	5.42		12	.078	.995	.10
.300	10	.034	.998	.04		4	.995	.588	8.24		11	.175	.987	.26
	9	.091	.995	.10	.360	12	.015	.999	.02		10	.333	.969	.62
	8	.205	.985	.30		11	.045	.998	.04		9	.540	.933	1.34
	7	.394	.961	.78		10	.112	.993	.14		8	.751	.872	2.56
	6	.639	.907	1.86		9	.237	.981	.38		7	.907	.785	4.30
	5	.865	.808	3.84		8	.429	.954	.92		6	.980	.680	6.40
	4	.978	.664	6.72		7	.660	.901	1.98		5	.998	.568	8.64
	3	.999	.500	10.00		6	.863	.812	3.76	.460	14	.014	.999	.02
.310	11	.014	.999	.02		5	.971	.691	6.18		13	.043	.998	.04
	10	.043	.998	.04		4	.998	.555	8.90		12	.109	.993	.14
	9	.110	.993	.14	.380	12	.024	.999	.02		11	.228	.982	.36
	8	.237	.981	.38		11	.067	.996	.08		10	.407	.958	.84
	7	.438	.953	.94		10	.154	.989	.22		9	.620	.914	1.72
	6	.684	.893	2.14		9	.305	.973	.54		8	.815	.846	3.08
	5	.892	.788	4.24		8	.513	.938	1.24		7	.939	.755	4.90
	4	.985	.643	7.14		7	.739	.874	2.52		6	.989	.651	6.98
.320	11	.018	.999	.02		6	.909	.777	4.46		5	.999	.543	9.14
	10	.053	.997	.06		5	.984	.656	6.88	.480	14	.022	.999	.02
	9	.130	.992	.06		4	.999	.526	9.48		13	.063	.996	.08
	8	.272	.977	.46	.400	13	.012	.999	.02		12	.147	.990	.20
	7	.483	.944	1.12		12	.037	.998	.04		11	.289	.974	.52
	6	.727	.878	2.44		11	.095	.994	.12		10	.484	.944	1.12
	5	.915	.768	4.64		10	.205	.984	.32		9	.695	.893	2.14
	4	.989	.624	7.52		9	.379	.962	.76		8	.867	.819	3.62
.330	11	.023	.999	.02		8	.598	.918	1.64		7	.962	.726	5.48
	10	.065	.997	.06		7	.807	.845	3.10		6	.994	.625	7.50
	9	.154	.990	.20		6	.942	.744	5.12	.500	14	.033	.998	.04

TABLE H-4 Continued **Finite Queuing Tables**

Population 20-30

X	M	D	F	L_q	X	M	D	F	L_q	X	M	D	F	L_q
	13	.088	.995	.10		14	.129	.991	.18		16	.272	.980	.40
	12	.194	.985	.30		13	.266	.978	.44		15	.487	.954	.92
	11	.358	.965	.70		12	.455	.952	.96		14	.703	.913	1.74
	10	.563	.929	1.42		11	.662	.908	1.84		13	.864	.859	2.82
	9	.764	.870	2.60		10	.835	.847	3.06		12	.952	.798	4.04
	8	.908	.791	4.18		9	.941	.772	4.56		11	.988	.733	5.34
	7	.977	.698	6.04		8	.986	.689	6.22		10	.998	.667	6.66
	6	.997	.600	8.00		7	.998	.603	7.94	.800	19	.014	.999	.02
.520	15	.015	.999	.02	.600	16	.023	.999	.02		18	.084	.996	.08
	14	.048	.997	.06		15	.072	.996	.08		17	.242	.984	.32
	13	.120	.992	.16		14	.171	.988	.24		16	.470	.959	.82
	12	.248	.979	.42		13	.331	.970	.60		15	.700	.920	1.60
	11	.432	.954	.92		12	.532	.938	1.24		14	.867	.869	2.62
	10	.641	.911	1.78		11	.732	.889	2.22		13	.955	.811	3.78
	9	.824	.846	3.08		10	.882	.824	3.52		12	.989	.750	5.00
	8	.939	.764	4.72		9	.962	.748	5.04		11	.998	.687	6.26
	7	.987	.672	6.56		8	.992	.666	6.68	.850	19	.046	.998	.04
	6	.998	.577	8.46		7	.999	.583	8.34		18	.201	.988	.24
.540	15	.023	.999	.02	.650	17	.017	.999	.02		17	.451	.965	.70
	14	.069	.996	.08		16	.061	.997	.06		16	.703	.927	1.46
	13	.161	.988	.24		15	.156	.989	.22		15	.877	.878	2.44
	12	.311	.972	.56		14	.314	.973	.54		14	.962	.823	3.54
	11	.509	.941	1.18		13	.518	.943	1.14		13	.991	.765	4.70
	10	.713	.891	2.18		12	.720	.898	2.04		12	.998	.706	5.88
	9	.873	.821	3.58		11	.872	.837	3.26	.900	19	.135	.994	.12
	8	.961	.738	5.24		10	.957	.767	4.66		18	.425	.972	.52
	7	.993	.648	7.04		9	.990	.692	6.16		17	.717	.935	1.30
	6	.999	.556	8.88		8	.998	.615	7.70		16	.898	.886	2.28
.560	15	.035	.998	.04	.700	17	.047	.998	.04		15	.973	.833	3.34
	14	.095	.994	.12		16	.137	.991	.18		14	.995	.778	4.44
	13	.209	.984	.32		15	.295	.976	.48		13	.999	.722	5.56
	12	.381	.963	.74		14	.503	.948	1.04	.950	19	.377	.981	.38
	11	.586	.926	1.48		13	.710	.905	1.90		18	.760	.943	1.14
	10	.778	.869	2.62		12	.866	.849	3.02		17	.939	.894	2.12
	9	.912	.796	4.08		11	.953	.783	4.34		16	.989	.842	3.16
	8	.976	.713	5.74		10	.988	.714	5.72		15	.999	.789	4.22
	7	.996	.625	7.50		9	.998	.643	7.14			Population 30		
.580	16	.015	.999	.02	.750	18	.031	.999	.02					
	15	.051	.997	.06		17	.113	.993	14	.004	1	.116	.999	.03

TABLE H-4 **Finite Queuing Tables**

Continued

Population 30

X	M	D	F	L_q	X	M	D	F	L_q	X	M	D	F	L_q
.007	1	.203	.998	.06		2	.286	.992	.24		2	.634	.951	1.47
.009	1	.260	.997	.09		1	.843	.899	3.03		1	.997	.616	11.52
.010	1	.289	.996	.12	.034	3	.083	.999	.03	.056	4	.086	.998	.06
.011	1	.317	.995	.15		2	.316	.990	.30		3	.267	.991	.27
.012	1	.346	.994	.18		1	.876	.877	3.69		2	.665	.944	1.68
.013	1	.374	.993	.21	.036	3	.095	.998	.06		1	.998	.595	12.15
.014	2	.067	.999	.03		2	.347	.988	.36	.058	4	.096	.998	.06
	1	.403	.991	.27		1	.905	.853	4.41		3	.288	.989	.33
.015	2	.076	.999	.03	.038	3	.109	.998	.06		2	.695	.936	1.92
	1	.431	.989	.33		2	.378	.986	.42		1	.999	.574	12.78
.016	2	.085	.999	.03		1	.929	.827	5.19	.060	5	.030	.999	.03
	1	.458	.987	.39	.040	3	.123	.997	.09		4	.106	.997	.09
.017	2	.095	.999	.03		2	.410	.983	.51		3	.310	.987	.39
	1	.486	.985	.45		1	.948	.800	6.00		2	.723	.927	2.19
.018	2	.105	.999	.03	.042	3	.138	.997	.09		1	.999	.555	13.35
	1	.513	.983	.51		2	.442	.980	.60	.062	5	.034	.999	.03
.019	2	.116	.999	.03		1	.963	.772	6.84		4	.117	.997	.09
	1	.541	.980	.60	.044	4	.040	.999	.03		3	.332	.986	.42
.020	2	.127	.998	.06		3	.154	.996	.12		2	.751	.918	2.46
	1	.567	.976	.72		2	.474	.977	.69	.064	5	.038	.999	.03
.021	2	.139	.998	.06		1	.974	.744	7.68		4	.128	.997	.09
	1	.594	.973	.81	.046	4	.046	.999	.03		3	.355	.984	.48
.022	2	.151	.998	.06		3	.171	.996	.12		2	.777	.908	2.76
	1	.620	.969	.93		2	.506	.972	.84	.066	5	.043	.999	.03
.023	2	.163	.997	.09		1	.982	.716	8.52		4	.140	.996	.12
	1	.645	.965	1.05	.048	4	.053	.999	.03		3	.378	.982	.54
.024	2	.175	.997	.09		3	.189	.995	.15		2	.802	.897	3.09
	1	.670	.960	1.20		2	.539	.968	.96	.068	5	.048	.999	.03
.025	2	.188	.996	.12		1	.988	.689	9.33		4	.153	.995	.15
	1	.694	.954	1.38	.050	4	.060	.999	.03		3	.402	.979	.63
.026	2	.201	.996	.12		3	.208	.994	.18		2	.825	.885	3.45
	1	.718	.948	1.56		2	.571	.963	1.11	.070	5	.054	.999	.03
.028	3	.051	.999	.03		1	.992	.663	10.11		4	.166	.995	.15
	2	.229	.995	.15	.052	4	.068	.999	.03		3	.426	.976	.72
	1	.763	.935	1.95		3	.227	.993	.21		2	.847	.873	3.81
.030	3	.060	.999	.03		2	.603	.957	1.29	.075	5	.069	.998	.06
	2	.257	.994	.18		1	.995	.639	10.83		4	.201	.993	.21
	1	.805	.918	2.46	.054	4	.077	.998	.06		3	.486	.969	.93
.032	3	.071	.999	.03		3	.247	.992	.24		2	.893	.840	4.80

TABLE H-4 **Finite Queuing Tables**

Continued

Population 30

X	M	D	F	L_q	X	M	D	F	L_q	X	M	D	F	L_q
.080	6	.027	.999	.03		5	.289	.985	.45		3	.986	.687	9.39
	5	.088	.998	.06		4	.570	.950	1.50	.150	9	.024	.999	.03
	4	.240	.990	.30		3	.890	.833	5.01		8	.065	.998	.06
	3	.547	.959	1.23		2	.998	.579	12.63		7	.155	.993	.21
	2	.929	.805	5.85	.120	7	.057	.998	.06		6	.322	.980	.60
.085	6	.036	.999	.03		6	.147	.994	.18		5	.580	.944	1.68
	5	.108	.997	.09		5	.327	.981	.57		4	.860	.849	4.53
	4	.282	.987	.39		4	.619	.939	1.83		3	.991	.665	10.05
	3	.607	.948	1.56		3	.918	.808	5.76	.155	9	.029	.999	.03
	2	.955	.768	6.96		2	.999	.555	13.35		8	.077	.997	.09
.090	6	.046	.999	.03	.125	8	.024	.999	.03		7	.177	.992	.24
	5	.132	.996	.12		7	.069	.998	.06		6	.357	.976	.72
	4	.326	.984	.48		6	.171	.993	.21		5	.622	.935	1.95
	3	.665	.934	1.98		5	.367	.977	.69		4	.887	.830	5.10
	2	.972	.732	8.04		4	.666	.927	2.19		3	.994	.644	10.68
.095	6	.057	.999	.03		3	.940	.783	6.51	.160	9	.036	.999	.03
	5	.158	.994	.18	.130	8	.030	.999	.03		8	.090	.997	.09
	4	.372	.979	.63		7	.083	.997	.09		7	.201	.990	.30
	3	.720	.918	2.46		6	.197	.991	.27		6	.394	.972	.84
	2	.984	.697	9.09		5	.409	.972	.84		5	.663	.924	2.28
.100	6	.071	.998	.06		4	.712	.914	2.58		4	.910	.811	5.67
	5	.187	.993	.21		3	.957	.758	7.26		3	.996	.624	11.28
	4	.421	.973	.81	.135	8	.037	.999	.03	.165	9	.043	.999	.03
	3	.771	.899	3.03		7	.098	.997	.09		8	.105	.996	.12
	2	.991	.664	10.08		6	.226	.989	.33		7	.227	.988	.36
.105	7	.030	.999	.03		5	.451	.966	1.02		6	.431	.967	.99
	6	.087	.997	.09		4	.754	.899	3.03		5	.702	.913	2.61
	5	.219	.991	.27		3	.970	.734	7.98		4	.930	.792	6.24
	4	.470	.967	.99	.140	8	.045	.999	.03		3	.997	.606	11.82
	3	.816	.879	3.63		7	.115	.996	.12	.170	10	.019	.999	.03
	2	.995	.634	10.98		6	.256	.987	.39		9	.051	.998	.06
.110	7	.038	.999	.03		5	.494	.960	1.20		8	.121	.995	.15
	6	.105	.997	.09		4	.793	.884	3.48		7	.254	.986	.42
	5	.253	.988	.36		3	.979	.710	8.70		6	.469	.961	1.17
	4	.520	.959	1.23	.145	8	.055	.998	.06		5	.739	.901	2.97
	3	.856	.857	4.29		7	.134	.995	.15		4	.946	.773	6.81
	2	.997	.605	11.85		6	.288	.984	.48		3	.998	.588	12.36
.115	7	.047	.999	.03		5	.537	.952	1.44	.180	10	.028	.999	.03
	6	.125	.996	.12		4	.828	.867	3.99		9	.070	.997	.09

TABLE H-4 **Finite Queuing Tables**
Continued

Population 30

X	M	D	F	L_q	X	M	D	F	L_q	X	M	D	F	L_q
	8	.158	.993	.21		10	.123	.994	.18		8	.676	.915	2.55
	7	.313	.980	.60		9	.242	.985	.45		7	.866	.841	4.77
	6	.546	.948	1.56		8	.423	.965	1.05		6	.970	.737	7.89
	5	.806	.874	3.78		7	.652	.923	2.31		5	.997	.617	11.49
	4	.969	.735	7.95		6	.864	.842	4.74	.280	14	.017	.999	.03
	3	.999	.555	13.35		5	.976	.721	8.37		13	.042	.998	.06
.190	10	.039	.999	.03		4	.999	.580	12.60		12	.093	.996	.12
	9	.094	.996	.12	.240	12	.031	.999	.03		11	.185	.989	.33
	8	.200	.990	.30		11	.074	.997	.09		10	.329	.976	.72
	7	.378	.973	.81		10	.155	.992	.24		9	.522	.949	1.53
	6	.621	.932	2.04		9	.291	.981	.57		8	.733	.898	3.06
	5	.862	.845	4.65		8	.487	.955	1.35		7	.901	.818	5.46
	4	.983	.699	9.03		7	.715	.905	2.85		7	.981	.712	8.64
.200	11	.021	.999	.03		6	.902	.816	5.52		5	.999	.595	12.15
	10	.054	.998	.06		5	.986	.693	9.21	.290	14	.023	.999	.03
	9	.123	.995	.15		4	.999	.556	13.32		13	.055	.998	.06
	8	.249	.985	.45	.250	13	.017	.999	.03		12	.117	.994	.18
	7	.446	.963	1.11		12	.042	.998	.06		11	.223	.986	.42
	6	.693	.913	2.61		11	.095	.996	.12		10	.382	.969	.93
	5	.905	.814	5.58		10	.192	.989	.33		9	.582	.937	1.89
	4	.991	.665	10.05		9	.345	.975	.75		8	.785	.880	3.60
.210	11	.030	.999	.03		8	.552	.944	1.68		7	.929	.795	6.15
	10	.073	.997	.09		7	.773	.885	3.45		6	.988	.688	9.36
	9	.157	.992	.24		6	.932	.789	6.33		5	.999	.575	12.75
	8	.303	.980	.60		5	.992	.666	10.02	.300	14	.031	.999	.03
	7	.515	.952	1.44	.260	13	.023	.999	.03		13	.071	.997	.09
	6	.758	.892	3.24		12	.056	.998	.06		12	.145	.992	.24
	5	.938	.782	6.54		11	.121	.994	.18		11	.266	.982	.54
	4	.995	.634	10.98		10	.233	.986	.42		10	.437	.962	1.14
.220	11	.041	.999	.03		9	.402	.967	.99		9	.641	.924	2.28
	10	.095	.996	.12		8	.616	.930	1.20		8	.830	.861	4.17
	9	.197	.989	.33		7	.823	.864	4.08		7	.950	.771	6.87
	8	.361	.974	.78		6	.954	.763	7.11		6	.993	.666	10.02
	7	.585	.938	1.86		5	.995	.641	10.77	.310	15	.017	.999	.03
	6	.816	.868	3.96	2.70	13	.032	.999	.03		14	.041	.998	.06
	5	.961	.751	7.47		12	.073	.997	.09		13	.090	.996	.12
	4	.998	.606	11.82		11	.151	.992	.24		12	.177	.990	.30
.230	12	.023	.999	.03		10	.279	.981	.57		11	.312	.977	.69
	11	.056	.998	.06		9	.462	.959	1.23		10	.494	.953	1.41

TABLE H-4 **Finite Queuing Tables**
Continued

Population 30

X	M	D	F	L_q	X	M	D	F	L_q	X	M	D	F	L_q
	9	.697	.909	2.73		12	.392	.967	.99		18	.041	.998	.06
	8	.869	.840	4.80		11	.578	.937	1.89		17	.087	.996	.12
	7	.966	.749	7.53		10	.762	.889	3.33		16	.167	.990	.30
	6	.996	.645	10.65		9	.902	.821	5.37		15	.288	.979	.63
.320	15	.023	.999	.03		8	.974	.738	7.86		14	.446	.960	1.20
	14	.054	.998	.06		7	.996	.648	10.56		13	.623	.929	2.13
	13	.113	.994	.18	.380	17	.020	.999	.03		12	.787	.883	3.51
	12	.213	.987	.39		16	.048	.998	.06		11	.906	.824	5.28
	11	.362	.971	.87		15	.101	.995	.15		10	.970	.755	7.35
	10	.552	.943	1.71		14	.191	.988	.36		9	.994	.681	9.57
	9	.748	.893	3.21		13	.324	.975	.75		8	.999	.606	11.82
	8	.901	.820	5.40		12	.496	.952	1.44	.460	19	.028	.999	.03
	7	.977	.727	8.19		11	.682	.914	2.58		18	.064	.997	.09
	6	.997	.625	11.25		10	.843	.857	4.29		17	.129	.993	.21
.330	15	.030	.999	.03		9	.945	.784	6.48		16	.232	.985	.45
	14	.068	.997	.09		8	.988	.701	8.97		15	.375	.970	.90
	13	.139	.993	.21		7	.999	.614	11.58		14	.545	.944	1.68
	12	.253	.983	.51	.400	17	.035	.999	.03		13	.717	.906	2.82
	11	.414	.965	1.05		16	.076	.996	.12		12	.857	.855	4.35
	10	.608	.931	2.07		15	.150	.992	.24		11	.945	.793	6.21
	9	.795	.876	3.72		14	.264	.982	.54		10	.985	.724	8.28
	8	.927	.799	6.03		13	.420	.964	1.08		9	.997	.652	10.44
	7	.985	.706	8.82		12	.601	.933	2.01	.480	20	.019	.999	.03
	6	.999	.606	11.82		11	.775	.886	3.42		19	.046	.998	.06
.340	16	.016	.999	.03		10	.903	.823	5.31		18	.098	.995	.15
	15	.040	.998	.06		9	.972	.748	7.56		17	.184	.989	.33
	14	.086	.996	.12		8	.995	.666	10.02		16	.310	.977	.69
	13	.169	.990	.30	.420	18	.024	.999	.03		15	.470	.957	1.29
	12	.296	.979	.63		17	.056	.997	.09		14	.643	.926	2.22
	11	.468	.957	1.29		16	.116	.994	.18		13	.799	.881	3.57
	10	.663	.918	2.46		15	.212	.986	.42		12	.910	.826	5.22
	9	.836	.858	4.26		14	.350	.972	.84		11	.970	.762	7.14
	8	.947	.778	6.66		13	.521	.948	1.56		10	.993	.694	9.18
	7	.990	.685	9.45		12	.700	.910	2.70		9	.999	.625	11.25
	6	.999	.588	12.36		11	.850	.856	4.32	.500	20	.032	.999	.03
.360	16	.029	.999	.03		10	.945	.789	6.33		19	.072	.997	.09
	15	.065	.997	.09		9	.986	.713	8.61		18	.143	.992	.24
	14	.132	.993	.21		8	.998	.635	10.95		17	.252	.983	.51
	13	.240	.984	.48	.440	19	.017	.999	.03		16	.398	.967	.99

TABLE H-4 **Finite Queuing Tables**
Continued

Population 30

X	M	D	F	L_q	X	M	D	F	L_q	X	M	D	F	L_q
	15	.568	.941	1.77		13	.972	.772	6.84		13	.999	.667	9.99
	14	.733	.904	2.88		12	.993	.714	8.58	.700	25	.039	.998	.06
	13	.865	.854	4.38		11	.999	.655	10.35		24	.096	.995	.15
	12	.947	.796	6.12	.580	23	.014	.999	.03		23	.196	.989	.33
	11	.985	.732	8.04		22	.038	.998	.06		22	.339	.977	.69
	10	.997	.667	9.99		21	.085	.996	.12		21	.511	.958	1.26
.520	21	.021	.999	.03		20	.167	.990	.30		20	.681	.930	2.10
	20	.051	.998	.06		19	.288	.980	.60		19	.821	.894	3.18
	19	.108	.994	.18		18	.443	.963	1.11		18	.916	.853	4.41
	18	.200	.988	.36		17	.612	.936	1.92		17	.967	.808	5.76
	17	.331	.975	.75		16	.766	.899	3.03		16	.990	.762	7.14
	16	.493	.954	1.38		15	.883	.854	4.38		15	.997	.714	8.58
	15	.663	.923	2.31		14	.953	.802	5.94	.750	26	.046	.998	.06
	14	.811	.880	3.60		13	.985	.746	7.62		25	.118	.994	.18
	13	.915	.827	5.19		12	.997	.690	9.30		24	.240	.986	.42
	12	.971	.767	6.99		11	.999	.632	11.04		23	.405	.972	.84
	11	.993	.705	8.85	.600	23	.024	.999	.03		22	.587	.950	1.50
	10	.999	.641	10.77		22	.059	.997	.09		21	.752	.920	2.40
.540	21	.035	.999	.03		21	.125	.993	.21		20	.873	.883	3.51
	20	.079	.996	.12		20	.230	.986	.42		19	.946	.842	4.74
	19	.155	.991	.27		19	.372	.972	.84		18	.981	.799	6.03
	18	.270	.981	.57		18	.538	.949	1.53		17	.995	.755	7.35
	17	.421	.965	1.05		17	.702	.918	2.46		16	.999	.711	8.67
	16	.590	.938	1.86		16	.837	.877	3.69	.800	27	.053	.998	.06
	15	.750	.901	2.97		15	.927	.829	5.13		26	.143	.993	.21
	14	.874	.854	4.38		14	.974	.776	6.72		25	.292	.984	.48
	13	.949	.799	6.03		13	.993	.722	8.34		24	.481	.966	1.02
	12	.985	.740	7.80		12	.999	.667	9.99		23	.670	.941	1.77
	11	.997	.679	9.63	.650	24	.031	.999	.03		22	.822	.909	2.73
	10	.999	.617	11.49		23	.076	.996	.12		21	.919	.872	3.84
.560	22	.023	.999	.03		22	.158	.991	.27		20	.970	.832	5.04
	21	.056	.997	.09		21	.281	.982	.54		19	.991	.791	6.27
	20	.117	.994	.18		20	.439	.965	1.05		18	.998	.750	7.50
	19	.215	.986	.42		19	.610	.940	1.80	.850	28	.055	.998	.06
	18	.352	.973	.81		18	.764	.906	2.82		27	.171	.993	.21
	17	.516	.952	1.44		17	.879	.865	4.05		26	.356	.981	.57
	16	.683	.920	2.40		16	.949	.818	5.46		25	.571	.960	1.20
	15	.824	.878	3.66		15	.983	.769	6.93		24	.760	.932	2.04
	14	.920	.828	5.16		14	.996	.718	8.46		23	.888	.899	3.03

TABLE H-4 **Finite Queuing Tables**
Continued

Population 30

X	M	D	F	L_q	X	M	D	F	L_q	X	D	F	L_q	
	22	.957	.862	4.14		26	.683	.953	1.41		28	.574	.973	.81
	21	.987	.823	5.31		25	.856	.923	2.31		27	.831	.945	1.65
	20	.997	.784	6.48		24	.947	.888	3.36		26	.951	.912	2.64
	19	.999	.745	7.65		23	.985	.852	4.44		25	.989	.877	3.69
.900	29	.047	.999	.03		22	.996	.815	5.55		24	.998	.842	4.74
	28	.200	.992	.24		21	.999	.778	6.66					
	27	.441	.977	.69	.950	29	.226	.993	.21					

TABLE H-5 **Table of Random Digits**

78466	83326	96589	88727	72655	49682	82338	28583	01522	11248
78722	47603	03477	29528	63956	01255	29840	32370	18032	82051
06401	87397	72898	32441	88861	71803	55626	77847	29925	76106
04754	14489	39420	94211	58042	43184	60977	74801	05931	73822
97118	06774	87743	60156	38037	16201	35137	54513	68023	34380
71923	49313	59713	95710	05975	64982	79253	93876	33707	84956
78870	77328	09637	67080	49168	75290	50175	34312	82593	76606
61208	17172	33187	92523	69895	28284	77956	45877	08044	58292
05033	24214	74232	33769	06304	54676	70026	41957	40112	66451
95983	13391	30369	51035	17042	11729	88647	70541	36026	23113
19946	55448	75049	24541	43007	11975	31797	05373	45893	25665
03580	67206	09635	84612	62611	86724	77411	99415	58901	86160
56823	49819	20283	22272	00114	92007	24369	00543	05417	92251
87633	31761	99865	31488	49947	06060	32083	47944	00449	06550
95152	10133	52693	22480	50336	49502	06296	76414	18358	05313
05639	24175	79438	92151	57602	03590	25465	54780	79098	73594
65927	55525	67270	22907	55097	63177	34119	94216	84861	10457
59005	29000	38395	80367	34112	41866	30170	84658	84441	03926
06626	42682	91522	45955	23263	09764	26824	82936	16813	13878
11306	02732	34189	04228	58541	72573	89071	58066	67159	29633
45143	56545	94617	42752	31209	14380	81477	36952	44934	97435
97612	87175	22613	84175	96413	83336	12408	89318	41713	90669
97035	62442	06940	45719	39918	60274	54353	54497	29789	82928
62498	00257	19179	06313	07900	46733	21413	63627	48734	92174
80306	19257	18690	54653	07263	19894	89909	76415	57246	02621
84114	84884	50129	68942	93264	72344	98794	16791	83861	32007
58437	88807	92141	88677	02864	02052	62843	21692	21373	29408
15702	53457	54258	47485	23399	71692	56806	70801	41548	94809
59966	41287	87001	26462	94000	28457	09469	80416	05897	87970
43641	05920	81346	02507	25349	93370	02064	62719	45740	62080
25501	50113	44600	87433	00683	79107	22315	42162	25516	98434
98294	08491	25251	26737	00071	45090	68628	64390	42684	94956
52582	89985	37863	60788	27412	47502	71577	13542	31077	13353
26510	83622	12546	00489	89304	15550	09482	07504	64588	92562
24755	71543	31667	83624	27085	65905	32386	30775	19689	41437
38399	88796	58856	18220	51016	04976	54062	49109	95563	48244
18889	87814	52232	58244	95206	05947	26622	01381	28744	38374
51774	89694	02654	63161	54622	31113	51160	29015	64730	07750
88375	37710	61619	69820	13131	90406	45206	06386	06398	68652
10416	70345	93307	87360	53452	61179	46845	91521	32430	74795
99258	03778	54674	51499	13659	36434	84760	76446	64026	97534
58923	18319	95092	11840	87646	85330	58143	42023	28972	30657
39407	41126	44469	78889	54462	38609	58555	69793	27258	11296
29372	70781	19554	95559	63088	35845	60162	21228	48296	05006
07287	76846	92658	21985	00872	11513	24443	44320	37737	97360
07089	02948	03699	71255	13944	86597	89052	88899	03553	42145
35757	37447	29860	04546	28742	27773	10215	09774	43426	22961
58797	70878	78167	91942	15108	37441	99254	27121	92358	94254
32281	97860	23029	61409	81887	02050	63060	45246	46312	30378
93531	08514	30244	34641	29820	72126	62419	93233	26537	21179

SOURCE. The Rand Corporation, *A Million Random Digits with 100,000 Normal Deviates.* The Free Press, Glencoe, Ill., 1955, pp. 180–183. (Reproduced with permission.)

INDEX